A GUIDE TO USEFUL WOODS OF THE WORLD

The cover shows an assortment of "colorwoods" of the world selected from the Wood Data Sheets in this book. The natural colors of wood represent every color of the rainbow not to mention black, white, brown and many variations and combinations of colors. A number of species of wood such as Gregg acacia (WDS 004, *Acacia greggi*), staghorn sumac (WDS 236, *Rhus typhina*) and black locust (WDS 237, *Robinia pseduocacacia*) produce even more spectacular color effects when viewed under ultraviolet light. The species shown on the cover, starting at 12 o'clock, represent the color:

1. red	WDS 035	pink ivory	*Berchemia zeyheri*
2. orange	WDS 051	Pernambuco	*Caesalpinia echinata*
3. yellow	WDS 050	boxwood	*Buxus sempervirens*
4. green	WDS 027	pawpaw	*Asimina triloba*
5. violet	WDS 092	kingwood	*Dalbergia cearensis*
6. black	WDS 096	African blackwood	*Dalbergia melanoxylon*
7. white	WDS 017	kukui	*Aleurites moluccana*
8. brown	WDS 002	mulga	*Acacia aneura*

A GUIDE TO USEFUL WOODS OF THE WORLD

Second Edition

Edited by
James H. Flynn, Jr.
and
Charles D. Holder

Forest Products Society
Madison, Wisconsin

Library of Congress Cataloging-in-Publication Data

A guide to useful woods of the world / edited by James H. Flynn, Jr. and Charles D. Holder.-- 2nd ed.

p. cm.

Includes bibliographical references and indexes.

ISBN 1-892529-15-7 (pbk.)

1. Wood. 2. Woodwork--Equipment and supplies. I. Flynn, James H. II. Holder, Charles D.

TA419.G83 2001

620.1'2--dc21 2001033458

ISBN 1-892529-15-7

Printed and bound in the United States of America.

01085000

DEDICATION

To the late *Harold Bailey Nogle*, a Texan, a professional newspaperman and a wood collector. The International Wood Collectors Society (IWCS) traces its heritage back to 1947 when the first meeting of the Society was held at the Nogles' Big Cow Creek Camp near Newton, Texas. Here, a small group of wood collectors worked out the details of the Society's mission, goals and aspirations, recognized its birth and called itself the "Wood Collectors Society". They had plenty of support and were encouraged by professionals such as Samuel J. Record and Robert W. Hess of the Yale University School of Forestry, Arthur Koehler of the Forest Products Laboratory and others. Today, over a half-century later, we can proudly salute Harold and his compatriots for their great sense of what it means to be a wood collector. Be they scientists in a laboratory delving into wood anatomy or humble wood crafters turning a bowl, there is a common link to all of them . . . a love of and admiration for wood and the trees from which it comes. We hope and trust that this latest *Guide to Useful Woods of the World* is an extension of Harold's dreams and it is respectfully dedicated to him.

TABLE OF CONTENTS

Preface ix

Acknowledgments xi

Introduction xv

Using The Book xvii

About The Editors xxi

Wood Data Sheets 1

Appendices
- Appendix A
  - *Biology and Taxonomy for Woodworkers* 563
- Appendix B
  - *Insights on Wood Toxicity* 571
- Appendix C
  - *Selected References* 587

Indices
- Common Name Index 591
- Family Name Index 607
- Scientific Name Index 615

PREFACE

Since 1947, the International Wood Collectors Society (IWCS), a not-for-profit association, has published the journal *World of Wood*, an outstanding source of data on woods of the world as well as a chronicle of the activities of the IWCS. In 1976, the *World of Wood* began featuring a column entitled Wood of the Month, which provided readers with numerous interesting facts about the wood species being highlighted, including its geographic range, woodworking properties, uses and supplies. In 1994, data on 201 woods that had been featured in Wood of the Month columns were gathered and published in the first edition of *A Guide to Useful Woods of the World.*

This all new edition of *A Guide to Useful Woods of the World* features 279 species of importance to woodworkers, wood collectors and others interested in wood. In addition to the substantially increased number of woods contained in this new edition, each Wood Data Sheet includes a line drawing of a key botanical feature, a photomicrograph of the end grain of the wood and a color photograph of a sample of the wood.

Woodworkers and others interested in wood will find this book to be an invaluable aid in assessing the characteristics and properties of commercially (or recreationally) useful woods and as a guide to their availability. Many of the attractive woods featured in this book are only available in certain geographical areas of the world. A distinct advantage that readers of this book will have is an understanding of the value of the network that membership in IWCS can afford, especially when it comes to locating supplies of hard-to-find species. Readers are encouraged to visit the IWCS website at www.woodcollectors.org and to become members of the Society in order to take advantage of the expertise that exists in this unique network of nearly 2,000 individuals throughout the world who are captivated by the beauty and versatility of wood.

Acknowledgments

The task of putting this book together was made much easier by many words of enthusiastic support from friends all over the globe. At times we were discouraged because data was difficult to gather and seemed to barely crawl into our offices. But, with time the information was collected and it began to look like a book was possible. So many helpful hands contributed a slice of data here and there that we are afraid we lost track of all those who helped. In spite of our not mentioning everyone individually, we are indeed grateful to all who contributed.

For the most part we relied on data and illustrations from government publications that are in the public domain. Most notable among these were the USDA's venerable book *Trees, The Yearbook of Agriculture, 1949* and *Tropical Timbers of the World*, Agriculture Handbook Number 607, to cite just two examples.

The botanical images placed on each Wood Data Sheet (WDS) became a unique problem because it was difficult to assemble them for some of the more exotic species. To the rescue came Dr. Mihaly Czako at the University of South Carolina who flooded us with photographs and drawings of the plants as e-mail attachments. From these we were able to make line drawings. We wish to thank Meg Palgrave, from Zimbabwe, for use of the botanical illustrations used and redrawn in WDS Nos. 13, 31, 35, 41 and 190. These images were in *Trees of South Africa*, Struik Publishers, Cape Town, South Africa.

The University of Chicago Press granted us permission to use several drawings from A *Field*

Guide to Families and Genera of Woody Plants of Northwest South America by Alwyn H. Gentry (a great field collector who, in the process, lost his life in an aircraft accident in South America). We also thank Michael Clayton, Department of Botany, Univeristy of Wisconsin–Madison for allowing us to use several of his photographs on the introductory pages.

The photomicrographs of the wood end grain were a cause of great concern until Dr. Bob Goldsack, along with his wife Gael, from Australia, stepped in and filled the gap. We can thank them for the outstanding photography in producing most of the high-quality black and white images. As we expected, we did not have all of the material to produce our own photomicrographs and we are thankful for the assistance of Dr. Jugo Ilic of the Forestry and Forestry Products Division of Australia's Commonwealth Industrial Scientific Research Office. He was instrumental in making available to us a goodly number of the valuable microscope images on wood microscope slides from the Dadswell Memorial Wood Collection. Dr. Eugene Dimitriadis, Camberwell, Victoria, Australia, helped tremendously in coordinating this effort Down Under. Again, fortune smiled upon us when Dr. Regis Miller of the U.S. Forest Service's Forest Products Laboratory supplied the ultimate shortages. Playing an equal role in this, we are indebted to Ernie Ives, Ipswitch, England, who day by day labored with a microtome and chemicals to make the microscope slides from which Bob made the photographic images. To Alan Curtis, Eugene, Oregon, and others for feeding wood to Ernie go our thanks as well. The nightmare of keeping track of all of the worldwide input of photomicrographs was made a bit easier when we recruited the services of a visiting scholar from Lima, Peru, Vania Rodriguez, who made a notebook of slide images for us. Added to the healthy mix of talent we cannot forget the help of Chris Flynn, Herndon, Virginia, in laboring with us to overcome the intricacies of computer programs in getting the images processed electronically. We were lucky indeed to enlist the services of James Lawton, Rochester, New York, in transferring the results of Bob Goldsack's work onto compact discs. For last minute help with the microscope and camera, we are indebted to marine biologist John Wojtowicz, Tiverton, Rhode Island, for dropping his work analyzing sand and jumping on our band wagon to process some of our photomicrographs.

Basic to the whole enterprise are the Wood Data Sheets which were formerly Wood of the Month columns in *World of Wood*, the journal of the International Wood Collectors Society. We sincerely thank Dorothy Kline, the wife of the late Max Kline, who gave us permission to use Max's work in an earlier edition of this book. Jon Arno and Alan Curtis, other contributors to these columns, graciously permitted us to use their work.

We wish to express thanks to Paul van Rijckevorsel, the Netherlands, who was occasionally consulted on botanical matters and did his best to be helpful. It was indeed a great privilege and pleasure to work with him.

We thank one and all and if we learned one thing in the process it was that we have a wonderful combination of skilled and ever-helpful members in the International Wood Collectors Society.

The opportunity for us to work with the staff of the Forest Products Society (FPS) was a unique and pleasurable experience. Both IWCS and FPS missions deal with forest products so we felt at home. The cordial working relationships with Arthur B. Brauner, FPS's Executive Vice President and the enthusiastic work of Susan Stamm, FPS's Special Publications Director, were instrumental in getting the book started on the right track, in the right direction and full speed ahead. We were impressed with Susan's obvious love of trees and their varied woods and tried, from time to time, to supply her with the odd piece of exotic tropical wood. We know that she enjoyed working with us as we did with her and regret that the project had to finally come to its end.

To our wives who constantly supported us, humoring us in times of despair and smiling at our boyish enthusiasm for the task, we are eternally grateful. Both Marys contributed more to the project than they will ever realize.

Jim Flynn
Vienna, Virginia, USA

Chuck Holder
Calgary, Alberta, Canada

INTRODUCTION

The publication of this book represents a combined effort of the **International Wood Collectors Society** and the **Forest Products Society**. Both of these not-for-profit Societies, independently founded in 1947, have been instrumental in describing, conserving and providing educational insights into the products of the world's forests.

The **International Wood Collectors Society (IWCS)** is dedicated to the advancement of information related to wood. The purposes of IWCS are:

1. To collect and disseminate all information considered pertinent and instructive to those interested in collecting and working with wood of woody plants from around the world.
2. To encourage the exchange of wood specimens by members from all parts of the world.
3. To encourage quality and creative craftsmanship with the innovative uses of wood.
4. To encourage the adoption of standard methods for wood sample collecting. This shall include specimen sizes, numbering systems, authenticity, ratings and other details relating to standardization and improvement of individual collections.
5. To assist in accurate identification and classification of wood specimens whenever information is not available from the source of the wood sample.

The **Forest Products Society (FPS)** is an international educational association founded to provide an information network for all segments of the forest products industry—from standing tree to finished product. FPS encourages the development and application of new technology and provides a link for technical interchange between industry and research through the collection and dissemination of the latest technical information.

The Wood Data Sheets (WDS) assembled in this edition were previously published in IWCS's journal *World of Wood* as Wood of the Month columns. The late Max Kline started these columns in 1976 and a continuum of authors has carried on this tradition. Each Wood Data Sheet lists the author. The wood described in the Wood Data Sheets does not necessarily cater to the needs of any particular user group such as the construction industry, cabinetmakers or craftspersons. Rather, they embrace a very wide range of wood that includes Australia's *grand dame*, mountain ash (WDS 116), to the humble, yet extremely attractive, lilac (WDS 255). Species that are included range from those that are becoming scarce, such as Puerto Rico's gommier (WDS 091), once the standard 2 by 4 on the island, to Caribbean pine (WDS 208), which is now planted worldwide on a sustainable yield basis. In this book, the reader is able to visualize the extremely wide span of wood-producing flora. No distinction is made between shrubs and trees as so many wood users have come to recognize the peculiar characteristics of wood regardless of the arboreal stature of the source. Albeit in shorter supply, the wood in our less spectacular shrubs has its own unique density and color attractiveness.

Every effort has been made to record the scientific names that are in current usage and in accordance with recognized taxonomic databases. For this purpose we have used, as the final authority, the USDA Agriculture Research Service's Germplasm Resources Information Network (GRIN). This website can be found at http://www.ars-grin.gov/npgs/tax/taxgenform.html. In the event data was not available in GRIN, we relied upon Index Kewensis. Because there can be considerable movement in the reclassification of plants resulting in scientific name changes, we have provided the synonyms of earlier names to enable the reader to associate these names with their current taxa.

USING THE BOOK

There are 279 species of wood covered herein and a Wood Data Sheet (WDS) number has been assigned to each one. The WDSs have been arranged in alphabetical order by scientific name and appear in numerical sequence by WDS number. A complete set of indices enables the reader to look up a common name or a scientific name and cross-reference it to a WDS. A Family Name Index is provided showing scientific and common names in family groups; the Scientific Name Index includes synonyms.

The standard international format for describing wood has been followed as closely as possible. We have added a color photograph of a wood sample as true to the original wood as was technically possible. Keep in mind, however, that wood is inherently different in terms of density and color even within the same species. The botanical illustrations serve as a reminder that the wood we use once grew as a tree. Because this book is a "wood" book and not published as a "tree" guide, the photomicrographs (PMGs) are of prime importance. The PMGs are magnified images of the wood end grain that were photographed from microscope slides. While there are persons who have vast experience in handling wood on a regular basis and can identify many woods by feel, color, smell and texture, they are rare indeed. Without herbarium material (essentially leaves, bark, flowers and fruit), the next best method for identifying wood is via wood anatomy. Use of the PMGs along with a 10X hand lens will greatly assist the reader in becoming acquainted with this process. Comparing the structure of the wood in question with the PMG is the beginning of a process that will enhance one's proficiency in wood identification.

The 279 photomicrographs were assembled from collectors in many parts of the world and may lack uniformity. We ask that this be overlooked. The vast majority of the images are at a nominal magnification of 35X. The exceptions to this are image numbers 004, 084, 131, 167, 189, 238 and 239, all of which are approximately 10X.

The header for each WDS contains the control number next to the scientific name and the preferred common name. For each species, the following information is provided:

SCIENTIFIC NAME: This name is recorded here as well as any synonyms of recent vintage. In addition, the derivation of both the genus and the specific epithet is given as many of these names provide interesting background data.

FAMILY: The family name is an important taxonomic lineage to enable the reader to associate various genera. For example, the oaks, genus *Quercus* are in the same family (Fagaceae) as are beeches, genus *Fagus*.

OTHER NAMES: A list of names other than the preferred one is noted. These names are cross-referenced in the Common Names Index.

DISTRIBUTION: The geographic area in which the species is endemic is listed as well as areas where the plant may have become naturalized.

THE TREE: A brief description of the tree is provided for the purpose of acquainting the reader with some of the features of the tree that would add greater understanding of the structure of the wood and the importance of the tree. A botanical illustration is provided.

THE TIMBER: The most important features of the wood are described to enable the reader to evaluate the particular species as to color, density, grain patterns, etc. A photomicrograph of the end grain of the wood is provided as well as a color photograph of the wood.

SEASONING: This all-important aspect of wood technology is discussed and shrinkage rates are included when available.

DURABILITY: Aspects of resistance to decay are discussed.

WORKABILITY: Anyone working with wood knows the variability of different species in terms of cutting, turning, finishing, etc., and these characteristics are discussed.

USES: Some woods are in abundant supply while others are extremely rare. A list of the major uses of the wood discussed is provided to give the reader some guidance as to the uses for the wood as related to its availability and characteristics.

SUPPLIES: This information is often hard to assemble, as there is no worldwide "clearing house" for specific woods in commercial trading circles. Many woods are attractive and available only in certain geographical areas and sources of supply must be independently cultivated. (This is one of the interesting and challenging benefits of membership in the International Wood Collectors Society.)

In addition to the Wood Data Sheets, two articles are included that cover subjects of interest to woodworkers: ***Biology and Taxonomy for Woodworkers*** and ***Insights on Wood Toxicity***. Both articles are intended to enhance the overall understanding and enjoyment of wood.

Biology and Taxonomy for Woodworkers discusses woody plants in the context of their overall classification by science within the Plant Kingdom. It also provides a brief overview of taxonomy, the branch of science that deals with finding, describing, defining, naming and classifying living things. Woodworkers and others interested in wood often ask, "How many different kinds of wood or species of trees or other woody plants are there?" It is a simple question with no readily known answer, but it is discussed and some broad estimates are provided. In addition, scientific and popular nomenclature for trees and wood are described along with examples of each for several kinds of trees.

Caution is indeed warranted when working with new woods. In ***Insights on Wood Toxicity*** wood toxicity is discussed in broad terms based on the possible dangers or potency. The hazardous substances found in or associated with cutting or handling wood are discussed in three main categories: irritants, sensitizers and poisons. In many respects, the problem of wood toxicity is one that requires individual management. The magnitude of the risk we face is not just a function of the wood species, but also our individual reactions. Although there is no one correct approach for all woodworkers, some sensible guidelines that are universally appropriate are suggested.

ABOUT THE EDITORS

JAMES H. FLYNN, JR.

Jim Flynn is a native of Fall River, Massachusetts and a retired Department of Defense Senior Operations Research Analyst. Having retired in 1976 after 36 years of service, he is now able to focus his energies on his life-long interest in wood.

While his interests in wood are varied and many, Jim undertook a highly specialized study of two Russian folk musical instruments, the balalaika and the gusli. Not content with just building them, he experimented with a variety of woods to determine the effect on the quality of sound produced. This work culminated in publishing *Building the Balalaika.* The book is a record of the history and dimensional data needed for the five sizes of the instrument and the only book printed in the English language. A similar manuscript was published on the gusli.

Jim is an Associate Editor of *World of Wood*, the journal of the International Wood Collectors Society. In this role he authors the column Wood of the Month. These columns became the first edition of A *Guide to Useful Woods of the World*, published by Jim in 1994. He does freelance work and has recently been engaged by the Discovery Channel in assisting in their book *Trees. Home Furniture*, a publication of The Taunton Press, features a series of his monographs on furniture timbers. Jim continues to add to his collection of 2,200 wood species and make more room for his extensive library of tree and wood related titles.

CHARLES D. HOLDER

Chuck Holder is a native of Alberta, Canada where he grew up on a farm and gained an early appreciation of nature. He went on to earn a BSc degree in Petroleum Engineering at the University of Alberta 1959 and then attended the University of Western Ontario, earning an MBA in 1962. After a 30-year career in industry as an engineer and energy economist, he retired in 1992.

Upon retirement Chuck converted a life-long hobby of woodworking into a custom woodcraft business. A keen amateur dendrologist and wood collector, he has written numerous articles on these subjects. In 1998 he developed the "Canadian Official Trees Wood Sample Kit" and, with the support of the Tree Canada Foundation and Canada's Department of Natural Resources, produced the booklet *Canada's Arboreal Emblems* describing Canada's official trees and their wood.

Chuck is a past Director of SAWS, the Southern Alberta Woodworkers Society, and has served as editor of *SAWSNEWS*. A member of the International Wood Collectors Society since 1993, he was elected Canadian Trustee in 1997 and Vice-President of the Society in 2000. He also serves IWCS on a number of committees and as an Associate Editor of *World of Wood*, the Society's journal.

WOOD DATA SHEETS

Abies balsamea

balsam fir

by Chuck Holder

SCIENTIFIC NAME

Abies balsamea. Derivation: The genus name *abies* is the classical Latin name for fir, probably derived from *abire* meaning to arise. The specific epithet *balsamea* is from the Latin *balsamum*, referring to the resinous pockets in the bark of balsam fir.

FAMILY

Pinaceae (*Abietaceae*), the pine family, a temperate Northern Hemisphere family of trees and shrubs with some nine genera and 210 species.

OTHER NAMES

Canada balsam, balsam, eastern fir, sapin baumier (French).

DISTRIBUTION

The growth range of balsam fir is from Alberta to Newfoundland and south to Wisconsin and New York.

THE TREE

Usually growing to 70 feet in height with a diameter of 24 inches, this medium-sized conifer develops a symmetrical, narrow, conical crown with branches extending nearly to the ground. The tree's maximum height is 160 feet and maximum diameter is 52 inches. The leaves, which are dark green above and whitish below, are needle-like, flattened, rounded at the tip and arranged in two ranks up to 1 inch long. The erect, dark purple and barrel-shaped cones are 2 to 5 inches long. After breaking up in September, a bare axis is left on the tree for several years. On young trees, the bark is smooth and pale gray with resin blisters. It becomes roughened and reddish-brown as the tree matures. Balsam fir may reach 200 years old. Balsam fir is the official tree of New Brunswick, Canada.

THE TIMBER

Balsam fir wood is soft, weak and somewhat brittle. It is white in color with no contrast between the heartwood and the sapwood. It is homogeneous, odorless and slightly resinous. The wood is straight grained and medium textured with moderate growth ring figure. It usually has a fairly large number of smallish knots. It weathers to a gray color with little sheen. The wood is fairly low in bending and compressive strength, in stiffness and in resistance to impact. Average reported specific gravity is 0.33 (ovendry weight/green volume), equivalent to an air-dried weight of 24 pcf.

SEASONING

Balsam fir seasons readily. It sometimes contains wet pockets that may result in defects during drying. Average reported shrinkage values (green to ovendry) are 2.9% radial, 6.9% tangential and 11.2% volumetric.

DURABILITY

Balsam fir is not very decay resistant but moderately resistant to impregnation with preservatives. It is also susceptible to butt rot, woolly adelgid and hemlock looper.

WORKABILITY

It glues readily, holds paint well but is considered below average in machining properties.

USES

Sold as SPF lumber, balsam fir is used in construction and particleboard and plywood manufacture. Large volumes are used for pulp. Resin from the blisters on young trees, refined and commercially known worldwide as Canada balsam, is used in cementing lenses and mounting specimens for observation with a microscope. Balsam fir is favored for Christmas trees due to its lengthy period of needle freshness.

SUPPLIES

Supplies are ample for this important commercial softwood.

Acacia aneura

mulga

by Max Kline

SCIENTIFIC NAME

Acacia aneura. Derivation: The genus name is Greek for thorn. The specific epithet is Latin for nerveless. Recent taxonomic studies in Australia have found that *A. aneura* was a mixture of closely related species. We will make note of this by adjusting the taxon to read *Acacia aneura (sens,lat.)* to indicate "in the broad sense".

FAMILY

Fabaceae or Leguminosae, the legume family; (*Mimosaceae*) the mimosa group.

OTHER NAMES

spearwood.

DISTRIBUTION

Mulga occurs mainly in the extremely dry interior of South Australia, extending into each of the adjoining states.

THE TREE

This tree varies from a small erect tree of 20 feet in height along waterways to a much-branched shrub in deep loamy soil. It grows on a variety of soils to produce a moderate amount of shade from the numerous erect branches. Foliage is gray, palatable to stock and extensively grazed throughout the pastoral arid regions.

THE TIMBER

Mulga is a coffee color or has reddish-brown alternating with golden brown stripes. The sapwood is a golden creamy yellow color. It has an extremely fine even texture and generally straight grain. The luster is generally low to medium, but it takes a high polish. The odor is distinct but not aromatic; taste is not distinct. Mulga is one of the hardest and heaviest woods known. Average specific gravity is 0.92 (ovendry weight/green volume), equivalent to an air-dried weight of 75 pcf. The wood is highly distinctive in appearance.

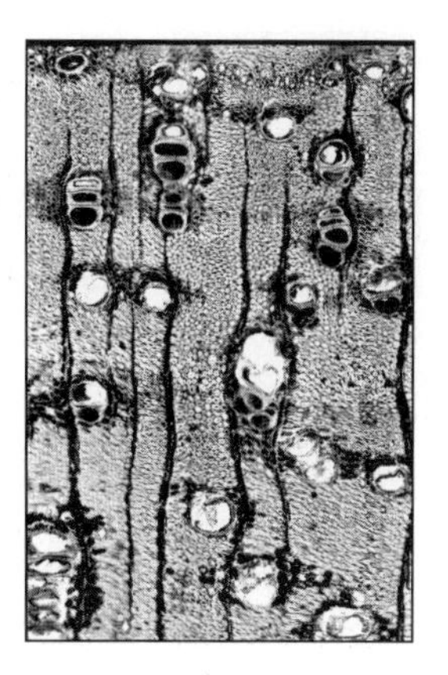

SEASONING

The bark is often stripped from the sawlogs for use in extracting tannin. These debarked logs require special efforts to prevent rapid drying of the wood; this is usually accomplished by spraying the logs with a fine mist of water. The rapid loss of moisture often results in major degradation of the wood from excessive splitting and checking. Shrinkage values are not available.

DURABILITY

The wood, especially the sapwood, is susceptible to attack by powder post beetles. Treatment with preservatives can remedy this problem.

WORKABILITY

Mulga turns well, finishes to a glossy surface and takes a high polish.

USES

Mulga is in demand for making ornamental woodenware such as trays, bowls, egg cups, boxes, candlesticks and carved articles. Mulga has been used by the Aborigines for shields, spears and boomerangs. It makes strong, durable fence posts and provides valuable fuel in the natural state and as charcoal.

SUPPLIES

This timber is very limited in supply beyond its native habitat as none is exported outside Australia. Wood collectors can obtain samples or small supplies from IWCS members living in South Australia.

Acacia auriculiformis

ear-pod wattle

by Jim Flynn

SCIENTIFIC NAME

Acacia auriculiformis. Derivation: The genus name is Greek for thorn. The specific epithet is Latin for ear-like appendage, which refers to the coiled and twisted shape of the seedpod. From this its preferred common name of ear-pod wattle is derived.

FAMILY

Fabaceae or Leguminosae, the legume family; (*Mimosaceae*) the mimosa group.

OTHER NAMES

Darwin black wattle. Australian wattles got their name when early settlers built wattle-and-daub huts using acacia saplings woven or "wattled" together and daubed with wet clay.

DISTRIBUTION

The tree is native to the savannas of Papua New Guinea, to the islands of the Torres Strait and to the northern areas of Australia. It has been naturalized in Florida.

THE TREE

The ear-pod wattle is a fast grower. It is evergreen with dense foliage and an open, spreading crown reaching heights of 100 feet or more with a trunk diameter of up to 2 feet. It is a colonizing species that provides the initial ground cover and shade for the establishment of rainforest species, but it is then unable to regenerate in the shade beneath the closed rainforest canopy. It is a very adaptable plant, and with practically no maintenance it will grow in a wide range of deep or shallow soils, including sand dunes, mica schist, compacted clay, limestone, podsols, laterite and lateritic soils. At Rum Junction, in Northern Australia, the tree grows on spoils from uranium mining. It is the only native woody plant adaptable enough to colonize these uranium spoil heaps. Like other wattles, ear-pod is short-lived having a life span between 10 and 15 years. It is easily raised from seed.

THE TIMBER

The heartwood of ear-pod wattle is dark and reddish-brown, and the sapwood is yellow. The wood is fine grained and of medium density having a specific gravity of 0.60 to 0.75 (ovendry weight/green volume), equivalent to an air-dried weight of 46 to 60 pcf. The tree has poor form for lumber because it grows large lateral branches that are near the ground, which limits its use as a timber crop. In Papua New Guinea, however, some straight-stemmed species have been found and seeds were obtained with the hope of growing a more useful timber stock.

SEASONING

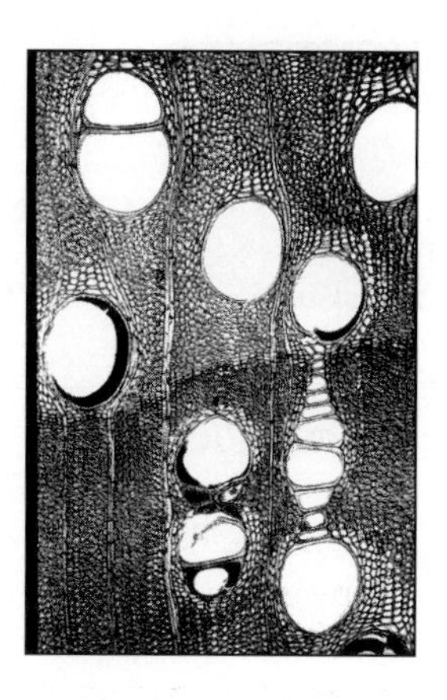

Average reported shrinkage values (green to ovendry) are 2.6% to 4.0% radial and 3.6% to 5.0% tangential.

DURABILITY

The tree is prone to attacks by woodborers and green ants. No data is available on the resistance of the wood to fungal decay.

WORKABILITY

The fine grain of this wood, which is often nicely figured, makes it very easy to work with hand and machine tools. It sands well and takes a good finish.

USES

The full utilization of this acacia has not yet been fully explored. It has known uses as a shelterbelt and shade tree but its potential for more useful applications needs further exploring. In the 1977 report, *Tropical Legumes, Resources for the Future*, the U.S. National Academy of Sciences suggested many avenues in which research could take advantage of the unusual characteristics of this tree. While the wood has been used in a limited way for furniture and paneling, its main use has been for paper pulp, and it is an outstanding fuelwood.

SUPPLIES

The wood is not available on the commercial market. Readers having a xylarium are able to obtain specimens from Australian friends as well as those in Florida who know the whereabouts of this tree.

Acacia greggii

Gregg acacia

by Max Kline

SCIENTIFIC NAME

Acacia greggii. Derivation: The genus name is Greek for thorn. The specific epithet is in honor of Josiah Gregg (1806 to 1850), an early American explorer who collected plants in the Southwest and Northern Mexico. A synonym is *A. greggi*.

FAMILY

Fabaceae or Leguminosae, the legume family; (*Mimosaceae*) the mimosa group.

OTHER NAMES

Gregg catclaw, devil's claw acacia, catclaw, paradise-flower, una-de-gato, wait-a-bit, Texas mimosa, catclaw acacia.

DISTRIBUTION

Southern, central and Trans-Pecos Texas, west to central New Mexico, northwestern Arizona, southwestern Utah, southeastern Nevada, and southeastern California, south to northern Mexico (lower California, Sonora, Chihuahua and Coahuila).

THE TREE

Gregg acacia ranges in height from 10 to 30 feet and has a trunk of 6 to 12 inches in diameter. The tree thrives in poor soils and is irregularly shaped with numerous open-spreading branches. It is called catclaw because of the many sharp-hooked spines, resembling the claws of a cat, on its twigs. The twigs are a light reddish-brown color minutely covered with fuzz. The fragrant yellow flowers furnish an excellent bee food, and the honey has a good flavor. Catclaw often grows in almost impenetrable thickets, and quails eat the seeds.

THE TIMBER

The heartwood is dark chocolate brown sometimes faintly tinged with purplish-red. Frequently it is a bright, clear brown streaked with yellow and red. The sapwood is usually a light cream or yellow color with a beautiful, dainty plain figure. The odor and taste are not distinct. The texture is medium to coarse. A standard 3- by 6- by 0.5-inch sample had a specific gravity of 0.78 (ovendry weight/green volume), equivalent to an air-dried weight of 62 pcf.

SEASONING

This timber air dries fairly well if the usual precautions are taken during the drying process. Shrinkage values are not available.

DURABILITY

Gregg acacia appears to be a durable wood.

WORKABILITY

It works well with power tools and has a "greasy" smooth feeling when working with it.

USES

The wood is used for fuel and small household articles. At one time, the Native Americans of Arizona and Mexico ground the seeds of the Gregg acacia into meal and ate them as mush or cakes. Since it is not available commercially, its uses have not been adequately explored. It should find valuable use for the hobbyist for turnery, small novelties and inlays. The home woodworker will be delightfully surprised when working with it.

SUPPLIES

Generally, Gregg acacia is only obtained by personal collecting by the woodworker or from a local resident in the growth range of the tree.

Acacia koa

koa

by Max Kline

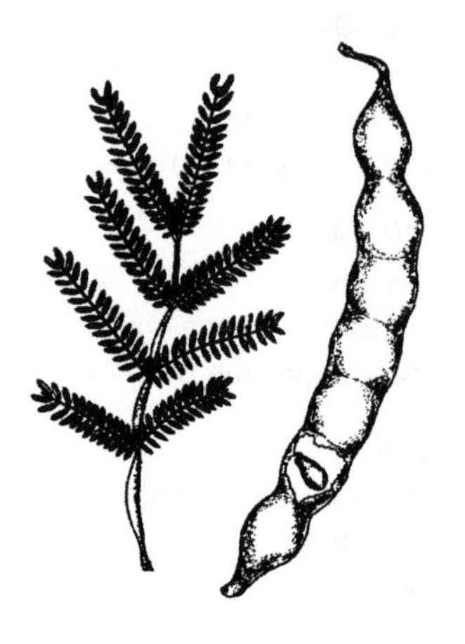

SCIENTIFIC NAME

Acacia koa. Derivation: The genus name is Greek for thorn. The specific epithet is an ancient Hawaiian epithet.

FAMILY

Fabaceae or Leguminosae, the legume family; (*Mimosaceae*) the mimosa group.

OTHER NAMES

koa-ka, Hawaiian mahogany.

DISTRIBUTION

On all of the larger islands of the Hawaiian groups.

THE TREE

Koa is not exacting in its requirements and adapts itself to almost any environment. It is most abundant in regions of heavy rainfall at elevations of 3,000 to 6,000 feet but can be found from a few hundred feet above sea level to almost the tops of the mountains. In heavy rain areas the tree can attain heights of 80 to 100 feet with a diameter of 3 to 4 feet. These trees produce the best straight-grained construction lumber, while those growing at higher elevations show the highest degree of attractive markings.

THE TIMBER

Koa is light to dark brown with a distinct golden luster sometimes with darker streaks. Odor and taste are not distinct. The texture of the wood is fine, and the grain is often curly and attractively figured. Average specific gravity is about 0.55 (ovendry weight/green volume), equivalent to an air-dried weight of 42 pcf.

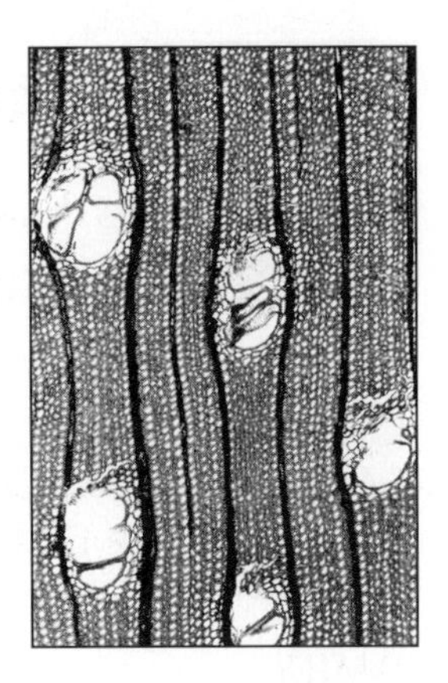

SEASONING

Koa seasons without splitting or warping, and veneer sheets remain flat. Average reported shrinkage values (green to ovendry) are 3.1% to 4.0% radial and 5.1% to 6.5% tangential.

DURABILITY

The seasoned wood is remarkably resistant to insect infestation and fungal decay. Standing trees, however, have many insect enemies.

WORKABILITY

Usually comparable to teak (*Tectona grandis* WDS 261), koa works easily taking a high lustrous finish, sanding smooth and taking glue well.

USES

In the past, Hawaiian natives made dugout canoes from large selected trees and took sea voyages of great distances. Today, koa is the best known wood of Hawaii. Ukuleles are made from koa wood because of its very fine resonant qualities and attractive appearance. Manufacturers of fine cabinets and high priced furniture in this and many other countries hold the wood in high esteem. Frequently used for veneer, the figured varieties are sought for turnery and carving.

SUPPLIES

Plantation-grown wood from many tropical and subtropical climates is now appearing in the commercial market at a competitive price.

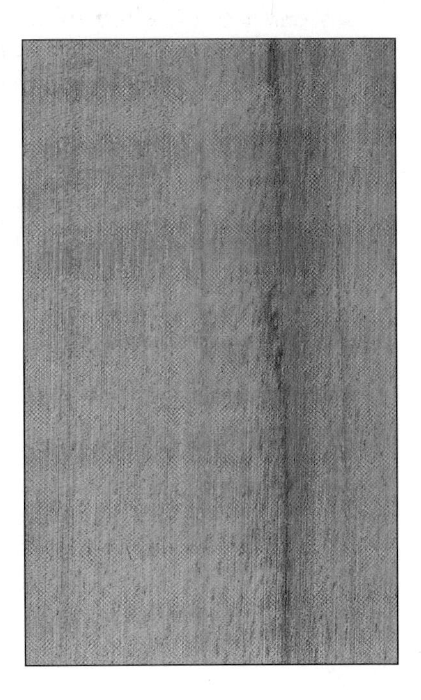

Acacia melanoxylon

Australian blackwood

by Max Kline

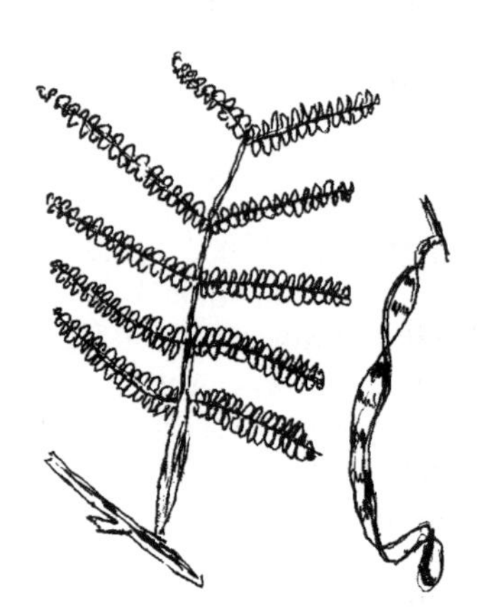

SCIENTIFIC NAME

Acacia melanoxylon. Derivation: The genus name is Greek for thorn. The specific epithet is Latin for black, referring to the color of the wood.

FAMILY

Fabaceae or Leguminosae, the legume family; (*Mimosaceae*) the mimosa group.

OTHER NAMES

Tasmanian blackwood, blackwood.

DISTRIBUTION

Blackwood grows in the mountain districts of southern New South Wales, Victoria, South Australia and is also found in Tasmania.

THE TREE

Normally reaching heights of 30 to 60 feet with a bole diameter up to 20 inches, blackwood varies from a small mountain shrub to heights of 110 feet making it one of Australia's largest acacias. Smaller trees in open situations are freely branched from the ground up, but larger trees have a well-developed clear bole providing ample opportunity for sawn lumber. There are 12 to 15 pairs of oblong leaflets on each of the two to five bipinnate leaves. The leaflets are oblong, 3 to 5 inches long, mid-green on the upper surface and paler below. The whitish or light yellow flowers are in clusters of globular heads. The fruits are legumes, small pods that are 2 to 4 inches long, mid-brown in color and when mature twist in up to two complete coils. The brownish-gray to very dark gray bark is hard and longitudinally furrowed reaching 2.5 inches thick at the base of large trees.

THE TIMBER

Blackwood is a dark golden brown or reddish-brown with dark brown or reddish-brown streaks. It is moderately hard and heavy and average reported specific gravity is 0.57 (ovendry weight/green volume), equivalent to an air-dried weight of 44 pcf. The luster is high. The texture is medium to fine, and the grain is usually straight but sometimes curly with an attractive figure. Odor and taste are not distinct. This timber is known to have relatively high strength properties, with impact resistance comparable to that of ash. It is also reputed to be good for steam bending purposes.

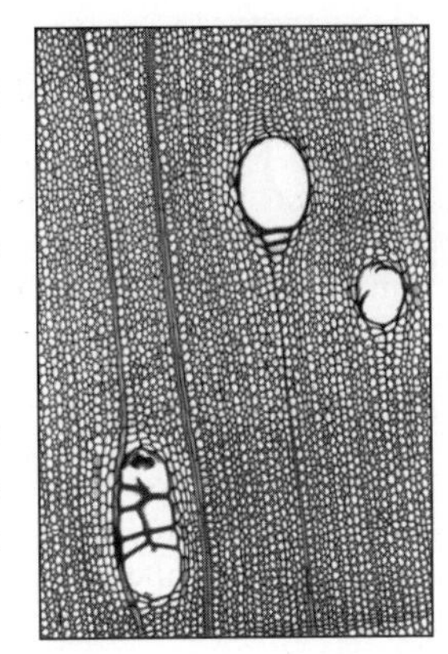

SEASONING

Care has to be taken to see that this species is properly seasoned to about 12% moisture content. It can be easily dried in 1-inch-thick boards without degrade, checking or cupping. Average reported shrinkage values (green to ovendry) are 3.4% radial and 9.0% tangential.

DURABILITY

It is reported in Australia that blackwood is susceptible to damage by powder post beetles and the common furniture beetle. It is, however, very resistant to preservative treatment so this method of improving durability cannot be used.

WORKABILITY

This timber works easily with hand and power tools, but the wood with curly grain requires careful techniques to obtain the finest surface. It takes nails and screws well and polishes to an excellent finish. Generally, glued pieces hold but there is some variability in this respect.

USES

Blackwood is one of the finest woods from Australia for cabinetry. The highly figured pieces make the most beautiful furniture, and it is used extensively in the construction of passenger rail cars, offices, billiard tables and all interior joinery. It is reported to have been successfully used for gunstocks and is excellent for carving and turnery. It can be sliced into veneers and used for paneling.

SUPPLIES

In Australia, supplies are adequate on a commercial scale. Since little is exported, other parts of the world find difficulty in securing blackwood, and the price is high. Wood collectors can obtain samples in small quantities through suppliers in Australia.

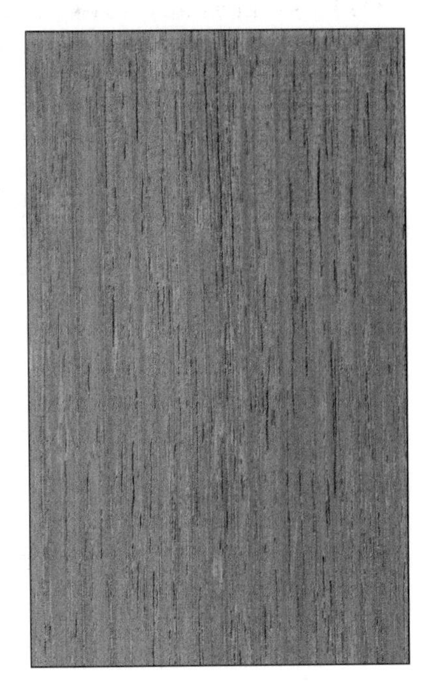

Acer negundo

boxelder

by Max Kline

SCIENTIFIC NAME

Acer negundo. Derivation: The genus name is the classical Latin name for maple. The specific epithet is from the Malayan common name *Vitex negundo*, negundo chastetree, in reference to the similarity of their compound leaves.

FAMILY

Aceraceae, the maple family.

OTHER NAMES

ash leafed maple, negundo maple, three leafed maple, cut leafed maple, Red River maple, Manitoba maple, box-elder, maple-ash.

DISTRIBUTION

Boxelder is found along the borders of streams and lakes from southern Quebec, west to Alberta, south to Florida, Texas and Colorado. It is also found in scattered areas of California and along the eastern slopes of the Rocky Mountains.

THE TREE

One of the maple species, boxelder is easily distinguished from other maples by its compound leaves with three and sometimes, although rarely, five leaflets per leaf as well as the olive-green twigs and large bead-like buds, which mature much later than all other native maples. It grows 30 to 70 feet in height with trunk diameters of 1 to 3.5 feet. Occasionally, a tree can reach a height of 75 feet. The short trunk usually divides into several irregular limbs, forming a well-rounded shapely crown. It grows rapidly, is easily transplanted and is short-lived. Because of its short life, frequent injury from wind and sleet and susceptibility to heart rot and an objectionable insect called boxelder bug, it is not considered a choice ornamental. Generic maple is the official national tree of Canada.

THE TIMBER

The heartwood is yellowish-brown merging rather gradually into the greenish-yellow sapwood. Sometimes the wood has coral red streaks due to the presence of a soluble pigment produced by a fungus. It is the lightest of the American maples. Average reported specific gravity is 0.42 (ovendry weight/green volume), equivalent to an air-dried weight of 31 pcf. It is soft, porous, close grained and weak. Occasionally, specimens are found with wavy or curly grain. This species is relatively low in strength properties.

SEASONING

There is no problem in drying boxelder as it seasons readily without degrade or checking. Average reported shrinkage values (green to ovendry) are 3.9% radial, 7.4% tangential and 14.8% volumetric.

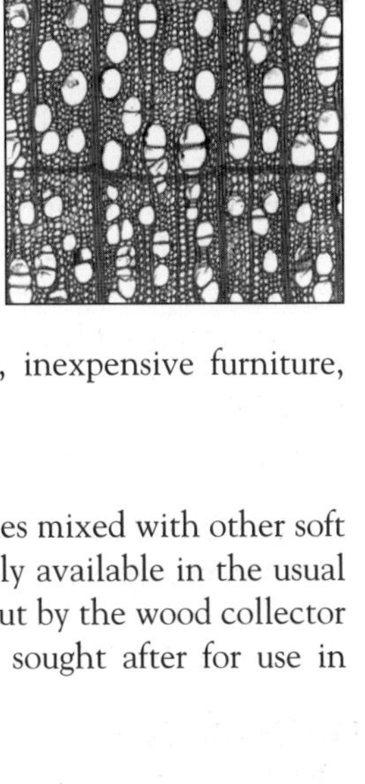

DURABILITY

Boxelder is not a durable timber due to its susceptibility to heart rot.

WORKABILITY

It is easily worked with machine and hand tools.

USES

Boxelder is used for boxes and crates, paper pulp, charcoal, inexpensive furniture, woodenware, fuel and cooperage.

SUPPLIES

Ordinarily boxelder is not cut for timber but sometimes becomes mixed with other soft maples in the trade. Thus, it cannot be classed as commercially available in the usual sense. There are plenty of trees in certain areas which can be cut by the wood collector for hobby use. The fungus red-streaked wood is particularly sought after for use in special cases by the woodworker.

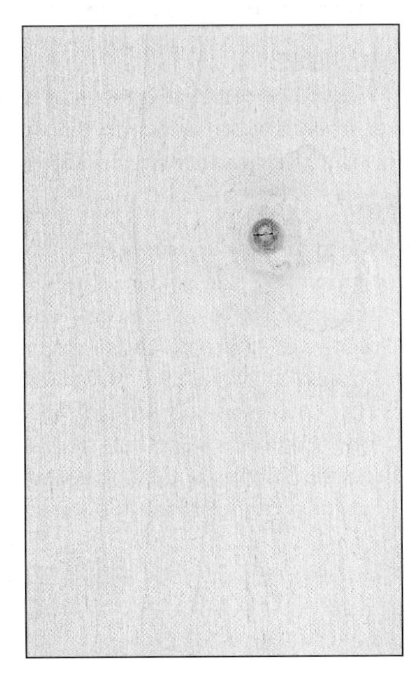

Acer pseudoplatanus

sycamore maple

by Max Kline

SCIENTIFIC NAME

Acer pseudoplatanus. Derivation: The genus name is the classical Latin name for maple. The specific epithet is from the Greek *pseudo* meaning false and *platan* meaning plane tree (*Platnanus* spp.).

FAMILY

Aceraceae, the maple family.

OTHER NAMES

plane tree, great maple, harewood, sycamore plane, sycamore, kaede, tokiwakaede (Japanese).

DISTRIBUTION

This tree has a fairly wide distribution in the temperate regions of Europe. Most of the timber for the market comes from Great Britain where it is a common tree popularly known as harewood. It is also found in Japan. [Ed. note: This species has been cultivated in North America for several centuries. It can be found in abundance along the northeastern Atlantic seaboard, as it is tolerant of saltwater spray.]

THE TREE

The sycamore maple grows to a height of more than 100 feet with a diameter of 5 feet. The straight, cylindrical bole can be 50 to 60 feet long. It is hardy and grows in forests and hedgerows thriving on all but the poorest soil.

THE TIMBER

When freshly cut, the wood is creamy white to yellowish-white, but tends to darken on exposure to pale golden brown. When the timber is allowed to season slowly in the open, the white color is not retained but stains to light brown and is then known as weathered sycamore. Sycamore maple can be dyed a silver-gray color by forcing the dye through the wood under hydraulic pressure; it is then sold on the market as harewood. The wood is usually straight grained and fine textured with a silky luster. Curly and wavy grain are sometimes found which is attractive when used as veneer. Average specific gravity is 0.50 (ovendry weight/green volume), equivalent to an air-dried weight of 38 pcf. Odor and taste are not distinct.

SEASONING

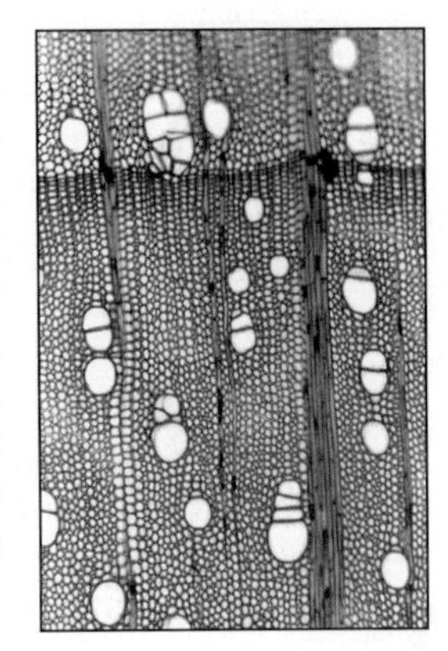

The greatest difficulty when seasoning is to prevent staining and discoloration. Inducing rapid drying immediately after converting to lumber minimizes this condition. Boards should be neither allowed to come in contact with each other nor piled in the usual way. The best methods are to edge pile or lean them separately against a wall for a week before normal piling. Great care is necessary in order to retain the original white color, which is second only to American holly (*Ilex opaca* WDS 147). While precise shrinkage values are not available, they are estimated to be within the range of those for red maple (*Acer rubrum* WDS 009).

DURABILITY

Sycamore maple is not a durable timber, tending to decay rapidly when left in the open. It is easy to impregnate with preservatives, and this property is utilized when dyeing the wood to sell as harewood. One company in London has been successful in achieving a fine, permanent silver-gray color.

WORKABILITY

It works easily with most tools provided they are sharp. When dull, tools will burn the wood quite readily. The figured material has a tendency to chip out when planed unless the cutting angle is reduced. It stains, paints and polishes well and can be glued satisfactorily.

USES

The white wood is in demand for dairy utensils, laundry and textile rollers, turnery, bobbins, brush handles and hat blocks. It has fine wearing qualities when used as flooring. The dyed silver-gray variety is very attractive for modern furniture, interior finish and cabinets. A strong fiddleback figure is in great demand for veneers for attractive craftwork, including turnery and marquetry.

SUPPLIES

Supplies outside Europe are scarce and costly but lumber may be available from the prominent importers. [Ed. note: An abundance of this wood can be found in many places in the United States where the species has naturalized, especially in the coastal section of New England. See Distribution.]

Acer rubrum

red maple

by Jon Arno

SCIENTIFIC NAME

Acer rubrum. Derivation: The genus name is the classical Latin name for maple. The specific epithet is Latin for red, which is the color of the flowers, petioles and autumnal foliage.

FAMILY

Aceraceae, the maple family.

OTHER NAMES

swamp maple, scarlet maple, water maple, soft maple.

DISTRIBUTION

Red maple is native to eastern North America from central Florida to Newfoundland and from eastern Texas to the Atlantic seaboard. It is most common on swampy sites, but also quickly seeds on cut or burned timberlands.

THE TREE

Red maple is not as shade tolerant or as long lived as the hard maples and, while it is capable of attaining heights of more than 100 feet and diameters of up to about 6 feet, it is usually displaced by the more dominant species on the better growing sites. Typical specimens reach about 70 feet with diameters of 18 to 24 inches. The leaves of red maple are similar in shape to those of sugar maple, but usually smaller in size with stems and veins which are noticeably more red in color. Red maple is the state tree of Rhode Island. Generic maple is the official national tree of Canada.

THE TIMBER

Like all of the maples, red maple is a fine textured, diffuse-porous wood with creamy white sapwood and light beige or tannish-brown heartwood. Sometimes, the heartwood exhibits a grayish-green hue. The configuration of anatomical features (rays and pores) makes it virtually indistinguishable from other maples (see Special Note). Although it is classified as a "soft" maple by the lumber trade along with silver maple (*A. saccharinum*) and bigleaf maple (*A. macrophyllum*), it is generally 5% to 7% heavier than these soft maples. With an average reported specific gravity of 0.49 (ovendry weight/green volume), equivalent to an air-dried weight of 37 pcf, it is slightly more dense than black cherry (*Prunus serotina* WDS 223) which has a specific gravity of 0.47, but still well below typical sugar maple (*A. saccharum* WDS 010, specific gravity of 0.56). Although it is slightly stronger than other soft maples and certainly rugged enough for most furniture applications, it is not as strong as sugar maple.

SEASONING

Red maple is easier to season and more stable than sugar maple. Nonetheless, care must be taken to ensure adequate airflow in the drying process to prevent blue staining. Average reported shrinkage values (green to ovendry) are 4.0% radial, 8.2% tangential and 12.6% volumetric.

DURABILITY

Like other maples, this species is not durable when exposed to the elements. It is susceptible to attack by fungi and, therefore, it is a very poor choice for exterior projects.

WORKABILITY

Red maple has virtually all of the favorable working characteristics of sugar maple. Since it is diffuse porous, it turns well, planes well and requires no fillers in preparing the surface to accept a glass smooth finish. In fact, because it is noticeably softer than sugar maple, it is easier to work and its lower volumetric shrinkage (even lower than black walnut) make it more reliably stable in use than sugar maple.

USES

Uses include turned articles, kitchen utensils, toys and novelties, crating, pallets, core stock, upholstered furniture framing and inexpensive cabinets.

SUPPLIES

Red maple is plentiful and inexpensive. Because red maple is sold as a "soft maple", it does not have the high demand or premium price of hard maple.

SPECIAL NOTE

Several technical references suggest there is a chemical test which may be used to differentiate red maple from sugar maple. When a solution of ferrous sulfate is applied to samples of these woods, red maple will turn dark blue in color, while sugar maple will turn green.

Acer saccharum

sugar maple

by Max Kline

SCIENTIFIC NAME

Acer saccharum. Derivation: The genus name is the classical Latin name for maple. The specific epithet is Latin for sugar, referring to the sweetish sap from which maple sugar is made.

FAMILY

Aceraceae, the maple family.

OTHER NAMES

hard maple, rock maple, sweet maple, black maple.

DISTRIBUTION

Sugar maple grows in every state east of the Great Plains, except Florida, South Carolina and Delaware but is most important in the New England states and the Great Lakes states, including the Canadian Great Lakes provinces.

THE TREE

A beautiful symmetrical tree, the sugar maple grows from 70 to 125 feet in height with a trunk 24 to 36 inches in diameter. Its leaves turn gorgeous shades of red and yellow in the autumn. Its name is derived from the delicious maple syrup and sugar obtained by boiling down the sap in the spring. Sugar maple is the state tree of New York, Vermont, West Virginia and Wisconsin. Generic maple is the official national tree of Canada.

THE TIMBER

The heartwood is a delicate, very light reddish-brown or very light tan, while the wide sapwood is nearly white. Sugar maple has a fine uniform texture and is very strong and hard with close straight grain. A small percentage of the trees produce special figures such as bird's eye, blistered, curly or fiddleback. Average reported specific gravity is 0.56 (ovendry weight/green volume), equivalent to an air-dried weight of 43 pcf. Because of the strength and stiffness of this maple species, it ranks as one of the more valuable hardwoods.

SEASONING

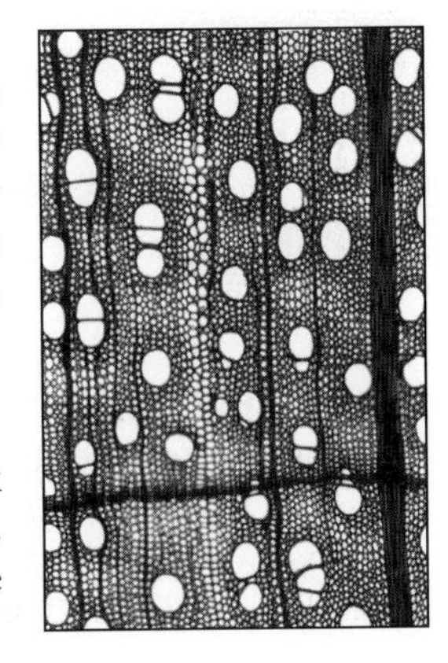

This timber can develop stain during the drying period unless fairly rapid drying is employed. Otherwise, it handles well by both kiln and air drying techniques and seasons with only slight degrade. Average reported shrinkage values (green to ovendry) are 4.8% radial, 9.9% tangential and 14.7% volumetric.

DURABILITY

Sugar maple is not resistant to either fungal or insect attack but is more durable than some of the other maple species. The fire resistant standards are higher than those of the average timber.

WORKABILITY

Sugar maple is well suited to all types of turnery, works well with tools and finishes very smoothly. Glue adheres well and nail and screw holding properties are good. Pre-boring is necessary on thin stock to prevent splitting. It takes stain readily and gives excellent results with paint or enamel.

USES

The finest dance floors, bowling alleys and bowling pins are made of sugar maple. It is also important in piano manufacture for making the part into which the pins for the strings are driven. A great deal of maple is used for furniture, paneling and cabinetry. A special use is for shoetrees and lasts. Wood with figured grain is highly prized for decorative cabinetry and in certain types of stringed musical instruments. It is a good timber for bending.

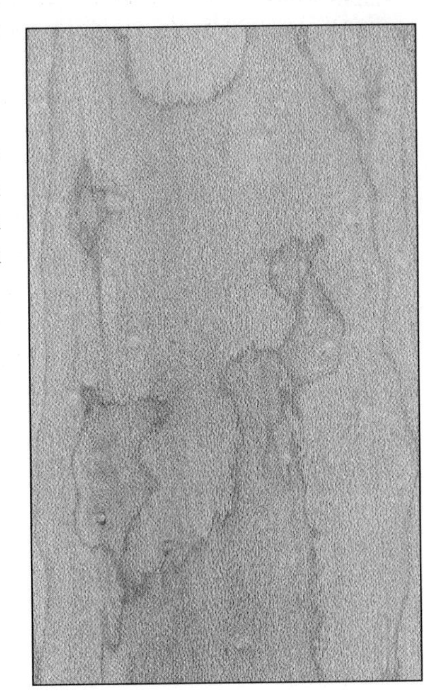

SUPPLIES

Sugar maple with plain straight grain is readily available from many lumber dealers at a medium price range. Wood with figured grains is much more costly and may be classified as rare in availability.

Aesculus glabra

Ohio buckeye

by Max Kline

SCIENTIFIC NAME

Aesculus glabra. Derivation: The genus name is the ancient Latin name of a European oak or other mast-bearing tree. The specific epithet is Latin for glabrous or hairless, referring to the foliage.

FAMILY

Hippocastanaceae, the horsechestnut family.

OTHER NAMES

stinking buckeye, fetid buckeye, buckeye.

DISTRIBUTION

From central Pennsylvania west to Nebraska, south to Alabama and Texas.

THE TREE

Ohio buckeye trees grow from 30 to 70 feet in height with a trunk seldom more than 2 feet in diameter. It is one of the first trees to leaf in the spring. The flowers and twigs have a disagreeable odor when crushed or broken. The large, smooth seeds resemble the eye of a buck deer, hence the name buckeye. Ohio buckeye is the state tree of Ohio.

THE TIMBER

The wood is almost white in color with occasional stain streaks that are light gray. It is odorless and tasteless when dry, but has a mildly unpleasant scent when fresh. The wood density is rather low, but for its weight it is strong and tough. Average specific gravity is 0.38 to 0.50 (ovendry weight/green volume), equivalent to an air-dried weight of 28 to 38 pcf. The texture is very fine with a straight to wavy grain. When cut radially, pronounced ripples can be seen. The luster is fairly high.

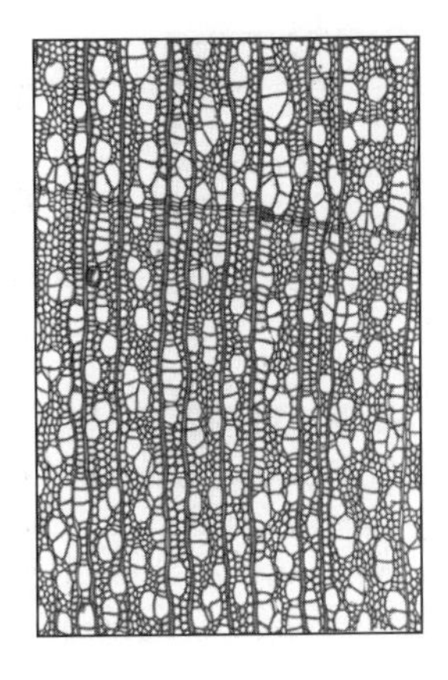

SEASONING

Ohio buckeye seasons well with little checking or cracking when kiln or air dried. Shrinkage values are not available.

DURABILITY

The wood has low resistance to decay. It is subject to a fungal disease, which can be controlled by spraying and disposing of all fallen infected leaves and fruit.

WORKABILITY

It is easily worked with all types of tools, but sometimes difficult to split in straight sticks due to twist. It finishes smoothly and presents a good surface for painting if desired. The stained variety should be treated with a natural finish.

USES

Some Ohio buckeye lumber is used in the manufacture of artificial limbs and splints because of its strength and light weight. It is also used for various types of woodenware, paper pulp, boxes and cases. The gray-stained material has a very unusual effect when used by woodworkers for turnery or carving and is often sought by wood collectors.

SUPPLIES

Ohio buckeye is rather limited on the commercial market, and it is often mixed with other species and not sold as buckeye. For the wood collector who prefers the stained wood with ripple figure for small projects, supplies are adequate.

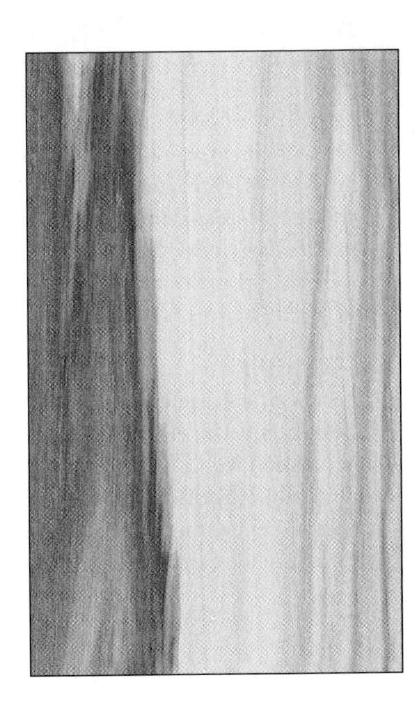

Afrocarpus falcata

podo

by Jim Flynn

SCIENTIFIC NAME

Afrocarpus falcata. Derivation: The genus name refers to the African origin of this tree as opposed to the *Podocarpus* species. The specific epithet is Latin meaning sickle-shaped, which refers to the leaves. A synonym is *Podocarpus falcatus*. There are some taxonomists that place this species in the genus *Nageia*.

FAMILY

Podocarpaceae, the podocarp family.

OTHER NAMES

East African yellowwood, yellowwood, wiri, mse, mushunga, musenene, sapta.

DISTRIBUTION

The tree is widely distributed in the highlands of East Africa, mainly in Kenya and south to South Africa.

THE TREE

Growing under various conditions in moist forests of southern Africa, in coastal swamp forests, in wooded ravines and in patches of mountain forests, this evergreen (gymnosperm) tree reaches 100 or more feet in height with diameters of 1.5 to 2.5 feet. The bark is thin, smooth and grayish-brown to dark brown in color. The small, sometimes sickle-shaped leaves are dark green, often with a grayish bloom. They are hard, leathery and narrow with a sharp pointed apex. Their margins are smooth and petioles short. The small cones are about 1.5 to 0.5 inches in length and produce only one seed. The tree is protected in South Africa and is becoming a popular garden plant.

THE TIMBER

The wood is a uniform light yellowish-brown with no clear distinction between the sapwood and the heartwood. It sometimes shows red streaks due to the presence of compression wood. The texture is very fine and even; the grain is straight. Growth rings are usually indistinct and resin ducts absent. Average specific gravity is 0.43 (ovendry weight/green volume), equivalent to an air-dried weight of 43 pcf. Because of the straight stems, this wood was once used for ship masts.

SEASONING

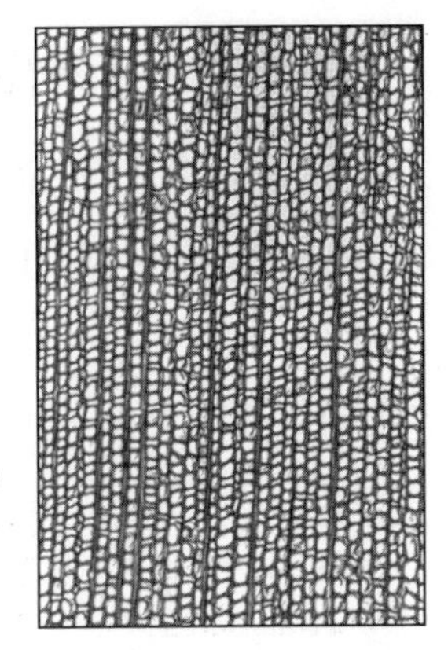

The wood has a long history of being unstable during seasoning because of its high degree of warping. Stickered piles that are heavily weighted tend to alleviate some of this problem. Average reported shrinkage values (green to ovendry) are 2.5% radial and 4.0% tangential. It seasons rapidly and is quite dimensionally stable when in service.

DURABILITY

The heartwood has low durability and is prone to damage from termites and other insects. It is easy to treat with preservatives, however, in both open-tank and pressure systems.

WORKABILITY

The wood is very easy to work with hand and machine tools and is similar to white pine. It takes an excellent finish. It shapes and turns well, glues satisfactorily and is easy to veneer. It has good steam bending qualities.

USES

Podo is a good, general, all-around useful wood. It is considered a very good substitute for the various species of soft pine but the price differential prohibits exploiting this. It is used for general construction, joinery, millwork, furniture components, boxes and crates, food containers and utility plywood. A friend of mine who spent many years in Africa knows a person who made a gunstock of podo and it twisted and warped terribly. On the other hand, he knew a person who made fine furniture from this wood and treasures it for its beauty.

SUPPLIES

Podo is probably not as plentiful as it was in the past because of the heavy demand and limited growth area. Evidence exists that it can be obtained on the commercial market, especially in Europe.

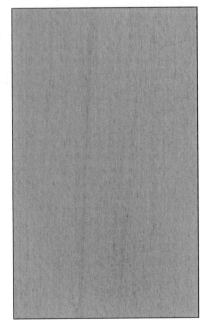

Afzelia quanzensis

chanfuta

by Jim Flynn

SCIENTIFIC NAME

Afzelia quanzensis. Derivation: The genus name is in honor of Adam Afzelius (1750 to 1837), a Swedish botanist. The specific epithet is in recognition of Cuanza, an African River. The name was originally misspelled and so remains in accordance with botanical nomenclature rules.

FAMILY

Fabaceae or Leguminosae, the legume family; (*Caesalpiniaceae*) the cassia group.

OTHER NAMES

pod-mahogany, mahogany bean, chanfuti, peulmahonia, afzelia, mkehli.

DISTRIBUTION

Several species of *Afzelia* can be found in the west, central and east regions of sub-Saharan Africa. The species A. *quanzensis* is distributed in both the lowland rain and dry forests and savannas in the coastal belt of eastern Africa from Kenya to South Africa.

THE TREE

This is a very large and spreading tree which reaches heights of 70 to 80 feet. Its straight bole, however, rarely exceeds 12 to 20 feet. In ideal growing conditions, its diameter can grow to 4 feet. The greenish-gray bark tends to flake in circular patches giving the tree a distinctive pitted effect. The alternate, compound leaves have four to six pairs of opposite or nearly opposite oblong-elliptic leaflets. The fruit is a large, flat, dark brown, thickly wooded pod. Embedded in the white pith inside the pod are six to ten distinctive dark blackish-brown seeds. The seeds are used for necklaces and sold as curios.

THE TIMBER

Chanfuta is light-colored when freshly sawn and turns rich reddish-brown when exposed to air. It has the appearance of a dense grade of a beautiful mahogany (*Swietenia* spp.). The texture is medium to coarse with interlocked grain. Wide boards can be obtained but the lengths are generally short. It lacks a characteristic taste or odor. The wood is related and somewhat similar to the *Intsia* spp. (WDS 148), a testimony to its beauty and usefulness. Average reported specific gravity is 0.67 (ovendry weight/green volume), equivalent to an air-dried weight of 52 pcf.

SEASONING

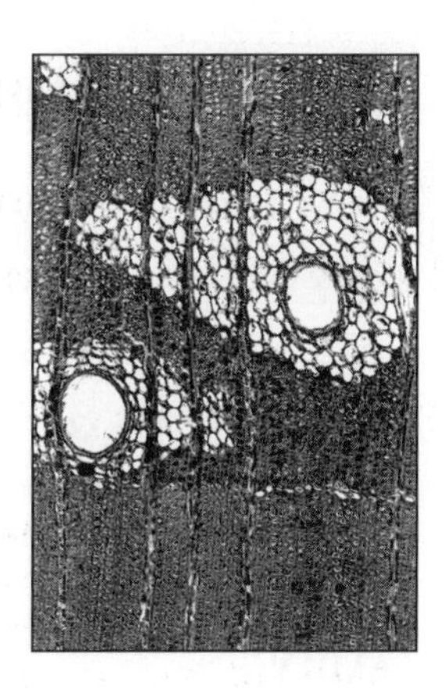

Chanfuta is stable when seasoned. It dries slowly with very little or no degrade. Average reported shrinkage values (green to 12% moisture content) are 1.0% radial and 1.5% tangential.

DURABILITY

The wood is very durable and moderately resistant to termite attacks. As with many other woods, the sapwood is vulnerable to powder post beetles. Its real glory is that it is resistant to the teredo in maritime applications. (The teredo, *Teredo navilis*, is a saltwater mollusk resembling a worm. It bores holes up to 1 inch in diameter through wood, but the holes never touch. It can honeycomb and destroy a 12-inch piling in 6 months or less.)

WORKABILITY

Because this wood is relatively hard, all tooling for its machining must be appropriate and cutting edges sharp and properly angled. Interlocked grain may be troublesome. It finishes well and takes a high polish.

USES

This strong and durable timber is used for upscale joinery in both indoor and outdoor applications. It is especially attractive when used for doors, window frames, floors and staircases. In its native locale, it is a favorite wood for highly ornamental carved doors and chests. Many of the Indian Ocean dhows are constructed of chanfuta.

SUPPLIES

Commercially, no distinction is made between the different species of *Afzelia* and they are marketed under the common name of chanfuta or one of its local names. Supplies seem to be adequate. Rio Rivuma, a private conservation group dedicated to establishing conservation practices in critical habitats, has been working with the *Mociboa da Praia Biosphere Reserve* in Mozambique to guide the harvesting of chanfuta (and other woods) to ensure that it is marketed in an environmentally responsible manner.

Agathis australis

kauri

by Max Kline

SCIENTIFIC NAME

Agathis australis. Derivation: The genus name is Greek for ball of twine. The specific epithet is Latin meaning the plant is "of the Southern Hemisphere".

FAMILY

Araucariaceae, the araucaria family.

OTHER NAMES

kauri-pine, New Zealand kauri, kaurikopal.

DISTRIBUTION

Agathis australis wood has a restricted distribution and is confined to the northern area of the North Island of New Zealand. Other species of kauri (not to be discussed) grow in Australia, Fiji, the Malay Peninsula, Borneo, the Philippines and other Pacific Islands. These related woods are all of inferior quality.

THE TREE

Kauri grows to an average height of 100 feet with an average diameter of about 3.5 feet. Specimens have been recorded which attained 200 feet in height and 17 feet in diameter with a straight, clear bole for 100 feet, classifying them as one of the true forest giants of the world. The tree constantly sheds its resinous, lead to ashy-gray colored bark in flakes causing mounds up to 5 feet deep around the base. Mature leaves are bright glossy green, up to 3 inches long, somewhat egg-shaped and blunt in the apex. Female cones are nearly round, 1.5 to 2 inches in diameter with a number of flat brown seeds similar to sunflower seeds.

THE TIMBER

This timber is a valuable softwood with fine straight grain, an even, silky texture and a lustrous surface. Normally a light biscuit color, where there is heavy infiltration of resin the wood may be yellowish or reddish-brown. Grain irregularities can produce an attractive mottling. There is no clear distinction between spring and summerwood. The odor is pleasant and agreeable when the wood is worked. Claims have been made that kauri is among the strongest of the world's softwoods. Average reported specific gravity is from 0.41 to 0.47 (ovendry weight/green volume), equivalent to an air-dried weight of 31 to 36 pcf.

SEASONING

Kauri is a wood that dries rather slowly. Average reported shrinkage values (green to ovendry) are 4.2% radial and 6.0% tangential. After kiln drying, this species is much better than normal in stability.

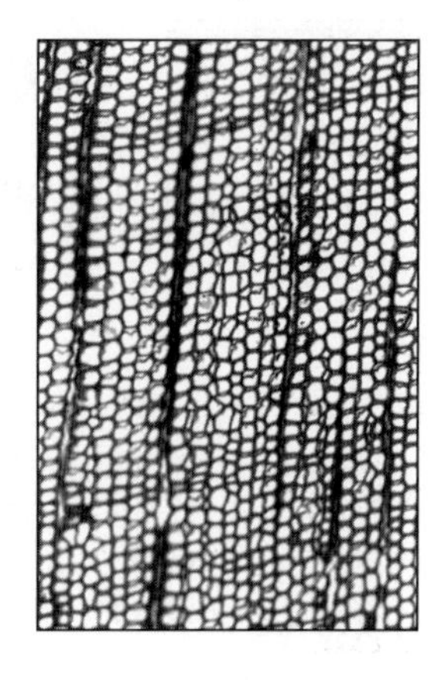

DURABILITY

For softwood, kauri has exceptionally high natural resistance to fungal attack, but is less satisfactory in regard to wood-boring pests. Felled logs are subject to severe damage from the kauri weevil when left in the forest. It is very durable under hot, wet conditions such as in vats for industrial processes.

WORKABILITY

This timber works easily by hand and machine tools with only a very slight dulling effect on the cutting edges. Worked edges remain sharp, and the wood sands well. Kauri may be used for turnery with good results. Nail and screw holding properties are good. Glue adheres well, and it responds satisfactorily to all finishing agents, including polish.

USES

Because of its outstanding qualities and the large sizes of clear material available, kauri was formerly one of the principal timbers used for vats and tanks and in ship building. Presently, the limited quantities are restricted to special uses in New Zealand, such as vats, wood machinery and boat building.

SUPPLIES

Kauri was exploited in the past, so currently only small-dimension timber is available and practically none is exported from New Zealand. Wood collectors can obtain samples from members in New Zealand.

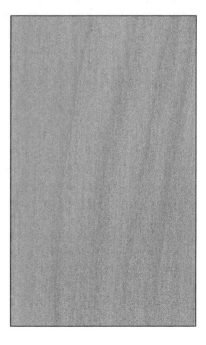

Ailanthus altissima

ailanthus

by Max Kline

SCIENTIFIC NAME

Ailanthus altissima. Derivation: The genus name is the Moluccan name *aylanto* meaning tree-of-heaven, referring to the height of the tree. The specific epithet is Latin for very tall.

FAMILY

Simaroubaceae, the ailanthus or quassia family.

OTHER NAMES

Chinese sumac, tree-of-heaven, paradise-tree, copal tree, heavenwood, A Tree Grows In Brooklyn, stinking chun.

DISTRIBUTION

Ailanthus is native to China but has become well established throughout the eastern United States, particularly in the north central states from Missouri to Pennsylvania. It has spread throughout the southern half of the United States to California.

THE TREE

The tree grows very rapidly in poor soil and environments and attains a height of 60 to 80 feet with a diameter of 2 to 2.5 feet. It can crowd out other more desirable trees if allowed to spread. The flowers and leaflets have a disagreeable odor. The alternate pinnate leaves are up to 24 inches long with 15 or more pairs of leaflets each 4.5 inches long and 2 inches wide ending in a tapered point. The greenish-yellow flowers with five or six petals are in the form of large panicles, and the fruit is a single winged seed 1.5 inches in length. The grayish-brown bark is streaked with gray.

THE TIMBER

The heartwood of ailanthus is pale yellow with numerous dark streaks, while the sapwood is rather wide and light creamy tan in color. It has conspicuous rays and is ring porous. Average reported specific gravity is 0.49 (ovendry weight/green volume), equivalent to an air-dried weight of 37 pcf. Odor is absent, and the taste is mildly to decidedly bitter. The grain is straight, like ash, but its annual rings are much wider. The texture is fine to rather coarse, and planed wood has a silky sheen.

SEASONING

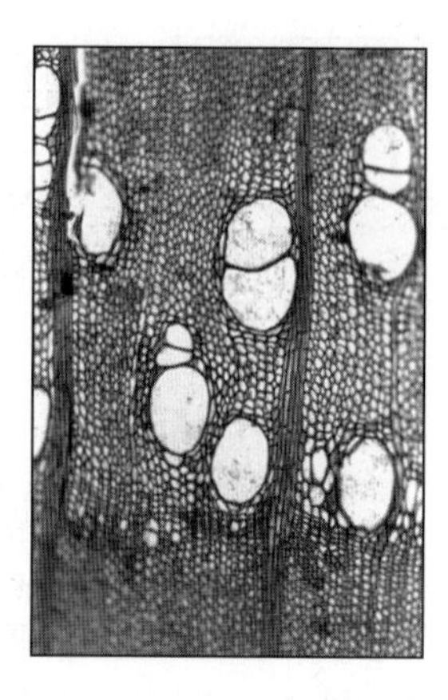

Usually there are no problems when air drying this timber. The moisture content stabilizes rather rapidly. Average reported volumetric shrinkage (green to ovendry) is 10.81%.

DURABILITY

Ailanthus is reputed to be resistant to insects, but not to decay fungi.

WORKABILITY

It is easily worked with tools, glues well and takes a good finish of any kind.

USES

Ailanthus is planted mainly for ornamental purposes and has very little, if any, commercial value except for fuel and for the home woodworker. The quartersawn timber is particularly attractive and should find more uses in the home workshop for novelties, where it is especially desirable for turnery projects.

SUPPLIES

This wood is not marketed commercially, but is abundant and can be cut by the collector. Many foresters consider it to be a serious forest weed and welcome its cutting and elimination.

Albizia lebbeck

lebbek

by Jon Arno

SCIENTIFIC NAME

Albizia lebbeck. Derivation: The genus name is in honor of Cavalier Filippo de' Albizzi, from the old and noble Italian family. Albizzi introduced this genus into Europe in 1749. The specific epithet is the Arabic common name. Like Queensland walnut (*Endiandra palmerstonii* WDS 110) in the laurel family, tigerwood (*Lovoa trichilioides* WDS 116) in the mahogany family and mansonia (*Mansonia altissima* WDS 175) in the Sterculiaceae (cacao) family, lebbek is one more walnut pretender and is not related to the true walnuts in the family Juglandaceae.

FAMILY

Fabaceae or Leguminosae, the legume family; (*Mimosaceae*) the mimosa group.

OTHER NAMES

woman's tongue, kokko, siris, East Indian walnut.

DISTRIBUTION

Lebbek originates in Southeast Asia, but has been so widely planted it is now pan tropic in distribution. Although not actively cultivated, lebbek has naturalized in southern Florida.

THE TREE

Not quite as robust as some albizias, lebbek is a medium-sized deciduous tree attaining a height of 20 to 40 feet, although it can reach a height of 90 feet with a maximum diameter of 3 feet. The tree is slender and erect with widely spreading branches that form a crown of 25 to 30 feet. Like most legumes, the tree produces a wind-rattling seed pod that is 4 to 8 inches long and 1 to 1.5 inches wide. The two to four pairs of leaves, each with many oblong leaflets that are 0.75 to 1.75 inches long and 0.5 inch wide, present a finely divided and fern-like appearance.

THE TIMBER

The sapwood is light beige-white contrasting sharply with the warm, reddish-brown heartwood. The grain is somewhat coarse textured and heavily flecked. The vessel flecks are longer than in American mahogany and they are not tiered, which gives the wood a tangential figure similar to the lauans, but with more surface luster. The grain is interlocked, producing a ribbon figure on radial surfaces. Because of its color and texture, an experienced woodworker might confuse lebbek with some of the so-called mahoganies, but should easily recognize that it is not a true walnut, both from its appearance (lack of pronounced figure) and lack of odor. Average reported specific gravity is 0.51 (ovendry weight/green volume), equivalent to an air-dried weight of 39 pcf.

SEASONING

Although shrinkage is about average, both surface checking and end splitting make lebbek moderately difficult to season. Average reported shrinkage values (green to ovendry) are 2.9% radial, 5.8% tangential and 9.6% volumetric.

DURABILITY

The heartwood is moderately durable.

WORKABILITY

With an average specific gravity of 0.51, lebbek and black walnut (*Juglans nigra* WDS 150) are virtually identical in terms of density, but that is where the similarity ends. Due to its interlocked grain, lebbek is difficult to machine. Although it will sand to a smooth, lustrous surface, the dust irritates the eyes and nose. Lebbek's radially cut ribbon-grained veneer is attractive. Since it is a tropical timber and lacks pronounced growth rings, the figure of both rotary cut and tangentially cut surfaces is bland in comparison to walnut. Overall, lebbek certainly should not be viewed as inferior or a poor choice relative to other imports for most woodworking projects. It is stronger and more stable, for example, than comparably dense lauans. It just is not in a class with walnut.

USES

Better known in Europe than in the United States, lebbek is used for cabinetry, joinery, furniture, decorative veneer and flooring.

SUPPLIES

Adequate. In fact, owing to the zeal with which foresters throughout the tropics have adopted this species and other *Albizias*, supplies may actually increase in the future. The price is moderate.

Aleurites moluccana

kukui

by Jim Flynn

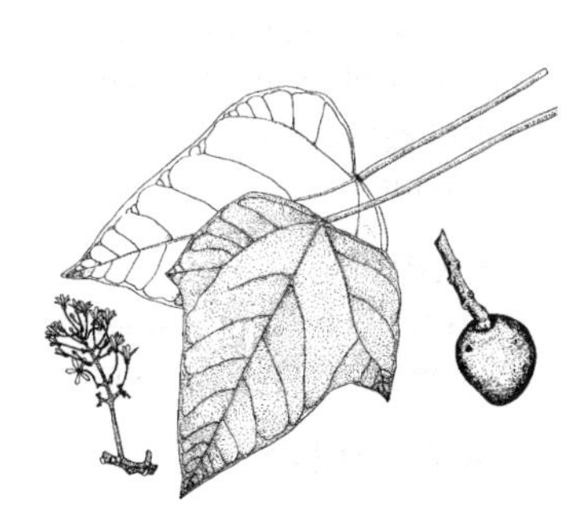

SCIENTIFIC NAME

Aleurites moluccana. Derivation: The genus name is Greek meaning "covered with a mealy powder". The specific epithet means the plant is "of the Molluccan Islands" in recognition of its native growth area.

FAMILY

Euphorbiaceae, the spurge family.

OTHER NAMES

Kukui is the official Hawaiian name. It is also known as candlenut-tree, candleberry, tung, varnishtree, lumbangtree, Indian-walnut, candlenut, noz da India, camirio, calumban, lumban, noyer de bancoul, Lichtnussbaum and noyer des Moloques.

DISTRIBUTION

Originally native to the South Pacific, kukui has spread widely through the Pacific Islands. It has been cultivated and has naturalized elsewhere in the tropics.

THE TREE

Kukui is the state tree of Hawaii. A partial quotation of the Legislative Resolution incident to this declaration is as follows: "WHEREAS, the multiplicity of its uses, to the ancient Hawaiians for light, fuel, medicine, dye and ornament and its continued value to the people of modern Hawaii, as well as the distinctive beauty of its light green foliage, which embellishes many of the slopes of our beloved mountains, causes the kukui tree to be especially treasured by the people of the Fiftieth State of the United States as an arboreal symbol of Hawaii". The trunk is often twisted and sprawling on the ground. When cramped in narrow gorges, however, the tree grows straight up to heights of 80 feet or more with diameters of 36 inches. The flowers are often used for leis. The fruit is a hard, woody, wrinkled and somewhat poisonous nut which is slightly smaller than an English walnut. Producing a valuable and useful oil, the nuts were often strung on the midribs of coconut leaves and ignited as candles. From this the common name candlenut tree is derived.

THE TIMBER

While the tree has been honored in Hawaii, there have been no such plaudits bestowed upon the timber. To be quite frank, there does not seem to be very much said in its favor. The timber is white often varying to creamy white with a variety of deeper tan and gray stripes coursing through it. It is rather light weight, quite similar to butternut (*Juglans cinerea* WDS 149), but without the pronounced grain patterns. The luster is low, and the texture varies from fine to coarse. There is neither discernible odor nor

taste. The specific gravity, calculated locally from a dry standard-sized wood sample, was 0.40, equivalent to an air-dried weight of 30 pcf.

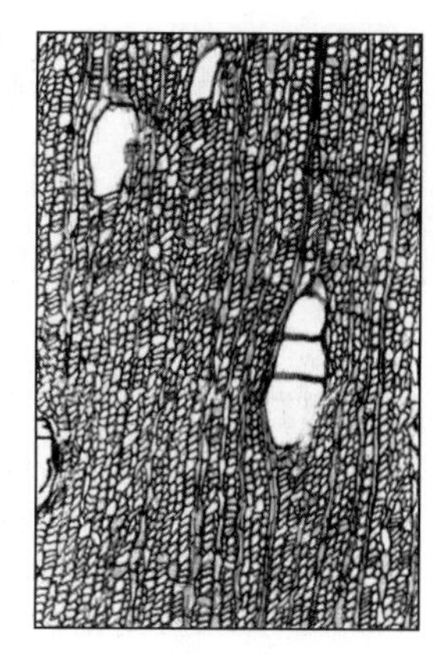

SEASONING

Kukui wood must be treated with an anti-fungicide as soon as it is cut to prevent blue stain from contaminating the timber. Common household bleach is a good treatment for this condition. Warping and checking have not been recorded but probably are known to wood users. Average reported shrinkage values (green to ovendry) are 4.5% radial and 7.0% tangential.

DURABILITY

This timber is not durable and should not be used in open areas exposed to moisture and ground contact. The effectiveness of preservative treatments is not known.

WORKABILITY

It is very easy to work because of its relatively low density and even grain structure. Assuming that the woodworker can acquire this timber without blue stain (which does not affect the quality of the wood), the wood can be run through shop machinery with ease. In spite of the softness of the wood, it can be sanded and finished acceptably. Gluing presents no problem.

USES

The timber was used for making canoes and floats for fishing nets. Products from the tree include flowers for leis and oil from the nuts. While these take precedence over uses for the timber, the wood appears ideal for carving, making lightweight models and in unexposed furniture applications.

SUPPLIES

This is not a commercial timber, but supplies are available where trees grow. Standard wood samples are relatively easy to acquire.

Alnus rubra

red alder

by Max Kline

SCIENTIFIC NAME

Alnus rubra. Derivation: The genus name is the classical Latin name for alder. The specific epithet is Latin for red, referring to the wood turning reddish-brown.

FAMILY

Betulaceae, the birch family.

OTHER NAMES

Oregon alder, western alder.

DISTRIBUTION

Along the Pacific coast from Alaska to central California.

THE TREE

One of the few northwestern hardwoods, red alder grows best along watercourses in fertile low lands and on slopes where moisture is abundant and rains are frequent. It grows to a height of 60 to 90 feet, but more frequently only to 35 to 40 feet with a diameter of 10 to 15 inches. Red alder is fast growing and reaches maturity in 50 years. It is often planted for roadside and home shade.

THE TIMBER

The sapwood and heartwood are not clearly defined. When freshly cut, the timber is almost white, turning light reddish-brown upon exposure to air. The grain is usually straight, while the texture is close and even. Average reported specific gravity is 0.37 (ovendry weight/green volume), equivalent to an air-dried weight of 27 pcf. While the coloring is delicate, the wood is without any outstanding figure. It is odorless and tasteless and has a low luster. Red alder is moderately strong for a lightweight wood but cannot be recommended where high strength is a requirement.

SEASONING

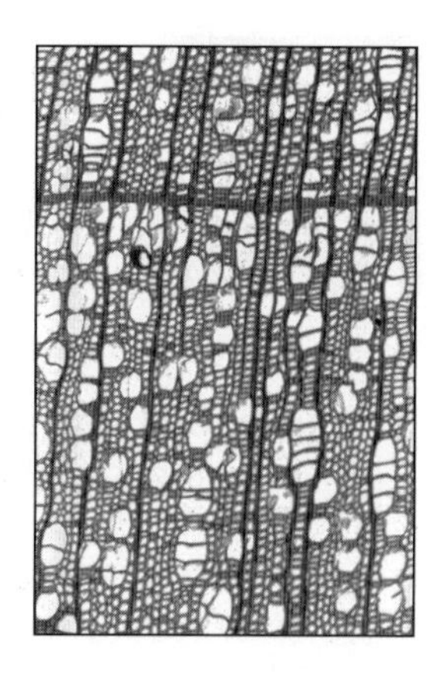

The rate of degrade during seasoning is small, and it may be kiln dried without undue trouble. Average reported shrinkage values (green to ovendry) are 4.4% radial, 7.3% tangential and 12.6% volumetric. Once properly seasoned, the stability factors are above average.

DURABILITY

Standing trees are often susceptible to an injurious fungus which causes heart rot. The timber is not durable in the ground and may also suffer damage from wood-boring pests and beetles. It stands up relatively well to underwater conditions.

WORKABILITY

Red alder is easily worked with tools, but they should be sharp to obtain the best surfaces. Nail and screw holding properties are not particularly good, but glue adheres well. In spite of its softness, it can be turned with satisfactory results and is a useful species for carving, either by hand or machine. It finishes quite easily, takes stain nicely and can be used to match other cabinet woods in color. Classified as one of the easiest commercial timbers to peel, red alder is often used for veneer. It is also adapted to uses where some elasticity is needed since it ranks better than most hardwoods in this respect.

USES

Since it holds its shape well after seasoning, an important use is as a core for valuable woods such as walnut or mahogany. Other uses include veneer, furniture of all types and cabinetry.

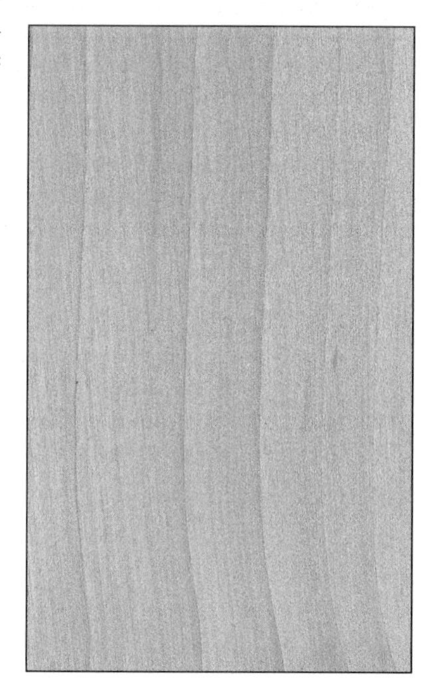

SUPPLIES

The veneer is scarce but the lumber is plentiful, especially on the west coast of the United States. The price range is in the lower bracket.

Alstonia scholaris

white cheesewood

by Jim Flynn

SCIENTIFIC NAME

Alstonia scholaris. Derivation: The genus name is in honor of Dr. Alston (1685 to 1760), a Professor of Botany at Edinburgh. The specific epithet describes the uniform method with which the branches are formed.

FAMILY

Apocynaceae, the dogbane family.

OTHER NAMES

Palimira alstonia, pulai, milky pine, milkwood, Ruk-attana, lettok, devil-tree, scholar-tree.

DISTRIBUTION

White cheesewood is a western Pacific species of *Alstonia* and is found in the Indo-Malayan region, Australia, Polynesia, India and Myanmar (Burma).

THE TREE

White cheesewood is an evergreen, which under favorable conditions grows to 50 feet or more in height with a diameter up to 4 feet. The main stem is fluted, and the base often buttressed. The tree exudes a large amount of milky juice from which one of its common names, milky pine, is derived. It is adaptable to many types of soils, sandy beach areas and on the fringes of rain forests. The leaves are simple and arranged in whorls. They are dark green above and pale green below. White cheesewood is proving to be a hardy and attractive garden tree, especially in the relatively hostile climate of Darwin, Australia.

THE TIMBER

White cheesewood is soft and even-grained with pinkish-brown heartwood that is often streaked with darker tones. The grayish-tan sapwood is clearly differentiated from the heartwood. The texture is moderately fine. Average reported specific gravity is 0.34 to 0.40 (ovendry weight/green volume), equivalent to an air-dried weight of 25 to 30 pcf.

SEASONING

Older reports from Myanmar indicate that this wood needs special care while seasoning to prevent stain. Other reports from Southeast Asian regions state that the wood dries easily with little or no degrade. The weather conditions under which the freshly cut wood is cured are a contributing factor to how well the wood dries. Average reported shrinkage values (green to ovendry) are 3.4 % radial and 6.1% tangential.

DURABILITY

White cheesewood is prone to stain, decay and powder post beetle attack.

WORKABILITY

Because of its favorable grain and density, it is easy to work with hand and machine tools. It turns well.

USES

In Myanmar, the wood is used for a variety of utilitarian purposes including coffins, blackboards, cheap furniture and carved items. Specific uses in other parts of the region include plywood, crates and general carpentry. One of its more interesting former uses was for boxes to ship bulk tea from Ceylon (Sri Lanka). The root wood of one of the species of *Alstonia* was used to make pith helmets.

SUPPLIES

White cheesewood is commercially traded in the southwestern Pacific region. Trade in worldwide commerce appears to be small. In 1993, trees of this species were destroyed in the devastation of Hurricane Andrew which struck the Fairchild Tropical Gardens in Florida. A limited amount of wood was salvaged for wood collectors.

Anacardium excelsum

espave

by Jon Arno

SCIENTIFIC NAME

Anacardium excelsum. Derivation: The genus name is Latin for heart-shaped, referring to the shape of the nut. The specific epithet infers tall.

FAMILY

Anacardiaceae, the cashew family.

OTHER NAMES

espavel, caju assu, caracoli, maranon.

DISTRIBUTION

Requiring well-drained soil, but preferring low elevations, espave is native to the Pacific coastal regions of Panama, Costa Rica, and Ecuador and along the Caribbean coasts of Venezuela and Colombia.

THE TREE

Usually of moderate size compared to some of the tropical timbers it grows near, espave can reach heights up to 150 feet with diameters to 5 feet, producing clear logs 60 feet long. Specimens like this require ideal growing conditions and when grown in the open, espave tends to branch out, producing a short bole and round crown. Generally, the tree reaches heights between 75 and 100 feet with diameters of 2.5 to 3 feet. Since it does not develop a large buttress, most of the lower bole is usable.

THE TIMBER

With an average specific gravity of 0.41 (ovendry weight/green volume), equivalent to an air-dried weight of 31 pcf, espave is about 10% less dense than bigleaf mahogany (*Swietenia macrophylla* WDS 254), but the wood is quite similar in appearance. The seasoned heartwood ranges from reddish-brown to amber brown, and flatsawn surfaces display dark, bold vessel lines. The grain is commonly interlocked and this, in combination with the natural luster and small, dark ray flecks, produces a beautiful ribbon-grained figure on quartersawn boards. The attractive heartwood contrasts sharply with the rather dull grayish-white sapwood, which can be as much as 10 inches wide. Espave is about two-thirds as strong as bigleaf mahogany in bending strength and crushing resistance so it is not a good choice for rugged applications.

SEASONING

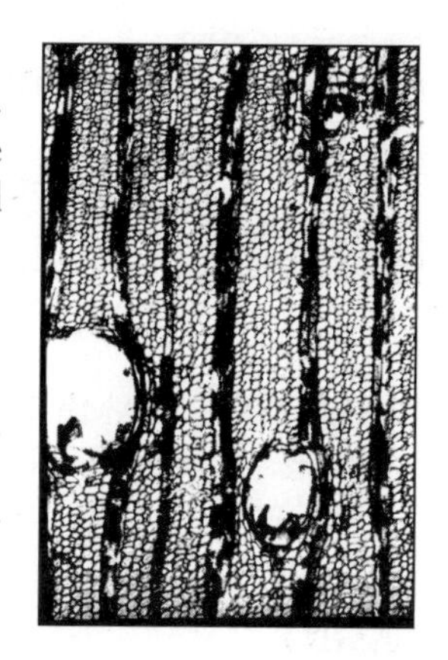

Espave is reported to be somewhat difficult to air dry. Average reported shrinkage values (green to ovendry) are 2.8% radial, 5.2% tangential and 8.4% volumetric. If dried too quickly it is prone to warping and checking.

DURABILITY

Durability tests are inconclusive. It appears to have resistance to some fungi but not to others. When dry, it is resistant to termite attack. Many members of Anacardiaceae produce toxic gums and oils, and this may account for espave's resistance to insect larvae. There does not appear to be any human risk even when it is used for kitchen utensils.

WORKABILITY

My experience is limited to one disappointing episode. Based on its mahogany-like appearance, low density, low shrinkage and botanical kinship to woods I like very much, such as goncalo alves (*Astronium fraxinifolium* WDS 030) and staghorn sumac (*Rhys typhina* WDS 236), I had great expectations. Although it was soft and easy to work, it was disappointingly stringy. My finest and sharpest tablesaw blades produced very frayed crosscuts, and the interlocked grain chipped badly under planes, chisels and router bits. Only with great care and very fine grit sandpaper was I able to cut the fuzzy surface close enough to be subdued by several coats of high-bodied varnish. The natural beauty of the wood prevailed in the finished project, but taming this species was a real struggle.

USES

Within its native range, it is a major, general purpose construction timber. Veneers are used for plywood, and solid stock is used for inexpensive furniture, boxes and crates and housewares such as bowls, spoons and cooking utensils. Because of its light weight and durability, it was favored by Indian tribes for dugout canoes. There are better choices for most cabinetry projects.

SUPPLIES

Unlike most tropical timbers, espave will grow in nearly pure stands on dry land in accessible terrain; therefore, it is relatively plentiful and easily harvested. Competing in applications with lauan, which is even more plentiful and less expensive, espave has not made an impact on the international market.

Andira inermis

partridge wood

by Max Kline

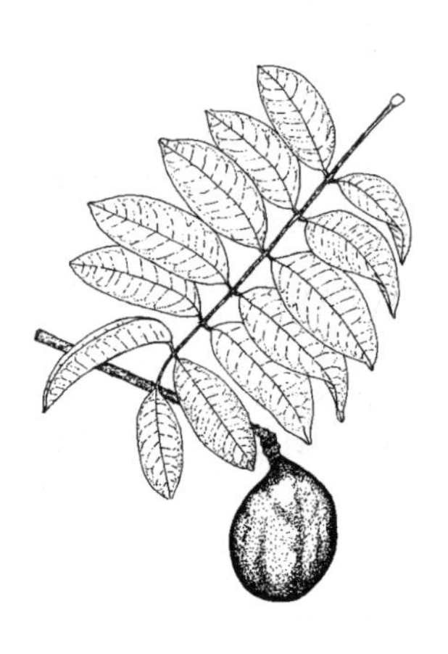

SCIENTIFIC NAME

Andira inermis. Derivation: The genus name is Brazilian. The specific epithet means that the tree is without thorns.

FAMILY

Fabaceae or Leguminosae, the legume family; (*Papilionaceae*) the pea or pulse group.

OTHER NAMES

red cabbage bark, moca, rode kabbes, macaya, pheasantwood, angelim, angelin, andira.

DISTRIBUTION

It occurs throughout the West Indies and from southern Mexico through Central America to northern South America and Brazil.

THE TREE

Partridge wood is an evergreen tree of moderate size, about 75 feet in height and 3 feet in diameter. Occasionally the tree can reach 100 feet in height. It grows under varying rainfall and soil conditions. The odd-pinnate leaves have several glabrous, oblong leaflets with leathery blades. The purplish, fragrant flowers or roseate are born at the terminal and contain many flowered panicles. The fruit is a drupe-like, indehiscent ovoid pod with one seed. The ragged bark has a disagreeable odor.

THE TIMBER

The heartwood is yellowish reddish-brown, sometimes very dark, and the conspicuous parenchyma striping suggests the markings of a partridge wing. The yellowish sapwood is not always sharply demarcated from the heartwood. The luster is low, and the odor and taste are not distinct. It is moderately hard and heavy. Average reported specific gravity is 0.64 (ovendry weight/green volume), equivalent to an air-dried weight of 50 pcf. The texture is quite coarse, and the grain is moderately irregular. Because of the parenchyma tissues, the timber has a distinctive figure. The timber is considered to be very tough and strong.

SEASONING

The wood air dries at a moderate rate with little degrade. Average reported shrinkage values (green to ovendry) are 4.6% radial, 9.8% tangential and 12.5% volumetric. During seasoning, the sapwood is susceptible to discoloration by fungi. After seasoning, partridge wood is dimensionally stable.

DURABILITY

The heartwood is resistant to termites. The sapwood is very vulnerable to attack by powder post beetles.

WORKABILITY

Due to the alternating bands of hard and soft tissue, the wood is not easy to work. It saws easily but is difficult to plane to a smooth surface. It works well on the lathe, holds nails and screws without problems and glues satisfactorily. It can be polished well after filling the open pores.

USES

Where strength is important, partridge wood is used for heavy construction, agricultural implements, house framing, exterior siding, sills and staircases. When beauty is the most sought after attribute, it is used for turnery, fancy articles, walking sticks, umbrella handles, furniture and cabinetry, parquet flooring and decorative veneer.

SUPPLIES

This is a fairly common timber tree, although only limited quantities are exported. When available beyond its native growth range, it is moderately high priced.

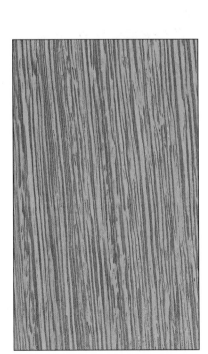

Aniba rosaeodora

pau rosa

by Max Kline

SCIENTIFIC NAME

Aniba rosaeodora. Derivation: The genus name is probably a local vernacular. The specific epithet is Latin for rose-smelling, referring to the wood. A synonym is *A. duckei.*

FAMILY

Lauraceae, the laurel family.

OTHER NAMES

Brazilian louro, lauro rosa.

DISTRIBUTION

Brazil.

THE TREE

Often reaching a height of 100 feet with a diameter up to 30 inches, clear bole lengths of 55 to 75 feet can be obtained. The base is mildly buttressed. The leaves and slash exude a pleasantly sweet-smelling odor. The entire, acute elliptic leaves have a distinctive yellowish-green sheen on the undersurface. The flowers are unusually small.

THE TIMBER

The timber is a uniform yellowish-green color or yellow with olive or pink streaks becoming light yellowish-brown with a greenish cast. The narrow sapwood is a light yellow color. The luster is medium. The texture is medium, and the grain is straight. The wood has a spicy odor which fades with age. The taste is not distinct. It is moderately hard and heavy. Average reported specific gravity is between 0.55 and 0.65 (ovendry weight/green volume), equivalent to an air-dried weight of 42 to 51 pcf.

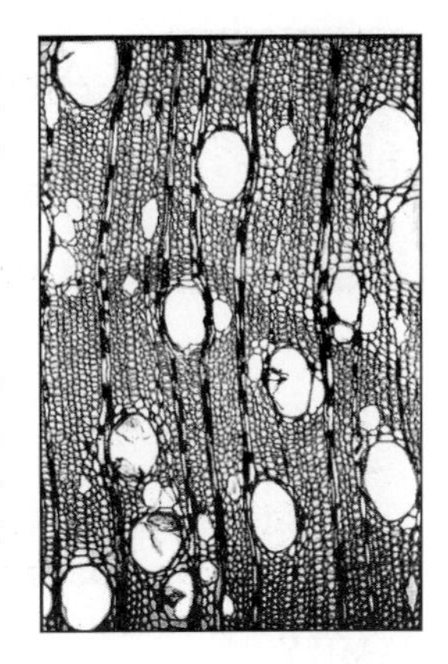

SEASONING

Pau rosa is moderately difficult to air dry. It dries at a moderate rate with slight warp and checking. Average reported shrinkage values (green to ovendry) are 4.7% radial, 7.0% tangential and 12.1% volumetric.

DURABILITY

This timber has an excellent reputation for resistance to decay. The heartwood is very durable to both white rot and brown rot fungi. It is known for its high resistance to moisture absorption and in this respect is comparable to teak (*Tectona grandis* WDS 261).

WORKABILITY

Pau rosa is easy to work with hand and machine tools and dresses to a smooth surface to give a satiny sheen.

USES

Wood distilled at the source with steam for about 1 to 1.5 hours has been used in the perfume industry. It is also used for furniture, turnery, inlay work, boat building, millwork, agricultural implements, crossties, flooring, tool handles, veneer and plywood.

SUPPLIES

Some pau rosa is imported into the United States and can be purchased from the largest importers, but it is generally not readily available.

Araucaria angustifolia

Parana-pine

by Jon Arno

SCIENTIFIC NAME

Araucaria angustifolia. Derivation: The genus name means the plant is "of Arauco", a Chilean province. The specific epithet means narrow leaf.

FAMILY

Araucariaceae, the araucaria family.

OTHER NAMES

pino parana, pino blanco.

DISTRIBUTION

Parana-pine is native to southern Brazil, Paraguay and parts of northern Argentina. As the name suggests, it is especially common in the state of Parana, Brazil.

THE TREE

When mature, Parana-pines range from 80 to 100 feet in height with diameters up to 5 feet. The tree's form is excellent for timber production, developing a tall, clear trunk topped by a flat crown with upturned branches.

THE TIMBER

The sapwood is a light yellowish-cream color similar to ponderosa pine, but with a slight grayish or "dirty" cast. The heartwood is light brownish-tan, occasionally exhibiting vivid, rusty red-colored streaks. With an average specific gravity of about 0.45 (ovendry weight/green volume), equivalent to an air-dried weight of 34 pcf, the density of the wood is similar to southern yellow pine (both loblolly and shortleaf pine which have a specific gravity of 0.47). The annual rings, however, while visible are far less bold than in the southern yellow pines, because the variance in density between earlywood and latewood is much less pronounced. In this respect, Parana-pine has a texture similar to eastern white pine (*Pinus strobus* WDS 212), but is as strong as southern yellow pine. Unlike most true pines, Parana-pine wood lacks a strong resinous odor and has a slightly lustrous surface. The wood is relatively hard and strong when compared to most softwoods.

SEASONING

Parana-pine is reported to be more difficult to dry than most softwoods, because it often contains compression wood which causes warping and checking. Average reported shrinkage values (green to ovendry) are 3.8% radial, 7.3% tangential and 11.6% volumetric.

DURABILITY

Nondurable, Parana-pine has low resistance to fungi and decay.

WORKABILITY

Although North America is blessed with a richer variety of softwoods than anywhere else in the world, when it comes to working characteristics Parana-pine presents a respectable challenge to the best available. In many respects, Parana-pine fills a void in the U.S. domestic lineup in that it has the strength and hardness of the yellow pines while offering the texture and easy working characteristics of the nicest of the white pines. Choice examples of Parana-pine heartwood, with bright red streaks on the soft brown background, can be very attractive. The wood often displays plentiful, but very small, tight knots which contribute to its attractive appearance without posing any significant functional problems. The grain is usually straight. Also, wide stock is available making it a good choice for larger projects in terms of convenience (minimum gluing). Parana-pine has excellent nail holding properties, dresses to a smooth surface and accepts virtually all glues and finishes without difficulty. In use stability is not outstanding but acceptable provided compression wood is avoided.

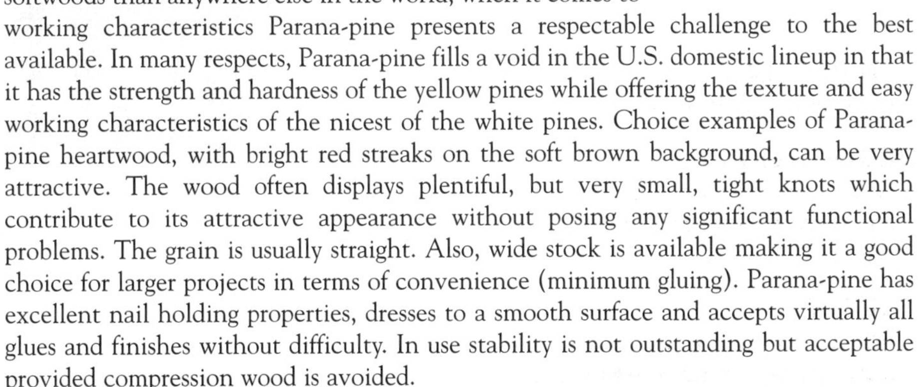

USES

The wood is used for construction lumber, trim, sash and doors, plywood veneer and furniture. The pulp is used for paper.

SUPPLIES

Parana-pine is plentiful, but availability in North America is somewhat limited. Shipping costs drive the price up to a level which makes it non-competitive with the U.S. domestic softwoods for utilitarian construction uses. Since it is not generally considered a wood for cabinetry, it is seldom stocked by dealers specializing in imported woods for the woodworker. When available, the price is moderate relative to other imports, but substantially higher than like grades of U.S. domestic softwoods.

Araucaria heterophylla

Norfolk-Island-pine

by Jim Flynn

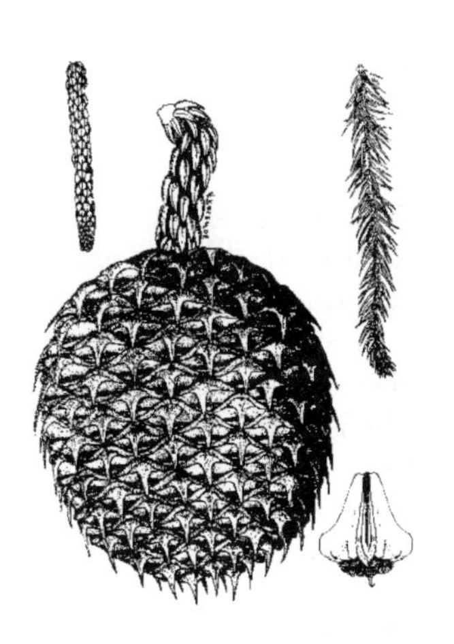

SCIENTIFIC NAME

Araucaria heterophylla. Derivation: The genus name means the plant is "of Arauco", a Chilean province where one of the species *A. araucana* (monkey puzzle) was found. The specific epithet refers to the two different types of foliage. A botanical synonym is *A. excelsa.*

FAMILY

Araucariaceae, the araucaria family.

OTHER NAMES

Araucaria (Spain, England), Christmas-tree (Belize), siete pisos (Cuba, Dominican Republic).

DISTRIBUTION

Norfolk-Island-pine is native only on Norfolk Island between New Zealand and New Caledonia in the South Pacific. It has been extensively planted in tropical and subtropical areas of the world primarily as an ornamental. In the 1902 edition of Bailey's *Cyclopedia of American* Horticulture, it is reported that 250,000 small Norfolk-Island-pines were sold in the United States from Ghent, Belgium, which was the world trade center for this conifer.

THE TREE

A conifer of magnificent symmetrical proportions, Norfolk-Island-pine grows tall and straight ending in a conical crown. As the species name indicates, the evergreen leaves are of 2 forms: the juvenile leaves are curved, narrow and triangular at the cross section; the mature leaves are a broad, curved pointed scale. The branches are horizontal in whorls of four to seven. Male and female cones do not appear on the same tree. The blackish bark is nearly smooth with white resin drops that taste like turpentine. In the most favorable climate, this tree can reach heights of 200 feet with diameters of 9 feet. In Florida, the tree can grow to a height of 60 feet with a diameter in excess of 2 feet.

THE TIMBER

The prime use of Norfolk-Island-pine has been as an ornamental. The creamy tannish-brown wood is straight grained, knotty, resinous and easy to work. It has characteristics similar to Scotch pine (*Pinus silvestris*). Average reported specific gravity is 0.42 (ovendry weight/green volume), equivalent to an air-dried weight of 31 pcf.

SEASONING

When freshly cut and properly sealed with a water-emulsion sealer, the wood will withstand normal checking until used. Average reported shrinkage values (green to ovendry) are 3.5% radial, 5.3% tangential and 8.9% volumetric.

DURABILITY

The timber is nondurable, and the sapwood is subject to blue stain.

WORKABILITY

A softwood with a tight, fairly straight grain, Norfolk-Island-pine is easy to work and takes a fine finish.

USES

The wood is useful for any application suitable to the softer species of pine. Some of the larger trees have been used for ship masts. Today, its prime use seems to be for turning bowls. Bowls turned from this wood exhibit interesting patterns due to the whorled branches that penetrate the stem to the pith.

SUPPLIES

The wood and bowl blanks of Norfolk-Island-pine are not generally available on the commercial market. Wood can be found where the tree grows in abundance as an ornamental, such as in Florida, and is frequently removed, replaced or storm damaged.

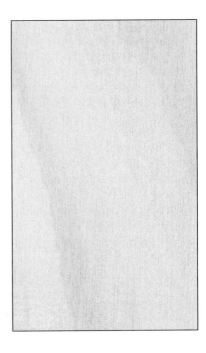

Arbutus menziesii

Pacific madrone

by Max Kline

SCIENTIFIC NAME

Arbutus menziesii. Derivation: The genus name is the classical Latin name *Arbutus unedo*, strawberry madrone, of southern Europe. The specific epithet is in honor of Archibald Menzies (1754 to 1842), a Scottish physician and naturalist who accompanied Captain George Vancouver on his voyage of discovery in the Northwest.

FAMILY

Ericaceae, the heath family.

OTHER NAMES

madrone, madrona, madrono, arbuti tree, coast madrone.

DISTRIBUTION

Southwestern British Columbia and southward through Washington, Oregon and California in the coastal mountains.

THE TREE

This evergreen tree is sometimes 60 to 80 feet in height with a diameter of 2 to 3 feet. Its orange-colored branches and shiny leaves immediately distinguish it from all other trees in its growing area. Dainty white flowers appear in large clusters and are very attractive against the evergreen leaves. The profuse, bright orange-red fruit ripens in the autumn adding to the overall beauty of the tree.

THE TIMBER

The heartwood is light pink to pale reddish-brown with occasional spots of deep red, and the sapwood is whitish or cream colored with a pinkish tinge. The timber is fine textured, heavy, moderately strong, dense and unusually brittle when dry. It resembles pear wood in texture and apple wood in color. Average reported specific gravity is 0.58 (ovendry weight/green volume), equivalent to an air-dried weight of 45 pcf.

SEASONING

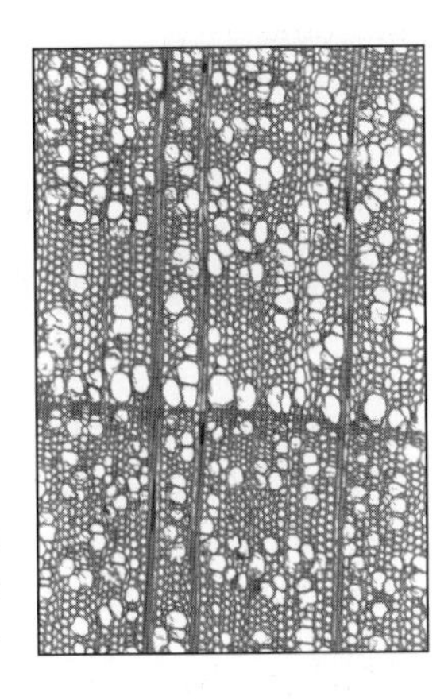

Pacific madrone is extremely difficulty to season as it warps and checks easily. Average reported shrinkage values (green to ovendry) are 5.6% radial, 12.4% tangential and 18.1% volumetric.

DURABILITY

It is not considered durable unless treated.

WORKABILITY

It is not easily glued, but works fairly well with all tools and polishes almost to the smoothness of American holly (*Ilex opaca* WDS 147). It is a beautiful wood and has a fine appearance when used to make furniture.

USES

As yet Pacific madrone does not have a great commercial importance as lumber. It is particularly suited for turned work, bowls, novelties and souvenirs. Small quantities are used for tool handles, mathematical instruments, furniture and inlay. The beautiful madrone burls are used for decorative veneers and making pipes. The charcoal obtained from the tree is one of the best sources for making gunpowder.

SUPPLIES

Lumber is available to a limited extent on the west coast of the United States. The burls are costly and scarce.

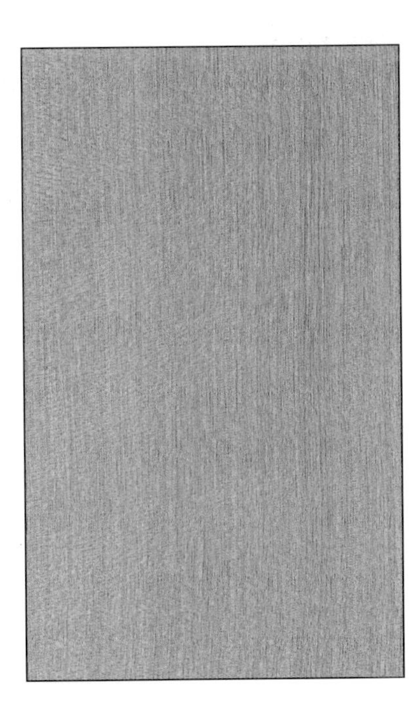

Arctostaphylos viscida

whiteleaf manzanita

by Max Kline

SCIENTIFIC NAME

Arctostaphylos viscida. Derivation: The genus name is Greek meaning bear and bunch of grapes. The specific epithet refers to the sticky twigs and fruits.

FAMILY

Ericaceae, the heath family.

OTHER NAMES

manzanita.

DISTRIBUTION

Foothills of southwestern Oregon, south in the coast ranges and the Sierra Nevada Mountains to central California.

THE TREE

Whiteleaf manzanita is a large evergreen shrub which occasionally grows to a tree height of 27 feet. It is noted that there are about 60 species of *Arctostaphylos* widely distributed in North America, mostly in the western parts of the United States and Mexico. The trunk is short and 4 to 10 inches in diameter. The leaves are leathery and somewhat oily. It is a prolific bloomer with a heavy crop of brownish-red berries that can be made into jelly. Flowers are typical of the heath family appearing in large panicles and are white in color and quite fragrant. The bark of the tree is reddish-brown.

THE TIMBER

The wood is hard, strong, heavy and fine textured. The heartwood is a reddish-brown to a predominantly red color, and the narrow sapwood is light brown. Because of the twisted character of growth in thickets, wood with attractive figures and grains can often be obtained. Average specific gravity is around 0.70 (ovendry weight/green volume), equivalent to an air-dried weight of 55 pcf.

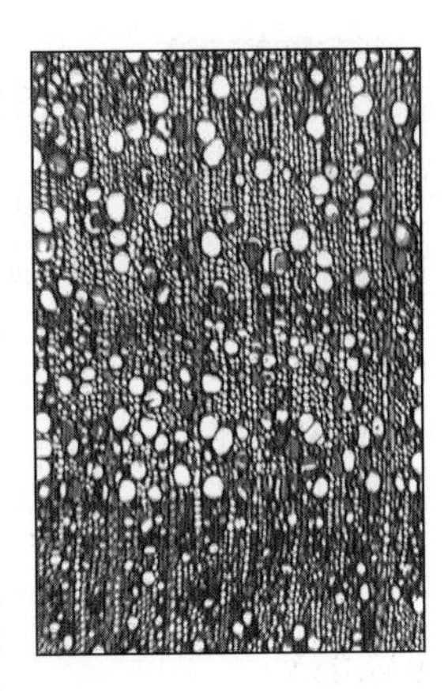

SEASONING

It is especially difficult to dry whiteleaf manzanita wood without severe splitting and checking. Shrinkage values are not available.

DURABILITY

There are no published reports on this characteristic. Use of the wood locally as fence posts, however, offers some testimony as to its endurance in contact with the ground.

WORKABILITY

The wood works well with machine tools, taking a high polish. It is a fine species for making novelties. It takes most finishes and can be readily glued, but nailing should not be attempted without pre-boring to prevent splitting.

USES

In growth areas, the wood is used for fuel and fence posts. Due to its pleasing color and figure, woodworkers can more advantageously use it, especially for turned novelties and projects. It is a favorite wood for pet bird perches.

SUPPLIES

Whiteleaf manzanita is usually only available in very limited quantities and not on a commercial scale. In most cases, woodworkers must know someone who has the wood for sale in order to obtain it.

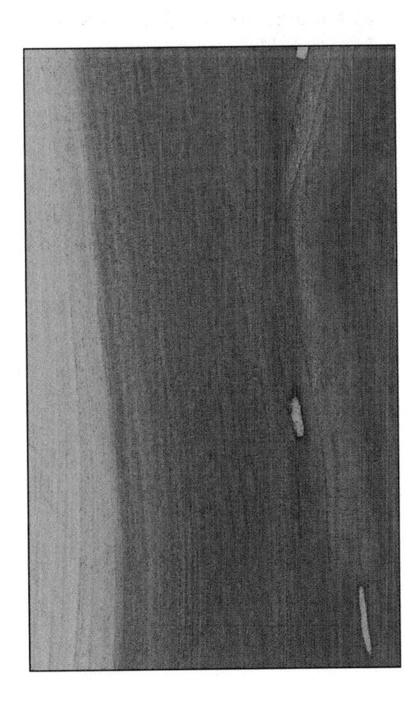

Asimina triloba

pawpaw

by Max Kline

SCIENTIFIC NAME

Asimina triloba. Derivation: The genus name is a Native American name translated from French. The specific epithet means three lobed, referring to the three sepals and two rows of three petals each.

FAMILY

Annonaceae, the annonia or custard-apple family.

OTHER NAMES

custard apple, false banana, jasmine, jasminier, fetid shrub, banana.

DISTRIBUTION

Found from western New York and New Jersey to southern Ontario, Michigan and Nebraska, south to Texas and Florida.

THE TREE

With its dense foliage and strange looking fruit, this tree seems more fitting for a tropical climate rather than a temperate zone. Reaching a maximum height of 40 feet with a diameter of 12 to 18 inches, pawpaw is a large shrub or small tree. It is a common undergrowth of hardwood forests, growing on rich soil, especially in river valleys. Its glossy leaves are 6 to 12 inches long, and the maroon flowers are nearly 2 inches long. Its banana-shaped fruit, which is 2 to 6 inches long with a smooth skin, a custard-like rich-flavored flesh and dark brown seeds, is relished by raccoons, opossums and foxes. [Ed. note: I *also* like it. jhf]

THE TIMBER

The wood of the pawpaw tree is greenish-yellow with white sapwood. It is light, soft, brittle, coarse textured and coarse grained. The rays are prominent, and when quartersawn the wood is extremely attractive. Pawpaw is not a strong wood and should not be used where this property is important. A standard 3- by 6- by 0.5-inch sample had a specific gravity of 0.43 (ovendry weight/green volume), equivalent to an air-dried weight of 32 pcf.

SEASONING

There are no particular problems when air drying pawpaw. While precise shrinkage values are not available, they are estimated to be within the range of those for ailanthus (*Ailanthus altissima* WDS 015).

DURABILITY

There is no information available on this characteristic other than to note that downed trees in the woods do not take long to disintegrate.

WORKABILITY

The timber works well with tools, but they must be sharp because of the soft texture of the wood. Staining is not recommended, but a clear, natural finish produces a pleasing result.

USES

The only commercial use for pawpaw is for fuel. The wood is recommended for woodworkers and is well adapted for turned novelties and all types of built-up designs because of its unique color.

SUPPLIES

Pawpaw is not available from commercial mills but must be collected by the individual for personal use.

Aspidosperma polyneuron

pink peroba

by Max Kline

SCIENTIFIC NAME

Aspidosperma polyneuron. Derivation: The genus name refers to the shielded seed. The specific epithet means many nerves, referring to the leaf veins. A synonym is *A. peroba*.

FAMILY

Apocynaceae, the dog bane family.

OTHER NAMES

red peroba, peroba rosa, palo rosa, amargosa, amarello.

DISTRIBUTION

Principally from the southeastern region of Brazil, also Argentina.

THE TREE

The average height of pink peroba is 90 feet with an average diameter of 2.5 feet. It may reach a height of 125 feet with a diameter of 4 to 5 feet. The trunk is straight and well formed. Pink peroba is often tapped for the reddish-orange latex it exudes. All of the leaves are alternate, elliptical and clustered on the ends of twigs. The flowers are white and usually very small. The distinctive fruit is a woody follicle, and the thin seeds are oval.

THE TIMBER

The yellowish sapwood is paler than, but not sharply distinguished from, the heartwood which is tan to rose-red and often streaked with purple or brown. The heartwood darkens on exposure. The luster is medium, and the taste is bitter. It is moderately hard and heavy and average reported specific gravity is about 0.65 (ovendry weight/green volume), equivalent to an air-dried weight of 51 pcf. The grain is straight to irregular, and the texture is fine. Pink peroba is normally a very strong wood but sometimes tends to be brittle. Strength is largely dependent on the grade of the timber, but in general is comparable to oak. When cross-grain is present, strength, especially in resistance to shock loads, is considerably reduced.

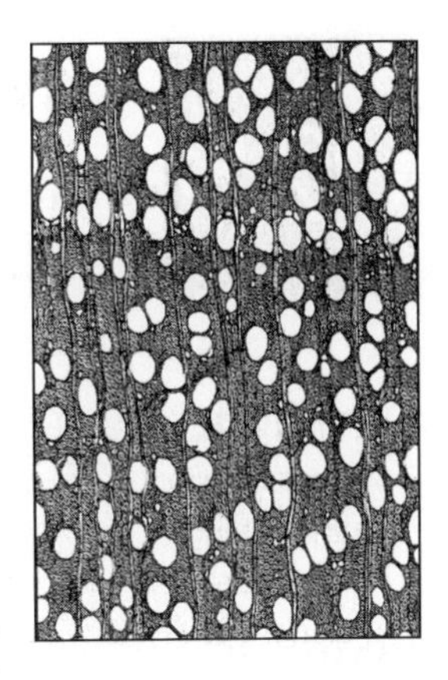

SEASONING

When kiln drying, little checking occurs but some distortion may develop. Average reported shrinkage values (green to ovendry) are 3.8% radial, 6.4% tangential and 11.6% volumetric.

DURABILITY

Occasionally there is a tendency for damage by borers, but in Brazil untreated pink peroba rail sleepers normally last 10 to 11 years.

WORKABILITY

If the grain is straight, pink peroba works easily and finishes smoothly. It does not cause undue dulling of tool edges. It stains and polishes well and can be glued satisfactorily.

USES

This timber is used for joinery, furniture, flooring and railway sleepers. It can be sliced for decorative veneer. It has been used successfully for ship decking and due to its acid resisting properties is used as staves for acid vat construction.

SUPPLIES

Pink peroba is available in adequate quantities and is considered to be in the moderate price range. In Brazil, entire buildings may be constructed of it. The quantity of pink peroba in Brazil is greater and more accessible than any other hardwood in the country.

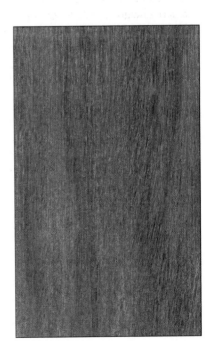

Astronium balansae

urunday

by Max Kline

SCIENTIFIC NAME

Astronium balansae. Derivation: The genus name is Greek for star. The specific epithet is Latin for balsam nut.

FAMILY

Anacardiaceae, the cashew family.

OTHER NAMES

aderno, ubatan, cuchi, urundel, aroeira.

DISTRIBUTION

Argentina, Paraguay, and most abundantly in Brazil.

THE TREE

Thriving best in semi-arid environments, the tree grows to a height of about 60 feet with a trunk diameter of 24 inches. The trunk is straight and spotted with many round, white patches on the bark. The small flowers are indistinct and a whitish-yellow color. The leaves are pinnately compound, 10 inches long, with five pairs of 1- by 0.5-inch elliptical leaflets.

THE TIMBER

The yellowish sapwood is comparatively thin and sharply demarcated from the uniformly cherry-red heartwood, which darkens on exposure. The luster is low to medium, and odor and taste are not distinct. The wood is very hard and heavy with an average specific gravity of 0.70 to 0.96 (ovendry weight/green volume), equivalent to an air-dried weight of 55 to 79 pcf. The grain is straight to wavy or irregular. The texture is fine. There is considerable resemblance to quebracho (*Schinopsis quebracho-colorado* WDS 246). The strength properties of urunday are very high.

SEASONING

The timber is difficult to season without checking. Average reported shrinkage values (green to ovendry) are 5.0% radial and 8.0% tangential.

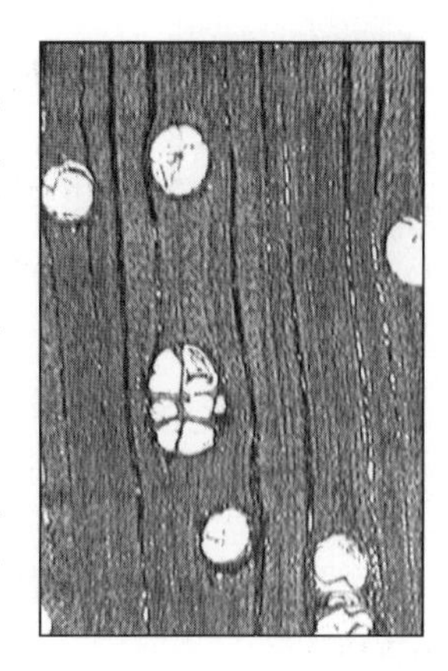

DURABILITY

Urunday is a very durable timber, which makes it usable for heavy construction purposes.

WORKABILITY

Depending on the piece of wood, workability can be fairly easy to difficult. The wood turns readily, finishes very smoothly and takes a high natural polish.

USES

Uses include railroad crossties, posts, pilings, bridge timbers, fuel and general construction. Finer grades can be used for knife handles.

SUPPLIES

Very little urunday is exported from South America since most of it is used locally. Occasionally a sample can be obtained for a wood collection.

Astronium fraxinifolium

goncalo alves

by Max Kline

SCIENTIFIC NAME

Astronium fraxinifolium. Derivation: The genus name is Greek for star. The specific epithet is Latin meaning resembling ash, which refers to the leaves.

FAMILY

Anacardiaceae, the cashew family.

OTHER NAMES

gateado, kingwood, mura, tigerwood, zorrowood.

DISTRIBUTION

Guyana, Colombia, Ecuador, Mexico, Peru, Guatemala, Honduras, El Salvador, Trinidad, Brazil.

THE TREE

Goncalo alves grows symmetrically to a height of 100 feet or more with diameters up to 3 feet. This strongly aromatic tree thrives best in a dry environment. The brown bark is rather thin but smooth with contrasting whitish and reddish patches. Trees in isolation from major growth areas, however, have a deeply ridged bark. The leaves are pinnately compound.

THE TIMBER

The color of goncalo alves is light golden-brown to reddish-brown with blackish-brown streaks of variable spacing. The odor and taste are not distinct, and the luster is dull to medium. The wood is very hard and heavy with an average specific gravity of 0.68 to 0.97 (ovendry weight/green volume), equivalent to an air-dried weight of 53 to 80 pcf. The grain is wavy, and the texture is fine. Sometimes the wood has a mottled figure resembling some rosewoods in the genus *Dalbergia*. The strength properties are very similar to those of flowering dogwood (*Cornus florida* WDS 088).

SEASONING

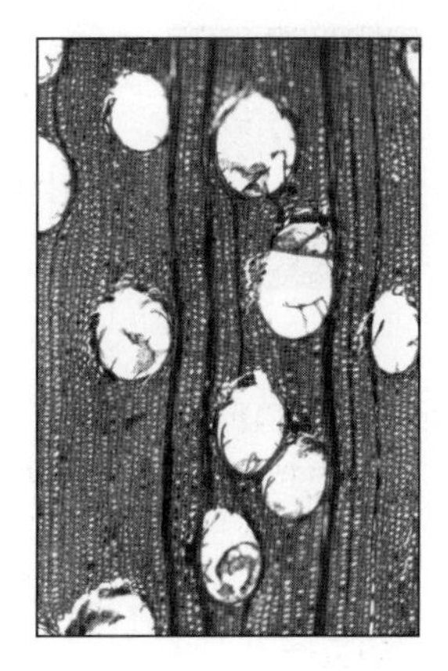

The wood should be seasoned slowly to avoid excessive warping and checking. Average reported shrinkage values (green to ovendry) are 4.5% radial and 7.0% tangential.

DURABILITY

Goncalo alves heartwood is very durable.

WORKABILITY

In general the wood is somewhat difficult to work, but workability varies considerably according to density. It finishes with a high natural polish, turns well and is noted for its durability.

USES

Goncalo alves is used for boat building, general construction, furniture and cabinetry, flooring, shutters and bobbins (as a dogwood substitute), veneer, plywood, turnery and for knife handles as a cocobolo (*Dalbergia retusa* WDS 098) substitute.

SUPPLIES

While the timber is abundant in its natural growth areas, it is limited in supply in the United States. When available commercially, it is considered a costly species.

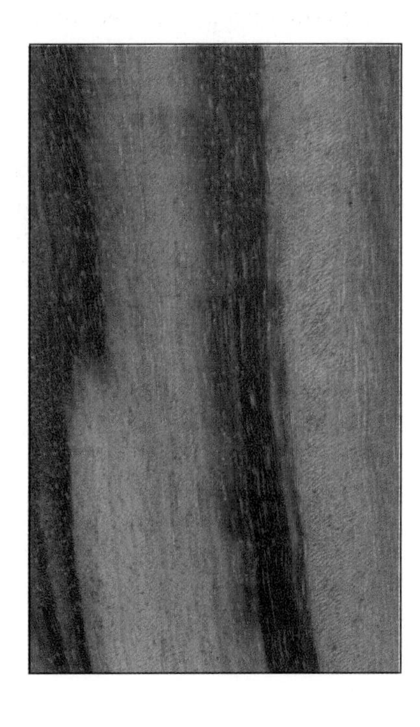

Baikiaea plurijuga

Rhodesian teak

by Jim Flynn

SCIENTIFIC NAME

Baikiaea plurijuga. Derivation: The genus name is in honor of Dr. William Balfour Baikie, a West African traveler and Royal Navy surgeon on the Niger Expedition of 1854 to 1857. The specific epithet is indicative of the many yokes (paired leaflets) in the pinnate leaf.

FAMILY

Fabaceae or Leguminosae, the legume family; (*Caesalpiniaceae*) the cassia group.

OTHER NAMES

Rhodesiese kiaat, gusi, umkusu, Zambesi redwood, umgusi, mukshi, mukusi.

DISTRIBUTION

This tree is a pure African species growing in Angola, Botswana, Namibia, Zambia and Zimbabwe. It dominates vast tracts of the Kalahari Desert and is often associated with other leguminous tree species.

THE TREE

Rhodesian teak is a much-branched tree growing to heights of 50 to 60 feet with clear boles 10 to 15 feet in height and diameters of 30 inches. On young trees, the bark is smooth but later becomes vertically fissured, cracked and brownish in color. The leaves are alternate and compound with four to five pairs of opposite leaflets. The very attractive, large flowers grow on large, strong axillary racemes that are 12 inches long. The fruit is a flattened, woody pod 2 inches wide by 5 inches long. From June to September, the pods split explosively and scatter seeds widely.

THE TIMBER

The heartwood is an attractive reddish-brown with prominent, irregular black streaks and flecks. The pale pinkish-brown sapwood is sharply demarcated from the heartwood. The texture is fine and even. The grain is straight or slightly interlocked. Luster is low, and the wood is without characteristic odor or taste. Like wood in the *Quercus* genus, Rhodesian teak will stain when moist and in contact with iron because of its tannin content. Average reported specific gravity is 0.73 (ovendry weight/green volume), equivalent to an air-dried weight of 58 pcf.

SEASONING

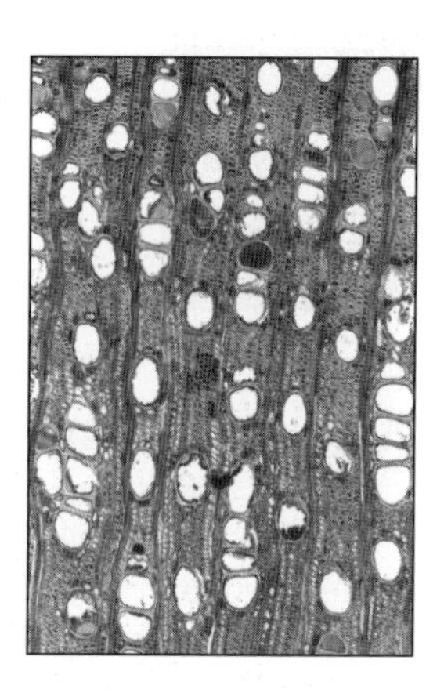

Rhodesian teak dries slowly with a good record in terms of degrade. Average reported shrinkage values (green to 12% moisture content) are 1.5% radial and 2.5% tangential. Care in stacking is essential. It is dimensionally stable in service.

DURABILITY

The durability of Rhodesian teak is excellent. It is moderately resistant to termite attacks. Sapwood is prone to attack by powder post beetles. The heartwood is extremely resistant to preservative treatment. Freshly cut logs must be handled quickly to prevent attack by the forest longhorn beetle. The Zimbabwe Forest Service reports that the wood is unaffected by fungi.

WORKABILITY

All reports indicate that the wood is difficult to saw. Machinery with adequate power and tungsten-tipped saw teeth is recommended because of the wood's high silica content and gumming of sawteeth when sawn green. Despite the interlocked and variable grain, it planes well to a lustrous, smooth finish. It is an excellent wood for turning. Gluing and finishing are without problems.

USES

A versatile wood, its uses include furniture, cabinetry, decorative flooring, turnery and carving, decorative veneer and store fittings. The timber is also used as mining timber and in railway sleepers.

SUPPLIES

The wood is available from commercial sources, and African suppliers state that the wood is in good supply. For members in Great Britain who cannot obtain this species, there is hope. They are advised to be on the alert in the event the flooring of the London Corn Exchange is ever replaced. During the rebuilding in 1952, a special grooved floor was designed to withstand the abrasion from the grain thrown to the floor by the merchants. Rhodesian teak was used for the parquet blocks because of its ability to withstand abrasion without undue wear.

Baillonella toxisperma

moabi

by Jon Arno

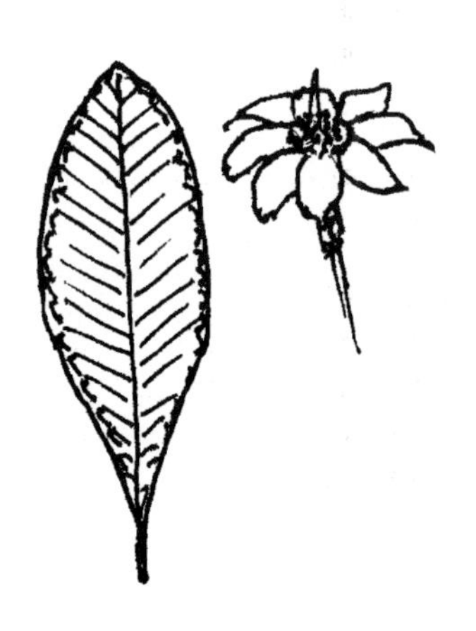

SCIENTIFIC NAME

Baillonella toxisperma. Derivation: The genus name is in honor of Frederick Manson Bailey (1827 to 1915), the Australian botanical author of *Queensland Flora*. The specific epithet is Latin for poison seed.

FAMILY

Sapotaceae, the sapodilla family.

OTHER NAMES

African pearwood, adza, njabi, dimpampi.

DISTRIBUTION

Moabi is native to the dense rain forests of Nigeria and southward into the Congo (formerly Zaire).

THE TREE

A big tree, moabi grows to heights of 200 feet with straight cylindrical boles as long as 100 feet. Trunk diameters range from 6 to 10 feet with some butt swelling on older trees.

THE TIMBER

Moabi is fine textured and often attractively figured. While the heartwood is virtually always of a reddish hue, its color can vary substantially from pinkish-brown to almost blood red. The sapwood is light pink or sometimes a soft grayish-tan, and the transition between sapwood and heartwood is relatively abrupt. Although the density of the wood shows considerable variation (specific gravity ranges between 0.65 and 0.77, ovendry weight/green volume, which is equivalent to an air-dried weight of 51 to 61 pcf), it is generally comparable to North American hickories and is considered to be hard. It is also roughly comparable to hickory in strength and elasticity.

SEASONING

This species dries very slowly, but is not difficult to season. Average volumetric shrinkage (green to ovendry) is 12.6%, which makes it comparable to black walnut (*Juglans nigra* WDS 150). With a radial shrinkage of 5.9% and a tangential shrinkage of 7.5%, drying stress is minimal, and warping is seldom a problem.

DURABILITY

Moabi is very durable and weathers exceptionally well. It is even suitable for marine applications.

WORKABILITY

Moabi is, on average, half again as dense as black walnut and, therefore, a bit of a challenge to work when using hand tools. Nonetheless, it machines predictably and is reported to have good steam bending properties. Its fine texture and low tendency to distort make it a good species for turning, and a glass smooth finish is easy to achieve even with light-bodied varnishes. On the down side, however, moabi contains high levels of silica, which rapidly dull cutters, and the dust is reported to be an irritant to mucous membranes. This wood has both great strength and beauty, but in large projects requiring any significant amount of hand shaping, a better choice might be makore (*Tieghemella heckelii* WDS 269). A fairly close relative of moabi, makore is similar in appearance, but substantially easier to work (see Special Note).

USES

Moabi is used for turning, carving, decorative veneers, high-quality cabinetry and furniture, flooring and paneling.

SUPPLIES

Because the species has a relatively extensive growth range, supplies are considered adequate, but it has long been exploited to supply the European market where it is more popular than it is in North America. When available in North America, the price is generally high, although not prohibitively so, considering it is normally used in small quantities for turnings and carvings.

SPECIAL NOTE

Tieghemella, a closely related genus, produces two similar and fairly popular woods: *T. heckelii* and *T. africana*, both from west central Africa. These species share most of moabi's favorable features, but are usually somewhat lighter in color and substantially less dense. Makore's average specific gravity, for example, is only 0.55, midway between that of black walnut and sugar maple, which is ideal for most cabinetry purposes.

Balfourodendron riedelianum

pau marfim

by Jon Arno

SCIENTIFIC NAME

Balfourodendron riedelianum. Derivation: The genus name is in honor of Isaac Bayley Balfour, a Professor of Botany at Glascow in 1879, at Oxford in 1884, and at Edinburgh in 1887 and possibly his father. The specific epithet is in honor of L. Riedel, a European plant collector in the 1800s.

FAMILY

Rutaceae, the rue or citrus family.

OTHER NAMES

guatambu, pau liso.

DISTRIBUTION

Pau marfim is native to South America from northern Argentina through Paraguay and into the state of Sao Paulo in southeastern Brazil.

THE TREE

With a maximum height of about 80 feet and diameters up to 30 inches, this tree is not large but produces a well formed bole yielding sawlogs up to 30 feet long.

THE TIMBER

The wood of pau marfim is dense, fine textured and creamy white in color. The heartwood may have a yellowish-brown tint, but there is little contrast between the heartwood and the sapwood. At first glance, the wood looks very much like sugar maple (*Acer saccharum* WDS 010) or yellow birch (*Betula alleghaniensis* WDS 037); however, it is harder and substantially more elastic than either of these U.S. domestic timbers. It is very strong with excellent wearing properties and, like the wood of its citrus relatives, is highly elastic. The specific gravity of pau marfim varies depending on where it grows. Reported specific gravity averages 0.73 (ovendry weight/green volume) for wood grown in Brazil, equivalent to an air-dried weight of 58 pcf, and specific gravity of wood from Argentina is 0.65, equivalent to an air-dried weight of 51 pcf.

SEASONING

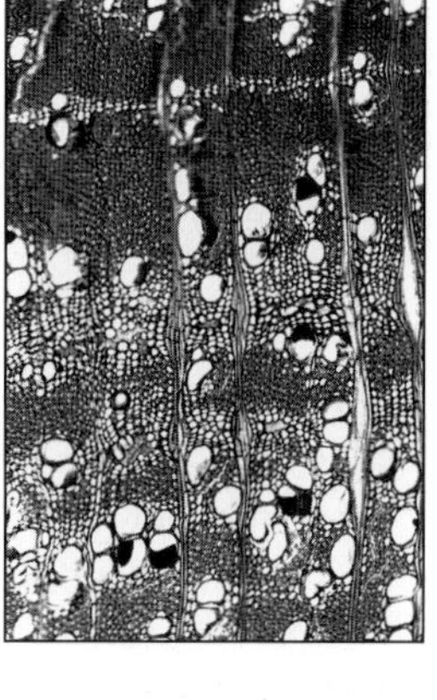

Although pau marfim has a rather high average reported volumetric shrinkage of 13.4% (green to ovendry), it is not especially difficult to kiln dry. Average reported radial shrinkage is 4.6% and tangential shrinkage is 8.8%. Like most dense, fine-textured woods, it is slow to air dry.

DURABILITY

Pau marfim is nondurable and not recommended for exterior projects.

WORKABILITY

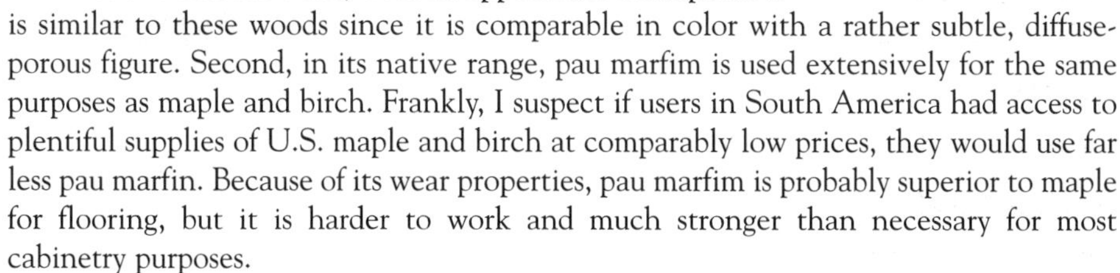

Pau marfim has been suggested as a substitute for maple or birch for two reasons. First, from an appearance standpoint it is similar to these woods since it is comparable in color with a rather subtle, diffuse-porous figure. Second, in its native range, pau marfim is used extensively for the same purposes as maple and birch. Frankly, I suspect if users in South America had access to plentiful supplies of U.S. maple and birch at comparably low prices, they would use far less pau marfin. Because of its wear properties, pau marfim is probably superior to maple for flooring, but it is harder to work and much stronger than necessary for most cabinetry purposes.

USES

Uses of pau marfim include cabinetry, furniture, flooring, turnery and tool handles.

SUPPLIES

Although most of the supply is consumed within its native markets, pau marfim is available in quantity. When carried by hardwood retailers in the United States, the price is moderate compared to other imports, but it is expensive relative to U.S. domestic maple or birch. Personally, I'm hard pressed to provide an argument for why it might be worth it.

Beilschmiedia tawa

tawa

by Max Kline

SCIENTIFIC NAME

Beilschmiedia tawa. Derivation: The genus name is name in honor of Karl Traugott Beilschmied (1793 to 1848), a German pharmacist and botanical author. The specific epithet is probably an old Maori name for the tree.

FAMILY

Lauraceae, the laurel family.

OTHER NAMES

Queensland walnut (U.S.).

DISTRIBUTION

Northeastern area of South Island and all of North Island, New Zealand.

THE TREE

Growing at sea level and altitudes of no more than 1,000 feet, the average size tawa tree is from 40 to 80 feet in height with a diameter of 1 to 4 feet. The slender branches are shorter than most New Zealand trees and have alternate leaves 2 to 4 inches in length and about 1 inch in width. The purplish-black, edible fruit looks like a small oblong plum and is covered with a grayish-blue down.

THE TIMBER

The wood is attractive, even colored and yellowish to grayish-brown with occasional streaks of darker brown. There is almost a complete lack of contrast between the sapwood and the heartwood. Planed surfaces of tawa are silky. The grain is almost invariably straight, and the texture is close and uniform. The wood has no distinctive smell. The wood is rather hard, brittle and easily split. Average reported specific gravity is 0.58 (ovendry weight/green volume), equivalent to an air-dried weight of 45 pcf. The strength of tawa is appropriate to its density, and therefore it is sufficiently strong for nearly all interior work.

SEASONING

The wood will air dry moderately quickly with only slight degrade. If care is taken, tawa can also be kiln dried with good results but a slight tendency to check may be observed. Average reported volumetric shrinkage (green to ovendry) is 11.4%.

DURABILITY

A certain amount of beetle attack may occur with the species, but it is usually not extensive. In general, resistance to infection is moderately good.

WORKABILITY

Sawing is easily accomplished with very little abrasive effect on cutting edges. The nail and screw holding properties are good, but thin stock shows some tendency to split during these operations so the wood should be pre-bored. If not adequately supported at the exit point of the tool, the timber will tend to chip out during drilling and mortising. Tawa responds satisfactorily to all types of finishes giving excellent results.

USES

A considerable amount of tawa is used in New Zealand for decorative woodwork, interior trim and furniture. In fact, IWCS member Bob Lynn reports using tawa for interior joinery in the new Parliament building in Wellington. Attractive plywood is manufactured from rotary cut tawa. A large quantity is used for staves in the cooperage industry.

SUPPLIES

Because of the distance from New Zealand to many of the major wood markets and the high cost of transportation, not much of this wood is exported for use outside New Zealand.

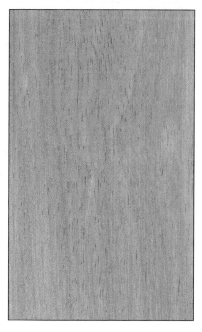

Berchemia zeyheri

pink ivory

by Max Kline

SCIENTIFIC NAME

Berchemia zeyheri. Derivation: The genus name is in honor of van Berchem (1720 to 1778), a French botanist. The specific epithet is in honor of Jean-Michel Zeyher (1770 to 1845), a botanist who collected plants in Africa with English botanist Joseph Burke (d. 1845). A synonym is *Rhamnus zeyheri*.

FAMILY

Rhamnaceae, the buckthorn family.

OTHER NAMES

red ivorywood, red ivory, umnini, umgoloti.

DISTRIBUTION

Zimbabwe, Mozambique, South Africa and scatterings in other parts of southern Africa.

THE TREE

Pink ivory is a deciduous tree with a spreading crown that varies in height from under 20 feet to over 50 feet. The boles are usually 7 to 9 inches in circumference. The bark is smooth and gray, but becomes darker when fully grown and rough near the base. The elliptic leaves are generally opposite, without hairs, 0.5 to 1.5 inches long and 0.75 to 1 inch wide. The flowers are small and greenish-yellow in color. The fruit is a small, ellipsoidal (0.5 by 0.75 inch) black berry very similar in appearance to the fruit of the buckthorn.

THE TIMBER

The wood is uniformly bright pink or pale red. The sapwood is almost white, and the pink heartwood, after long exposure, tends to become orange or orange-brown in color. The luster is low, and odor and taste are not distinct. It is hard and heavy, with an average specific gravity of 0.78 (ovendry weight/green volume), equivalent to an air-dried weight of 62 pcf. The grain is straight to irregular, while the texture is very fine. The rays are so close together they are not easily seen. This timber is reported to be very strong and stiff.

SEASONING

Pink ivory seasons very slowly and needs some care to prevent checking. Shrinkage values are not available.

DURABILITY

The Zimbabwe Forestry Commission reports that this wood has "excellent" durability. No other information was reported.

WORKABILITY

It is difficult to work with hand tools, but is an excellent wood for turnery and carving. Pink ivory takes a high polish.

USES

Pink ivory cannot be considered a commercial timber because the trees are so scattered exploitation would be too costly a process. The small quantities that are felled are used for fancy articles, inlaid work, small turned goods and carving.

SUPPLIES

Pink ivory is one of the rarest wood species to obtain. When available from special sources in small quantities, it commands a exceedingly high price and is usually sold by weight.

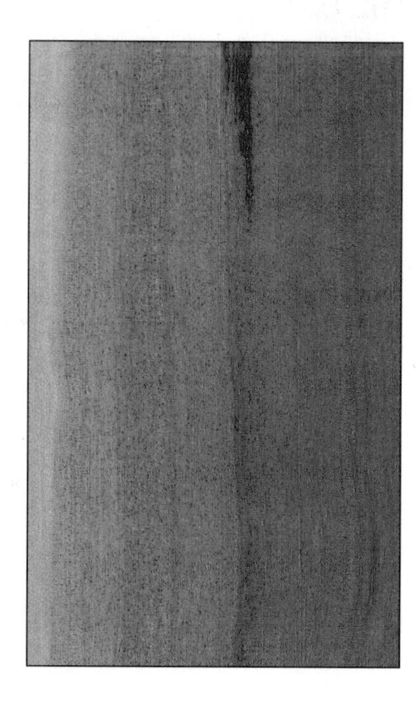

Bertholletia excelsa

Brazilnut

by Jon Arno

SCIENTIFIC NAME

Bertholletia excelsa. Derivation: The genus name is in honor of Louis Claude Berthollet (1748 to 1831), a French chemist. The specific epithet is Latin for tall or high.

FAMILY

Lecythidaceae, the Brazilnut family.

OTHER NAMES

castanheiro, juvia, tetoka, turury, tocary, castana del maranon.

DISTRIBUTION

Brazilnut is native to the Amazon River basin and rain forests along its tributaries in Brazil, Peru, Colombia and Venezuela.

THE TREE

Brazilnut is one of the largest trees in the Amazon reaching heights of 150 feet or more with diameters in excess of 6 feet. The bole is of excellent form for timber production. The bright yellow flowers and large (6 inches wide and 15 to 20 inches long) leathery leaves make it an attractive ornamental.

THE TIMBER

As a result of the conditions within its extensive range, the wood density and color vary. When freshly cut, the heartwood is pinkish-brown, becoming softer amber-brown after prolonged exposure. The sapwood is yellowish-tan and sharply demarcated from the heartwood. Although interlocked grain produces a ribbon effect on radial surfaces, the normal figure is plain and monotonous. With medium texture, medium to low luster and no taste or odor, Brazilnut is neither memorable nor easily identified by any unique attribute. The wood is moderately strong and compares favorably to bigleaf mahogany (*Swietenia macrophylla* WDS 254) in crushing strength but is substantially more elastic and more resistant to abrasion. Average reported specific gravity is 0.59 (ovendry weight/green volume), equivalent to an air-dried weight of 46 pcf.

SEASONING

Brazilnut air dries quickly with little or no degrade. Average reported shrinkage values (green to ovendry) are 3.9% radial, 8.3% tangential and 11.2% volumetric.

DURABILITY

The wood is very durable with respect to fungi attack.

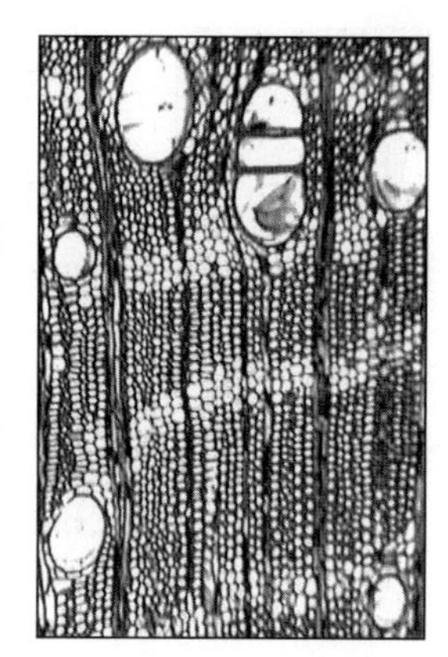

WORKABILITY

Though just slightly denser and comparable in working characteristics to sugar maple (*Acer saccharum* WDS 010), brazilnut is not as finely textured and may clog saw blades due to high gum concentrations. Brazilnut undergoes about 25% less volumetric shrinkage and is more stable than sugar maple. It is also more elastic, steam bends well, glues easily and polishes without difficulty. It is a good, general purpose timber.

USES

Because of its decay resistance, it is used for ship decking, water tanks, railroad ties and exterior construction. Interior uses include flooring, furniture and cabinetry. The nuts are valuable as a table delicacy and the oil that can be extracted and used as a substitute for olive oil. In its native habitat, the fibrous inner bark is sometimes used for caulking boats.

SUPPLIES

Because of the value of its nut crop, the tree is common throughout northern South America, but its exploitation as a timber seems wasteful. The wood is not so unique or showy as to command a high price on the international market and is not commonly stocked by retailers. It is capable, however, of producing clear lumber of immense width and length, and if you can find some at a reasonable price it is handy for large projects.

SPECIAL NOTE

Interestingly, the hard-shelled seeds grow in a cluster (as many as 12 to 24) encased in a heavy, woody capsule about the size of a softball. Foraging can be quite dangerous since the capsule does not release the nuts while affixed to the tree. Instead, the entire capsule, weighing about 1 pound, breaks off the tree and silently plummets to the ground. Falling from heights of 150 feet, it reaches the forest floor with skull crushing force, and fatalities among gatherers of the nuts are not uncommon. The tree also absorbs strontium which becomes concentrated in the endosperm of the seed. This led to concern in the early 1960s that the nuts were becoming radioactive as a result of the atmospheric testing of nuclear weapons though the radioactivity never posed a problem when working with the wood.

Betula alleghaniensis

yellow birch

by Chuck Holder

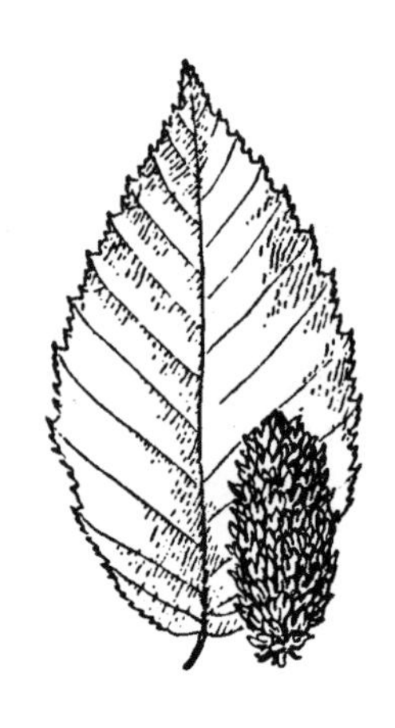

SCIENTIFIC NAME

Betula alleghaniensis. Derivation: The genus name is classical Latin for birch. The specific epithet means the plant is "of the Allegheny Mountains" in recognition of its native geographical range. A synonym is *B. lutea*.

FAMILY

Betulaceae, the birch family, a Northern Hemisphere family of trees and shrubs with some six genera and 100 species.

OTHER NAMES

grey birch, silver birch, swamp birch, hard birch, curly birch, boleau jaune.

DISTRIBUTION

The growth range of yellow birch extends from the southeast corner of Manitoba to and throughout Canada's Atlantic provinces and the northeastern United States where it grows with western larch (*Larix occidentalis* WDS 159) and black ash (*Fraxinus nigra* WDS 123).

THE TREE

The largest of the eastern birches, yellow birch is a medium-sized hardwood tree which normally grows 80 feet in height with a trunk diameter of 2 feet. Maximum growth is 114 feet in height and 4.5 feet in diameter. Yellow birch has an irregularly rounded crown and well formed bole. The doubly serrate leaves are 3 to 4.5 inches long and 1 to 2 inches wide. They are ovate to oblong with an acute apex and are dark green above and pale yellowish-green below. Its fruit is an ovoid, short-stalked, erect catkin. The yellowish golden-gray to bronze bark peels horizontally into thin papery strips and breaks into reddish-brown fissures and plates in maturity. The papery bark curls are highly flammable, even when wet. Both the bark and the twigs have a wintergreen taste. Typically, the tree matures in 150 years and lives to 300 years. Yellow birch is the official tree of Quebec, Canada.

THE TIMBER

The diffuse-porous heartwood is light to dark golden-brown to light reddish-brown. The sapwood is whitish, pale yellow or light reddish-brown. Hard, often wavy grained and strong, the wood has an average reported specific gravity of 0.55 (ovendry weight/green volume), equivalent to an air-dried weight of 42 pcf. The fine, uniform textured wood has high impact resistance.

SEASONING

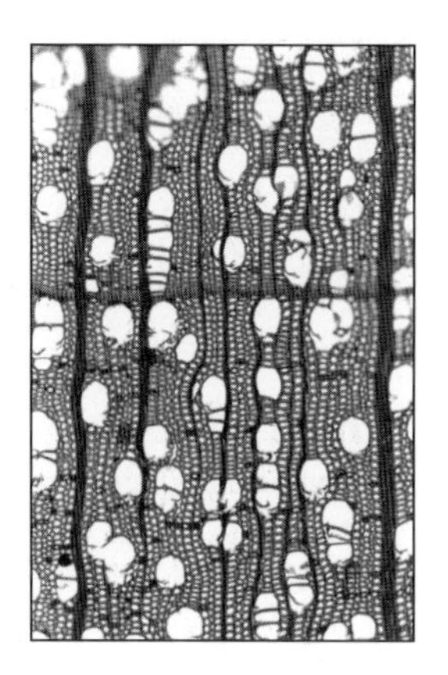

The wood seasons slowly with little degrade. Average reported shrinkage values (green to ovendry) are 7.3% radial, 9.5% tangential and 16.8% volumetric.

DURABILITY

Yellow birch is not very resistant to decay in moist service conditions, but can be readily treated with preservatives.

WORKABILITY

Yellow birch bends fairly well but requires some care when gluing. It has good nail and screw holding properties and splitting resistance. It takes a smooth finish and has good machining properties. I have seen some beautifully crafted and finished chairs made of yellow birch.

USES

Used extensively for furniture, flooring, doors and cabinetry, yellow birch is also in demand for veneer and plywood. It is an important source of hardwood lumber for eastern Canada and the northeastern United States.

SUPPLIES

Yellow birch is widely available commercially.

Betula lenta

sweet birch

by Jim Flynn

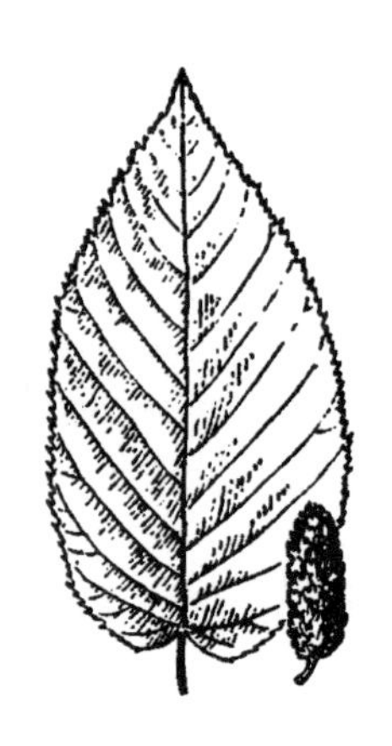

SCIENTIFIC NAME

Betula lenta. Derivation: The genus name is classical Latin for birch. The specific epithet *lenta* means flexible or tough referring to the twigs.

FAMILY

Betulaceae, the birch family.

OTHER NAMES

black birch, cherry birch, red birch, mahogany birch.

DISTRIBUTION

The growth range extends from southern Maine, west to New York, New Jersey and eastern Ohio, south mostly in the mountain areas, including Pennsylvania, West Virginia, Virginia and Kentucky to the western portions of North Carolina, the extreme northwest South Carolina, northern Georgia, Alabama and eastern Tennessee. In Canada, it is found in limited numbers in eastern Ontario and southwestern Quebec.

THE TREE

Sweet birch is a medium-sized tree attaining heights of 50 to 60 feet with diameters of 18 to 24 inches and a long clear bole. It grows best in the ancient forest loam and can almost always be found near mountain streams at the head of coves where its deep, smooth, mahogany-red bark stands out. The tree can also be found on rocky, boulder-strewn sites in spite of its need for nourishment. On poor soil, it is apt to be shrubby and have a stunted appearance. The numerous branches spread out from the stem at a wide angle. The twigs and smaller branches droop at the ends. The roots are deep and wide spreading. It seldom grows in pure stands and is found scattered amongst white pine, hemlock, yellow birch, black cherry, white ash, basswood and yellow poplar. Sweet birch is frequently confused with yellow birch (*B. alleghaniensis* WDS 037), but can be distinguished by its fine-toothed leaves with scalloped or heart-shaped bases and by the non-peeling blackish-red bark. The bark and twigs have a wintergreen taste.

THE TIMBER

Sweet birch has a light-colored sapwood, and the dark brown heartwood is often tinged with red. The wood has no distinct taste or odor. This diffuse-porous wood is strong with a very good shock-resisting ability. The texture is fine and uniform. Average reported specific gravity is 0.60 (ovendry weight/green volume), equivalent to an air-dried weight of 46 pcf.

SEASONING

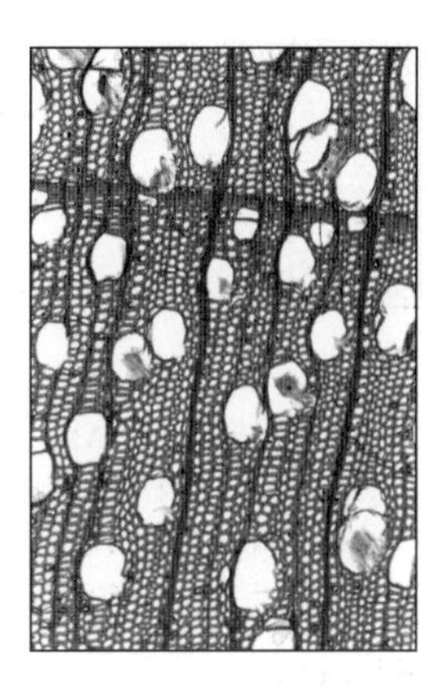

It is hard to air dry because of its tendency to warp. Average reported shrinkage values (green to ovendry) are 6.5% radial, 9.0% tangential and 15.6% volumetric.

DURABILITY

Sweet birch is not durable when in contact with moist ground or when used in conditions where moisture is a problem.

WORKABILITY

Sweet birch has very good machining, gluing and finishing qualities. It turns well and, if splitting can be avoided, it has good nail holding capacity. Once dry, it is relatively stable and holds its shape well.

USES

One of the principal uses of this wood is veneer and plywood. Because it resembles maple, it is often used interchangeably in the furniture industry. It has many other uses including paneling, flooring, woodenware and butcher blocks. One of the old-time mountaineer uses was birch beer. The recipe in *A Natural History of Trees of the Eastern and Central North America* by Donald Culross Peattie is: "Tap the tree as the sugar maple is tapped, in spring when the sap is rising and the buds are just swelling, jug the sap and throw in a handful of shelled corn" and natural fermentation will finish the job.

SUPPLIES

In North America, the three birches of highest commercial importance are yellow birch, white birch (*B. papyrifera* WDS 039) and sweet birch. It is difficult to identify the specific species when confronted with the wood alone. White birch can be distinguished from the others on the basis of weight because it is much lighter. A lumber dealer may have between three and five different species combined in stacks labeled birch. It is essential when seeking a specific species of birch that it be identified before it is harvested. Supplies are plentiful.

Betula papyrifera

white birch

by Chuck Holder

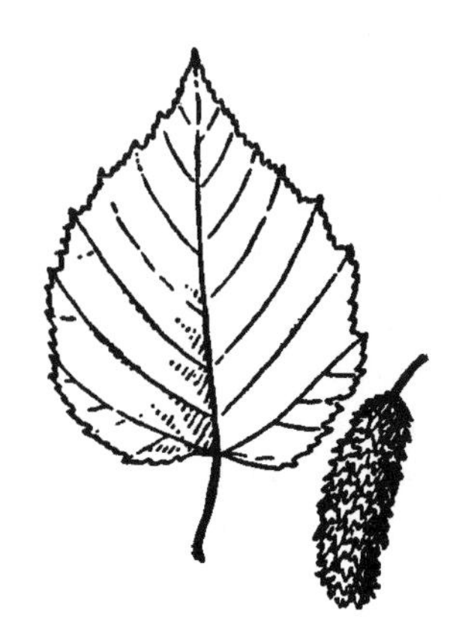

SCIENTIFIC NAME

Betula papyifera. Derivation: The genus name is classical Latin for birch. The specific epithet *papyifera* is Latin for paper-bearing meaning with a papery bark. It is derived from the Greek *papyros* meaning paper reed and Latin *ferre* meaning to bear.

FAMILY

Betulaceae, the birch family.

OTHER NAMES

paper birch, canoe birch, silver birch, bouleau a papier.

DISTRIBUTION

The most widely distributed of native birches, white birch grows in every region of Canada and most of the northern tier states of the United States. It reaches as far south as West Virginia and North Carolina.

THE TREE

White birch is a medium-sized deciduous tree with a pyramidal crown that grows 70 feet in height with a diameter of 2 feet. Maximum growth is 120 feet in height and 6 feet in diameter. The irregularly rounded, coarsely double serrate leaves are ovate to oval, 2 to 4 inches long by 1 inch wide or larger, with an open 0.5- to 2-inch margin and are dull dark green above and pale yellowish-green below turning light yellow in the autumn. The fruit is a seed catkin which is 1.5 to 2 inches long and held erect. In young trees, the bark is dark brown but as the tree matures the bark turns a chalky white separating into thin papery strips. In old trees, the bark is blackish and fissured near the base. There are a number of western varieties of white birch, including Alaska birch, mountain white birch, Kenai birch and northwestern paper birch. White birch is the official tree of Saskatchewan, Canada and the state tree of New Hampshire.

THE TIMBER

Fine and uniformly textured with no odor, the wood is diffuse porous and creamy white with a pale brown core. It is moderately heavy. Average reported specific gravity is 0.48 (ovendry weight/green volume), equivalent to an air-dried weight of 36 pcf. The wood of white birch is somewhat weaker than yellow birch (*B. alleghaniensis* WDS 037) or eastern birch and is much lower in resistance to suddenly applied loads.

SEASONING

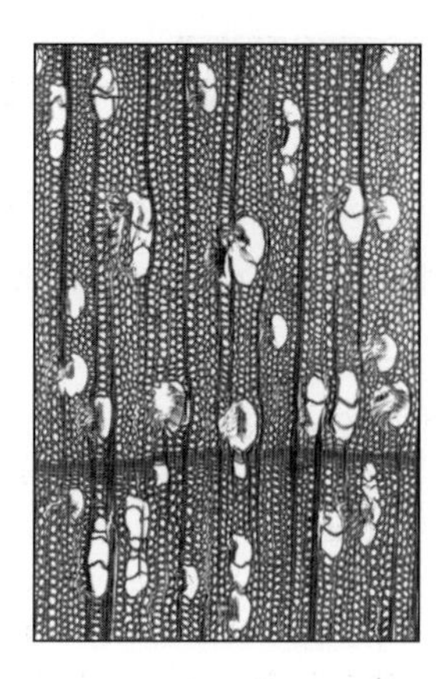

White birch seasons well but has relatively high shrinkage. Average reported shrinkage values (green to ovendry) are 6.3% radial, 8.6% tangential and 16.2% volumetric.

DURABILITY

It is not very resistant to decay. Spalting is readily induced.

WORKABILITY

White birch glues satisfactorily and has good machining properties. It takes a good finish and has good nail and screw holding properties.

USES

White birch is a popular ornamental and is a valued source of hardwood timber. In thin curls, the bark makes excellent fire starter, and the wood is premium firewood. Uses include veneer, plywood, interior finish, furniture, woodenware, toys, dowels, pallets and crates. It is also used for pulpwood. The tough pliable bark has long been used for making canoes and ornaments.

SUPPLIES

Available as western birch, supplies are ample at reasonable prices.

Bischofia javanica

bishopwood

by Jim Flynn

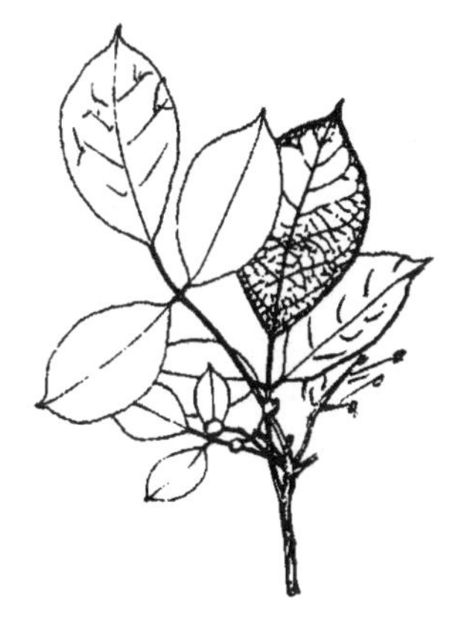

SCIENTIFIC NAME

Bischofia javanica. Derivation: The genus name is in honor of G. W. Bischoff (1797 to 1854), a German botanist, lexicographer and glossographer in Heidelberg. The specific epithet means the plant is "of Java".

FAMILY

Euphorbiaceae, the spurge family.

OTHER NAMES

Java cedar, gintungan, paniala, aukkyu, ye-padauk, nhoi, term, toog, tuai, koka, tayokthe.

DISTRIBUTION

Native to India, central China and Malaysia eastward into the southeast Pacific, it has been cultivated in warmer climates including Africa, Hawaii and Florida.

THE TREE

Because the species has a very wide geographical growth range, there are apt to be large variances in its morphological characteristics. Typically, the tree has a large and dense crown and reaches heights of 100 feet with diameters of 36 inches or more. The main stem is not buttressed. The rough bark of this deciduous tree is dark gray tinged with brown, and it exfoliates in angular scales. It is a fast grower with a compound leaf consisting of three ovate to oblong-ovate leaflets 2 to 7 inches long. It exudes a colorless latex sap. The trees grow best in shady ravines and swamps and can be found on hillsides up to 4,000 feet in elevation if the site is damp.

THE TIMBER

Bishopwood has a pink to brick or reddish-brown sapwood that is up to 1.5 inches wide and is sharply demarcated from the purplish or dark reddish-brown heartwood. Mature trees rarely produce sawlogs larger than 25 feet because of the tree's branching habits. The texture of the wood is medium coarse, and the grain can be either straight or interlocked and is occasionally curly. It is slightly lustrous. When freshly cut, it has a strong vinegar odor. Depending upon its growth range, there is a wide variation in the density of this wood. Average reported specific gravity ranges from 0.45 to 0.71 averaging about 0.56 (ovendry weight/green volume), equivalent to an air-dried weight of 34 to 56 pcf, with an average of 43 pcf.

SEASONING

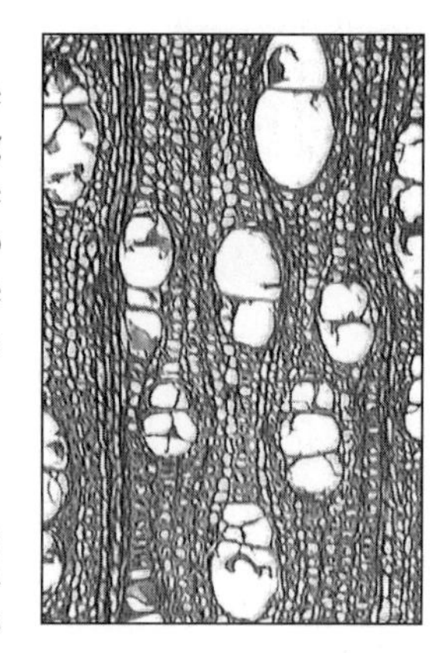

Bishopwood is extremely difficult to season. Reports indicate that the species is subject to severe warping and checking during seasoning. No advice is given as to how best overcome this fault. Timber harvested in Malaysia has been reported to be less susceptible to degrade. Average reported shrinkage values (green to ovendry) are 4.4% radial and 9.8% tangential.

DURABILITY

The timber is susceptible to termites, pinhole borers, longhorn beetles and marine borer attacks. In tropical climates it is only moderately durable when used outdoors under a roof or other covering but is very durable in temperate zones. With preservatives the wood is durable when used underground in such applications as piles and railroad sleepers and is especially durable when used in freshwater applications.

WORKABILITY

In spite of its relative hardness, bishopwood saws and machines well and can be worked to a very smooth finish. Some chipping on the radial surface can be expected when planing. Sanding and finishing operations are without problems. The wood appears to be similar to jarrah (*Eucalyptus marginata* WDS 115) in workability. Bishopwood will darken with age.

USES

Contrary to an early 1800's report in the Indian *Tea Gazette* that the wood was too heavy for tea boxes, it was used for some of Rajah's coffins. Current usage includes indoor construction work, flooring and some furniture. There are several cautionary notes that the wood is not suitable for manufacturing plywood.

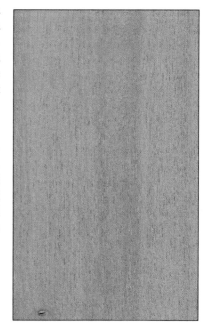

SUPPLIES

Bishopwood is not readily available on the commercial market except in areas of heavy growth. The Fiji Department of Forestry states the wood is normally retailed in mixtures labeled "mixed hardwoods". Because the trees are widely planted in many parts of the world, however, the wood is available if collectors seek it.

Brachylaena huillensis

muhuhu

by Max Kline

SCIENTIFIC NAME

Brachylaena huillensis. Derivation: The genus name *brachy* is Greek meaning short and *laena* meaning cloak, referring to the ring of bracts around the base of the flowers. The specific epithet is in reference to a geographical name Huila, in Angola. A synonym is *B. hutchinsii.*

FAMILY

Compositae, the composite or daisy family; also Asteraceae.

OTHER NAMES

muhugu, mubuubu, watho, mvumvo, mshenzi.

DISTRIBUTION

In the semi-evergreen and lowland dry forests of the coast of tropical East Africa, also common in particular parts of Tanzania and Kenya.

THE TREE

Muhuhu is a small tree that grows 80 to 90 feet in height with a diameter of 2 feet. The bark if often ingrown. Because the tree is often misshapen, it is difficult to obtain timber in large dimensions.

THE TIMBER

Muhuhu is dark yellowish-brown with darker streaks and a fine, even texture and interlocked grain. Average reported specific gravity is 0.75 (ovendry weight/green volume), equivalent to an air-dried weight of 60 pcf. Muhuhu resembles sandalwood (*Santalum album* WDS 242) from its marked scent that is pleasant and spicy when the wood is fresh. Machining dry wood also brings out the same odor. Muhuhu does not have outstanding strength relative to bending, stiffness or resistance to applied loads. Its ability to resist indentation, however, makes it a very good wood to use to withstand abrasion.

SEASONING

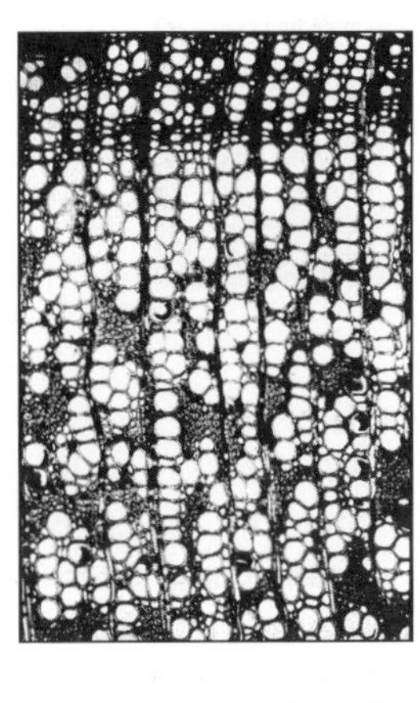

As with all dense woods, care must be exercised to slowly dry this species to prevent checks and end splitting. Average reported shrinkage values (green to 12% moisture content) are 2.0% radial and 3.0% tangential. Once seasoned, the timber is noteworthy for very small movement under fluctuating humidity conditions.

DURABILITY

The heartwood is extremely durable and is reputed to be resistant to both termite and marine borer attack.

WORKABILITY

This timber is rather difficult to work and requires pre-boring when nailing. It also has a tendency to overheat during sawing operations unless the proper saw is used. It turns well but is quite brittle.

USES

Due to its abrasion resistance and even wear, muhuhu is a good heavy-duty flooring timber. Its attractive appearance makes it suitable for floors in hotels and public buildings where there is heavy pedestrian traffic. In Africa it is used for fence posts, piles, railway sleepers, carving, turnery, doors and window frames. Aromatic oil is extracted from the wood for use in perfumes and toilet preparations. Small quantities of the wood have been exported to India as a substitute for sandalwood for cremation purposes.

SUPPLIES

The majority of this wood is available as flooring and in small-dimension lumber at relatively high prices.

Brosimum guianense

snakewood

by Max Kline

SCIENTIFIC NAME

Brosimum guianense. Derivation: The genus name is Greek for edible fruit. The specific epithet is Latin meaning the plant is "of Guyana". Synonyms are *Piratinera guianensis* and *B. aubletii*.

FAMILY

Moraceae, the mulberry family.

OTHER NAMES

letterwood, leopardwood, tortoiseshell wood, specklewood.

DISTRIBUTION

Panama, French Guiana, Suriname, Guyana and the Amazon region of Brazil.

THE TREE

This tree grows to a height of 80 feet with a long cylindrical bole, sometimes over 30 inches in diameter above the root, and is covered with a smooth bark containing a thick, sticky, white latex.

THE TIMBER

The heartwood is slow-growing and a log that is 15 inches long may have heartwood that is only 1 to 4 inches thick. The sapwood is of no commercial value and is always discarded. The heartwood is dark red or reddish-brown with conspicuous, irregular black speckles or stripes resembling the spotted skin of certain snakes. The dark areas are a result of variations in the color of the gummy deposits that fill the cell cavities. Sometimes the wood is plain with no markings. It has a fine texture and is highly lustrous. The grain is straight. The odor and taste are not distinct. The wood is very hard and heavy, and average reported specific gravity is 0.82 to 1.10 (ovendry weight/green volume), equivalent to an air-dried weight of 66 to 93 pcf. Snakewood is strong but brittle.

SEASONING

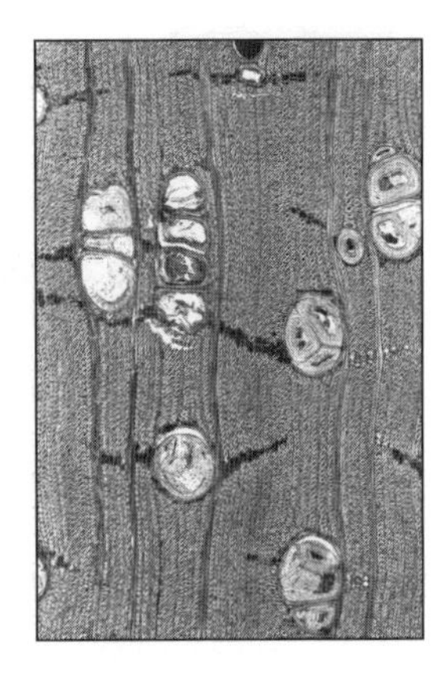

Snakewood should be dried carefully and in small pieces. In a wood of such high density, an excessive moisture gradient may easily build up during seasoning causing case hardening. Shrinkage is reported to be high.

DURABILITY

The heartwood is very resistant to decay and insects, while the sapwood is readily attacked.

WORKABILITY

Snakewood is inclined to be splintery and splits rather easily. Because of its hardness and density, the wood is difficult to work. It is hard to cut but turns well, finishes smoothly and polishes beautifully. Pre-boring is necessary when using nails or screws. Gluing is very difficult.

USES

Because of its limited availability and size, snakewood is used for walking sticks and is considered one of the finest of all woods for this purpose. Other uses include umbrella handles, drumsticks, fishing rod butts and miscellaneous fancy articles. To take advantage of unusual figure, larger pieces are sometimes cut into veneer for cabinetry.

SUPPLIES

Due to the slow formation of heartwood, snakewood is only available in very small quantities. In the past, it was imported in the form of small logs which were 7 feet long and 2 to 8 inches in diameter with all of the sapwood removed in the forest at the time of cutting. It is sold by weight and is one of the most expensive woods on the commercial market. There have been very few imports into the United States, and this highly prized wood has become essentially nonexistent for wood collectors.

Brosimum rubescens

satine

by Max Kline

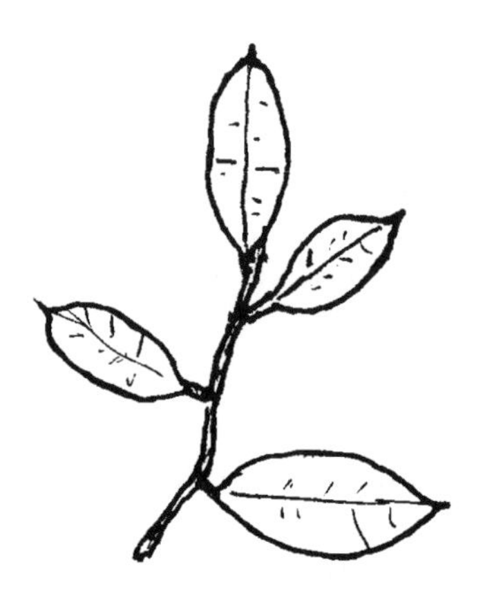

SCIENTIFIC NAME

Brosimum rubescens. Derivation: The genus name is Greek for edible fruit. The specific epithet is Latin for light red, referring to the color of the wood. A synonym is *B. paraense*.

FAMILY

Moraceae, the mulberry family.

OTHER NAMES

bloodwood, Brazil redwood, cardinalwood, muirapiranga, pau rainha, palo de sangre, satine urbane.

DISTRIBUTION

French Guiana, Brazil, Peru, Panama and Venezuela.

THE TREE

Satine is a rather tall tree growing in the upland rain forest, most frequently on sandy soils. It grows to a height of 120 feet or more with a trunk diameter in the range of 30 to 40 inches. The straight, cylindrical boles are mostly clear for 75 feet or more. They are without buttresses, and the smooth bark exudes a white latex. The leaves are elliptic in shape with a broad base. The fruit is similar in appearance to a fig but has a single large seed.

THE TIMBER

The color of satine heartwood is various shades of rich, lustrous red and yellow overlaid with a golden sheen. The wide sapwood is yellowish-white and sharply demarcated from the heartwood. Average specific gravity ranges from 0.71 to 0.82 (ovendry weight/green volume), equivalent to an air-dried weight of 56 to 66 pcf. The sapwood is lighter and variable in weight. The texture is rather fine to coarse, and the grain is straight to variable. The luster is high and golden. The wood is odorless and tasteless.

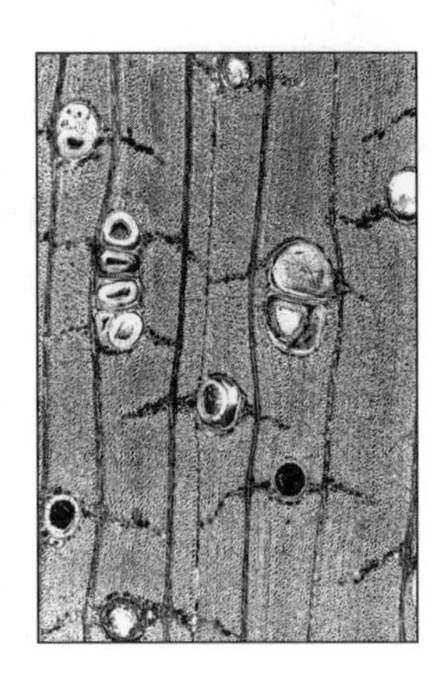

SEASONING

Satine lumber air dries rapidly and easily with little or no degrade. Material containing tension wood, however, will be prone to warp. Average reported shrinkage values (green to ovendry) are 4% radial and 6% tangential.

DURABILITY

The heartwood is very resistant to decay and insects, while the sapwood is readily attacked.

WORKABILITY

Although hard, the wood is not difficult to work, glues well and takes a high natural polish. It finishes very smoothly.

USES

Satine is used for veneer, furniture, marquetry inlays, turnery, cabinetry, bows and fishing rods. The heartwood has very limited commercial use due to the small size of this portion of the tree. The logs obtainable rarely exceed 8 inches in diameter. The much wider sapwood finds expanded uses but does not have the beautiful and variegated color of the heartwood.

SUPPLIES

Satine is scarce and only found occasionally on a commercial basis in the United States from the most important lumber importers. For this reason, its cost is regarded as high. The best grades come from French Guiana, and the wood from Brazil and Peru usually has inferior color characteristics.

Broussonetia papyrifera

paper-mulberry

by Jim Flynn

SCIENTIFIC NAME

Broussonetia papyrifera. Derivation: The genus name is in honor of Auguste Broussonet (1761 to 1807), a physician and naturalist of Montpellier, France. The specific epithet means paper-bearing, referring to the use of the inner bark in making paper.

FAMILY

Moraceae, the mulberry family.

OTHER NAMES

wauke, po'a'aha, tapa, kappa, broussonet.

DISTRIBUTION

Paper-mulberry is native to China, Japan, Thailand and Myanmar (Burma). It has been widely planted as an ornamental, especially in the temperate zone, and has naturalized in many places.

THE TREE

Paper-mulberry is a fast-growing tree reaching heights of 50 feet or more. Its rounded crown is supported on an often large and twisted trunk. The trunk of an old tree is often a mass of gnarled and wrinkled formations which appear grotesque. Because the tree produces many suckers, its branches are tangled hopelessly. The leaf shape is variable, some being lobed and others not. The trees are unisexual with male and female flowers appearing on different plants. The tough and fibrous bark is soaked and pounded into a fabric, which functions as paper or cloth. It once had universal use in Polynesia for white togas and simple loincloths. This tree should not be confused with *Morus alba* (white mulberry), which is cultivated for the silkworm industry.

THE TIMBER

Although paper-mulberry grows worldwide, there is little published data on the characteristics of the wood. Many references pass it off with such statements as "There is little commercial use for the wood." Nevertheless, it seems that we should be aware of this timber and perhaps encourage its use. At an IWCS meeting in Florida, fairly large saw logs of paper-mulberry were run through a wood miser producing clear slabs in the 20-inch-wide range. It has a density equivalent to white pine (*Pinus strobus* WDS 212) which has a specific gravity (ovendry weight/green volume) of 0.34, equivalent to an air-dried weight of 25 pcf. The wood is ring porous, and the sapwood is a light tan. On quartersawn surfaces of heartwood, there is a pleasing array of light tan to brown stripes. Depending on the growth area, there is wide variation in wood color and density. An added feature is that dark reddish-orange wood develops around insect holes and sites of trauma.

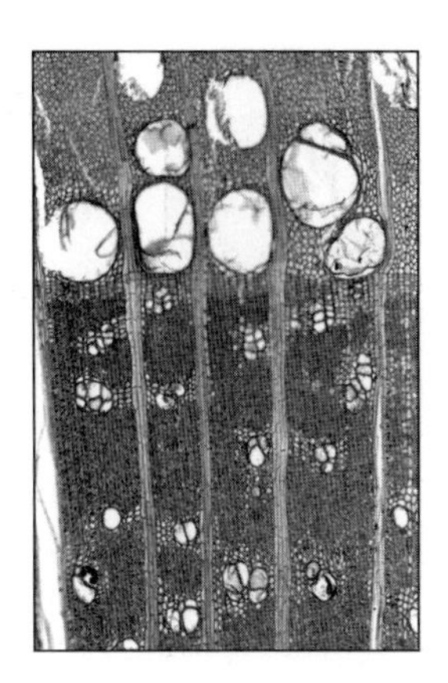

SEASONING

Shrinkage values are not available. Because the main stem of this tree may or may not be clear of branches nor the growth straight, it would seem that there would be a wide range of seasoning problems. Slabbed wood should be cut and stacked with weights while it is slowly seasoned.

DURABILITY

Again, there is an absence of published data on this characteristic. From observing some saw logs, it appears that the timber would not serve well in applications where moisture would be a factor.

WORKABILITY

There is every reason to believe that paper-mulberry will have the same working characteristics as soft white pine. The wood glues and finishes well. When processing a piece with cross-grain that was cut from a distorted sawlog, a noticeable problem was the tendency to chip when using a thickness planer with dull blades.

USES

Uses remain obscure, but paper-mulberry makes a nice addition to a wood collection.

SUPPLIES

Samples are obtainable by careful watching, waiting and trading.

Brya ebenus

cocuswood

by Max Kline

SCIENTIFIC NAME

Brya ebenus. Derivation: The genus name is from the Greek *byron* meaning low growing. The specific epithet is Latin for ebony black.

FAMILY

Fabaceae or Leguminosae, the legume family; (*Papilionaceae*) the pea or pulse group.

OTHER NAMES

brown ebony, green ebony, Jamaican ebony, granadillo, torchwood, cocus, tropical American ebony.

DISTRIBUTION

Cocuswood is only found in Jamaica and Cuba.

THE TREE

This is a small tree which grows no more than 25 feet in height and 8 inches in diameter. The branches are armed with short, sharp prickles. The leaves are deciduous, and the yellow or orange flowers are arranged in sub-terminal clusters.

THE TIMBER

The heartwood is yellow to rich brown, almost black, variegated or striped, sometimes with an olive hue. When fresh the heartwood is sharply demarcated from the yellowish sapwood, which is 0.5 to 1 inch thick. The wood has a waxy appearance, and it is very hard, heavy, compact, tough and strong. Average specific gravity ranges from 0.85 to 0.95 (ovendry weight/green volume), equivalent to an air-dried weight of 69 to 78 pcf. The texture is very fine and uniform with a natural luster. The grain is straight to wavy, and the timber is considered to be very strong.

SEASONING

Cocuswood is difficult to season without checking but, once dried, it is dimensionally stable. Shrinkage values are not available.

DURABILITY

Cocuswood timber is highly durable.

WORKABILITY

Cocuswood is not difficult to work, despite its heavy and hard properties. It finishes very smoothly with a high natural luster.

USES

This species was formerly used for making musical instruments, such as flutes, clarinets and oboes but has been largely replaced by African blackwood (*Dalbergia melanoxylon* WDS 096). Other uses have included handles for table cutlery, brush backs, inlay and articles of turnery for which it is especially suited due to its handsome appearance and excellent turning properties.

SUPPLIES

Sometimes logs 8 feet long and 3 to 6 inches in diameter are exported from Cuba and Jamaica. The sapwood is left on to protect the heartwood from checking and splitting. The sapwood has no other value. As lumber, the wood is scarce, and the price is very high.

Bucida buceras

jucaro

by Jon Arno

SCIENTIFIC NAME

Bucida buceras. Derivation: Both the genus name and specific epithet refer to ox horn because the horn-like galls on the fruit, which are caused by a mite, resemble ox horns.

FAMILY

Combretaceae, the combretum family.

OTHER NAMES

oxhorn bucida, gri-gri, grignon, ucar, black-olive, bucida.

DISTRIBUTION

Jucaro is native to the islands of the West Indies and along the coast of Central and South America from southern Mexico to French Guiana. Owing to its attractive appearance and its ability to withstand salt spray, it has been used for landscaping beach properties and is now quite common in southern Florida where it is better known as oxhorn bucida.

THE TREE

Capable of reaching heights in excess of 100 feet with diameters up to 5 feet, jucaro can produce an excellent, straight bole which yields good saw logs. Unfortunately, when grown in the open, it branches out quickly. The leaves tend to cluster toward the tip of the branches, giving it an interesting appearance in landscape settings. The flowers form on slender spikes and develop into small, dry, egg-shaped fruits, which turn reddish-brown and have a leathery texture when mature.

THE TIMBER

With an average reported specific gravity of 0.93 (ovendry weight/green volume), equivalent to an air-dried weight of 76 pcf, jucaro is a very hard, heavy and a somewhat fine-textured wood. The transition between the yellowish-green sapwood and the brownish, drab olive heartwood is rather gradual. Interlocked grain is common, and radially cut (quartersawn) boards often possess a very attractive ribbon-striped figure which is further enhanced by the way the wood's high surface luster reflects light. When first cut, jucaro has a faint creosote-like scent, but once dry, the wood is without odor or taste. It is very strong, superior to Brazilian rosewood (*Dalbergia nigra* WDS 097) in bending strength, with excellent resistance to abrasion.

SEASONING

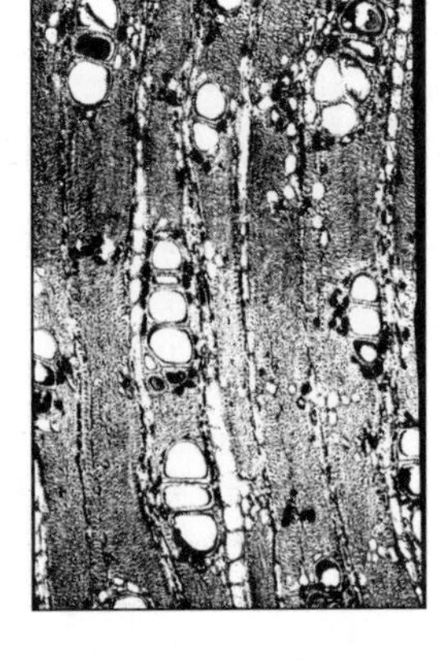

Given its extremely high density, this wood dries slowly, but it is not difficult to season. Average reported shrinkage values (green to ovendry) are 4.4% radial, 7.9% tangential and 12.2% volumetric.

DURABILITY

Jucaro weathers exceptionally well, is durable in contact with the ground but is not resistant to marine borers.

WORKABILITY

Jucaro's working qualities are punishing. By comparison, hickory (*Carya* spp.), black locust (*Robinia pseudoacacia* WDS 237) and even eastern hophornbeam (*Ostrya virginiana* WDS 194) seem soft. Nonetheless, for patient woodworkers with sharp tools, the end results can be rewarding because jucaro polishes beautifully and has great stability. Although hard, it shapes, turns and machines well, with the only exception being a slight tendency for the interlocked grain to tear out when planing. A smooth, even surface is best achieved with hand scrapers and fine grit sandpaper. The wood's translucent luster and soft green highlights give it a rich look when used for small, decorative pieces, but its structural properties also make it a good choice for rugged utilitarian items such as chopping blocks, mallet heads, wood clamp jaws and workbench tops.

USES

Heavy construction timbers, fence posts, poles, railroad ties and flooring represent this wood's primary commercial applications. Some, however, is used for making charcoal, and the bark and fruit are used in tanning leather.

SUPPLIES

Jucaro is fairly plentiful, but seldom seen in lumber form on the international market. Its extreme hardness is a negative in most general construction applications. In addition, although almost as dense as lignumvitae (*Guaiacum officinale* WDS 132), jucaro cannot compete with lignumvitae in its traditional use for making marine bearings. Jucaro lacks lignumvitae's natural, self-lubricating qualities and is not resistant to marine borers. Jucaro can best be obtained by foraging.

Bulnesia arborea

verawood

by Jon Arno

SCIENTIFIC NAME

Bulnesia arborea. Derivation: The genus name was assigned by Claude Gay, the French botanist, in his book *Flora for Chili* (ca. 1850) to honor Manuel Bulnes, President of Chile (1841 to 1851), and the Chilean town of Bulnes. The specific epithet is Latin meaning tree-like.

FAMILY

Zygophyllaceae, the caltrop family.

OTHER NAMES

Maracaibo lignumvitae, guayacan, bera cuchivaro.

DISTRIBUTION

Verawood is native to the coastal areas of Colombia and Venezuela. As one of its common names suggests, it is plentiful in the Lake Maracaibo region. Despite the fact that its range tends to be coastal, this species prefers higher ground and drier soils.

THE TREE

While small relative to most tropical timbers, verawood may occasionally reach heights of up to 100 feet but typical specimens average 40 to 50 feet in height. Under ideal growing conditions, the bole is straight, yielding good saw logs, even though the diameter seldom exceeds 2 feet. Genetically, this species is perhaps the best timber producer in the caltrop family, but the family's other genera offer little competition.

THE TIMBER

With an average reported specific gravity of 1.00 (ovendry weight/green volume), equivalent to an air-dried weight of 83 pcf, verawood is about as heavy as water. While this ranks it among the heaviest woods in the world, it is slightly less dense than its close relative lignumvitae (*Guaiacum officinale* WDS 132) which has a specific gravity of 1.05. Being extremely fine textured and oily to the touch, this species is similar in appearance to lignumvitae, but often exhibits a brighter green color and more attractive figure with dark brown stripes. It is extremely strong and is especially resistant to abrasion. Because of its high density, it is exceptionally rigid and not particularly elastic. Technically, this defines the wood as brittle, but compared to other woods it does not break easily.

SEASONING

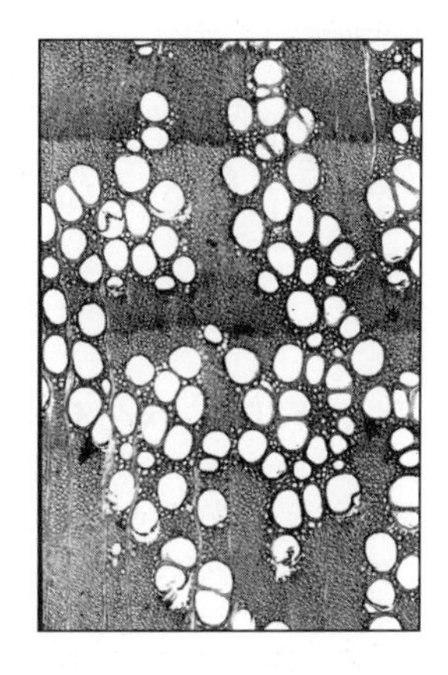

Because it is dense and drenched with oil and resins, verawood air dries very slowly. Shrinkage values are not available.

DURABILITY

This species is one of the most durable woods in the world. When used as pilings and fence posts, verawood will outlast most treated timbers.

WORKABILITY

The working characteristics of verawood are comparable to the softer metals than to virtually any other wood, except for its relative lignumvitae. Fasteners of any kind must literally be "fitted" with nearly exact pilot holes. The wood resists adhesives and penetrating finishes, but it polishes so well that no additional finish is necessary. Although it is exceptionally hard on tools, verawood produces nice results when turned on the lathe. When turned, it is very important that the wood is well seasoned or it may distort or check after the work is completed. Verawood has an exceptionally strong and pleasant scent especially when sanded, since the heat caused by friction releases some of its volatiles.

USES

Because of verawood's self-lubricating properties, it is used in many of the same industrial applications as lignumvitae, such as for bushings, collars and rollers. Lignumvitae is generally preferred for marine use as propeller shaft bushings where its ever so slight advantage in terms of density makes it more resistant to wear. Verawood is an excellent choice for mallet heads, splitting wedges or dowel pins that must withstand constant wear. Within its native range, some verawood is used for fence posts, but its price on the international market makes it too costly for such ordinary, high-volume applications. This is a species to be used sparingly when the project calls for its unique combination of incredible strength and durability.

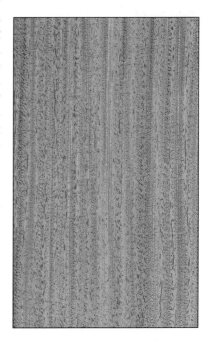

SUPPLIES

Currently, verawood is a little more plentiful than lignumvitae, but both are becoming scarce and expensive. Being extremely slow growers, both species require literally centuries to achieve full maturity.

Burkea africana

makarati

by Jim Flynn

SCIENTIFIC NAME

Burkea africana. Derivation: The genus name is in honor of Joseph Burke (d. 1845), a British botanist who collected plants in Africa with Jean-Michel Zeyher (1770 to 1845). The specific epithet means the plant is "of Africa".

FAMILY

Fabaceae or Leguminosae, the legume family; (*Caesalpiniaceae*) the cassia group.

OTHER NAMES

burkea, false ash, mukarati, mgando, mkarati, msangala, red syringa, rooisering, muscarala, umnondo.

DISTRIBUTION

Widely distributed in dry savanna forests in tropical Africa with heavy concentrations in northeastern South Africa and Zimbabwe.

THE TREE

Greatly influenced by climate and soil conditions, makarati is a small to medium-sized tree which reaches heights of 40 to 70 feet with diameters of 1 to 2 feet. The alternate leaves are pinnately compound, and the leaflets are distinctly broad. New shoots are densely ferruginous tomentose. The scented flowers are white and grow in long pendulous spikes. The long, flat oval seedpods are indehiscent and persistent with wing-like margins. Keith Coates Palgrave in *Trees of Southern Africa* reports the trees are host to a species of caterpillar which, when roasted and dried, are relished by Africans. This epicurean delight is called *mukarati*, which is translated to "the tree with caterpillars". The Zezuru people in Zimbabwe prepare an extract from the bark, chew it and use it as a poultice on septic sores.

THE TIMBER

Technical literature is inconsistent about the color of the wood. Some indicate that the wood has narrow, yellowish sapwood with a dark brown to reddish-brown heartwood, while other authoritative sources indicate to the contrary. The growth habitat and timber size will influence descriptions of color. A 3-inch diameter stem provided to me by the Zimbabwe Forest Service is a rather bright yellow with just a trace of heartwood. The color rapidly oxidizes to a deep reddish-black after being cut. The texture is fine and even. The interlocked grain produces an attractive ribbon figure. Average reported specific gravity is 0.60 to 0.80 (ovendry weight/green volume), equivalent to an air-dried weight of 46 to 64 pcf.

SEASONING

Makarati timber has a good seasoning record. It dries rather rapidly with little warping or splitting. When dry, it has good dimensional stability. Average reported shrinkage values (green to 12% moisture content) are 1.2% radial and 2.1% tangential.

DURABILITY

The heartwood is rated as very durable and is immune to borer, fungal and termite attack.

WORKABILITY

Despite its hardness, makarati machines cleanly and easily. It planes to a lustrous and smooth finish but care needs to be exercised when machining along interlocked grain. Working the wood with hand tools is a bit more difficult. Pre-boring is advised when using nails or screws. The wood takes a high gloss varnish very well and glues firmly.

USES

Due to the hardness of the wood, it has very good wearability thus rendering it an outstanding material for flooring. It is also used for small pieces of furniture, joinery, artwork and even mining timbers.

SUPPLIES

Zimbabwe, one of the major timber exporting countries, states that supplies are abundant. It does not appear, however, to be abundant in all parts of the world. When needed, wood samples and larger supplies can be obtained from the commercial market.

Bursera simaruba

gumbo-limbo

by Jon Arno

SCIENTIFIC NAME

Bursera simaruba. Derivation: The genus name is in honor of Joachim Burser (1593 to 1639), a German botanist and physician. The specific epithet was derived from the Carib Indian name *simarouba* for another tree. This is in reference to another word and it is used as a generic name.

FAMILY

Burseraceae, the bursera or torchwood family.

OTHER NAMES

almacigo, turpentine tree, carate, chaca.

DISTRIBUTION

Gumbo-limbo's natural growth range extends from east central Florida through the West Indies and eastern Mexico into northern South America. Like most members of Burseraceae, this species is sensitive to cold, but is otherwise very adaptive to a broad range of soil types and growing conditions.

THE TREE

While capable of reaching heights of over 90 feet with diameters in excess of 3 feet, gumbo-limbo is best characterized as a medium-sized tree that is normally 50 to 60 feet in height and seldom over 2 feet in diameter. The leaves are pinnate, but the tree's most striking feature is its reddish-brown, birch-like bark which peels off to expose the greenish inner bark. Generally gnarly and broad crowned, this species makes an interesting shade tree for landscaping purposes, but is a rather poor lumber producer.

THE TIMBER

Because gumbo-limbo tolerates widely different growing conditions, the wood can be quite variable. Its average reported specific gravity ranges between 0.30 and 0.38 (ovendry weight/green volume), equivalent to an air-dried weight of 22 to 28 pcf, which puts it in a category with some of the softer North American hardwoods, such as trembling aspen (*Populus tremuloides* WDS 219) and eastern cottonwood (*Populus deltoids* WDS 218). Like eastern cottonwood, it has a medium to fine texture, but its yellowish-beige heartwood is slightly darker and more lustrous. The sapwood tends to be creamy white just inside the bark and gradually darkens as it blends into the heartwood. It is a weak wood and compares closely with trembling aspen.

SEASONING

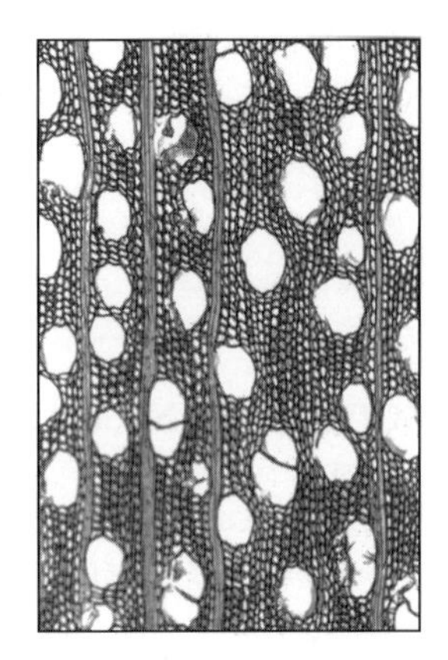

Gumbo-limbo air dries quickly with very little checking or distortion, but it is extremely susceptible to blue staining. [Ed. note: I have used household bleach to remove blue stains when the wood is freshly cut.] Average reported shrinkage values (green to ovendry) are 2.6% radial, 4.2% tangential and 7.3% volumetric.

DURABILITY

Durability is very poor. Gumbo-limbo ranks as the filet mignon in fungi circles and is also well liked by termites and powder post beetles.

WORKABILITY

While gumbo-limbo lacks the pleasant figure and warm tan color of butternut (*Juglans cinerea* WDS 149), these two woods are remarkably similar in terms of working characteristics. Both tend to become a little fuzzy when sanded, but are otherwise exceptionally easy to work. For such a soft wood, gumbo-limbo turns well and accepts both screws and nails without splitting. Although the wood has a faint, resinous scent when sanded, it is surprisingly non-aromatic considering its botanical kinship to frankincense and myrrh.

USES

The primary uses of gumbo-limbo include interior construction, crating, boxes, construction plywoods, corestock and match sticks. Some is harvested for charcoal production, and its resins and oils have been used for centuries as incense and in making varnish, adhesives and medicines. In the Yucatan, young boys are seen carving novelties with this wood while you wait.

SUPPLIES

While this species is plentiful, it is not a major timber in international commerce. Occasionally, it may be found on the U.S. market in the form of low-grade plywood (plyscore), mostly of Mexican manufacture, but as lumber, its lack of strength and its poor weathering characteristics limit its value in the building trade.

Buxus sempervirens

boxwood

by Max Kline

SCIENTIFIC NAME

Buxus sempervirens. Derivation: The genus name is ancient Latin for the box tree. The specific epithet is Latin for evergreen.

FAMILY

Buxaceae, the box family.

OTHER NAMES

Turkey boxwood, Abyssinian boxwood, Persian box, Circassian box.

DISTRIBUTION

Southern Europe, Western Asia and North America.

THE TREE

The boxwood tree grows to a height of 20 or 30 feet. It is often of shrubby growth, and large sizes are never found. While specimens with diameters of 12 inches have been recorded, diameters over 8 inches are rare. It grows on chalk and limestone soils in England.

THE TIMBER

The color of boxwood is uniform light yellow, and the heartwood and sapwood are barely distinguishable. The wood is hard and heavy with an average specific gravity of 0.70 to 0.87 (ovendry weight/green volume), equivalent to an air-dried weight of 55 to 71 pcf. The grain is straight to very irregular, and the texture is extremely fine and uniform. The odor is not distinct, and the luster is low to medium. Boxwood is recognized as perhaps the best species in the world for wood engraving.

SEASONING

During seasoning, slow drying is essential as the wood is subject to splitting. It is also liable to develop minute surface checks. It was customary to cut boxwood into small billets, which were stored and allowed to dry in boxes of sawdust. With care it can be kiln dried with satisfactory results. Average reported shrinkage values (green to ovendry) are 1.9% radial and 4.0% tangential. Once seasoned, boxwood has good stability.

DURABILITY

Considering the uses to which boxwood will be put, durability is not likely to be an important factor. Its resistance to fungal or insect attack is not high, and it is not a timber that would be worth the investment of a preservative treatment.

WORKABILITY

Boxwood is rather hard to work, but there are no unusual dulling effect to tools so special equipment is not required for working this timber. Surfacing wood with highly irregular grain can present some difficulty. In this case, sanding is required to produce a good surface. Pre-boring is necessary to prevent splitting when using nails or screws. Boxwood can be glued satisfactorily and is generally finished with a clear finish to preserve its unique color.

USES

Since boxwood is only available in small sizes, its uses are rather limited and generally confined to turnery, engraving, inlays, mathematical scales, musical instruments, shuttles, rulers and mallet heads. A unique use is for tool handles that have to withstand an exceptional amount of hammering. It is an excellent species for both plain and ornamental turnery.

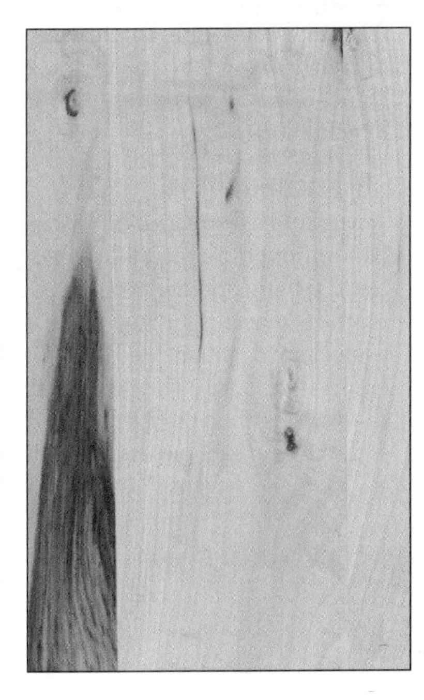

SUPPLIES

As supplies are quite limited, the price is quite high. In recent years related species are being used as a substitute for genuine boxwood. These include *Buxus balearica* known as Balearic boxwood and Cape boxwood (*Buxus macowani*) from South Africa. Other substitutes not botanically related are Knysna boxwood (*Gonioma kamassi*), Maracaibo boxwood (*Gossypiospermum praecox*) and San Domingo boxwood (*Phyllostylon brasiliensis*).

Caesalpinia echinata

Pernambuco

by Max Kline

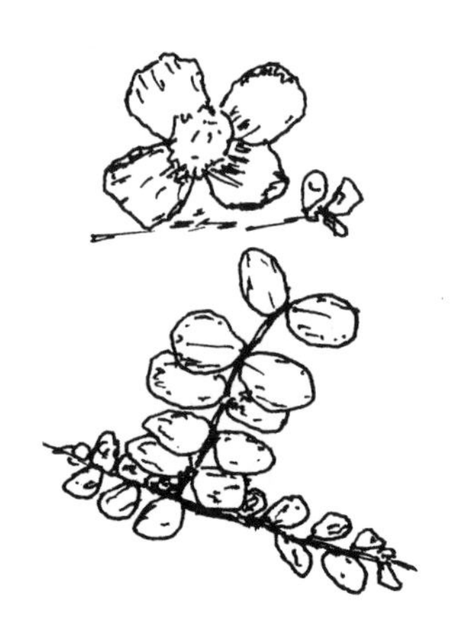

SCIENTIFIC NAME

Caesalpinia echinata. Derivation: The genus name is in honor of Andrea Cesalpino (Caesalpini) (1519 to 1603), an Italian physician and botanist. The specific epithet is Greek for prickly. A synonym is *Guilandina echinata*.

FAMILY

Fabaceae or Leguminosae, the legume family; (*Caesalpiniaceae*) the cassia group.

OTHER NAMES

Bahia wood, Braziletto, Brazilwood, pau Brasil, Fernambuco.

DISTRIBUTION

Coastal forests of eastern Brazil.

THE TREE

Although commonly much smaller, Pernambuco can sometimes be more than 100 feet in height with a symmetrical bole as much as 3 feet in diameter which is free of branches for 50 to 60 feet. The seeds are contained in a brown, bristly pod that is 2 to 3 inches long.

THE TIMBER

The nearly white sapwood is sharply demarcated from the heartwood, which is a fairly uniform bright orange to orange-red when fresh but turns deep red on exposure. The luster is high, and the taste and odor are not distinct. It is a very hard and heavy wood. Average specific gravity ranges from 0.71 to 1.05 (ovendry weight/green volume), equivalent to an air-dried weight of 56 to 88 pcf. The texture of Pernambuco is usually fine and uniform. The grain is straight to irregular. The deeply colored heartwood of this wood is characteristic of a group of about 20 species that previously were important as a source of red dye in the world trade.

SEASONING

Much care is required, and Pernambuco dries especially slowly. Proper precautions, however, ensure satisfactory results. Shrinkage values are not available.

DURABILITY

Pernambuco wood is highly resistant to decay.

WORKABILITY

The wood is not difficult to work and finishes very smoothly with a high natural polish.

USES

Currently, carefully selected pieces are used almost exclusively for violin bows. In fact, the great musicians will consider no other wood for bows because of the unique, strong, resilient spring found in the species. Other uses include parquet flooring, furniture and some turnery.

SUPPLIES

There is ample Pernambuco available for violin bows. In 1623, when it was in great demand as a dyewood, a monopoly was placed on it to prevent its exploitation.

Caesalpina granadillo

partridge wood

by Max Kline

SCIENTIFIC NAME

Caesalpina granadillo. Derivation: The genus name is in honor of Andrea Cesalpina (Caesalpini) (1519 to 1603), an Italian physician and botanist. The specific epithet is from the Spanish word *grana* indicating the ripening season for the fruit.

FAMILY

Fabaceae or Leguminosae, the legume family; (*Caesalpiniaceae*) the cassia group.

OTHER NAMES

coffeewood, brown ebony, ebano, granadillo, guayacan, Maracaibo ebony.

DISTRIBUTION

Mexico, Nicaragua, Colombia, Venezuela, Paraguay, Argentina and Brazil.

THE TREE

Mature trees are 50 to 75 feet in height. The trunks have unusually smooth bark and can grow to 36 inches in diameter and be free of branches for 35 feet. This tree grows best in dry areas. Occasionally spiny, the leaflets on the bipinnate leaves are always oblong to elliptic and small and numerous. The spiked flowers are yellow. The fruit pods are narrowly oblong, irregularly flattened and indehiscent.

THE TIMBER

The heartwood is coffee brown or dark red to chocolate brown or nearly black with fine parenchyma striping sharply demarcating the heartwood from the yellowish or pinkish-white sapwood. The luster is low to medium, and odor and taste are not distinct. The wood is exceedingly hard, heavy and strong with an average specific gravity of 0.95 (ovendry weight/green volume), equivalent to an air-dried weight of 78 pcf. The texture is medium, and the grain is fairly straight to very irregular. The wood has exceptional strength properties.

SEASONING

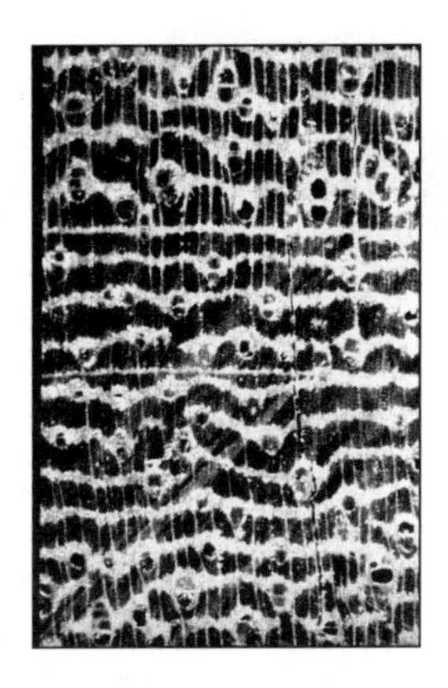

To prevent fine checks, partridge wood requires care when seasoning. Average reported shrinkage values (green to ovendry) are 5% radial and 8% tangential.

DURABILITY

The species is highly durable, which is beneficial when it is used for railway crossties, marine piling and in harbor works.

WORKABILITY

Partridge wood is relatively difficult to work and has a tendency to chip out. A good smooth finish is achievable when using a lathe.

USES

The wood is used in its native location for general construction, marine piling, bridge construction, crossties, fence posts, agricultural implements and vehicles. Exported timber has been used chiefly for umbrella handles.

SUPPLIES

Very little partridge wood is exported from South America, therefore it is difficult to obtain. Occasionally, the large importers will receive a shipment. Primarily used for ornamental purposes, it is of limited commercial use which adds to its scarcity in the United States and Europe. When available, it is expensive but worth using by the hobbyist.

Callitris glaucophylla

white cypress pine

by Jim Flynn

SCIENTIFIC NAME

Callitris glaucophylla. Derivation: The genus name is Greek meaning beautiful. The specific epithet is Greek meaning "leaves covered with whitish bloom".

FAMILY

Cupressaceae, the cypress family.

OTHER NAMES

Murray River pine, Murray River cypress, cypress-pine, Murray pine, cypress, western cypress.

DISTRIBUTION

Native to all of Australia and the main source of commercial timber supplied by New South Wales and Queensland. It has been planted extensively for shelterbelts in tropical and semi-tropical regions, including Florida.

THE TREE

There are at least 17 species of *Callitris*. White cypress pine thrives best in areas of moderate rainfall. The original, pure stands of these trees in New South Wales and Queensland have been harvested, and the land is now devoted to growing wheat. It is an evergreen with scale-like leaves arranged in alternating whorls of three, which sheath the needle-like green branchlets. The tiny leaves are a mere 0.03 to 0.125 inch long and 0.016 inch wide. Depending on the soil and growth site, it can grow to 100 feet with a circumference in excess of 3 feet. Usually, the main stem is straight and slender with rough, furrowed brown or gray bark. Trees of this genus are similar in appearance to eastern redcedar (*Juniperus virginiana* WDS 152). When young, the tree is sensitive to fire but older trees, with thickened bark, tolerate light grass fires. In some areas, growth of young white cypress pines hinders the growth of pasture plants. Not seen as commercially attractive, these scrub-like trees are considered a pest.

THE TIMBER

The wide sapwood is paler than the tan to dark brown heartwood, but the most notable attribute of white cypress pine is the pleasant, spice-like odor. The wood has a greasy feel. The texture is fine, lustrous and even, and the grain runs straight in larger sawlogs. Smaller billets that have grown crooked are apt to produce interesting, wavy patterns in the heartwood and bordering sapwood. Small knots present no problems. Average reported specific gravity is 0.58 (ovendry weight/green volume), equivalent to an air-dried weight of 45 pcf.

SEASONING

Even large pieces air dry rapidly, providing care is taken when stickering and weighting the stack. It must be protected from direct sun and weather. Kiln-dried timber may contain stress areas around knots. The wood is often used in the green or partially dried condition. Average reported shrinkage values (green to ovendry) are 2.1% radial, 2.8% tangential and 4.0% volumetric.

DURABILITY

Highly durable, wood in contact with the ground will not materially degrade for up to 25 years. It is highly resistant to attack by both decay fungus and insects, including termites and marine organisms.

WORKABILITY

Do not expect this wood to work like those in the *Pinus* genus as it is a bit heavier. Surprisingly, it works easily with both hand tools and woodworking machinery. There is apt to be some tearing near knots that can be dressed after planing. No problems are encountered in gluing or finishing. A disadvantage is that the wood is extremely brittle, requiring special attention to nailing and pre-drilling for screws is recommended. The timber is a dramatic burner and produces a fire almost as readily as southern yellow pine (*Pinus* spp.).

USES

White cypress pine is a versatile, useful wood. Because of its durability, it is a prime candidate for applications such as boat building, poles, construction timber and piling. It is also used for turnery, interior trim, furniture, boxes and crates, and light flooring.

SUPPLIES

This wood is available in the world market. The wood can be found in central Florida, where the trees are used as wind and sand shelterbelts. Supplies of white cypress pine go largely unnoticed since it is similar in appearance to eastern redcedar (in form, not the wood).

Calocedrus decurrens

incense-cedar

by Jim Flynn

SCIENTIFIC NAME

Calocedrus decurrens. Derivation: The genus name *calo* is Greek meaning drop or tear which refers to the trickling of resin and *cedrus* meaning cedar. The specific epithet refers to the leaf margin running gradually into the stem. A synonym is *Libocedrus decurrens*.

FAMILY

Cupressaceae, the cypress family.

OTHER NAMES

California incense-cedar.

DISTRIBUTION

This tree has a rather circuitous range but generally can be found in the mountainous regions of western Oregon, in the higher elevations of the coast ranges and the Sierra Nevadas, south to California and into western Nevada. Pockets of growth can be found elsewhere in these states as well as Baja California. It seldom grows in pure stands.

THE TREE

John Charles Fremont on February 3 and 10, 1844, in *A Report of the Exploring Expedition to Oregon and North California in the Years 1843-1844*, wrote the following statements about *Libocedrus decurrens*: ". . . a lofty cedar which here made its first appearance the usual height was 120 to 130 feet, and one that was nearby was 6 feet in diameter and, . . . The forest here has a noble appearance the tall cedar is abundant its greatest height being 120 feet, and circumference 20, three or four feet above the ground . . ." With Kit Carson as the guide, the expedition was trying desperately to get across the snow covered mountains to Sutters Fort via what is now Alpine County, California, with Freemont as the expedition's botanist. Incense-cedar can be recognized by its prominent tapering trunk, widely buttressed base and the cinnamon-brown, deeply furrowed and ridged bark, which is 2 or more inches thick at the base. The scale-like evergreen leaves are 0.125 to 0.5 inch long and arranged in four ranks on the twig, in alternating pairs. Those on the sides of the twigs are keeled and glandular on the back, gradually narrowed toward the tip and nearly cover the top and bottom pair which are flattened and abruptly pointed. The droopy cones are light brown and 0.75 to 1 inch long. The oldest of these trees is close to 1,000 years.

THE TIMBER

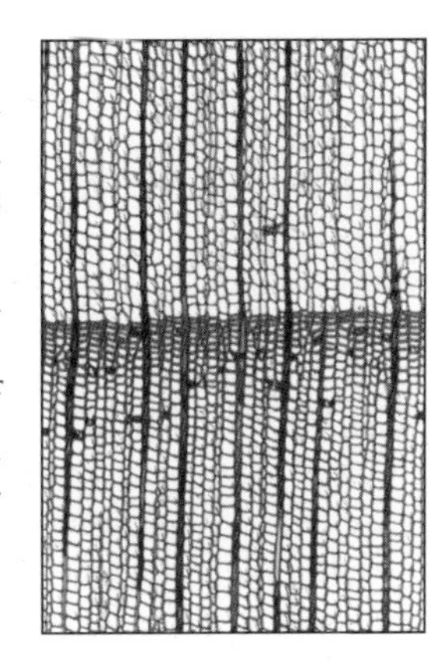

The sapwood is white or cream-colored, and the heartwood is light brown often tinged with red. It has a fine, uniform texture and a spicy odor. Average reported specific gravity is 0.35 (ovendry weight/green volume), equivalent to an air-dried weight of 26 pcf. It is moderately low in the strength properties, low in shock resistance, soft and low in stiffness. Most of the lumber is pecky in that it contains pockets of disintegrated wood caused by advanced stages of localized decay in the living tree. Decay in these areas is arrested after the wood has been cut and seasoned.

SEASONING

This wood seasons very well. Average reported shrinkage values (green to ovendry) are 3.3% radial, 5.9% tangential and 7.7% volumetric.

DURABILITY

The heartwood is extremely durable making the wood ideally suited for applications in moist areas or outdoors.

WORKABILITY

Because it is a light and low-density wood, incense-cedar is extremely easy to work with machine and hand tools. It glues well. Because of its stability after it is dry, it is rated as one of the best woods to paint and finish.

USES

Incense-cedar is best used where a durable wood is required, including fence posts, poles, shingles and railroad ties. A principal use is in the manufacture of pencils because of the ease with which the pencil can be sharpened either by turning or whittling. This suggests a good wood for carvers. It is also used in manufacturing woodenware, novelties, millwork (both indoor and outdoor applications), chests, furniture, veneer and plywood.

SUPPLIES

Supplies are adequate. Because the trees do not grow in dense stands, they are harvested with other species and then separated. The trees used in the pencil industry are often harvested individually.

Calophyllum brasiliense

Maria

by Jim Flynn

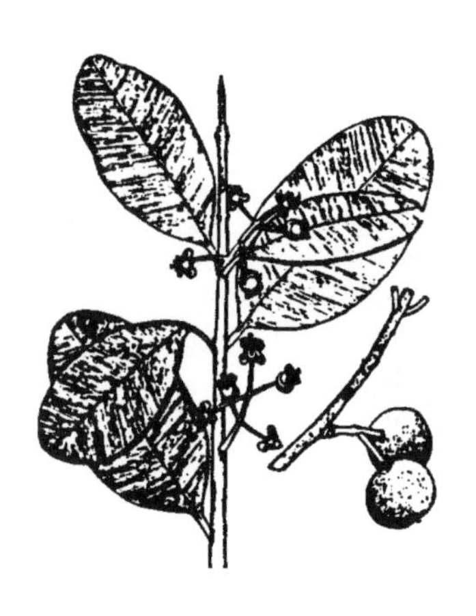

SCIENTIFIC NAME

Calophyllum brasiliense. Derivation: The genus name is derived from the Greek *kalos* meaning beautiful and *phyllon* meaning leaf. The specific epithet means the plant is "of Brazil". A synonym is C. *lucidum*.

FAMILY

Guttiferae, the mangosteen family; also Clusiaceae.

OTHER NAMES

Santa-Maria, lagarto caspi, jacareuba, bari, leche de Maria, calaba tree, aceite Maria, edabalii, kurahara, balsamaria, false-mamey, guanandi, varilla, oscuro, calambuca.

DISTRIBUTION

Maria grows throughout the West Indies and from Mexico southward through Central America and into northern South America. It is found on all types of soils, from wet and humid to very dry. It is naturalized in Bermuda and has been introduced in southern Florida.

THE TREE

Maria is a medium-sized tropical evergreen. Frequently used for reforestry, it is easily established on almost all soils. It is tolerant of saltwater spray and forms a dense crown with small, fragrant flowers that make it a popular shade tree. The leaves are opposite, elliptical and dark green with many parallel, lateral veins. A medicinal yellow sap exudes from broken leaves, twigs and incisions in the bark. The seeds, when pressed, produce an oil suitable for illuminating purposes. The bark has many diamond-shaped fissures. Under favorable conditions, the tree grows to a height of 100 to 150 feet with an unbuttressed bole 3 to 6 feet in diameter.

THE TIMBER

Maria wood is widely used in the tropics where a strong, moderately durable timber and general utility wood is needed. The heartwood varies from yellowish-pink to reddish-brown while the sapwood is generally lighter in color. Usually, Maria has interlocked, wavy grain. Average reported specific gravity is 0.51 (ovendry weight/green volume), equivalent to an air-dried weight of 39 pcf. The luster is low, and the texture is medium and fairly uniform. Early records from timber cutters in British Honduras (Belize) indicate that they recognized three classes of Maria after the trees were felled: white, red and dark. The first two were floatable while the third was not, suggesting that there were several varieties of the species some considerably more dense than others.

SEASONING

The wood is moderately difficult to air dry and tends to warp severely. Kiln drying is suggested for wood used indoors. Average reported shrinkage values (green to ovendry) are 4.6% radial, 8.0% tangential and 13.6% volumetric.

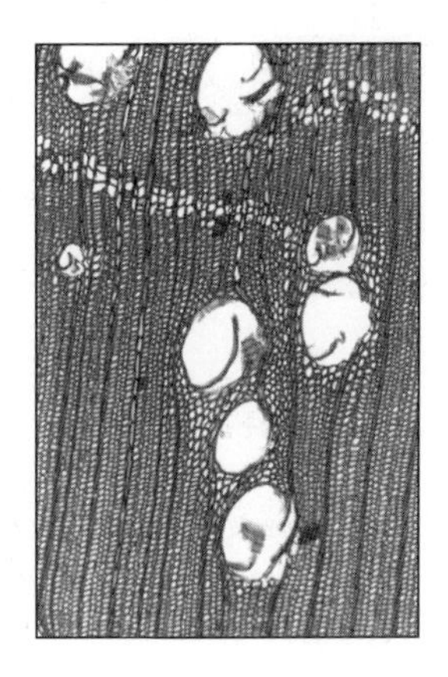

DURABILITY

Generally, the heartwood is rated as durable to moderately durable with respect to decay resistance and is very susceptible to attack by dry-wood termites and marine borers. The heartwood resists preservative impregnation by both pressure and non-pressure systems. The sapwood, however, can be impregnated with preservatives successfully.

WORKABILITY

The wood is moderately easy to work, and good surfaces can be attained if machinery is properly set up. Planing is the most difficult process as torn and chipped grain can result if knife blades are not properly adjusted. The best machining can be done when the wood is at 6% to 7% moisture content. Nail and screw holding properties are good, and there are no special problems when gluing, staining or finishing.

USES

Maria is widely used in the tropics for general construction, flocring, furniture, boat construction and is a favored general utility wood. It is particularly valuable in the furniture industry because dimensional stability in manufactured items is lower than in conventionally used domestic woods. The quartersawn surfaces are considered suitable as face veneer for plywood used in boat construction.

SUPPLIES

Stocks are occasionally available commercially, but it seems most harvested supplies are used in the growth area.

Calycophyllum candidissimum

degame

by Max Kline

SCIENTIFIC NAME

Calycophyllum candidissimum. Derivation: The genus name is derived from the Greek *kalyx* for calyx and *phylum* meaning divided, referring to the flowers. The specific epithet is Latin meaning very white or hoary.

FAMILY

Rubiaceae, the madder family.

OTHER NAMES

lemonwood, lancewood.

DISTRIBUTION

Degame is found in Cuba, Central America, Colombia and Venezuela.

THE TREE

Degame is a small to medium-sized tree which grows 40 to 60 feet in height with a diameter of 8 to 20 inches. When in bloom, it is very beautiful with large white calyx lobes.

THE TIMBER

The color of the extremely broad sapwood is almost white to brownish-white, and the well defined heartwood has a variegated dark-brown tint. Unlike most timbers, the heartwood is of no commercial importance but significant use is made of the sapwood. It has no distinctive odor or taste. The luster is low, and the texture is very fine. The grain may vary from straight to markedly irregular. Average reported specific gravity is 0.67 (ovendry weight/green volume), equivalent to an air-dried weight of 52 pcf. Degame is similar in strength, toughness and resilience to hickory (*Carya* spp.).

SEASONING

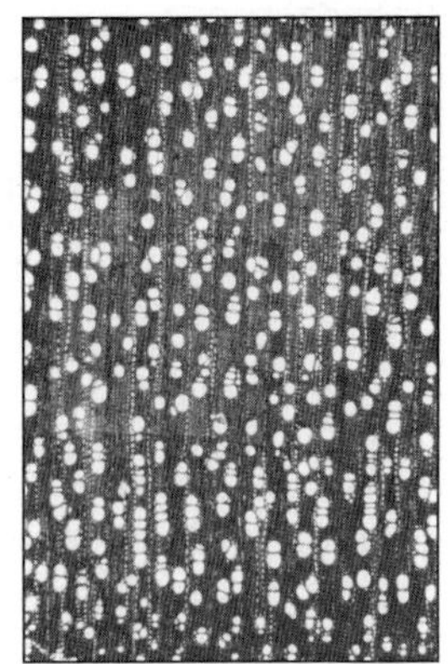

The difficulty when seasoning will depend largely on the type of grain present in the stock. If noticeably irregular, a certain amount of twisting will occur but straight-grained stock presents no problems. In general, the timber behaves reasonably well when both kiln and air dried. Degrade from checking is normally only slight. Average reported shrinkage values (green to ovendry) are 4.8% radial, 8.6% tangential and 13.2% volumetric. Once properly seasoned, the wood holds its shape quite well.

DURABILITY

This wood will readily succumb to fungal attack of all types, but does show reasonable resistance to borer-type pests. Since the wood is not used in applications where durability is a factor, preservatives are not likely to be used.

WORKABILITY

Generally workability with hand or power tools is satisfactory. A good surface can be obtained on the planer, and sawing can be done cleanly. Glue adheres satisfactorily, and nail and screw holding properties are good with little tendency to split during nailing. The response to stains and finishes is very good.

USES

One of the specialized uses for degame is in the manufacture of bows for archery. Other uses include agricultural implements, fishing rods, tool handles, rules and scales (as a substitute for boxwood [*Buxus sempervirens* WDS 050]), carving and turnery, skewers and shuttles.

SUPPLIES

Because Cuba is the chief exporter of this wood, supplies are almost non-existent resulting in a high cost for this species.

Cananga odorata

ylang-ylang

by Jim Flynn

SCIENTIFIC NAME

Cananga odorata. Derivation: The genus name is a Latinized version of the Malayan vernacular for these tropical trees, one of which yields the perfume known as Macassar oil. The specific epithet is Latin for fragrant smelling, referring to the delicate and attractive odor of the flowers. *Canangium odoratum* is a synonym for this taxa. (See Supplies for further information on scientific names.)

FAMILY

Annonaceae, the annonia or custard-apple family.

OTHER NAMES

There are many ways to spell the preferred name. Other names include ilang-ilang, ylang-ylang-tree, cananga, cadmia, perfume tree, canang odorant and ylang-ylangbaum.

DISTRIBUTION

This is a widely spread species. In tropical Asia, it can be found in Cambodia, Indonesia, Laos, Malaysia, Myanmar (Burma), Papua New Guinea, the Philippines, Thailand and Vietnam. It flourishes in Australia. The tree is also widely cultivated in the tropics.

THE TREE

Often, the olfactory nerves can detect the presence of this tree before it is seen. The unusually striking and long-lasting yellow flowers, 3 to 5 inches wide, are renown for their attractive smell. In the past, distilled oil from the flowers was used to perfume coconut oil for men's hairdressing and sold as Macassar oil. Ylang-ylang is a tall and narrow tree growing to heights of about 100 feet with diameters up to 30 inches. The base of the trunk is slightly buttressed, and the boles are straight and cylindrical producing very good sawlogs.

THE TIMBER

There is no distinctive color separation between the heartwood and the sapwood. The wood is a light yellow often with a greenish or pinkish tint. When freshly planed, it does not appear to be very lustrous but it does have an attractive silver figure on the radial surface. The texture is coarse and uneven. The grain is generally straight. There is neither distinctive smell nor taste to the wood. Average reported specific gravity is 0.30 (ovendry weight/green volume), equivalent to an air-dried weight of 22 pcf. As can be seen in the photomicrograph or with a hand lens and a sample of the wood, a distinctive feature is the ladder-like cross bars.

SEASONING

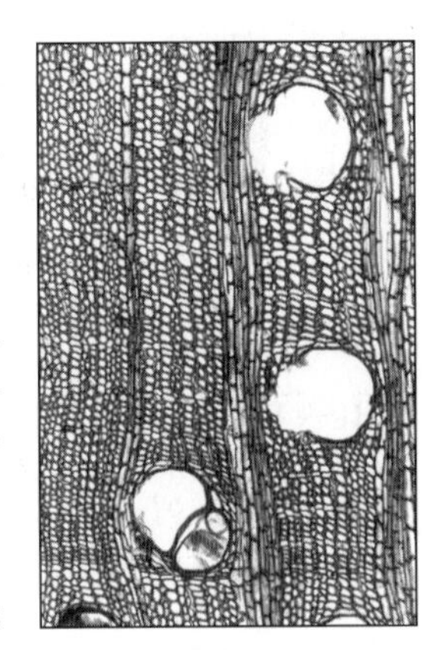

This timber is easy to season, and very little stock is lost due to degrade. Average reported shrinkage values (green to ovendry) are 3.3% radial and 8.0% tangential.

DURABILITY

This is not a durable timber and should not be used when contact with the ground is expected. It is very perishable and vulnerable to termite attack. There is no information available related to the effectiveness of preservatives.

WORKABILITY

Ylang-ylang timber is easy to work. It has an excellent density, luster and grain arrangement permitting machining operations to run smoothly. It sands, finishes and glues well.

USES

Earlier uses for ylang-ylang included carving drums and the hollowing of trunks for canoes. It is now considered a good wood for turnery, small boxes and crates, wood shoes, fishnet floats and small craft items.

SUPPLIES

There does not seem to be any commercial supplies of this timber available in the United States. Wood samples, however, are not in short supply. Worthy of note is the practice of lumping wood with similar working characteristics under a common commercial name. In consulting the reference publication, *Malayan Forest Records No. 25*, published in 1993 by the Malaysia Forest Research Institute, this species, C. *odorata*, is marketed as a "light hardwood" under the Malayan common name mempisang. Timbers so identified consist of at least 28 different species in 11 genera and several families; the chances of identifying and obtaining a piece of ylang-ylang from this mix would be small.

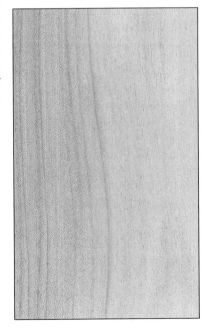

Caragana arborescens

caragana

by Chuck Holder

SCIENTIFIC NAME

Caragana arborescens. Derivation: The genus name *Caragana* is a Latinization of the Mongolian name for the plant Caragan. The specific epithet is Latin meaning tree-like.

FAMILY

Fabaceae or Leguminosae, the legume family; (*Papilionaceae*) the pea or pulse group.

OTHER NAMES

common caragana, pea-shrub, pea-tree, Siberian pea-shrub, Siberian pea-tree.

DISTRIBUTION

Some 20 species of caragana are known. It is widely distributed in its native range in northeast Asia. It was brought to Western Canada in the 1880's for use as a field shelterbelt and garden ornamental. It is extremely hardy and now common on the Canadian prairies and some adjacent northern tier states.

THE TREE

Caragana is a deciduous, fast growing, upright, spreading shrub or small tree. When left untended it will attain a height of 20 feet with major stems growing up to 5 inches in diameter. These are rare, however, and a height of 12 to 14 feet with a diameter of 2 to 3 inches is much more common. Its multi-stalked stems attain a spread of up to 15 feet. The bark is olive green and bronze to gray with horizontal markings. Covering the branches in small clusters on short shoots, the fragrant flowers resemble yellow sweet peas in the spring. The leaves, 1.5 to 3 inches long, are alternate and pinnately compound on a central stalk with no terminal leaflet. Divided into four to six pairs, the bright green, oval leaflets each 0.5 to 1 inch long are short pointed, toothless and stalkless. The fruit pod is 1.6 to 2 inches long and splits open with a twist and loud pop when ripe. Stipules at the base of each leaf become small spines. A terminal bud is present with chaff-like scales. The twigs are green dwarf shoots borne on previous year's branchlets.

THE TIMBER

Very little has been published on the characteristics of caragana timber. The wood of caragana is variable in texture and color. The sapwood is light cream to yellow and exhibits a light-sensitive sheen. The heartwood can have unusual patterns in brown, purple and rust. It is often punky in older stems and sometimes is difficult to distinguish from the sapwood. Solid heartwood or sapwood seems to be about as hard as black walnut (*Juglans nigra* WDS 150). Dry samples I own have a specific gravity of 0.61, equivalent to an air-dried weight of 38 pcf.

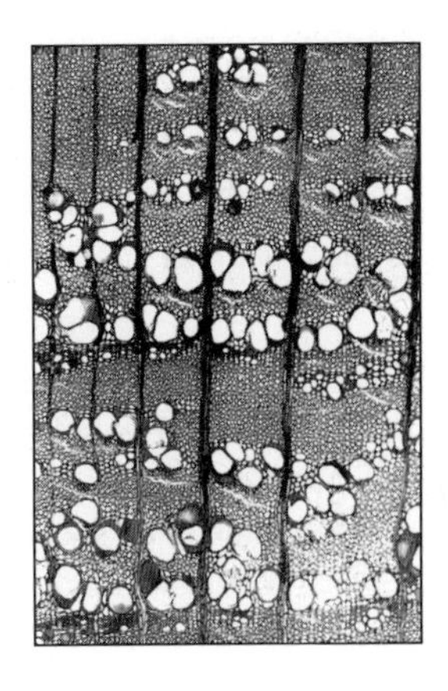

SEASONING

Fresh cut stems will split quickly when drying and care must be taken to coat ends immediately after cutting. Stems 2 inches or more in diameter require up to 3 years to air dry to 8% moisture content. Microwave drying with a turntable at the low heat level works well. Shrinkage values are not available.

DURABILITY

No information is available.

WORKABILITY

Solid, non-punky stock works well, accepts glue well and can be brought to a fine finish. Dry caragana is stable in use. When sorting through dry, seasoned randomly selected stems, Harold Biswanger, a fellow IWCS member, reports a success rate of about 1 in 10 to find solid, turnable stock with interesting heart.

USES

Due to its small size, uses of caragana are limited to small craft items such as turnings, jewelry and boxes. It can be edge glued to make larger blanks.

SUPPLIES

Caragana is not available commercially. It is plentiful on the Canadian prairies and is worth seeking for craftwork.

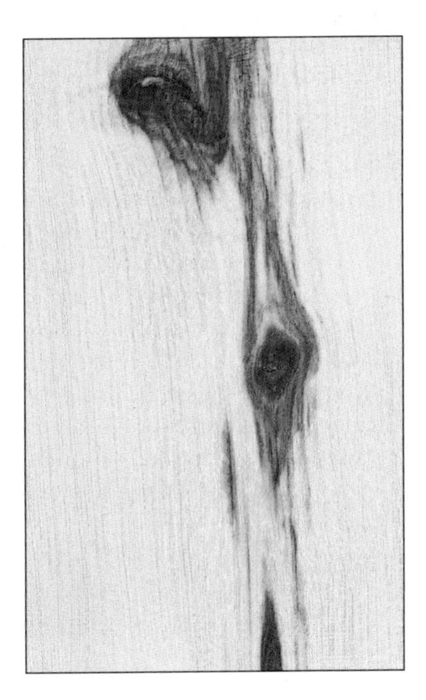

Carapa guianensis

andiroba

by Jon Arno

SCIENTIFIC NAME

Carapa guianensis. Derivation: The genus name is a local name. The specific epithet is Latin meaning the plant is "of Guyana".

FAMILY

Meliaceae, the mahogany family.

OTHER NAMES

crabwood, krapa, cedro macho, bateo, mazabalo.

DISTRIBUTION

Andiroba has a very expansive range, extending from Cuba and other islands in the West Indies, southward through Central America into Brazil, Colombia and Peru. Thriving in moist soils, it is predominantly a lowland species, but will grow at higher elevations where moisture is plentiful.

THE TREE

Andiroba is capable of growing to a height of 170 feet. Because the buttress is less pronounced than with many tropical species, it has an excellent form for timber production. Occasionally, clear, straight logs 50 feet long with diameters of 5 to 6 feet are harvested.

THE TIMBER

Because andiroba tolerates a rather wide range of climates, the properties of the wood can be highly variable. With an average reported specific gravity of 0.56 (ovendry weight/green volume), equivalent to an air-dried weight of 43 pcf, it is generally heavier than bigleaf mahogany (*Swietenia macrophylla* WDS 254) with a specific gravity of 0.45. Since bigleaf mahogany is also quite variable, these two woods are easy to confuse. Andiroba is usually finer textured and less figured. On average, andiroba is substantially stronger than bigleaf mahogany, especially in terms of elasticity. The heartwood is light salmon to reddish-brown when freshly cut, becoming darker when dry. The odor is very variable. The sapwood is pinkish, turning pale brown or grayish and is not always sharply demarcated from the heartwood.

SEASONING

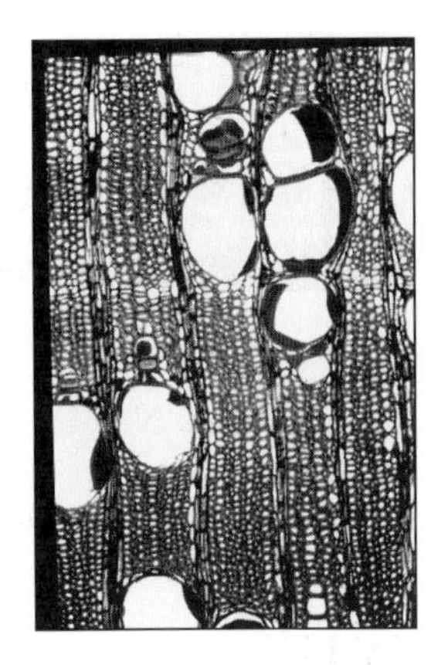

Andiroba is more difficult to season than bigleaf mahogany. With an reported average volumetric shrinkage of 10.4% (green to ovendry), andiroba experiences about a third again as much shrinkage as bigleaf mahogany (7.8%). Also, warping can be a problem because andiroba has a greater than 2:1 ratio between its tangential shrinkage of 7.6% and radial shrinkage of 3.1%. While these statistics tend to put andiroba in an unfavorable light, remember that bigleaf mahogany ranks among the most stable wood for cabinetry use in the world. Actually, andiroba has greater dimensional stability than many other esteemed tropical timbers.

DURABILITY

Comparable to bigleaf mahogany, andiroba is slightly more susceptible to attack by termites and powder post beetles.

WORKABILITY

Owing to its greater average density, andiroba is slightly more difficult to work than either bigleaf or African mahogany (*Khaya* spp. WDS 155). Although mahogany-like in appearance, its working characteristics are not unlike those of sugar maple (*Acer saccharum* WDS 010) or cherry. It turns well and fewer coats of varnish are required to establish a smooth surface than is the case with mahogany.

USES

Uses include furniture, cabinetry, interior trim, flooring, plywood veneer and turnery.

SUPPLIES

Substantially more plentiful than bigleaf mahogany, andiroba has a greater propensity to grow in pure stands and is still plentiful enough to be used as a construction timber in some areas of its native range. Although somewhat less available through hardwood suppliers both in the United States and Europe than are the true mahoganies (*Swietenia* and *Khaya*), when available, it is generally less expensive.

Cardwellia sublimis

lacewood

by Max Kline

SCIENTIFIC NAME

Cardwellia sublimis. Derivation: The genus name is in honor of Lord Edward Cardwell (1813 to 1886), a friend of Charles Darwin and the British Secretary of War under Gladstone. The specific epithet means high or lofty.

FAMILY

Proteaceae, the protea family.

OTHER NAMES

silky oak, selano, Australian silky oak, northern silky oak.

DISTRIBUTION

Found in Queensland, Australia, especially in the northern coastal areas.

THE TREE

This tree is tall and straight reaching heights of 130 feet with an unbuttressed trunk which can grow to 7 feet in diameter. The slightly flaky bark is a biscuit-brown color and rather nondescript. Adult leaves are pinnate and comprised of 5 to 12 oblong, spirally arranged leaflets each 1 to 2 inches wide and 2.5 to 5 inches long. There is no terminal leaflet. The flowers are terminal accumulations of spike-like racemes that bloom from October to December. The fruits are oval, woody follicles 2.5 to 4 inches long and 1.5 to 2.5 inches in diameter containing several flat, brown winged seeds.

THE TIMBER

The heartwood is pale pink to pinkish-brown with a distinct silvery sheen. Upon exposure, the wood turns brown. The odor and taste are not distinct. The texture is coarse, and the grain is straight. The most prominent surface feature of lacewood is the large rays which result in a small, flaky grain pattern that is very attractive for decorative purposes. Lacewood is reported to be light and soft, yet firm, strong and tough. Average specific gravity is 0.44 (ovendry weight/green volume), equivalent to an air-dried weight of 33 pcf. It is reputed to be a very good timber for steam bending purposes.

SEASONING

Kiln-dried lacewood can cup rather easily. The best drying procedure is to reduce the drying rate and to use a final high humidity treatment. Average tangential shrinkage (green to 12% moisture content) is 5%.

DURABILITY

This timber is only moderately durable and is occasionally damaged by beetles.

WORKABILITY

The wood works easily with hand and machine tools having relatively little dulling effect on their cutting edges. The large rays tend to crumble which requires care when working to produce a smooth surface that is not rough or fibrous. Quartersawn material usually produces some pick up in the planer unless the knives are kept very sharp. The timber has good nail and screw holding properties and glues easily. It also stains readily and finishes satisfactorily.

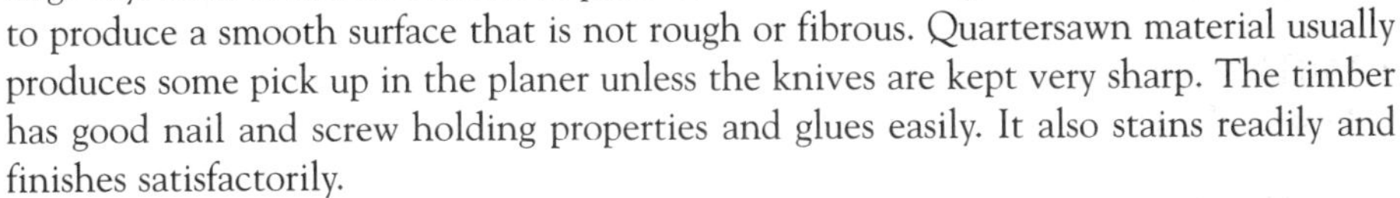

USES

Lacewood is a highly ornamental wood used for cabinetry and furniture and for paneling and trim in fine residences and banks. It possesses a high resistance to wear and is suitable for heavy traffic when used as a flooring timber. The home woodworker will find it to be a very attractive wood for turnery and carving projects where a striking and unusual appearance is desired.

SUPPLIES

There are ample supplies in Australia for local consumption, but quantities available in Europe and the United States are very limited due to the high shipping cost. The boards sold are usually 1 inch thick and about 14 feet long. The price is high.

Carpinus caroliniana

American hornbeam

by Jim Flynn

SCIENTIFIC NAME

Carpinus caroliniana. Derivation: The genus name is the classical Latin name. The specific epithet means the plant is "of the Carolinas" in recognition of where it was found.

FAMILY

Betulaceae, the birch family; (*Corylaceae, Carpinaceae*).

OTHER NAMES

blue beech (preferred in Canada), musclewood, ironwood, water beech. In Mexico: lechillo, palo silo, palo barranco, ico.

DISTRIBUTION

With 26 known species of *Carpinus*, only one is native to North America. American hornbeam is native to most of the eastern United States and extends into Canada in southwest Quebec and southeast Ontario. Its western limit is just beyond the Mississippi River from north-central Minnesota to the Missouri River, ranging southwestward into the Ozark and Ouachita Mountains and eastern Texas. It grows throughout most of the South but is absent from the Mississippi River bottomland south of Missouri, the lowermost Gulf coastal plain, and the southern two-thirds of Florida. It is not found in the New Jersey Pine Barrens, much of Long Island, Cape Cod, northern and eastern Maine and the White and Adirondack Mountains. It is found in central and southern Mexico, Guatemala and western Honduras.

THE TREE

American hornbeam is a small, slow-growing short-lived tree thriving best in the understory of eastern forests along the banks of mountain streams. The main stem is often short, crooked and covered with a thin, smooth grayish bark not unlike that of American beech (*Fagus grandifolia* WDS 119) except that the bark is ridged and resembles the muscles of arms of the human anatomy. Heights up to 25 to 30 feet are normal with some specimens growing to 40 feet or more. Diameters of the main stem are usually 4 to 12 inches. The alternate, deciduous, egg-shaped, finely toothed leaves are 2 to 4 inches long and 1 inch or more wide. Male and female flowers are in separate catkins. The fruit is a small, winged nut edible by numerous forest creatures.

THE TIMBER

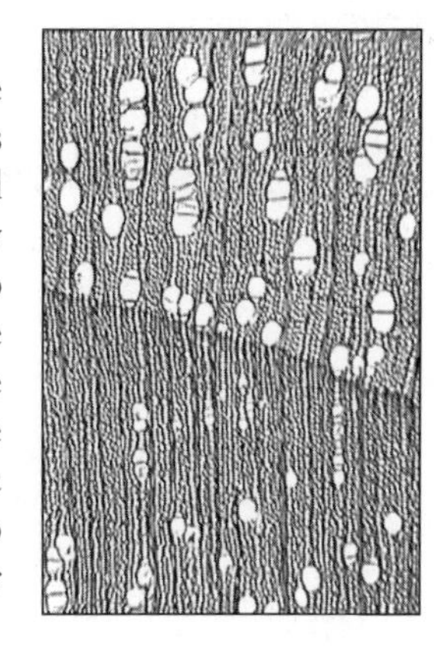

There is an inherent beauty in wood from trees that struggle for existence and become short, crooked and thus overlooked. Even unfinished, the timber is smooth and highly lustrous with subtle patterns. The sapwood is nearly white, while the heartwood is pale yellow and often varies to brownish-white with deeper brown streaks. Adding to the wood's beauty are the numerous small patterns that are formed where small branches grew. The wood is diffuse porous with small indistinct pores. Average reported specific gravity is 0.58 (ovendry weight/green volume), equivalent to an air-dried weight of 45 pcf. There is no distinctive odor or taste.

SEASONING

The wood is difficult to season because it is prone to warping. Average reported shrinkage values (green to ovendry) are 5.7% radial, 11.4% tangential and 19.1% volumetric. If slab-cut, the pieces must be stickered, the pile heavily weighted and dried very slowly.

DURABILITY

The wood is not durable unless treated with preservatives.

WORKABILITY

Working with wood from trees that are not tall or straight requires more time, patience and care. This is especially true in planing operations since grain changes are often abrupt. The wood deserves this extra time. Because it is so dense and smooth, care in clamping surfaces being glued is cautioned lest too much glue be squeezed from the joints. The wood sands and scrapes to an excellent surface and could very well be finished with just one coat of wax.

USES

The wood is best used for items requiring heft and strength such as mallets, tool handles, wedges and other small items. It is an excellent wood for small craft items and turnery.

SUPPLIES

The very wide distribution of this tree in the eastern United States, lower eastern Canada and parts of Central America as well as related species in England and Europe suggest that the wood is readily available in non-commercial venues.

Carya illinoinensis

pecan

by Max Kline

SCIENTIFIC NAME

Carya illinoinensis. Derivation: The genus name is Greek for nut. The specific epithet is Latin meaning the plant is "of Illinois". A synonym is C. *illinoensis*.

FAMILY

Juglandaceae, the walnut family.

OTHER NAMES

pecan nut, pecanier.

DISTRIBUTION

From Indiana to the southwestern corner of Wisconsin, eastern Iowa and Kansas, southward to Alabama, Texas and Mexico.

THE TREE

The pecan tree is the largest member of the hickory genus. It has a massive trunk and reaches heights of 160 to 170 feet with diameters of 6 to 7 feet. Very long-lived, trees of 350 years have been known to exist. The feather-like leaves consist of 9 to 17 leaflets that are 1 to 3 inches wide and 3 to 8 inches long. They are lance-shaped and curved and taper to a long pointed tip. The fruits are 1.25 to 2 inches long and also pointed at the tip. The husks are thin and split in quarters to expose the cylinder-shaped, sweet tasting nut. Pecan is the state tree of Texas.

THE TIMBER

The heartwood of pecan is pale brown to reddish-brown, while the sapwood is almost white, sometimes tinged with brown. Average reported specific gravity is 0.60 (ovendry weight/green volume), equivalent to an air-dried weight of 46 pcf. It is strong, stiff, brittle and very high in shock resistance. The rays are quite conspicuous. Pecan wood is noted for its strength properties and is comparable to other hickories in this regard.

SEASONING

Pecan, when kiln dried, seasons well. Average reported shrinkage values (green to ovendry) are 4.9% radial, 8.9% tangential and 13.6% volumetric.

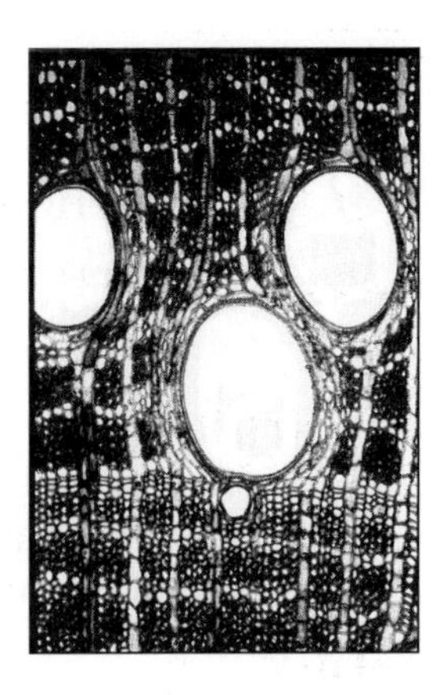

DURABILITY

Pecan is subject to attack by the hickory bark beetle and can easily be damaged by frost. It is seldom damaged by fire since it commonly prefers a moist habitat.

WORKABILITY

The wood requires careful machining but may be planed readily and turned for many uses where strength is important. To achieve the smoothest surface, it must be meticulously sanded.

USES

Common uses include furniture, handles, baseball bats, flooring and veneer for interior trim. It is also used for smoking meats because of the pleasing flavor imparted by the smoke. It is an excellent wood for fireplace fuel. The nuts are remarkable for their high food value and probably the most important product of the tree.

SUPPLIES

Lumber, veneers and plywood from pecan trees are available at a moderate cost on the commercial market.

Carya myristiciformis

nutmeg hickory

by Jon Arno

SCIENTIFIC NAME

Carya myristiciformis. Derivation: The genus name is Greek for nut. The specific epithet *myristica* indicates the nut is similar in shape to a nutmeg.

FAMILY

Juglandaceae, the walnut family.

OTHER NAMES

swamp hickory, pecan.

DISTRIBUTION

Nutmeg hickory grows from north central Texas to South Carolina. Preferring moist soil, its distribution is sporadic and most plentiful along the Red River in Texas, Oklahoma, Arkansas and Louisiana.

THE TREE

Identification can be a problem in that its appearance is very similar to both water hickory (C. *aquatica*) and black hickory (C. *texana*). Normally, nutmeg hickory is a larger tree capable of reaching heights up to 100 feet (water hickory seldom exceeds 80 feet and black hickory is even shorter). Nutmeg hickory generally has fewer leaflets (seven to nine) versus water hickory (seven to fourteen). The nut is small and tasty whereas the nut of water hickory is bitter. The scaly bark is a lighter reddish-brown than black hickory, which derives its name from its deeply furrowed nearly black bark. All three of these minor species of pecan are easily distinguished from true pecan (C. *illinoinensis* WDS 062) in that true pecan's leaflets are broader, more coarsely toothed and less hairy. The nuts of true pecan are much larger and more elliptical.

THE TIMBER

With an average reported specific gravity of 0.56 (ovendry weight/green volume), equivalent to an air-dried weight of 43 pcf, it is identical in density to sugar maple (*Acer saccharum* WDS 010). While hard and heavy relative to most cabinetry woods, it is about 7% less dense than true pecan. The heartwood is reddish-brown which contrasts sharply with the almost white sapwood. Pecans are semi-ring porous and relatively coarse textured which provides for an attractive but subtle figure on flatsawn surfaces. Close examination of the end grain with a hand lens reveals faint wavy bands of parenchyma cells forming fine, concentric, white lines (a primary feature in identifying the *Carya* species). Unlike their cousins on the walnut side of the family, the hickories have no distinct odor. Strength is outstanding and it is very hard and elastic.

SEASONING

Prone to checking and warping, nutmeg hickory is somewhat slow and difficult to season. It is susceptible to blue staining if airflow is not adequate to quickly remove surface moisture early in the drying process. While precise shrinkage values are not available, they are estimated to be within the range of those for pecan (*Carya illinoinesis* WDS 062).

DURABILITY

Durability is poor, and it is substantially less resistant to decay than walnut.

WORKABILITY

While nutmeg hickory is the softest wood in the pecan group and this group represents the softer hickories, it is still quite hard. When used in quantity, expect blades to dull quickly and be prepared to expend more effort in hand sanding, mortising for hinges and carving. Pecan is the traditional choice for furniture of the French Provincial design but if you have access to nutmeg hickory, it is definitely the "lazyman's choice". Like all hickories, it performs quite well on the lathe and polishes to a soft luster that is ideal for oil finishes. Due to its high shrinkage, it is important that it is thoroughly seasoned or turnings may distort and check.

USES

Uses include furniture, cabinetry, flooring, tool handles, sporting goods and fuel. It is especially good for smoking meats. Veneer is used for decorative paneling. The nuts are edible, but due to their smaller size and harder shell, they are not highly prized by commercial growers.

SUPPLIES

Owing to its sporadic range and that it is not planted for nut production, it is the least plentiful of the hickories. Since the wood is marketed with other members of the pecan group, this scarcity commands no premium. In order to ensure you are getting nutmeg hickory you may have to cut your own or follow the log from stump to kiln.

Carya ovata

shagbark hickory

by Max Kline

SCIENTIFIC NAME

Carya ovata. Derivation: The genus name is Greek for nut. The specific epithet is Latin for ovate or egg-shaped, referring to the fruit.

FAMILY

Juglandaceae, the walnut family.

OTHER NAMES

Carolina hickory, red heart hickory, shell bark, white hickory, upland hickory.

DISTRIBUTION

Extends from southern Maine, westward to southeastern Minnesota, southward to eastern Texas, eastward to western Georgia, and from there northward to New Hampshire, except in the coastal plains of the Carolinas.

THE TREE

Reaching heights of 120 to 140 feet with diameters of 20 to 30 inches, shagbark hickory is a distinctly American tree because only one other hickory exists beyond our shores. The bark is attached at the middle of a section but free at the ends, giving the trunk a shaggy appearance. The nuts are edible and delicious. Shagbark hickory is seldom found in pure stands, but is widely scattered among other hardwoods in the eastern United States.

THE TIMBER

Shagbark hickory is the hardest, heaviest and strongest North American wood in common everyday use. The heartwood is brown to reddish-brown. The very wide and nearly white sapwood is considered more valuable than the heartwood. The luster is medium, and odor and taste are not distinct. The texture is rather coarse, and the grain is usually straight but sometimes wavy or irregular. Average reported specific gravity is 0.64 (ovendry weight/green volume), equivalent to an air-dried weight of 50 pcf. Elasticity is outstanding. No other commercial wood has the combination of strength, toughness and elasticity, and no other North American hardwood could adequately substitute for shagbark hickory in the event of a supply shortage.

SEASONING

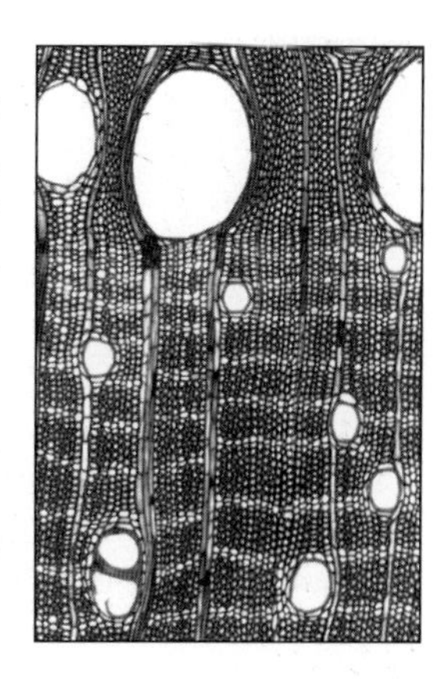

Shagbark hickory requires careful seasoning to prevent splitting, checking, warping and other defects due to its high degree of shrinkage while drying. Average reported shrinkage values (green to ovendry) are 7.0% radial, 10.5% tangential and 16.7% volumetric.

DURABILITY

Shagbark hickory's resistance to decay is poor. It is attacked by various insects, but suffers the greatest harm from the hickory bark beetle. It is not durable when in contact with the soil.

WORKABILITY

The working properties are excellent provided tools are sharp. It is hard to nail without splitting, but finishes very smoothly. No finish is used in many applications, such as tool handles, but it will finish satisfactorily with varnish or oil stains. Bending properties are excellent.

USES

Of all hardwoods, shagbark hickory is the best wood for handles for axes, hammers, hatchets, and picks and nearly 80% is used for this purpose. In the past, large quantities were used in making spokes and rims of wheels, single-trees and buggy shafts. Today increasing quantities are used for athletic goods, such as skis or golf clubs. Other minor uses include ladder rungs, inexpensive archery bows and agricultural parts.

SUPPLIES

The lumber is available commercially at medium to high prices.

Castanea dentata

American chestnut

by Max Kline

SCIENTIFIC NAME

Castanea dentata. Derivation: The genus name is the classical Greek and Latin name for chestnut. The specific epithet is Latin for toothed, which refers to the leaves. A synonym is C. *americana*.

FAMILY

Fagaceae, the beech family.

OTHER NAMES

chestnut, sweet chestnut, o-heh-yah-tah (Native American).

DISTRIBUTION

From Maine to southern Ontario, Michigan and Illinois, south to Florida and Mississippi.

THE TREE

The chestnut blight first appeared in 1904 in New York City and spread rapidly. By the late 1930's, the entire range had become affected and trees were dying. The largest trees were found in western North Carolina and eastern Tennessee. Today only small sprouting shoots are seen from old roots and stumps. The American chestnut tree has a short trunk with branches forming a broad tree 60 to 90 feet high and a trunk diameter of 2 to 4 feet. Sweet, edible nuts develop in prickly burs 2 to 2.5 inches in diameter that split open when ripe in the autumn to produce two to five nuts each.

THE TIMBER

American chestnut heartwood is reddish-brown with lighter-colored sapwood that becomes darker with age. It has large pores and splits easily. The texture is coarse, and the grain is straight. The luster is low to medium. Average reported specific gravity is 0.40 (ovendry weight/green volume), equivalent to an air-dried weight of 30 pcf.

SEASONING

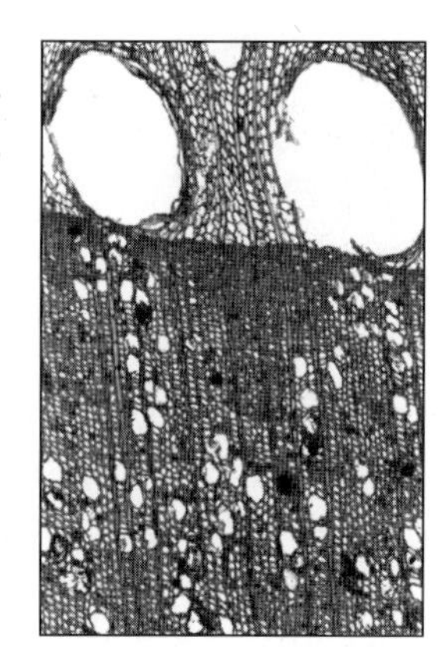

American chestnut can easily be kiln dried without checking or warping. Average reported shrinkage values (green to ovendry) are 3.4% radial, 6.7% tangential and 11.6% volumetric.

DURABILITY

This wood has the ability to resist attack by wood-destroying fungi and for this reason is particularly suited for fence posts in contact with the ground.

WORKABILITY

American chestnut is easily worked with all types of tools and is also easily glued. Nailing will cause splitting, but when holes are pre-bored, the nails stay in place well.

USES

In the past, American chestnut had a wide variety of uses from the manufacture of caskets, furniture, musical instruments, boxes and crates, woodenware, novelties and core wood for veneer to structural purposes including interior trim, fence posts, shingles, piling and railroad ties. The bark was an important source of tannin for the leather industry, and one cord of wood produced as much as 700 pounds of 25% extract. The spent wood chips were used for paper pulp. The delicious nuts were highly prized by our pilgrim ancestors, but are no longer obtainable from American trees. Today, the "wormy" type of American chestnut is used for picture frames and furniture.

SUPPLIES

The commercial supply of American chestnut has disappeared since the blight has eradicated the tree. Scientists have been working to develop a blight-resistant strain. The dead trees were soon attacked by small insects, which burrowed small round holes throughout the tree, giving the wood its "wormy" appearance. [Ed. note: In 1990, while hiking in a secluded area of the Shenandoah National Park in the Blue Ridge Mountains of Virginia, I found a grove of young chestnut trees large enough to produce edible nuts. They were roasted by the evening campfire. jhf]

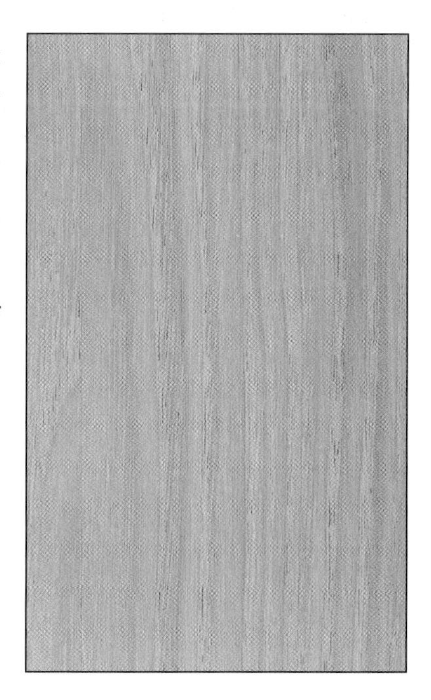

Castanospermum australe

blackbean

by Max Kline

SCIENTIFIC NAME

Castanospermum australe. Derivation: The genus name is Latin for chestnut seed, which refers to the similar taste of blackbean. The specific epithet is Latin meaning the plant is "of the Southern Hemisphere".

FAMILY

Fabaceae or Leguminosae, the legume family; (*Papilionaceae*) the pea or pulse group.

OTHER NAMES

bean tree, Moreton Bay chestnut.

DISTRIBUTION

Sparsely distributed in New South Wales and Queensland.

THE TREE

This is a moisture-loving tree which grows in Australia where rainfall is heavy. It can reach a height of 120 feet with a diameter of 40 to 47 inches. The bole is not prominently buttressed. The fruit is pea-pod shaped.

THE TIMBER

Blackbean is one of the most celebrated Austrailian woods for cabinetry. The color is medium brown to dark chocolate brown, sometimes with darker streaks. The grain is usually straight but may be slightly interlocked. Odor and taste are not distinct. The texture is rather coarse and uneven, and the wood has a greasy feel. The attractive appearance of finished wood owes its beauty to its color and pronounced vessel lines. It is moderately strong and stiff and inclined to be brittle. Average reported specific gravity is 0.57 (ovendry weight/green volume), equivalent to an air-dried weight of 44 pcf. It is unsuitable for steam bending.

SEASONING

This timber seasons slowly with a tendency to collapse, honeycomb and split when kiln dried. To minimize collapse, it is recommended that it be slowly air dried under cover and then put in the kiln. Average reported shrinkage values (green to ovendry) are 2.1% radial and 3.5% tangential.

DURABILITY

Blackbean is known to be of high durability even when exposed to extremes of moisture and dryness. Beetles, termites and other borers are not attracted to the heartwood.

WORKABILITY

Due to the mineral deposits in the wood, blackbean is somewhat difficult to cut on woodworking machinery found in most shops. Maximum performance is best attained with industrial equipment. It handles fairly well with knives which makes it desirable for turnery. It sometimes shows variable gluing properties due to its greasy character. It finishes beautifully and takes a high and lustrous natural polish. Its freedom from checking is held in high esteem by wood carvers. It has reasonably good nail and screw holding properties.

USES

Blackbean timber is one of the finest Australian woods for decorative purposes. It is used mainly for up-scale furniture, paneling, carving and fancy articles. Because of its good insulating properties, it is suitable for switchboards and electrical fittings. It is also much less flammable than most woods.

SUPPLIES

Blackbean veneer can be purchased at a high price. Usually lumber has to be obtained directly from IWCS members living in Australia.

Casuarina spp.

casuarina

by Jon Arno

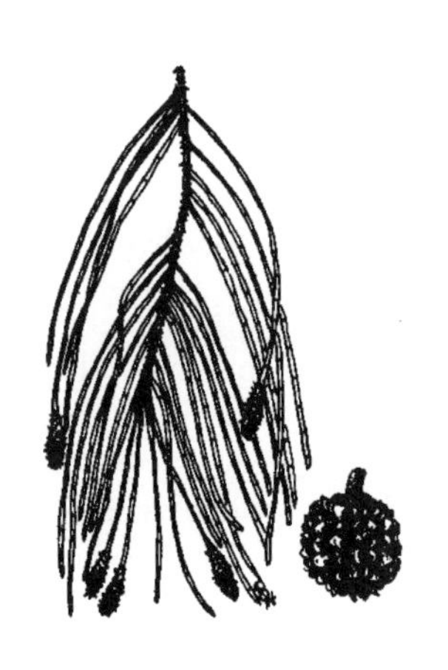

SCIENTIFIC NAME

Casuarina spp. Derivation: The genus name *Casuarina* is in honor of the large and flightless cassowary bird in the Casuarius family. The fine, long leaves of this tree are characteristically segmented and needle-like similar to the feathers of the bird.

FAMILY

Casuarinaceae, the casuarina or beefwood family.

OTHER NAMES

she-oak, beefwood, Australian pine, horsetail tree, South-Sea ironwood, aru, ru, surra, agoho.

DISTRIBUTION

Although most plentiful in Australia, some species are native to Africa, India, Indochina, the Philippines and various islands of Polynesia. Because casuarina will grow in beach sand and seems immune to salt spray, several species have been introduced into other tropical and semi-tropical regions around the world. The species C. *equisetifolia*, known as Australian pine, is now quite plentiful in southern Florida.

THE TREE

The larger species of this genus reach heights between 120 and 150 feet with diameters of 2 to 2.5 feet. In general appearance, the trees look very much like conifers with very small, scale-like leaves borne on branchlets, which, from a distance, look like pine needles. While the tree may be helpful in preventing beach erosion, it is messy since it sheds its branchlets, along with its fruit (a woody, cone-like structure). Both can be painful when stepped on with bare feet.

THE TIMBER

The heartwood is usually reddish-tan, but in mature trees is darker and rather vivid reddish-brown. The sapwood is a soft, creamy beige color that is clearly demarcated from the heartwood. The wood is medium to fine textured with a low luster. The rays are conspicuous on the radial surface giving it a faint resemblance to oak. When freshly cut, the wood has a fairly strong, turpentine-like odor. It is hard, heavy and very strong. Average reported specific gravity is 0.83 (ovendry weight/green volume), equivalent to an air-dried weight of 67 pcf.

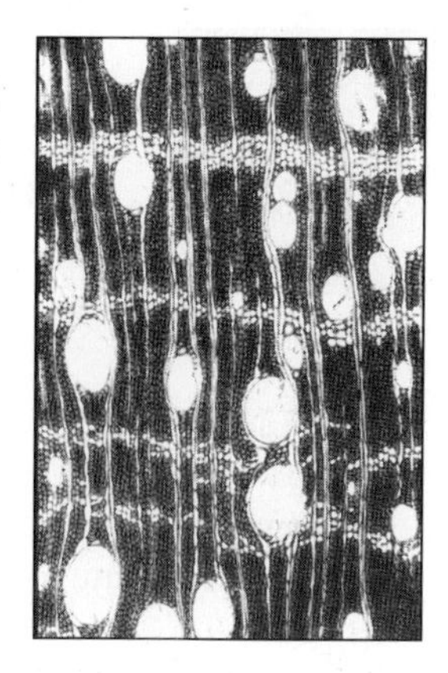

SEASONING

Casuarina is difficult to dry and very prone to warping and checking. Average reported shrinkage values (green to ovendry) are 6.4% radial, 11.7% tangential and 17.6% volumetric.

DURABILITY

The heartwood is reported to be non-durable.

WORKABILITY

Working with casuarina can be unpleasant. Although this wood finishes well and quartersawn stock has an attractive, ray-dominated figure, it is simply a much harder wood than is necessary for most furniture applications. Also, its high shrinkage makes it unsuitable to any project requiring dimensional stability.

USES

Uses include interior rough construction, posts, poles and tool handles. Some wood is used for making charcoal, and the bark can be used in tanning leather.

SUPPLIES

Supplies are plentiful both within its native range and now in warmer regions throughout the world. Casuarina is not commonly carried by retail hardwood suppliers in North America. It is an easy wood to acquire either by foraging or by trading with IWCS members in Florida or southern California.

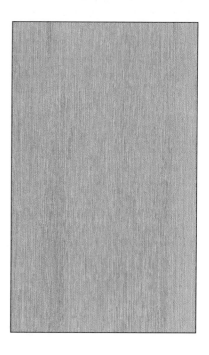

Catalpa speciosa

northern catalpa

by Jon Arno

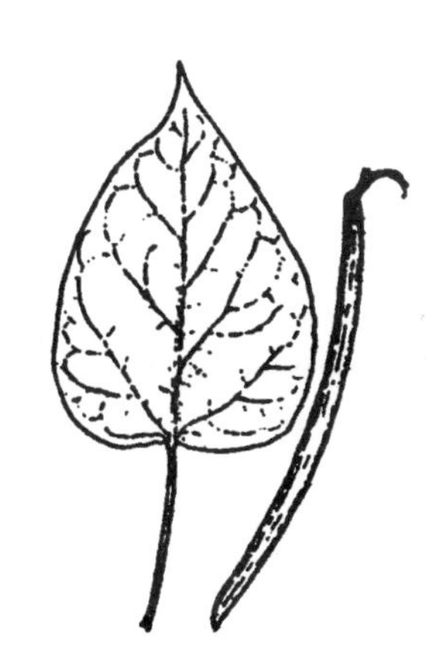

SCIENTIFIC NAME

Catalpa speciosa. Derivation: The genus name is a Native American name. The specific epithet is Latin for showy, referring to the clusters of large flowers.

FAMILY

Bignoniaceae, the bignonia or trumpet creeper family.

OTHER NAMES

Indian bear, cigar tree.

DISTRIBUTION

Originally, the native range of northern catalpa consisted of a narrow, arc-shaped band scarcely 100 miles wide extending down the Ohio and Mississippi River basins from about Louisville, Kentucky to Memphis, Tennessee. As a popular landscaping ornamental, however, it has been widely planted and is now well established throughout the midwest and southern Canada.

THE TREE

Northern catalpa is capable of reaching a height of 100 feet or more with diameters of up to 5 feet. It is a pretty tree with large heart-shaped leaves and orchid-like flowers that bloom in late spring. The fruit is a long, slender capsule which looks like a giant green bean. The capsule turns tobacco brown by the time the leaves fall, hence, the tree's common name, cigar tree.

THE TIMBER

Northern catalpa is ring porous, producing a somewhat racy tangential figure very similar to ash. The heartwood is a light gray-brown or buff color often with very attractive dark brown or purple streaks. The heartwood contrasts sharply with the light cream, almost white, sapwood. The tree quickly converts sapwood to heartwood and, in logs cut from mature trees, all but the outer two or three annual rings will be heartwood. Average reported specific gravity is 0.38 (ovendry weight/green volume), equivalent to an air-dried weight of 28 pcf. The wood is soft, weak and brittle and should not be considered for applications where either strength or wear properties are critical. Northern catalpa has a unique and somewhat less than pleasant creosote-like odor.

SEASONING

This species air dries quickly and is exceptionally easy to season. The end grain, however, should be thoroughly coated to prevent checking. Average reported shrinkage values (green to ovendry) are 2.5% radial, 4.9% tangential and 7.3% volumetric.

DURABILITY

Northern catalpa is surprisingly durable when in contact with the soil. It also produces a very attractive silver-gray patina when exposed to weather.

WORKABILITY

For a ring-porous, hardwood (deciduous) species, working with northern catalpa is a joy. It is almost as soft and easy to cut and shape as eastern white pine (*Pinus strobus* WDS 212) and, therefore, performs well when used for turned articles, even though it is far less dense than most of the species preferred for lathe work. Its natural, buff-brown color and attractive highlight allow the woodworker to achieve an acceptable, light "fruitwood" finish without staining, although a perfectly smooth surface will require filling. When sanding, care must be taken to use a rigid, unpadded block behind the sandpaper or the porous earlywood will disintegrate, leaving ripples in the surface. These minor inconveniences are certainly sufferable considering the beautiful results that can be achieved.

USES

Owing to its remarkable durability, northern catalpa and its close relative southern catalpa, *C. bignonioides*, are used for fence posts and rails. Sadly, much of it is also used for crating. On a limited basis, some is used for turned articles, carving and cabinetry, but it is far less popular than it ought to be.

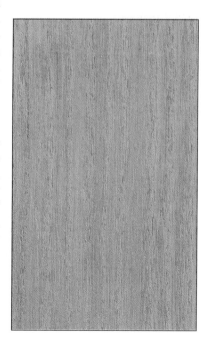

SUPPLIES

Northern catalpa is seldom available in large quantities, because mills do not have access to pure stands where it can be cut and processed in volume. Due to this on-again/off-again supply and the lack of awareness on the part of woodworkers, it does not command a premium price. It can be bought at the prevailing price for "random hardwoods" throughout much of North America.

Cecropia peltata

trumpet-tree

by Jim Flynn

SCIENTIFIC NAME

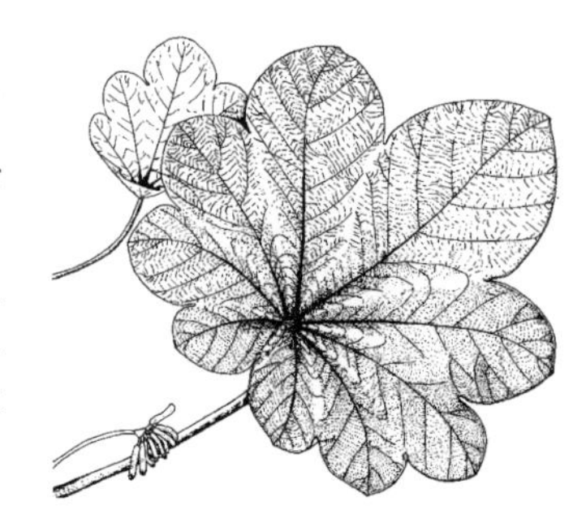

Cecropia peltata. Derivation: The genus name is from Greek mythology. Cecropius pertains to Cecrops, the founder and first King of Attica, portrayed as half man and half dragon whose legs were fabled to be snakes. The connection with the plant name is in reference to the long snake-like flowers. The specific epithet means a shield-shaped leaf. There are about 80 species of *Cecropia*.

FAMILY

Cecropiaceae, the cecropia family.

OTHER NAMES

cecropia, trumpet-wood, yagrumo (Cuba, Venezuela), guarumo (Central America), boessi papaja (Surinam), imbauba (Brazil), cetico and tacuna (Peru), ambahu (Argentina), yagrumo (also spelled yagrumbo), hembra (Puerto Rico) and trompy (West Indies). There are many other names but the few listed illustrate the wide growth range.

DISTRIBUTION

Throughout tropical America, it is abundant in open areas and in virgin and cutover forests often growing in pure stands. No other single plant has had so large a part in giving Central American vegetation its characteristic appearance.

THE TREE

Trumpet-tree is a medium sized, rapidly growing species that reaches heights of 70 feet or more in ideal surroundings. More commonly, it grows to 40 feet with trunk diameters of 8 to 12 inches when mature. The tree invades open areas that have been disturbed by cutting and propagates naturally. The large (12 to 20 inches), alternate, evergreen, mildly lobed leaves grow in clusters at the ends of hollow branches. The gray bark is smooth and thin with narrow rings and large leaf scars at the nodes. The inner bark is pinkish, slightly bitter and exudes watery latex. The stem and branches are mostly hollow and were used by early inhabitants to make trumpets hence its name. In many growth areas, the hollow stem and branches are inhabited by stinging ants. The ants live in the internodes and feed upon glycogen produced by the tree.

THE TIMBER

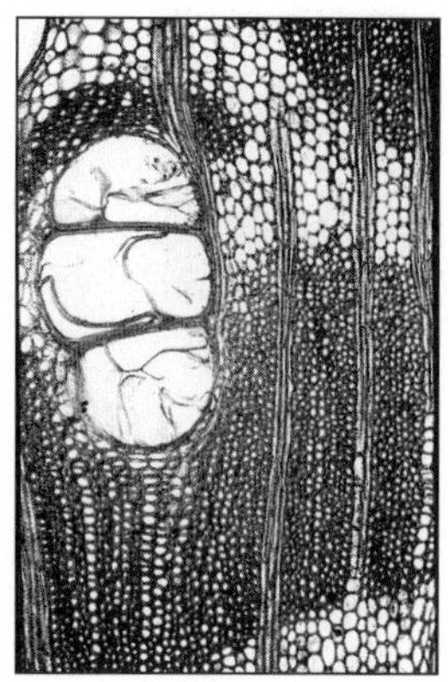

The wood is rather plain. It is a pale brown or oatmeal color when exposed to the air and light, and there is no distinction between sapwood and heartwood. The wood is fairly lustrous and has a coarse texture with generally straight grain. There is no distinctive taste or odor. Average reported specific gravity is 0.26 to 0.34 (ovendry weight/green volume), equivalent to an air-dried weight of 19 to 25 pcf. The wood resembles North American black cottonwood both in density and mechanical properties.

SEASONING

The wood air dries rapidly, with little checking, but warps severely. Average reported shrinkage values (green to ovendry) are 2.0% radial, 6.2% tangential and 8.3% volumetric. It is easily kiln dried.

DURABILITY

Prone to blue stain, susceptible to fungal decay, termites and a variety of other insects, the wood is not durable. If incised or if there is high end-grain exposure, the wood will accept preservatives using either pressure-vacuum or open tank systems.

WORKABILITY

Seasoned wood is very easy to saw and machine. Because of its low density, the surfaces tend to tear and fuzz in shaping and turning operations. It can be planed and sanded without problems but is very difficult to finish with varnish and lacquer. There are no problems with gluing. Nailing is readily done and screws hold well.

USES

Uses include plywood core stock, particleboard, match sticks, boxes and crates and light furniture. In Puerto Rico it is used in making "Puertorrican cuatros" a guitar-like musical instrument. The hollow branches are often split and used for rain gutters, and unsplit sections are used as pipes.

SUPPLIES

There is no significant commercial market for this timber in the United States because of the wealth of other hardwood species with similar characteristics. It is listed in the USDA's Agriculture Handbook *Tropical Timbers of the World* which indicates the wood is important in world trade. There are sufficient quantities available to wood sample collectors.

Cedrela odorata

Spanish-cedar

by Alan B. Curtis

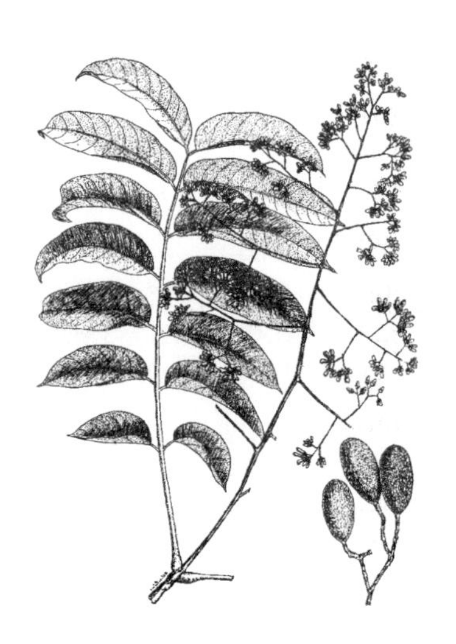

SCIENTIFIC NAME

Cedrela odorata. Derivation: The genus name is derived from the Latin *cedrus* meaning cedar from the similarity in the appearance and fragrance of the wood. The specific epithet is Latin for fragrant smelling. A synonym is C. *mexicana*.

FAMILY

Meliaceae, the mahogany family.

OTHER NAMES

cedro (both Spanish and Portuguese), cigar-box cedar.

DISTRIBUTION

All countries in Central and South America, except Chile; also the islands of the West Indies. It is occasionally planted as far north as Florida.

THE TREE

Spanish-cedar can grow to heights of more than 100 feet with straight trunks of 3 to 6 feet in diameter. Average-sized trees are considerably smaller. Sometimes, the base of the trunk is buttressed. Although the tree grows in seasonally dry forests and rain forests, it only grows where drainage is good.

THE TIMBER

Spanish-cedar heartwood is pinkish to reddish-brown and has prominent growth rings. The color darkens slightly on exposure to light. Timber from fast-grown trees is paler in color and lighter in weight than that from slowly grown trees. The wood is soft. Its texture is fine and even, and the grain is straight. It has a lustrous appearance and a pleasant, distinctive cedary odor that is especially noticeable when the wood is freshly surfaced. Occasionally trees produce a highly resinous timber. The wood may have a bitter taste. It is classified as semi-ring porous. The wood is light weight with an average reported specific gravity of 0.40 (ovendry weight/green volume), equivalent to an air-dried weight of 30 pcf.

SEASONING

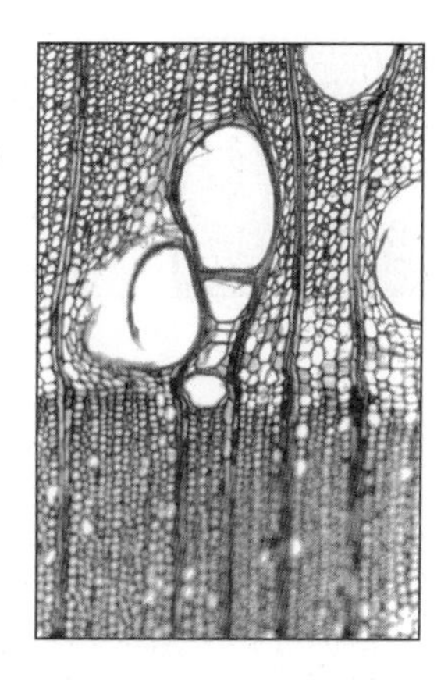

The wood air dries rapidly without warping or splitting. Average reported shrinkage values (green to ovendry) are 4.2% radial, 6.3% tangential and 10.3% volumetric.

DURABILITY

It is highly resistant to decay and insect attack.

WORKABILITY

Easy to work and finish, Spanish-cedar takes a smooth polish. The wood has good dimensional stability when manufactured. In proportion to its weight, it is strong.

USES

In the past, Spanish-cedar was exported to Europe and North America for cigar boxes. Indians made dugout canoes from the trunks of this tree. Currently, Spanish-cedar is the most important local timber for domestic use in tropical America. Spanish-cedar is also a preferred wood for furniture, cabinetry, doors and windows, interior trim and carved figures. Because the wood is resistant to insects and aromatic, it is used for clothes chests. It is a favorite timber for the hulls of light racing boats and is used in making musical instruments. The wood is also used to make veneer and plywood.

SUPPLIES

Although Spanish-cedar is becoming scarce on the world market, it is usually available from specialty lumber dealers at reasonable prices. In its native range, the trees are often planted for shade along highways, in parks and in coffee and cacao plantations. Such trees, if harvested, yield little lumber due to their poor form and many limbs. Spanish-cedar has not done well in plantations due to the presence of a shoot borer insect, which deforms the young trunks. If this problem can be overcome, the trees, which grow rapidly, could provide adequate timber for the future.

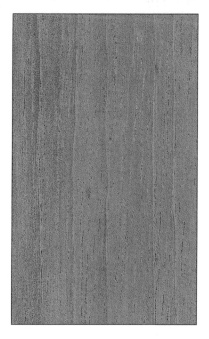

SPECIAL NOTE

In appearance and color, the wood of Spanish-cedar resembles bigleaf mahogany (*Swietenia macrophylla* WDS 254) and can be easily confused with it. The distinguishing feature of Spanish-cedar is the pleasant cedary odor from fresh-cut surfaces. The use of the name cedar can also be confusing since this tree is a hardwood (angiosperm) and not a softwood or conifer (gymnosperm) as are the temperate climate cedars.

Ceiba pentandra

ceiba

by Jon Arno

SCIENTIFIC NAME

Ceiba pentandra. Derivation: The genus name is the Mexican name ceibo, the fruit of this tree. The specific epithet is Latin referring to the flower with five petals. A synonym is *Bombax pentandrum*.

FAMILY

Bombacaceae, the bombax family.

OTHER NAMES

kapoktree, silk-cottontree, white silk-cottontree, bonga, pochota, sumauma, fromagier.

DISTRIBUTION

Widespread throughout the tropics including west Africa, Malaysia and Indonesia, ceiba is plentiful from central Mexico to Brazil. A tree of open, dry areas, it will quickly establish on abandoned farmland. It is grown on plantations for the kapok fiber in its seed capsules.

THE TREE

Mature specimens can reach heights of 150 feet and maximum diameters in excess of 6 feet. Mounted on a widespread buttress, the trunk is often slightly barrel-shaped. The smooth, almost skin-like bark is protected by cone-shaped protrusions, reminiscent of the spikes on a medieval war mace, and this, in combination with the tree's huge size, gives it a striking appearance. Because it favors open environments, it is usually shorter and tends to form a broad crown composed of heavy, contorted branches, leaving a relatively short, usable log between the buttress and the crown.

THE TIMBER

The heartwood is pinkish-white to ashy brown and is not clearly distinguished from the sapwood. It is normally straight grained, course textured, without luster and extremely soft. With an average reported specific gravity of 0.25 (ovendry weight/green volume), equivalent to an air-dried weight of 18 pcf, ceiba is almost twice as heavy as lightweight balsa (*Ochroma pyramidale* WDS 189), but it is still very soft compared to American basswood (*Tilia americana* WDS 270). Despite its greater density, ceiba is only slightly stronger than balsa in bending and crushing strength.

SEASONING

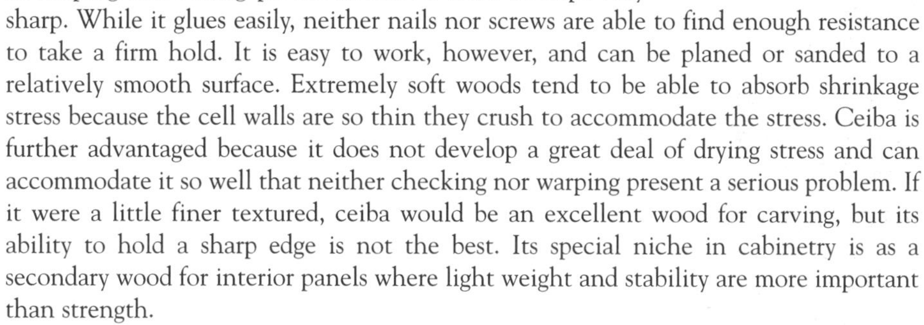

Ceiba is easy to air or kiln dry. Average reported shrinkage values (green to ovendry) are 2.1% radial, 4.1% tangential and 7.7% volumetric. Microbial staining can be a problem, and the drying process should begin as soon after cutting as possible.

DURABILITY

Timber decays very quickly when exposed to the elements. The tree is often attacked by insect larvae.

WORKABILITY

Ceiba tends to tear out or leave fuzzy edges when crosscut or in shaping and boring processes. Blades must be kept very sharp. While it glues easily, neither nails nor screws are able to find enough resistance to take a firm hold. It is easy to work, however, and can be planed or sanded to a relatively smooth surface. Extremely soft woods tend to be able to absorb shrinkage stress because the cell walls are so thin they crush to accommodate the stress. Ceiba is further advantaged because it does not develop a great deal of drying stress and can accommodate it so well that neither checking nor warping present a serious problem. If it were a little finer textured, ceiba would be an excellent wood for carving, but its ability to hold a sharp edge is not the best. Its special niche in cabinetry is as a secondary wood for interior panels where light weight and stability are more important than strength.

USES

Ceiba is used for core stock, plywood veneer, crating and packaging. The wood pulp is used for paper. Kapok fiber is an important commodity for stuffing, flotation and insulation applications. The seeds yield oil for the soap and cosmetics industries. The gum has medicinal value, and the fibrous, inner bark is used for making rope. Ceiba was sacred to the Mayans and worshipped as the "tree of life". It was also a favored wood for rafts and dugout canoes.

SUPPLIES

Ceiba is plentiful. It regenerates quickly and yields many products, but is not a major timber in international commerce. The wood is hampered by mediocrity in that it is not light enough to compete with balsa, but is too soft and weak to be an important construction timber or cabinetry wood.

Celtis occidentalis

hackberry

by Max Kline

SCIENTIFIC NAME

Celtis occidentalis. Derivation: The genus name is the classical Latin name for a species of lotus. The specific epithet means the plant is "of the Western Hemisphere".

FAMILY

Ulmaceae, the elm family.

OTHER NAMES

bastard elm, hacktree, hoop ash, nettle tree, sugar-berry tree.

DISTRIBUTION

Hackberry grows from southern Quebec to southern Manitoba and North Dakota, south to North Carolina, Georgia, Alabama, Oklahoma and Texas.

THE TREE

Usually a small tree of 30 to 50 feet in height, hackberry can reach heights of 130 feet with a diameter of 2 to 3 feet. The hackberry tree is not shapely but is quite rugged. It exudes a sticky gum, and tannin may be extracted from the bark. The fruit resembles a black cherry, has a large pit and is eaten by birds.

THE TIMBER

The wide sapwood is pale yellow to grayish or greenish-yellow, frequently discolored with blue sapstain. The heartwood, when present, is yellowish-gray to light brown streaked with yellow. The grain is coarse and sometimes interlocked. The wood is fairly uniform in texture. It is without characteristic odor or taste. Hackberry is moderately heavy with an average reported specific gravity of 0.49 (ovendry weight/green volume), equivalent to an air-dried weight of 37 pcf. It is usually sold with lower grades of ash and elm, which it resembles.

SEASONING

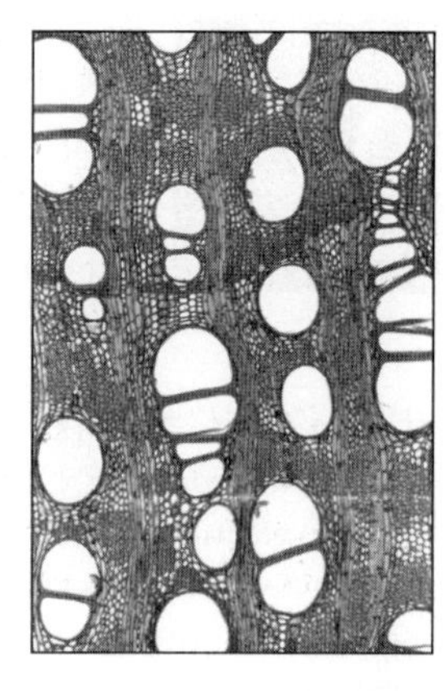

Hackberry is rather difficult to season because the boles are not usually straight. Average reported shrinkage values (green to ovendry) are 4.8% radial, 8.9% tangential and 13.8% volumetric.

DURABILITY

Hackberry is not durable when in contact with the soil and is frequently badly damaged by wood-boring pests.

WORKABILITY

This timber works fairly well with any type of tool. Because of its usually straight and uniform grain, it can be successfully carved. It stains and finishes easily and has an attractive appearance when finished in its natural color. It is not strong enough or available in sufficient commercial quantities for building construction.

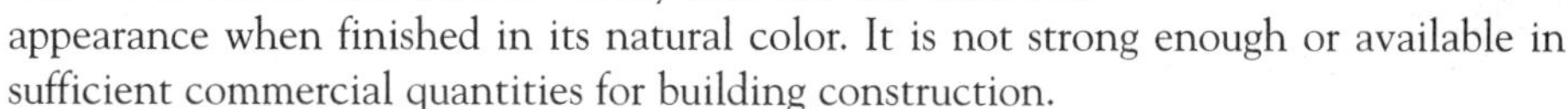

USES

It is used mainly in the manufacture of farm implements, shipping crates and boxes. It is used to a limited extent for carving, furniture and cabinetry. Other uses are similar to those of ash and elm.

SUPPLIES

Hackberry veneer is plentiful either quartered, sliced or rotary cut, and lumber is available. The price range is medium to expensive.

Centrolobium spp.

canarywood

by Jim Flynn

SCIENTIFIC NAME

Centrolobium spp. Derivation: The genus name is Latin meaning "many lobes on leaves". Because the anatomy of the wood of this genus is considered fairly consistent and distinctive and at least three of its species have been recently advertised on the commercial market as canarywood, it was decided to group these woods into one Wood Data Sheet. Canarywood may likely be C. *robustum*, C. *tomentosum*, C. *paraense*, C. *orinocense* or C. *ochroxylon*.

FAMILY

Fabaceae or Leguminosae, the legume family; (*Papilionaceae*) the pea or pulse group.

OTHER NAMES

This is quite complicated. Practically every locality where this tree can be found has applied one or more distinctive common names to the genus without regard to the species. Commercially, the wood is marketed under only a few names; the most popular names are canarywood, arariba, porcupinewood and putumuju. In their book *Timbers of the New World*, Record and Hess list 42 common names for these species.

DISTRIBUTION

These species occur irregularly from Panama to Ecuador and southern Brazil.

THE TREE

These species are large trees growing to a height of 100 feet or more with a trunk diameter of 30 to 50 inches. The large imparipinnate leaves bear 7 to 17 opposite to alternate leaflets. The yellow or purplish flowers are borne in terminal panicles. The large samara-like indehiscent pod contains one to three seeds and resembles a chestnut bur with the wing of a gigantic maple seed attached to it.

THE TIMBER

The heartwood, which is clearly demarcated from the yellowish sapwood, is bright yellow or orange, typically variegated and sometimes "rainbow-hued". In time, this color usually changes to red or brown. The texture and luster varies. Likewise, the grain varies from straight to irregular. Some species have no odor or taste while others may have a distinctive odor and a faint taste. The average reported specific gravity varies from 0.61 to 0.69 (ovendry weight/green volume), equivalent to an air-dried weight of 47 to 54 pcf.

SEASONING

The wood dries at a moderate rate with little or no warp and low shrinkage. Average reported shrinkage values (green to ovendry) are 2.4% radial, 5.6% tangential and 8.4% volumetric. Canarywood is dimensionally stable when in use.

DURABILITY

Canarywood is very durable. It is reported to be highly resistant to attack by decay fungi, termites and other insects as well as teredo marine borers. It does not take impregnation with preservatives very well.

WORKABILITY

The wood of these species is easy to work with all types of tools. It sands and finishes well. Hand cabinet scrapers do well to smooth out the occasional fuzzy spots.

USES

Because of its attractive grain and stability, canarywood is used for the manufacture of fine furniture and cabinetry. It also has many other applications including heavy construction timbers, decorative veneer and flooring as well as ship components such as planking, keel, decking and trim.

SUPPLIES

Canarywood is occasionally seen on the commercial market at a fairly expensive price.

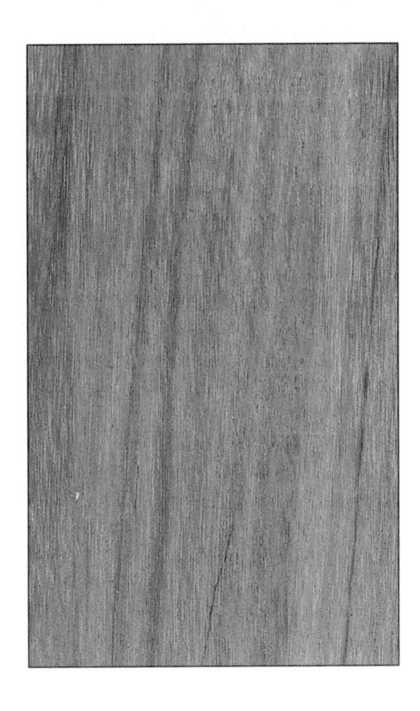

Ceratopetalum apetalum

coachwood

by Max Kline

SCIENTIFIC NAME

Ceratopetalum apetalum. Derivation: The genus name is Latin meaning horned petal. The specific epithet is Latin for petal.

FAMILY

Cunioniaceae, the cunonia family.

OTHER NAMES

scented satinwood, rose mahogany.

DISTRIBUTION

In the coastal range forests of northeastern New South Wales, from the latitudes of Sydney northwards to the Queensland border in Australia.

THE TREE

Coachwood is a medium-sized tree, shortly buttressed, reaching a height of 100 feet with a diameter of about 2.5 feet. The bole is straight and slender. The bark is smooth, whitish-gray and spotted with darker shades due to the growth of lichens. The leaves are opposite, broad lanceolate with serrated margins from 1 to 10 inches long and 0.5 to 3 inches wide. The five-lobed flowers are cream to white and turn pink when fully grown. The fruit is a thin, woody, egg-shaped capsule containing a single seed.

THE TIMBER

The heartwood is light brown to pinkish-brown, darkening on exposure. It is not well defined from the heartwood. The grain is fine and straight. The texture is fine and even. The luster is medium, and the taste is not distinct. The delicate figure can be seen on flatsawn surfaces. Average specific gravity is 0.51 (ovendry weight/green volume), equivalent to an air-dried weight of 39 pcf. The wood has a very pleasant scent which is described as being like caramel or freshly mown hay. The strength to weight ratio is high.

SEASONING

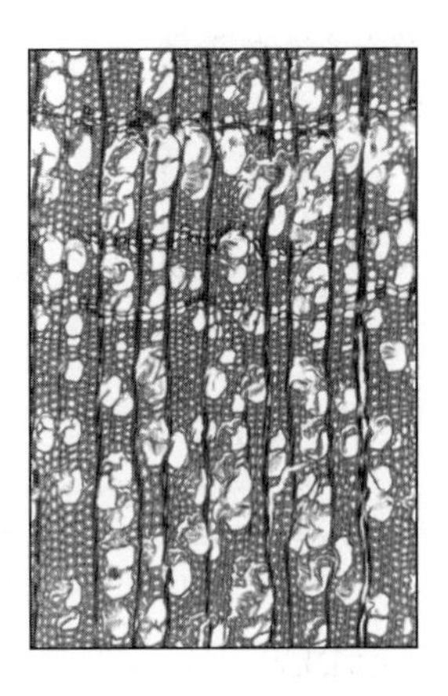

The thinner dimensions of coachwood will air dry comparatively quickly, but thick sections tend to split, warp and collapse. Shrinkage values are not available.

DURABILITY

This timber is immune to attack by powder post beetles. It is not resistant to fungal attack. To be used successfully for exterior work, it must be treated with preservatives. Treatment markedly enhances its durability.

WORKABILITY

Coachwood is easily worked to a smooth, silky finish with hand or machine tools. It turns well and is a very good wood to use when carving. It needs to be pre-bored before nailing as it has a tendency to split. Gluing is satisfactory, and the response to all types of finishes is good. It has excellent peeling properties, enabling the manufacture of rotary veneers. It cannot be recommended for steam bending purposes.

USES

It is confined to indoor use including furniture, cabinetry, joinery, turnery such as brushes, brooms, handles, dowels, plywood, sporting goods, corestock and laminated panels. It is the best Australian wood for rifle stocks.

SUPPLIES

The availability of coachwood is very limited, and the price is high.

Cercis canadensis

eastern redbud

by Max Kline

SCIENTIFIC NAME

Cercis canadensis. Derivation: The genus name is Greek for weaver's shuttle, referring to the fruit. The specific epithet is Latin meaning the plant is "of Canada" when French Canada extended down the Mississippi Valley.

FAMILY

Fabaceae or Leguminosae, the legume family; (*Caesalpiniaceae*) the cassia group.

OTHER NAMES

Judas-tree, June bud, American redbud.

DISTRIBUTION

From southern New York and Connecticut west to Iowa, south to northeastern Texas and east to northern Florida.

THE TREE

Eastern redbud is a small tree, 25 to 50 feet in height with a trunk 6 to 12 inches in diameter. Because of its exquisite beauty in early spring when the bright pink flowers appear in numerous clusters along the branches from its crown to its trunk, it is planted as an ornamental. The bark is reddish-brown, and the leaves are heart-shaped and turn from dark green to bright yellow in the autumn. It is called the Judas-tree because it resembles the Asiatic species which legend claims to be the tree from which Judas hanged himself. Eastern redbud is the state tree of Oklahoma.

THE TIMBER

The heartwood of eastern redbud is light olive-brown, often with dark streaks, with a golden luster, deepening to russet-brown. The narrow sapwood is white and sharply demarcated from the heartwood. The wood is heavy, moderately hard, not strong and coarse grained. Average specific gravity is 0.61 (ovendry weight/green volume), equivalent to an air-dried weight of 47 pcf. It is odorless and tasteless with a medium texture.

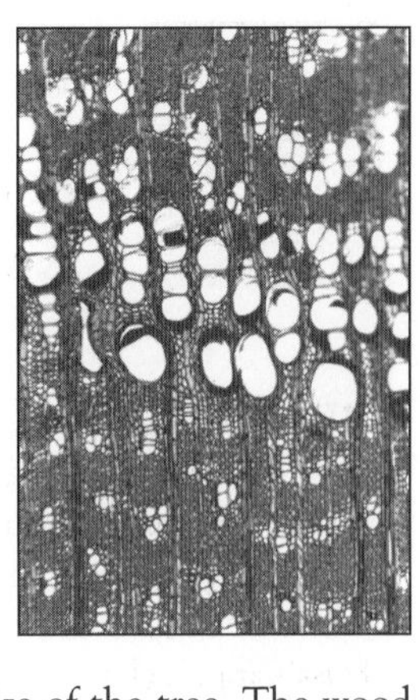

SEASONING

This wood can be air dried quite readily. Shrinkage values are not available.

DURABILITY

The wood is highly durable.

WORKABILITY

Eastern redbud is easily worked with tools and takes a good finish of any kind. It tends to split when nailing but glues well.

USES

Primarily, eastern redbud is used as an ornamental. The lumber has little commercial importance due to the limited size of the tree. The wood is attractive for novelties, however, and a very pleasing effect may be obtained by working it to achieve the maximum benefit from the grain and figure. It is used on a small scale for small cabinetry work and turnery.

SUPPLIES

Limited small dimensions are sometimes available to the home woodworker.

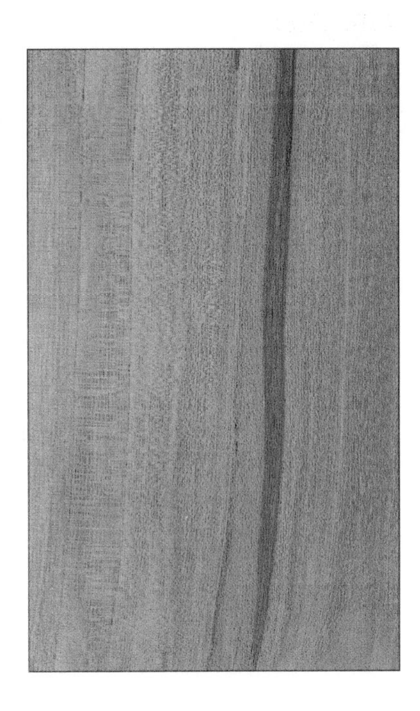

Cercocarpus spp.

mountain-mahoganies

by Max Kline

SCIENTIFIC NAME

Cercocarpus spp. Derivation: The genus name is for tail and fruit referring to the long-tail hairy fruit. Since several species of mountain mahogany are known, they will be treated as a group in this Wood Data Sheet, but descriptive information will deal specifically with the largest tree, *Cercocarpus ledifolius* (curlleaf cercocarpus). Also covered are: C. *betuloides* (birchleaf cercocarpus), C. *breviflorus* (hairy cercocarpus) and C. *montanus* (alderleaf cercocarpus).

FAMILY

Rosaceae, the rose family.

OTHER NAMES

hardtack, lentisco, ramon, sweetbrush.

DISTRIBUTION

Mountain-mahoganies are found in the dry interior and mountainous regions of western North America from Oregon and Montana to Oaxaca, Mexico.

THE TREE

Usually these species are large shrubs only 15 to 25 feet in height. Occasionally they can grow to a round-topped tree 40 feet in height with a diameter of 30 inches. The trunk is short, crooked and deformed with irregular limbs. The edges of the thick evergreen leaves curl downward and underneath are covered with minute hairs. The tree is slow growing and found at elevations of 4,000 to 10,000 feet.

THE TIMBER

The wood of mountain-mahogany is very hard, dense, brittle and heavy usually with straight grain. The heartwood is cherry red, reddish-brown or chocolate-brown, often with lighter or darker shades intermingled and is sharply demarcated from the narrow, yellowish sapwood. Average specific gravity is 0.81 (ovendry weight/green volume), equivalent to an air-dried weight of 65 pcf. It has a medium luster and is without distinctive odor and taste. The texture is fine.

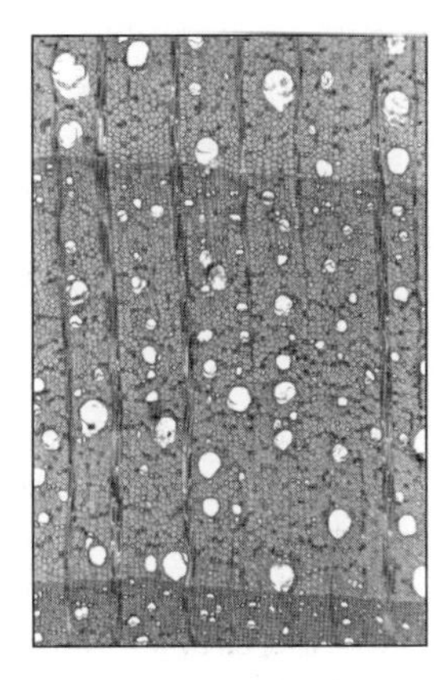

SEASONING

The mountain-mahoganies are extremely prone to checking and warping during drying. Shrinkage values are not available.

DURABILITY

The heartwood is very resistant to decay.

WORKABILITY

It is rather difficult to work with hand tools but will machine readily. It takes a beautiful, highly polished finish.

USES

The leaves of the mountain-mahoganies are a favorite food of deer since they are evergreen. The wood makes an outstanding fuel, burning for a long time with an intense heat. Because of its durability, it is utilized in its growth area for fence posts. The timber has no other commercial use but lends itself admirably to the making of unusual novelties by the home woodworker. Souvenirs made from mountain-mahogany are sometimes offered to the tourist trade.

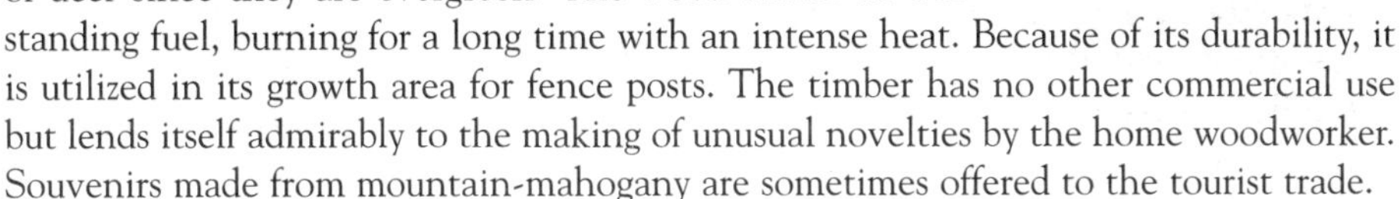

SUPPLIES

Because this species is found only sparsely scattered at high altitudes, it is difficult and expensive to obtain. Although rare, wood collectors seem to have small samples available for exchange with other members.

Chamaecyparis lawsoniana

Port-Orford-cedar

by Jim Flynn

SCIENTIFIC NAME

Chamaecyparis lawsoniana. Derivation: The genus name is the Greek name for lavender-cotton or ground cypress, a dwarf shrubby Old World composite which resembles a dwarf cypress. The specific epithet is in honor of Peter Lawson (1794 to 1873) and sons, nurserymen of Edinburgh who introduced this species into cultivation in England.

FAMILY

Cupressaceae, the cypress family.

OTHER NAMES

Port-Orford white-cedar, Oregon-cedar, Lawson cypress.

DISTRIBUTION

This tree, in dwindling supply, grows naturally in a narrow belt near the Pacific coastline in southwest Oregon and northwest California. It needs a mild climate with an abundance of rain, thus its range is limited. It has been grown successfully in New Zealand and Europe.

THE TREE

Port-Orford-cedar is a tree of great beauty which ranges from 125 to 180 feet in height with a diameter from 3.5 to 6 feet. Its reddish-brown bark is very thick and deeply furrowed into large, flat, fibrous ridges. The twigs are slender and flattened. The bright green leaves are 0.0625 inch long and pale and glandular underneath. The cones are about 0.375 inch in diameter. Currently the tree is under serious attack by juniper scales, spruce mites and fungus. It is believed that the fungus, which is now decimating the forests, came from Japan and extreme measures are being taken to prevent further spreading.

THE TIMBER

The fine-grained wood is faint yellowish-white with tinges of red. When freshly cut, it has a distinctive ginger-like aroma that is overpowering in confined spaces. Average reported specific gravity is 0.43 (ovendry weight/green volume), equivalent to an air-dried weight of 32 pcf.

SEASONING

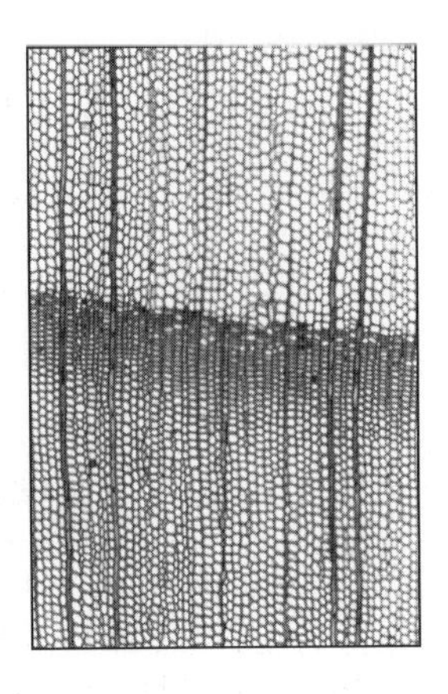

There are no known seasoning problems with this timber. Average reported shrinkage values (green to ovendry) are 4.6% radial and 6.9% tangential.

DURABILITY

Port-Orford-cedar is extremely durable. In the past, it was used for making caskets and large quantities were exported to the Far East for this purpose. It was also used as storage battery separators because of its resistance to acid. Serving as a testimony to its value in a marine environment, Sir Thomas Lipton, in the early part of the last century, had his Cup-challenger yachts built of Port-Orford-cedar.

WORKABILITY

Because of its even grain and medium density, Port-Orford-cedar is a woodworker's dream. It cuts easily and can be trimmed, carved and shaped with hand tools with little effort. It takes a high polish and stains and paints extremely well.

USES

Because of its limited supply, the former uses for railway ties, mining timbers and venetian blinds have been largely curtailed. It remains a premier wood for cabinetry as well as for arrows. To some extent, the wood is utilized as soundboards for stringed musical instruments.

SUPPLIES

Supplies are very limited and therefore quite expensive. Limited supplies can be found where the tree grows, but this is mostly culled from prime grade saw logs that are exported to Japan.

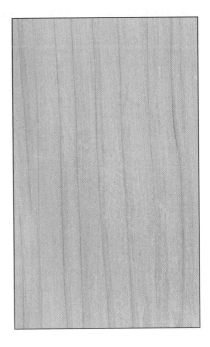

Chamaecyparis nootkatensis

Alaska-cedar

by Jim Flynn

SCIENTIFIC NAME

Chamaecyparis nootkatensis. Derivation: The genus name is the Greek name for lavender-cotton or ground cypress, a dwarf shrubby Old World composite which resembles a dwarf cypress. The specific epithet means the plant is "of Nootka Sound", Vancouver Island, British Columbia, in recognition of where it was discovered.

FAMILY

Cupressaceae, the cypress family.

OTHER NAMES

Alaska yellow-cedar, yellow-cedar, Nootka cypress, Sitka cypress, yellow cypress, Alaska cypress.

DISTRIBUTION

This is a tree of the North American Pacific coast. It can be found from south Alaska, in western British Columbia and in the mountains of western Washington and Oregon. Also in southeastern British Columbia, northeastern Oregon in the Blue Mountains and in the extreme northern portion of California in the Siskiyou Mountains.

THE TREE

Reaching heights of 75 to 80 feet or more with a diameter of 2 to 3 or more feet, Alaska-cedar has an open, narrowly conical crown with drooping branches. Trees growing in dense forest areas often are branchless from the ground to about 30 feet or more making for knot-free saw logs. The main trunk is frequently not straight and has one or two slight bends. Because of its thin bark, it seldom survives a forest fire. The scale-like leaves have very distinctive sharp spreading points and are harsh and prickly to the touch. The cones, which ripen in late September or early October, are a deep russet-brown with a conspicuous bloom. Under each cone scale, two to four seeds are borne. The tree grows very slowly.

THE TIMBER

Considering that Alaska-cedar is a softwood, its weight is high with an average reported specific gravity of 0.42 (ovendry weight/green volume), equivalent to an air-dried density of 31 pcf. It is aromatic and sulfur yellow in color. The wood is finely textured.

SEASONING

There are no known unusual seasoning problems with Alaska-cedar. Average reported shrinkage values (green to ovendry) are 2.8% radial, 6.0% tangential and 9.2% volumetric.

DURABILITY

Alaska-cedar is highly resistant to decay. Logs that are known to have lain in the forest for decades have been found to be perfectly sound. Immersion in salt water does not harm the wood.

WORKABILITY

The wood is easily worked and takes a fine finish. Unfinished, it wears smooth with use.

USES

The wood is not overly plentiful so it finds special uses befitting its high quality. Early generations of Native Americans on the northwest coast used it for paddles to propel their western redcedar (*Thuja plicata* WDS 268) canoes. It is a superb wood for interior finishes and cabinetry. In boat building, it is most useful for decking, railings and interior paneling. While the wood has excellent tone, it has limited use in stringed instruments because of its color and the relatively indistinct growth lines.

SUPPLIES

Because U.S. domestic demand is low, it is cut and exported in the round directly to Japan. The wood is very expensive. Most wood dealers on the west coast stock it. A great place to acquire some of the wood is amongst the driftwood on the beaches of Oregon. In most cases, permission should be sought from either state or local land owners to gain access to the beaches and remove the wood.

Chlorocardium rodiei

greenheart

by Max Kline

SCIENTIFIC NAME

Chlorocardium rodiei. Derivation: The genus name is Latin for green heart. The specific epithet is in honor of Cecil Rhodes (Rodi in Italian) (1853 to 1902), an English colonial capitalist and administrator in South Africa. Synonyms are *Ocotea rodiei* and *Nectandra rodiei*.

FAMILY

Lauraceae, the laurel family.

OTHER NAMES

bebeere, demerara, bibira, sepira, sipiroe.

DISTRIBUTION

Guyana, Suriname, French Guiana and northern Brazil.

THE TREE

Greenheart is a large evergreen tree 75 to 125 feet in height with an average diameter of 3 feet. It has a straight, cylindrical trunk that is free from buttresses. The bole is 50 to 80 feet in length.

THE TIMBER

Greenheart wood has outstanding strength, straight grain, a fine and even texture and freedom from knots and other defects. The heartwood is light to dark olive green, sometimes marked with brown or black streaks. The sapwood is pale yellow or greenish gradually blending into the heartwood. Average reported specific gravity is 0.80 to 0.91 (ovendry weight/green volume), equivalent to an air-dried weight of 64 to 74 pcf. It has no odor or taste. In comparison to oak, greenheart is twice as hard and more than twice as strong in bending and in compression along the grain. It is also more than twice as stiff and twice as high in resistance to shock loads.

SEASONING

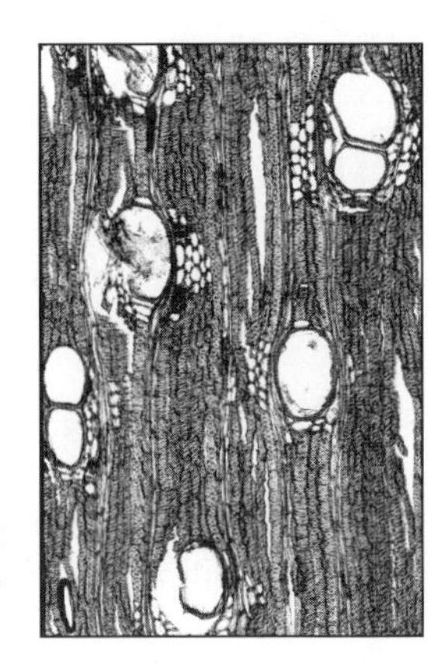

Fresh cut timber can have a very high moisture content and requires careful seasoning. With thick timbers, it is best to partially air dry before final kiln drying. Distortion is not serious, but checking and splitting tends to be severe during seasoning. Average reported shrinkage values (green to ovendry) are 8.8% radial, 9.6% tangential and 17.1% volumetric.

DURABILITY

Greenheart is little affected by fungal attack and is almost impervious to shipworm damage, especially in saltwater environments.

WORKABILITY

This timber is fairly hard to work with hand or machine tools and has a dulling effect on tool edges. The fine dust formed when working with these species affects some operators by irritating the mucous membrane. Pre-boring is necessary when using screws because of the hardness of the wood and its tendency to split. Most finishing treatments can be used satisfactorily, and gluing properties are generally good. It is not a timber recommended for steam bending processes or turnery. Large timbers burn quite slowly because of the high density of the wood.

USES

Most greenheart is used in its growth area for construction work including docks, ships, pilings, flooring, piers, sluice gates, bench slats, picket fences, wheel spokes, house posts and boardwalks. A unique use is in the manufacture of fishing rods since it will bend to a great extent before breaking.

SUPPLIES

In the past, the principal market was Europe with only small supplies coming into the United States. It is only available as lumber from a few of the leading importers and is considered expensive.

Chloroxylon swietenia

Ceylon satinwood

by Max Kline

SCIENTIFIC NAME

Chloroxylon swietenia. Derivation: The genus name is from the Greek *chloros* meaning yellow green and *pherin* meaning to carry, referring to the yellowish-green inflorescences. The specific epithet is in honor of Gerard von Swieten (1700 to 1772), a German physician.

FAMILY

Rutaceae, the rue or citrus family.

OTHER NAMES

East Indian satinwood, buruta, flowered satinwood.

DISTRIBUTION

Sri Lanka, southern India.

THE TREE

Ceylon satinwood grows to a maximum height of 60 feet but the average height is between 45 to 50 feet with a diameter of 12 inches. In Sri Lanka, specimens with diameters of 3 feet are sometimes found. The bole is straight and cylindrical and about 10 feet long. The tree is common in the drier, deciduous forests of India and Sri Lanka.

THE TIMBER

Ceylon satinwood is pale to golden yellow, frequently with darker streaks. Although somewhat lighter in color, the sapwood is rarely clearly distinguishable from the heartwood. The odor is fragrant, and the taste is not distinct. The luster is high and satiny. The grain is mostly wavy producing a narrow ribbon figure. The texture is close and uniform. Average reported specific gravity is 0.80 (ovendry weight/green volume), equivalent to an air-dried weight of 64 pcf. While not a major consideration for the purposes for which this timber is likely to be used, it does have a high strength factor.

SEASONING

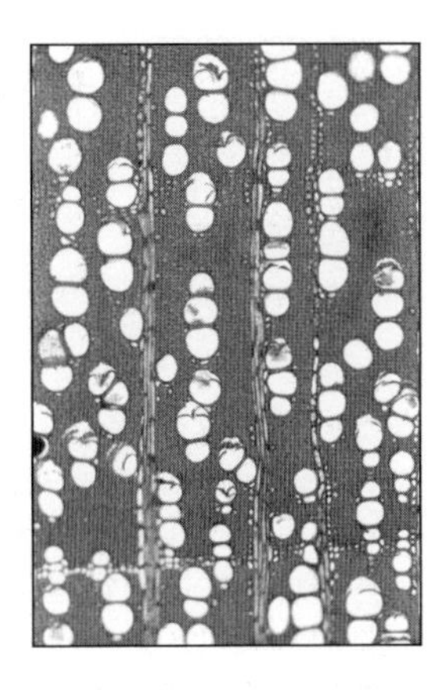

Ceylon satinwood timber has a decided tendency to surface checking and to a lesser degree to twist and warp. Average reported shrinkage values (green to ovendry) are 5.5% radial and 7.1% tangential. Seasoning in log form has given good results and girdled trees show less checking. The timber tends to dry rather slowly so kiln drying will speed up the drying process. Once seasoned, dimensional stability is good.

DURABILITY

Resistance to all types of fungal attack is high. A certain amount of beetle damage may occur with seasoned stock, but it is neither extensive nor prominent.

WORKABILITY

The timber is fairly difficult to work with hand tools and moderately hard to saw and machine. It has a dulling effect on tools. Due to its high density, the wood must be held firmly in machinery to prevent chattering. It turns excellently. Nail and screw holding properties are good, and glue adheres satisfactorily. Ceylon satinwood responds very well to stain and finish treatments.

USES

Primarily Ceylon satinwood is used for cabinetry. It is also used for tool handles, brush backs, flooring, veneer and plywood and for carving and turnery for fancy products. In veneer form, it is used for inlays and decorative purposes.

SUPPLIES

Due to limited supplies, Ceylon satinwood is considered an expensive timber in lumber form.

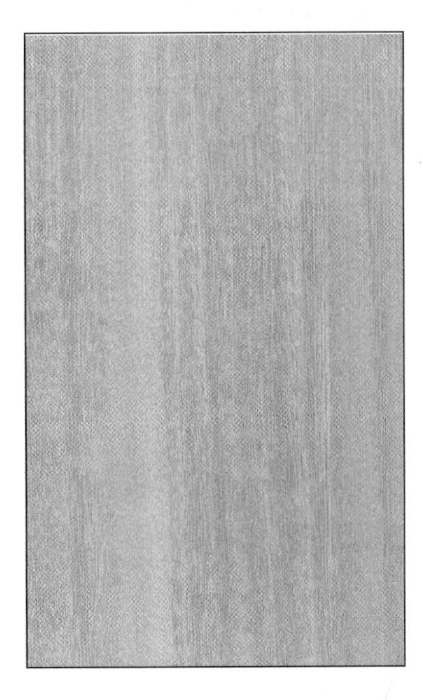

Chrysolepis chrysophylla

giant chinkapin

by Jim Flynn

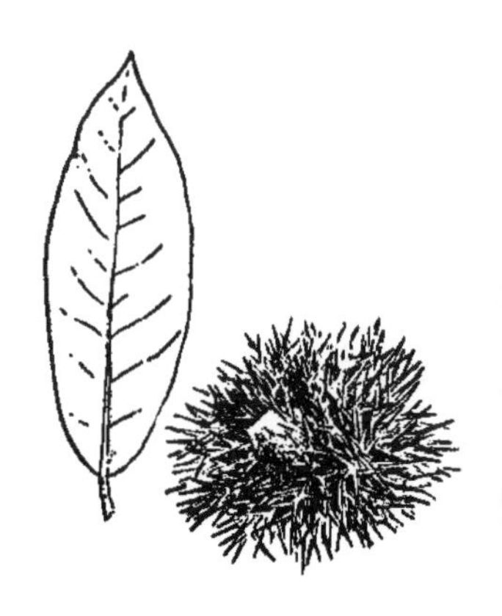

SCIENTIFIC NAME

Chrysolepis chrysophylla. Derivation: Both the genus name and the specific epithet mean golden scaled referring to the thick coating of minute golden scales on new shoots and the underside of the leaves. A shrub-like chinkapin is identified as C. *chrysophylla* var. *minor*. A synonym is *Castanopsis chrysophylla*.

FAMILY

Fagaceae, the beech family. Giant chinkapin is closely related to the chestnuts and oaks in the genera *Castanea* and *Quercus*, respectively, which are also in this family.

OTHER NAMES

goldenleafed chinkapin, golden chinkapin, evergreen-chinkapin, western chinkapin and chinkapin.

DISTRIBUTION

It can be found in the Pacific coast region of the United States from southwest Washington to western Oregon and in the coast ranges to central California. It is endemic to the Sierra Nevadas of central California. The majority of the other 30 species of *Castanopsis* are native to tropical and sub-tropical mountains of Asia and may be found as ornamentals elsewhere.

THE TREE

When approaching old age at 400 to 500 years, giant chinkapin can be 115 feet or more in height yet be dwarfed by redwoods (*Sequoia sempervirens* WDS 247) which are scattered throughout its range. Occasionally, it is found in pure stands. When fully grown, giant chinkapin develops a fluted trunk and is free of branches from one-half to two-thirds of its length. The bark of young trees is thin, smooth and dark gray. As the tree matures, the bark, which consists of very wide plates that have a reddish-brown exterior and are brilliant white within, becomes thicker (0.75 to 1.5 inches or more) and slightly scarred. The thick, leathery evergreen leaves persist about 3 years and are shiny yellow-green on top. Underneath they are coated with minute golden yellow scales that extend to the leaf stems. The leaves are 2.5 to 3.5 inches long and on vigorous shoots extend 4 to 6 inches. The flowers open in early summer and throughout the season into mid-winter. The fruit matures in the autumn of the second season when the spiny burs split open into four divisions to liberate an edible nut.

THE TIMBER

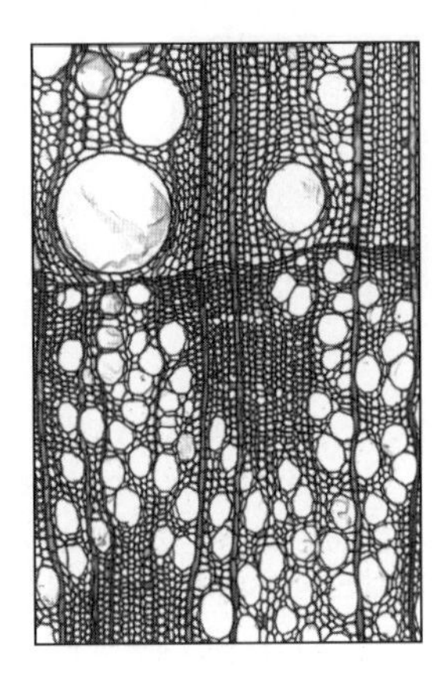

The narrow sapwood is light brown with a pinkish cast and is barely distinguishable from the heartwood. The grain is fairly straight. There is neither characteristic taste nor odor. Average specific gravity is 0.42 (ovendry weight/green volume), equivalent to an air-dried weight of 31 pcf. The wood is ring porous with conspicuous growth rings. The rays are very fine and barely visible with a hand lens.

SEASONING

According to the U.S. Forest Service, giant chinkapin is one of the most difficult of the U.S. hardwoods to cure. It tends to check badly. Shrinkage values are not available.

DURABILITY

The timber is not durable under conditions favorable to fungi.

WORKABILITY

Properly seasoned, the wood has excellent working qualities. Machining, gluing, sanding and finishing are straightforward and easy to perform. The wood holds a nice crisp edge. If supplies were plentiful, it would rate high on the scale of favorite woods.

USES

References from the turn of the century indicate the primary use was for heating and campfires, which is a shame. Currently, there is a small market for use in furniture, cabinetry, paneling and surface veneer. Other uses, such as turnery and craftwork, are surely unrecorded.

SUPPLIES

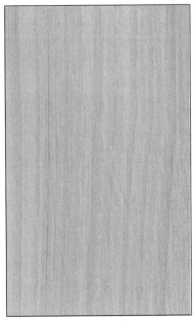

Sawlogs make ideal candidates for the mill, but so does its redwood neighbor. Timber merchants in growth areas know the value even though limited quantities prevent it from becoming a widely known commercial timber. In addition, transportation costs to mills capable of processing the timber often prohibit it from being extracted from the forest. Therefore, very little is on the commercial market. Wood can be obtained in the growth area with a little searching. IWCS wood sample suppliers have access to limited supplies.

Cinnamomum camphora

camphor-tree

by Max Kline

SCIENTIFIC NAME

Cinnamomum camphora. Derivation: The genus name is the ancient name, coming into Latin through Hebrew and Greek. The specific epithet is an ancient name for camphor. A synonym is *Laurus camphora.*

FAMILY

Lauraceae, the laurel family.

OTHER NAMES

kusonoki, ohez, kalingag, dalchini, kayu.

DISTRIBUTION

Japan, Taiwan, China, and southward to Australia. [Ed. Note: Widely planted in tropical and sub-tropical regions, it may be found in southern parts of the United States but not of the stature as found in the native area.]

THE TREE

Camphor-tree is medium sized generally 60 to 100 feet in height with a straight, cylindrical bole 40 feet in length. Trunk diameters range from 2 to 4 feet. The leathery evergreen leaves are 2 to 5 inches long and have a distinct camphor-like smell when broken. The very small (0.1 inch) yellow flowers are produced in the axils of the leaves. The globe-shaped fruit is equally small measuring only 0.4 inch.

THE TIMBER

Camphor-tree timber is light yellowish-brown or light pinkish or reddish-brown usually with darker streaks. The heartwood and sapwood are not sharply demarcated. The grain is straight, interlocked or wavy with a medium coarse to fine texture. The typically pleasant and spicy odor is of camphor or anise oil. Average reported specific gravity varies from 0.35 to 0.50 (ovendry weight/green volume), equivalent to an air-dried weight of 26 to 38 pcf. The luster is high, and taste is not distinctive.

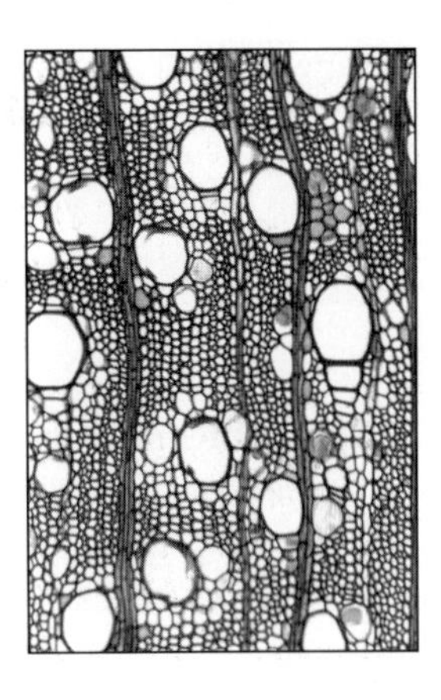

SEASONING

The timber will air dry with little or no degrade, although warping is sometimes encountered. Average reported volumetric shrinkage (green to ovendry) is 7.4%. After drying, the wood is quite dimensionally stable in use.

DURABILITY

Camphor-tree is reported to be durable in contact with the ground, except when attacked by termites.

WORKABILITY

This light and soft wood works well both with hand and machine tools and saws easily. It finishes smoothly with a high degree of luster.

USES

In the Orient, camphor-tree is used for trunks, chests and coffins. Before the advent of synthetic camphor, the tree was the source of the camphor of commerce, which was obtained by destructive distillation. Figured veneer is used for fine cabinetry.

SUPPLIES

Camphor-tree is only available on the world market in very small quantities, almost entirely in the form of veneer. Planted trees, however, are sometimes cut when grown in tropical or subtropical parts of the world, for example in Florida, where wood collectors can often obtain specimens.

Cladrastis lutea

yellowwood

by Max Kline

SCIENTIFIC NAME

Cladrastis lutea. Derivation: The genus name is Greek for branch and little. The specific epithet is Latin for the color yellow. A synonym is C. *kentukea*.

FAMILY

Fabaceae or Leguminosae, the legume family; (*Papilionaceae*) the pea or pulse group.

OTHER NAMES

gopherwood, yellow ash, yellow locust, virgilia, American yellowwood.

DISTRIBUTION

Extreme western North Carolina, western slopes of the high mountains of Tennessee and Kentucky, northern Alabama, southern Missouri and northern Arkansas. Due to its use as an ornamental, it is often planted over a much wider range than found naturally.

THE TREE

The yellowwood tree is a medium-sized deciduous tree, sometimes 50 to 60 feet in height with a wide spreading crown and a trunk 18 to 24 inches in diameter with smooth bark. Because of its delicate white flowers that hang in loose clusters 12 to 14 inches long, yellowwood is often planted as an ornamental.

THE TIMBER

The heartwood is bright yellow, becoming brownish upon exposure. The sapwood is nearly white. The wood is hard with an average specific gravity of 0.47 to 0.48 (ovendry weight/green volume), equivalent to an air-dried weight of 40 pcf. Taste is not distinctive. The luster is rather high, and the texture is moderately coarse. The grain is straight to irregular. This timber is quite tough and strong.

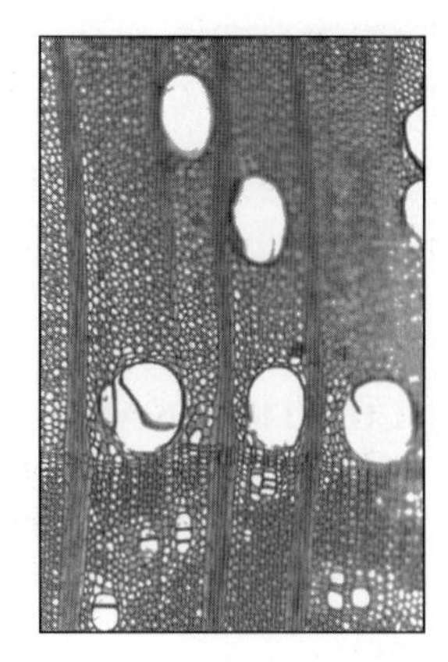

SEASONING

Yellowwood can be air dried, but has a tendency to check moderately. Shrinkage values are not available. After seasoning, it is dimensionally stable in use.

DURABILITY

The wood is of medium durability.

WORKABILITY

Yellowwood is not difficult to work and finishes smoothly. The bright yellow color that makes the wood attractive, however, turns to a dull brown color with age. Knowing this, it is best to stain the wood before applying the final finish.

USES

Due to its scarcity, the timber is not used often. Occasionally it has been made into gun stocks. The heartwood yields a yellow dye. Yellowwood makes an attractive contrast to woods of other colors when used in laminated craftwork.

SUPPLIES

This wood is not commercially available, but samples for wood collections are sometimes obtainable from other IWCS members living in the growth area of the tree.

Cocos nucifera

coconut

by Jim Flynn

SCIENTIFIC NAME

Cocos nucifera. Derivation: The genus name *cocos* is a Portuguese word meaning monkey, referring to the three pores or "eyes" that resembles a monkey's face on a coconut that has had the outer husk removed. The specific epithet means nut-bearing.

FAMILY

Palmae, the palm family; also Arecaceae.

OTHER NAMES

Coconut seems to be the worldwide common name.

DISTRIBUTION

This is a tropical tree growing worldwide in moist areas especially near salt water. It probably had its origin in the vast stretches of the Pacific Ocean. Reports that it was native in the New World in pre-Colombian time have been rejected. Because the fruit is not affected by saltwater immersion, ocean currents have carried them far and wide thus naturalizing trees in widely separated venues. In Florida, the large population of coconut palms is attributable to the wreck of a ship near Fort Walton. The cargo of coconuts was lost and washed ashore in 1878. It was estimated that, as a result, in 10 years there were over 500,000 trees growing in Florida.

THE TREE

The coconut is ranked as one of the 10 most important trees in the world. The stately fashion in which the tree grows, often to a height of 100 feet or more, is familiar to almost everyone and symbolizes tropical paradises. The graceful, leaning and swaying trunk grows clear to the top, which is crowned with long, pinnate leaves and clusters of coconuts. They are covered with a light brown fibrous husk from 8 to 12 inches long, which often fall to the ground when one is not expecting it. The nuts are rich in oil and find use in making margarine and other cooking products including the delicious sprinkling on top of richly applied cake frosting. Unfortunately, many trees in Florida and elsewhere are being killed by a lethal yellowing disease caused by a microplasm. A dwarf variety from Malaysia is replacing them.

THE TIMBER

The wood in the palm family is considerably different from that of our familiar softwood and hardwoods. The stems contain scattered vascular strands embedded in soft tissues. The wood on the outer perimeter of the trunk is much denser than that in the center, which is sometimes very fuzzy or pithy. Average specific gravity is about 0.56 (ovendry weight/green volume), equivalent to an air-dried weight of 43 pcf.

SEASONING

There does not appear to be any problem in the transition from a wet to a dry condition if care is taken to dry slowly and stacks are weighted. A number of slabs I obtained from trees in Florida showed a tendency to warp slightly. Shrinkage values are not available. The timber can be treated with preservatives.

DURABILITY

Timber in contact with the ground has a poor durability record. It is not susceptible to attack by powder post beetles.

WORKABILITY

Slab cutting and surface planing old logs is rather difficult because of the mixed density of the wood. It is difficult to turn but sands very well. It cannot be peeled for veneer. Nailing often results in splitting. The key to its workability is in shaping it first with sharp cutting tools and concentrating on a sanding operation to work the wood to a smooth surface. It finishes well and produces a unique and pleasing pattern.

USES

In spite of its poor durability rating, coconut is used for posts in the tropics. It was used in many places as defense revetments and bunkers on the islands of the Pacific during WWII. When dry, it makes good firewood. Uses include walking sticks, light flooring, turnery and miscellaneous joinery. Woodworkers find the unusual composition of the wood useful in making attractive novelties such as small boxes and jewelry.

SUPPLIES

There is evidence that demand for coconut wood is increasing and limited quantities are appearing on the commercial market. It is certainly abundant in many parts of the tropics. Contact with IWCS members in the growth areas should produce results.

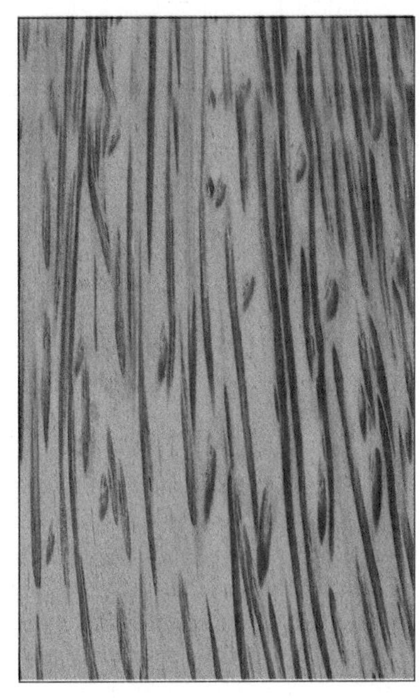

Cordia dodecandra

sericote

by Alan B. Curtis

SCIENTIFIC NAME

Cordia dodecandra. Derivation: The genus name is in honor of Euricius Cordus (1486 to 1535) and his son, German physicians and botanists. The specific epithet is Latin for the 12 stamens.

FAMILY

Boraginaceae, the borage family.

OTHER NAMES

ziricote.

DISTRIBUTION

This tree is native to southern Mexico (Yucatan Peninsula, westward into the state of Chiapas), Belize and Guatemala.

THE TREE

Sericote grows throughout tropical deciduous forests, usually at low elevations but occasionally to elevations of 1,500 feet in the mountains. It is not a rainforest inhabitant but does need at least 40 inches of rainfall annually. In an undisturbed mature forest, it may be one of the dominant trees with a straight trunk 60 to 90 feet in height and occasionally diameters of 30 inches. Over much of its range, however, the tree is considerably smaller in size. The simple, oval-shaped leaves have a very rough surface. The showy, orange-red tubular flowers grow in clusters, and each flower measures 2 inches across its petals. The fruit is egg-shaped, nearly 2 inches long, yellowish and edible. The tree flowers and fruits throughout the year.

THE TIMBER

Sericote wood can be immediately recognized by its heartwood which is various shades of black with irregular wavy black streaks, lines and variegations on a tan background. The black markings may curve, run diagonally, or may be at nearly right angles to the main axis of the tree. This author does not know of any other timber (except a few other species of *Cordia*) which produces such attractive and distinctive markings. The white sapwood is often left on the boards to provide a handsome contrast to the black heartwood. The wood is hard, heavy and strong. It has a more or less oily or waxy appearance, a medium luster, straight grain, and attractive ray flecks on the tangential surface. The texture is fine to medium and not always uniform. When freshly cut, the wood has a slight odor. The specific gravity is 0.63 to 0.84 (ovendry weight/green volume), equivalent to an air-dried weight of 49 to 68 pcf.

SEASONING

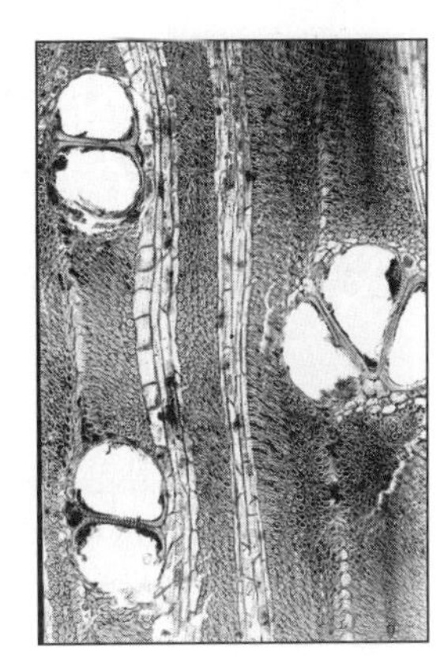

Sericote is difficult to dry as it readily develops surface checks and end splits. Average reported shrinkage values (green to ovendry) are 4.2% radial and 6.2% tangential. The wood air dries slowly.

DURABILITY

The wood is moderately durable when exposed to the weather.

WORKABILITY

Sericote is not difficult to work and carves and turns nicely. It finishes very smoothly and attractively and takes a high polish. The wood is stable in use.

USES

Uses include fine furniture, cabinetry, turnings, rifle butts, doors, paneling, flooring and face veneer on plywood. Sometimes it is cultivated for its fruit, which can be eaten raw or made into a conserve. In the past, the bark and wood were used to make a cough syrup and the very rough leaves were used as sandpaper. Today, the tree is sometimes used for street and yard plantings.

SUPPLIES

Supplies are very limited due to scattered occurrence and the small size attained by most trees. In its native area, the wood is usually available at local sawmills, although the quality may be poor. Standing dead and down trees in the forest provide good wood, and pieces can sometimes be purchased from local residents. The tree is a slow-grower and has not been tried in plantations. Small quantities of sericote lumber are occasionally imported into the United States and sold at a very high price.

Cordia gerascanthus

canalete

by Max Kline

SCIENTIFIC NAME

Cordia gerascanthus. Derivation: The genus name is in honor of Euricius Cordus (1486 to 1535) and his son, German physicians and botanists. The specific epithet refers to the many prickles on the leaf.

FAMILY

Boraginaceae, the borage family.

OTHER NAMES

amapa, baria, ziricote, canaletta, lauro pardo, lauro Negro.

DISTRIBUTION

Florida, West Indies, Central America and southward to Brazil and Argentina.

THE TREE

Canalete is a small to large tree, sometimes growing to 100 feet in height. It is found in tropical dry zones and sparsely distributed over wide areas. The entire, elliptical, alternate leaves are 4 inches long and 3 inches wide with acuminate tips and acute bases. The white flowers are small usually with seven petals on a tubular base.

THE TIMBER

The heartwood is reddish-brown to dark brown with irregular dark brown or blackish streaks. It is sharply demarcated from the grayish or yellowish sapwood. The odor is mildly fragrant. Taste is not distinct, and the luster is medium. It has an oily or waxy appearance with a fine to medium texture. Average reported specific gravity ranges from 0.63 to 0.84 (ovendry weight/green volume), equivalent to an air-dried weight of 49 to 68 pcf. The grain varies from straight to wavy.

SEASONING

The wood is difficult to dry. It readily develops surface checking and end splitting, but once dried it holds its shape well. Average reported shrinkage values (green to ovendry) are 4.0% radial, 7.4% tangential and 11.6% volumetric.

DURABILITY

The durability of canalete is rated high.

WORKABILITY

Canalete is a readily worked timber, finishing very smoothly with a very attractive appearance.

USES

Canalete is highly esteemed in its growth range for joinery, furniture and house construction. Exported timbers are used for brush backs, turned articles, cabinetry, veneer, rifle stocks, handles and general millwork.

SUPPLIES

Since the timber is quite scarce, the availability is very limited. When found, it is expensive.

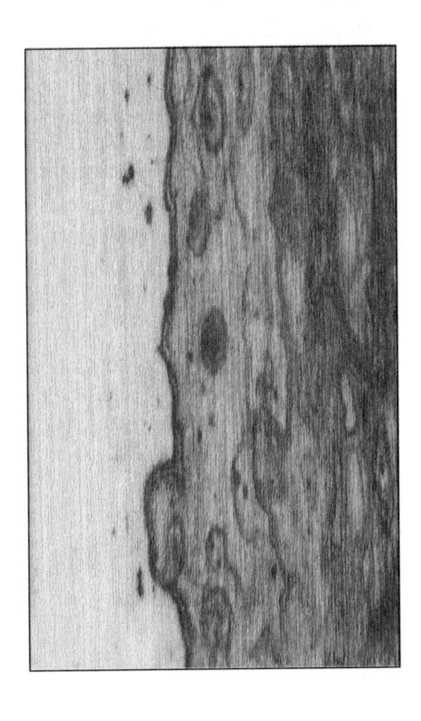

Cordia goeldiana

freijo

by Max Kline

SCIENTIFIC NAME

Cordia goeldiana. Derivation: The genus name is in honor of Euricius Cordus (1486 to 1535) and his son, German physicians and botanists. The specific epithet is in honor of Emil A. Goeldi (1859 to 1917), a naturalist famous for Goeldi's monkey, Callimico goeldi.

FAMILY

Boraginaceae, the borage family.

OTHER NAMES

South American walnut, Jenny wood, cordia wood.

DISTRIBUTION

Occurs in the tropical forests of Brazil, mainly in the Amazon basin.

THE TREE

Freijo is of medium size, frequently 40 to 60 feet in height, with diameters of 18 to 24 inches. It grows in moist soil but is absent from river locations which are subject to flooding.

THE TIMBER

The heartwood is yellowish to brown sometimes marked with darker streaks. The darker colored heartwood is distinct but not usually sharply demarcated from the lighter sapwood. The luster of good quality lumber is rich and golden when viewed in proper lighting. The taste is not distinctive, and the odor is spice-scented. The texture is uniform but coarse, and the grain is generally straight. In general, the wood resembles teak (*Tectona grandis* WDS 261) in appearance. In strength, freijo is not essentially different from teak, except it is superior in toughness. The bending properties are suitable for moderate radii of curvature. Average reported specific gravity varies from 0.44 to 0.52 (ovendry weight/green volume), equivalent to an air-dried weight of 33 to 40 pcf.

SEASONING

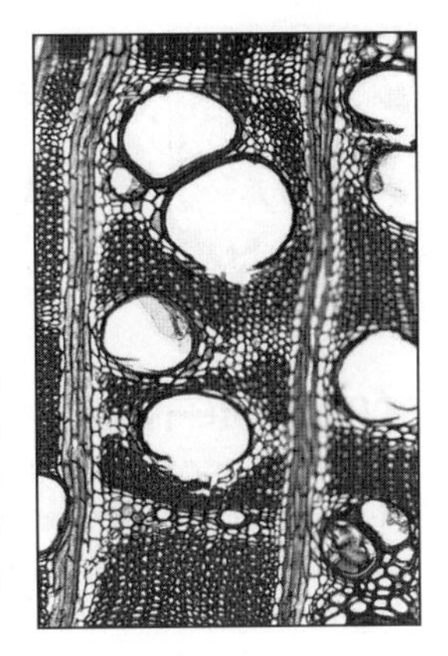

This lumber can be air or kiln dried with only minimal end splitting and practically no distortion. Average reported shrinkage values (green to ovendry) are 3.4% radial, 7.1% tangential and 9.2% volumetric.

DURABILITY

Freijo is reported to be very durable upon exposure to white or brown-rot fungi, but a certain amount of pinhole woodborer damage can occur. The wood has good weathering resistance and absorbs moisture at a moderate rate. The heartwood is not receptive to preservative treatment.

WORKABILITY

Since it is rather brittle, freijo tends to split when nailing, and the use of thin-gauge nails coupled with pre-boring is advisable. Nail and screw holding properties are good, and glue adheres satisfactorily. It is not a recommended species for turnery. Keen-edged tools are essential for good results when working the wood. Stains dry evenly, but a fair amount of filling is needed prior to polishing, after which a good glaze can be obtained with a beautiful luster.

USES

This timber is essentially a cabinet wood and used for furniture and interior joinery. In some cases, it has been a substitute for teak in shipbuilding. In the cooperage trade, it has been employed for wine casks and barrels. In Brazil, it is used for general construction and flooring.

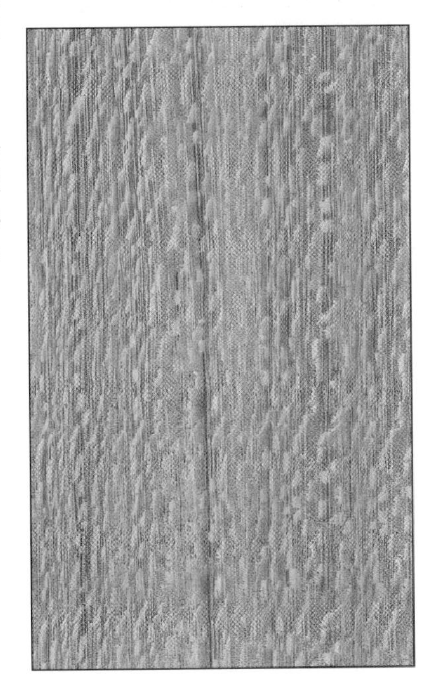

SUPPLIES

Supplies of freijo are adequate, but only very limited quantities reach the United States and Europe. Therefore, when available, the price is higher than expected for a timber of this quality.

Cornus florida

flowering dogwood

by Max Kline

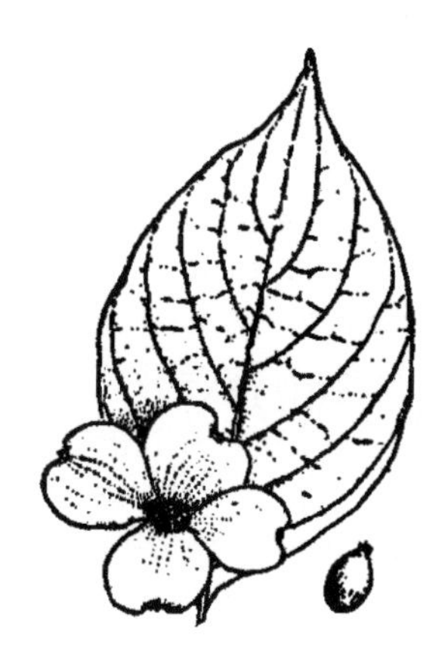

SCIENTIFIC NAME

Cornus florida. Derivation: The genus name is Latin for a species of Cornelian-cherry in Europe. It is derived from the word horn, which refers to the hardness of the wood. The specific epithet means flowering, referring to the showy petal-like bracts.

FAMILY

Cornaceae, the dogwood family; also Corylaceae.

OTHER NAMES

arrow-wood, cornel, false boxwood, Florida dogwood, boxwood, white cornel.

DISTRIBUTION

From eastern Texas to central Florida northward throughout the Mississippi Valley and the southern Appalachians to southern Missouri and Michigan, lower Ontario, central New York and southern Maine. It is also found on the uplands of northern Mexico.

THE TREE

The flowering dogwood is best known for its large, attractive, white, pinkish and rose colored blossoms which appear in the spring and for its brilliant scarlet-colored autumnal foliage. It is a slow-growing tree seldom more than 40 feet in height and has a short trunk 8 to 16 inches in diameter. The reddish-brown bark is only about 0.25 inch thick and has the texture of alligator hide. The tree thrives best in rich well-drained soils or along the banks of streams. Flowering dogwood is the state tree of Virginia.

THE TIMBER

The rather small heartwood of flowering dogwood is reddish-brown to light chocolate in color, sometimes streaked with mottled white lines. The wide sapwood has a creamy white to pinkish cast. The wood texture is fine and uniform. The interlocked grain is very fine, hard and compact. Average reported specific gravity is 0.64 (ovendry weight/green volume), equivalent to an air-dried weight of 50 pcf. This wood is classified among the best for strength, firmness and hardness.

SEASONING

Flowering dogwood is sometimes difficult to air dry without checking unless the process is done slowly under controlled conditions. Average reported shrinkage values (green to ovendry) are 7.1% radial, 11.3% tangential and 19.9% volumetric.

DURABILITY

There are no known uses for flowering dogwood in situations exposed to fungi and insects, consequently, no reports of durability have been found.

WORKABILITY

Flowering dogwood finishes with glossy smoothness. Even though it is a hard wood, it can be sawn, turned and planed with ease because of its close-grained characteristics.

USES

Probably 90% of all flowering dogwood was used commercially for the manufacture of shuttles for textile weaving. Its hard, close-grained features have little wearing effect upon the threads. It is also used for spool and bobbin heads, small pulleys, skewers, golf club heads and tool handles. Native Americans made a scarlet dye from the bark, and the early U.S. colonists made a medicine for the treatment of fevers.

SUPPLIES

Because of the scarcity of supplies of clear wood, flowering dogwood is very valuable and sometimes sold by the pound. It has not been introduced to a great extent into the home workshop but it does have important advantages where a hard, smooth wood of pleasing color is desired. Wood collectors have no problem in procuring samples for their collections, but the availability as lumber is scarce and it is high priced.

Cotinus obovatus

American smoketree

by Max Kline

SCIENTIFIC NAME

Cotinus obovatus. Derivation: The genus name is from the Greek *kotinos*, the name for olive probably in reference to the shape and size of the fruit. The specific epithet is Latin for obovate, referring to the shape of the leaves.

FAMILY

Anacardiaceae, the cashew family.

OTHER NAMES

chittamwood, mist tree, yellowwood, smoketree.

DISTRIBUTION

American smoketree grows in southeastern Tennessee, northern Alabama southward into northcentral Georgia, southwestern Missouri, the northwestern tip of Arkansas, in a small section of northeastern Oklahoma, and in a small area of south-central Texas. As a cultivated ornamental, it is planted in many states farther north.

THE TREE

At its best, American smoketree rarely exceeds 30 feet in height with a trunk 12 to 14 inches in diameter. The small, yellow or purplish flowers are most attractive. The oval-shaped leaves, 4 to 6 inches long and 2 to 3 inches wide, turn brilliant shades of orange and scarlet in the autumn. The billowy mass of green, yellow, and red has the appearance of a hazy puff of smoke and from this the common name smoketree is derived.

THE TIMBER

The wood is moderately soft, light and close grained. The heartwood is bright, clear, rich orange to light yellow with a narrow, nearly white sapwood. It has an attractive figure when plainsawn. The luster is fairly high, and odor and taste are not distinct. The texture is rather coarse and uneven. A standard 3- by 6- by 0.5-inch sample had a specific gravity of 0.50 (ovendry weight/green volume), equivalent to an air-dried weight of 38 pcf.

SEASONING

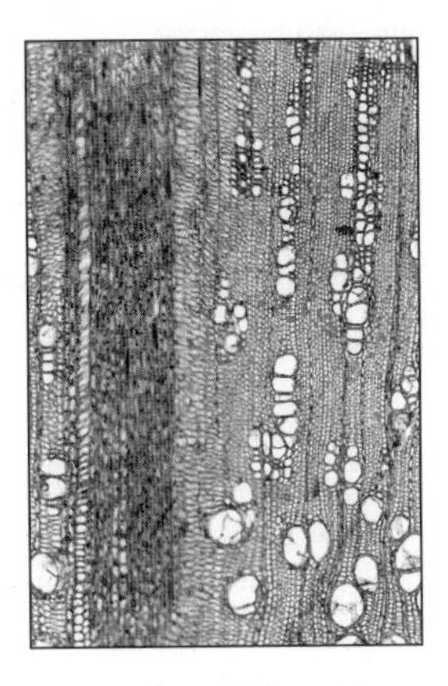

Because American smoketree is not a commercial wood, kiln drying is not used. Air drying of wood by the hobbyist is easily accomplished. Shrinkage values are not available.

DURABILITY

American smoketree is highly resistant to decay when in contact with the soil leading to its use, where available, for fence posts.

WORKABILITY

American smoketree is easily worked with tools, but tends to split when nailing. It holds screws well and is easily glued. In order to highlight and preserve the natural color of the wood, it is best finished with a product that does not contain a stain as an additive.

USES

In its growth range, it is used for fence posts and a dye is obtained from the rich orange heartwood. There are no commercial uses. Home woodworkers, however, should utilize this wood for its unusual color.

SUPPLIES

Since the tree is not widely distributed, it may be classified as rare. The only supply source is wood collectors who have access to trees.

Cyrilla racemiflora

swamp cyrilla

by Jim Flynn

SCIENTIFIC NAME

Cyrilla racemiflora. Derivation: The genus name is in honor of Domenico Cirillo (1734 to 1799), an Italian physician, botanist and patriot. The specific epithet is Latin and refers to flowers growing in clusters.

FAMILY

Cyrillaceae, the cryilla family. In this family there are two genera, *Cyrilla* and *Cliftonia.* The preferred common name for *Cliftonia* is buckwheat-tree. Both genera, however, are known in parts of the southern United States as titi.

OTHER NAMES

leatherwood, swamp leatherwood, southern leatherwood, red titi, white titi, black titi, swamp ironwood, he-huckleberry, palo colorado (Puerto Rico).

DISTRIBUTION

In the United States, along the coastal plain from Virginia to Texas. In the West Indies, in the mountains from Cuba to Puerto Rico and the Lesser Antilles. Then along the Atlantic coast of Central America from Belize to Nicaragua, and in northern South America from Guyana to Venezuela, Colombia and Brazil.

THE TREE

In the southeastern United States, swamp cyrilla will grow to as much as 30 feet in height along river banks and in swamps and lowland depressions. In other parts of its range, it grows taller. Trunk diameters range from 8 to 12 inches. In Puerto Rico, the tree grows to its maximum height of 60 feet with diameters of as great as 3 feet. It has attractive and graceful white flowers in elongated clusters made of many tiny flowers. Tiny fruits (0.1 inch long) make the tree easily identifiable. The bright, simple, alternate, green leaves turn orange in autumn. The tree often is called leatherwood because the base of the trunk is spongy and pliable.

THE TIMBER

The heartwood of swamp cyrilla is light brown but may vary in some locations to dark reddish-brown. The darker woods are rather oily. The sapwood is lighter and not well demarcated. The luster is low, and the wood is hard. Average specific gravity is about 0.70 (ovendry weight/green volume), equivalent to an air-dried weight of 55 pcf. The texture is fine and uniform. In appearance, swamp cyrilla is similar to maple and birch.

SEASONING

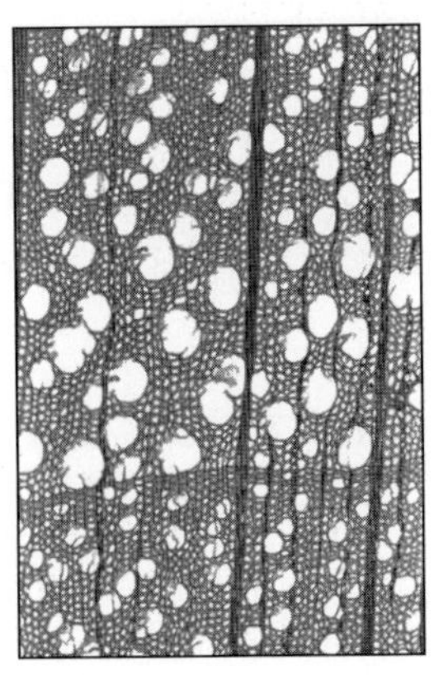

The wood is slow to dry. It also has a marked tendency toward warping, which is one of the main reasons why the wood is not popular. Shrinkage values are not available.

DURABILITY

Swamp cyrilla is susceptible to dry-wood termites but, otherwise, is durable. Most large trees are found with hollow trunks, a point to be considered in deciding its durability. Because this tree is not commercially marketed, there may be little data on the subject.

WORKABILITY

When properly seasoned, the wood is easily worked. It planes well and does not split when using screws. It can be bored, turned and finished with few problems.

USES

In Cuba, the darker wood is used to make fine furniture. Swamp cyrilla is suitable in applications where a hard wood is needed such as tool handles, small furniture and turnery. It is also an excellent fuel.

SUPPLIES

There is no record of this wood being available on the commercial market. It is used most often by woodworkers in the area where it grows and can be found readily.

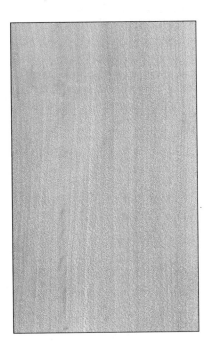

Dacryodes excelsa

gommier

by Jim Flynn

SCIENTIFIC NAME

Dacryodes excelsa. Derivation: The genus name is Greek for tears (as in crying). The specific epithet is Latin for tall or high. We can assume that this nomenclature refers to the "high and lofty" tree and the amber resin it exudes.

FAMILY

Burseraceae, the bursera or torchwood family.

OTHER NAMES

candle tree, tabanuco, tabonuco, gommier blanc, gommier montagne, bois cochon.

DISTRIBUTION

Puerto Rico and Lesser Antilles from St. Kitts to Grenada. Generally, the species can be found growing in small groups along upper slopes. On the island of Dominica, it forms almost pure stands at higher elevations.

THE TREE

Gommier is a very large erect tree, rising above the forest canopy and easily recognized on the mountainsides by its size and dark evergreen foliage. Early botanists referred to the species as "the most majestic tree of Puerto Rico". The tree grows to heights of 100 feet or more with a large trunk 3 to 5 feet in diameter. The short, broad buttresses are similar in appearance to an elephant's foot. It has a smooth whitish bark that peels off in thick flakes and exudes streaks of fragrant whitish resin when cut. This resin is an amber liquid when fresh and becomes white and hard after exposure to the air. The flammable exudate is used locally for caulking boats, candles and torches and for medicinal uses. The alternate pinnate leaves consist of five to seven elliptic leaflets that are 2.5 to 5 inches long and 1.75 to 3 inches wide. The small greenish flowers grow in clusters on lateral branches 3 to 8 inches long. The fruit is similar in appearance to an olive and has one seed.

THE TIMBER

The narrow sapwood is grayish and the heartwood, when first cut, is uniform brown with a pinkish cast but turns pinkish-brown when exposed. The wood is moderately heavy, moderately hard, tough and strong, of fine to medium uniform texture with wavy, interlocked grain and ripple marks but lacks growth rings. It has high luster, and there is no odor or taste. Average reported specific gravity is 0.52 (ovendry weight/green volume), equivalent to an air-dried weight of 40 pcf.

SEASONING

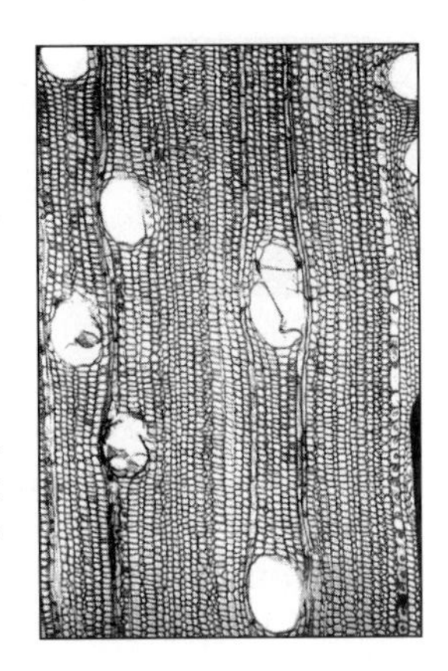

The wood seasons very rapidly with only minor degrade in the form of slight warping and end checking but no apparent surface checking. Average reported shrinkage values (green to ovendry) are 4.1% radial, 6.4% tangential and 10.5% volumetric.

DURABILITY

The heartwood is only slightly resistant to attack by fungi when exposed to the ground and is very susceptible to attack by dry-wood termites. It is not resistant to marine borer attacks.

WORKABILITY

Exercise caution with your cutting tools. The U.S. Forest Service reports that the wood has extremely high silica content. Nevertheless, it is a moderately good wood for machining and cuts and saws easily. Planing, shaping, sanding, gluing and finishing are all rated satisfactory. Turning and boring operations are only rated as fair.

USES

Once believed to be unlimited in availability, this species was much used in Puerto Rico. In furniture, it was used as a substitute for mahogany. It serves well for cabinetry, interior trim, general construction, box and crate manufacture and for shingles and small boats.

SUPPLIES

Because this species has only a mediocre growth rate and is not readily transplanted, supplies have become limited and actions are being taken to protect current stands. Since the native growth range of this tree is in the heart of the winter vacation area in the sunny Caribbean, wood-collecting tourists should stop by a local *tienda de madera* (lumber yard) and inquire as to the availability of tabonuco which is the common name in Puerto Rico or for gommier in the other Islands.

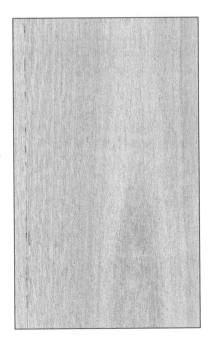

Dalbergia cearensis

kingwood

by Max Kline

SCIENTIFIC NAME

Dalbergia cearensis. Derivation: The genus name is in honor of Nicholas and C. G. Dalberg, Swedish botanists about 1730 to 1820. The specific epithet means the plant is "of Ceara" in Brazil.

FAMILY

Fabaceae or Leguminosae, the legume family; (*Papilionaceae*) the pea or pulse group.

OTHER NAMES

violetwood, violete.

DISTRIBUTION

This tree is characteristic of northeastern Brazilian caatinga vegetation, occurring from Ceara and southern Piaui to southern Bahia and possibly extending to northern Minas Gerais.

THE TREE

Kingwood is a small to medium-sized tree growing to a maximum of about 50 feet in height with diameters that rarely exceed 10 inches. The branches and leaves are glabrous. The leaves are five to seven foliate non-persistent stipules, and the leaflets are 2 inches long and 1 inch wide. The fruit is a flat legume.

THE TIMBER

Streaked with varying lighter and darker lines of golden yellow, kingwood timber is a rich, violet-brown color sometimes with shades almost to black. The scent is fragrant, but scarcely noticeable in dry material. It has a bright luster with a fine, uniform texture, although the dark layers are slightly harder than the others. It is hard and strong, but brittle. Average specific gravity is 0.92 (ovendry weight/green volume), equivalent to an air-dried weight of 75 pcf. The grain varies from straight to finely wavy. The taste is not distinct.

SEASONING

Kingwood is exported in the form of small logs 3 to 6 feet long and 3 to 8 inches in diameter, without sapwood. Some checking takes place during shipment, but once dried and manufactured it stays in place with little expansion. Shrinkage values are not available.

DURABILITY

While specific data is not available, the wood is reported to be highly durable.

WORKABILITY

The wood is not difficult to work with sharp tools, finishing to an exceptionally smooth surface with a natural waxy polish. When first used, this is a beautiful wood and becomes more so as it tones with age, developing a metallic sheen. It is well named, as it is truly the wood of kings.

USES

Because of the small size of the kingwood tree, the wood has always been restricted to the use of veneer in inlays and marquetry with occasional use as lumber for articles of turnery.

SUPPLIES

Since the exportation of kingwood is very limited, it is quite scarce on the general market. If available, it can only be bought as a complete log on a weight basis. Wood in the form of veneer is found more often than lumber. The price might be considered to be among the highest for any commercial wood species.

Dalbergia cochinchinensis

flamewood

by Jim Flynn

SCIENTIFIC NAME

Dalbergia cochinchinensis. Derivation: The genus name is in honor of Nicholas and C. G. Dalberg, Swedish botanists about 1730 to 1820. The specific epithet means the plant is "of Cochin China".

FAMILY

Fabaceae or Leguminosae, the legume family; (*Papilionaceae*) the pea or pulse group.

OTHER NAMES

The present commercial name, flamewood, is an attempt to emphasize the color and grain of the wood

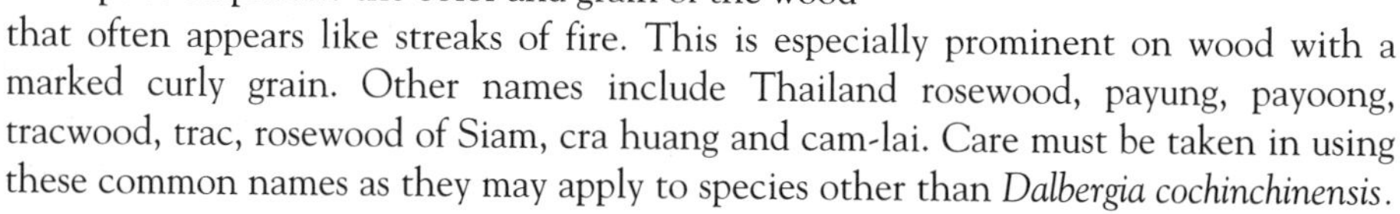

that often appears like streaks of fire. This is especially prominent on wood with a marked curly grain. Other names include Thailand rosewood, payung, payoong, tracwood, trac, rosewood of Siam, cra huang and cam-lai. Care must be taken in using these common names as they may apply to species other than *Dalbergia cochinchinensis*.

DISTRIBUTION

This plant species is found scattered in mixed deciduous forests and dry evergreen forests in Vietnam, Laos, Kampuchea, Thailand and Myanmar (Burma).

THE TREE

Growing to heights of 80 feet with trunk diameters up to 2.5 feet, flamewood trees are apt to be fluted at the base. The bark is gray, smooth, fibrous and much lenticeled. The leaves are pinnate, 6 to 8 inches long with seven to nine large, subopposite ovate-acute glabrous leaflets.

THE TIMBER

Flamewood is considered a "luxury timber" as indeed it is. The wood has a well defined whitish sapwood. The heartwood varies from a light rose-purple to burgundy streaked with brown and black. The texture varies from fine to medium. There is a slight vinegary odor detectable when working the wood. The cells contain a red, shiny gummy substance that can be observed with a magnifying glass. The irregular growth rings add to the beauty of the wood and are visible on quartersawn surfaces. The wood will darken with age. Average specific gravity is 0.82 (ovendry weight/green volume), equivalent to an air-dried weight of 66 pcf.

SEASONING

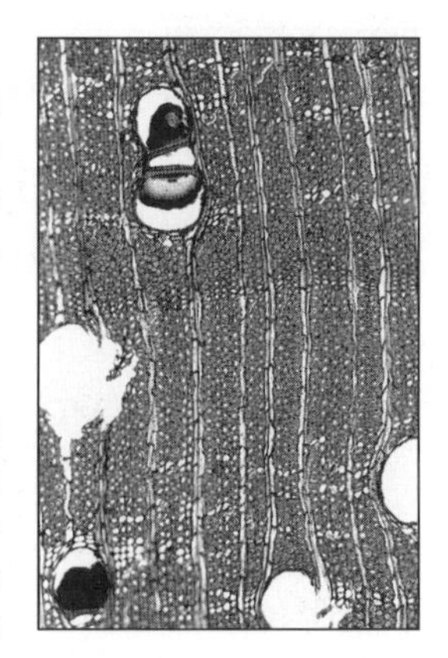

This timber seasons best when in log form. It will season well when kiln dried, if it is done slowly. End splitting and surface checking may occur. It has a low shrinkage level and average reported shrinkage values (green to ovendry) are 3.0% radial and 6.0% tangential.

DURABILITY

The durability of flamewood is rated high. Estimates are that it will survive at least 25 years in contact with the ground and in exposed conditions. It is susceptible to attack by powder post beetles. The heartwood is extremely resistant to preservation treatment. Because this is a costly, high-class furniture wood, it is seldom used in unprotected environments.

WORKABILITY

Although flamewood timber is hard, very strong, at times brittle, and very dense, it is not too difficult to machine. The key is to use saws with short and stiff teeth. The wood will hold nails and screws, but pre-boring may be necessary. Gluing may be done satisfactorily if care is taken to make sure that the adjoining surfaces are freshly milled, dust-free and, if necessary, wiped with a solvent to remove surface oil. An extremely lustrous glass-like surface can be obtained by finish-sanding with 1600-grit paper. A hand-rubbed penetrating oil finish is suggested. Dust may be a concern for dermatitis.

USES

This wood is used for high-class furniture, especially in China. Other uses include heavy-duty flooring, veneer, musical instruments, carvings, turnery and precision instruments.

SUPPLIES

The wood is available from dealers who specialize in unusual and exotic forest products. It is often sold as a substitute for East Indian rosewood (*Dalbergia latifolia* WDS 095) and Brazilian rosewood (*D. nigra* WDS 097).

Dalbergia decipularis

tulipwood

by Max Kline

SCIENTIFIC NAME

Dalbergia decipularis. Derivation: The genus is name is in honor of Nicholas and C. G. Dalberg, Swedish botanists about 1730 to 1820. The specific epithet *descipere* is Latin meaning deceptive, referring to the limited growth of the tree.

FAMILY

Fabaceae or Leguminosae, the legume family; (*Papilionaceae*) the pea or pulse group.

OTHER NAMES

boise de rose, Brazilian pinkwood, pau rosa.

DISTRIBUTION

Northeastern Brazil.

THE TREE

The tulipwood tree is small, growing only to 30 to 40 feet in height with diameters up to 16 inches. It has an irregular trunk, slender branchlets and odd-pinnate leaves generally with seven leaflets. The small, white flowers are only about 0.25 inch. The fruit is a pod about 2 inches long that holds a single seed.

THE TIMBER

The heartwood has a straw-colored background irregularly striped in shades of yellow, rose, pink and violet. The sapwood is solid yellow. It has a mildly fragrant scent when worked, but the odor does not persist after machining. The luster is rather high, and it is very hard and strong with a specific gravity ranging from 0.71 to 0.85 (ovendry weight/green volume), equivalent to an air-dried density of 56 to 69 pcf. The texture is rather fine, and the grain is straight to wavy.

SEASONING

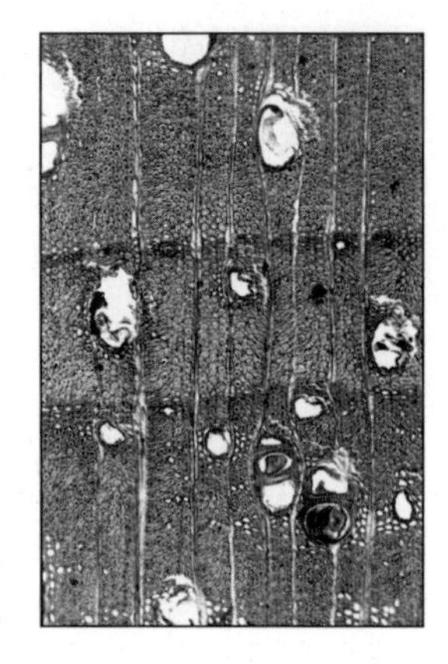

The wood is not usually available in widths sufficient to warrant details on shrinkage.

DURABILITY

Reports on this characteristic do not seem to be available with the exception of typical statements such as "The wood is highly durable".

WORKABILITY

Tulipwood is not easy to work as it is inclined to be splintery. It does take a high natural polish. It has been known to cabinetmakers for many years and was a favorite for French furniture, especially of the Empire Period. The color fades on exposure, as does the scent emanating during machining operations.

USES

Present day uses are limited to inlay for veneer, marquetry, brush backs, turnery and bandings.

SUPPLIES

The timber is exported from Brazil in the form of round logs without sapwood that are usually under 6 feet in length and from 2 to 8 inches in diameter. Often sold by the pound, tulipwood is costly in the United States. Veneer is readily available from supply houses.

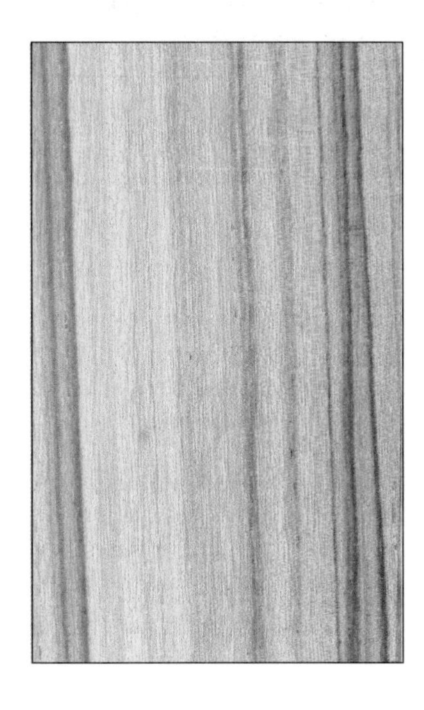

Dalbergia latifolia

East Indian rosewood

by Max Kline

SCIENTIFIC NAME

Dalbergia latifolia. Derivation: The genus name is in honor of Nicholas and C. G. Dalberg, Swedish botanists about 1730 to 1820. The specific epithet is Latin for broad-leafed.

FAMILY

Fabaceae or Leguminosae, the legume family; (*Papilionaceae*) the pea or pulse group.

OTHER NAMES

Bombay blackwood, shisham, sitsal, malabar.

DISTRIBUTION

East Indian rosewood occurs in most of India except the northwest. Its grows best in southern India, where it is found scattered in the dry, deciduous forests.

THE TREE

East Indian rosewood varies considerably in size, from 1 to 5 feet in diameter, according to locality. Generally it is moderately sized, reaching a height up to 80 feet with a straight, clean, cylindrical bole of 40 feet maximum. It will grow up to altitudes of 3,500 feet. The trees are found in forests along with teak (*Tectona grandis* WDS 261) and bamboo. The gray bark is 0.5 inch thick with irregular short cracks, exfoliating in thin, fibrous longitudinal flakes. The opposite, entire, oval leaves with acute bases are generally 2.5 inches long by 1.5 inches wide with short petioles. The pods remain on the tree in the hot season and fall only when the rains have begun. It is sometimes planted as a shade tree in coffee plantations.

THE TIMBER

The purple-brown heartwood has a handsome figuring of darker streaks and is well defined from the sapwood. The sapwood is whitish-yellow often with a purple tinge. The odor is faintly aromatic, and the luster is low to medium. The grain is commonly interlocked producing a narrow ribbon figure on quartersawn surfaces. The texture is uniform and moderately coarse. Average reported specific gravity is 0.70 (ovendry weight/green volume), equivalent to an air-dried weight of 55 pcf. This heavy timber has high strength properties. When seasoned, it is about 2.5 times as hard, 15% stiffer and about 25% stronger in bending and compression along the grain than oak.

SEASONING

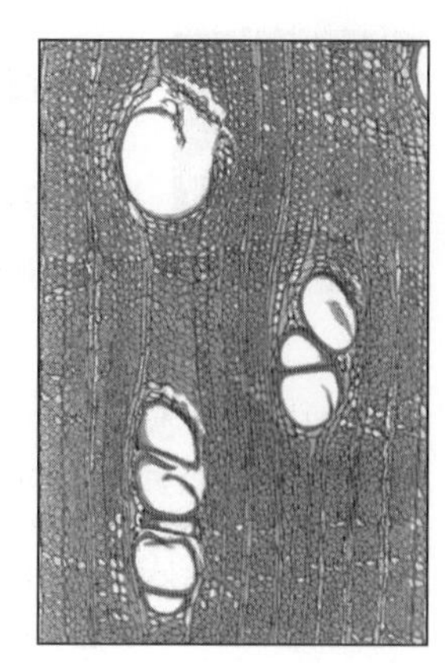

East Indian rosewood air dries fairly rapidly with a slight amount of surface checking. This may be minimized by the use of slower drying rates and end coatings. Warping does not present a problem. Average reported shrinkage values (green to ovendry) are 2.7% radial, 5.8% tangential and 8.5% volumetric. It can also be kiln dried satisfactorily, but should be done slowly to avoid end splitting.

DURABILITY

The heartwood is quite resistant to all types of fungal attack and is reported to be the most resistant timber to termites. The sapwood is susceptible to beetle attack.

WORKABILITY

East Indian rosewood is moderately hard to work by hand and offers a fair resistance to cutting by machines. It has a dulling effect on cutting edges. Slower than normal speeds should be used for mortising and drilling. Nail and screw holding properties are good, and glue adheres satisfactorily. It is capable of a high polish, but requires filling prior to finishing. A dull or waxed finish gives the most effective results.

USES

Since this timber has a handsome appearance, its primary use is for decorative and ornamental purposes, such as furniture, cabinetry and paneling, especially in the veneer form. It is also suitable for musical instruments, brush backs, combs and inlay work. In India, it has been used for construction purposes and for boat building and agricultural implements. It is unsuitable for plywood manufacture.

SUPPLIES

Since the demand for East Indian rosewood is increasing, supplies are rather limited. It is imported in log form (10 to 36 inches in diameter) and much of it is cut into veneer. [Ed. note: *Dalbergia latifolia* should not be confused with *D. sissoo*. The latter is also a native of India and is abundant in Florida. It bears the common name of Indian rosewood as opposed to East Indian rosewood and is a much inferior wood.]

Dalbergia melanoxylon

African blackwood

by Max Kline

SCIENTIFIC NAME

Dalbergia melanoxylon. Derivation: The genus name is in honor of Nicholas and C. G. Dalberg, Swedish botanists about 1730 to 1820. The specific epithet means black, referring to the color of the wood.

FAMILY

Fabaceae or Leguminosae, the legume family; (*Papilionaceae*) the pea or pulse group.

OTHER NAMES

Congowood, Mozambique ebony, grenadillo, Senegal ebony, Cape Damson ebony, mpingo.

DISTRIBUTION

This tree has an extensive range on the African continent. It can be found in the savanna regions of the Sudan southward to Mozambique, then westward to Angola and northward to Nigeria and Senegal.

THE TREE

Not a large tree in most respects, African blackwood normally reaches a height of 25 feet. Errant trees, however, can grow as high as 50 feet in ideal surroundings. Common throughout its range, African blackwood grows in medium and low altitudes and thrives on termite mounds. The bole, seldom over 24 inches in diameter, is rarely cylindrical and is often fluted. The smooth bark is pale gray and peels in strips when mature. The compound leaf has alternating leaflets from 0.5 to 2 inches long. The fruit is a thin and fragile oblong pod which holds one to two seeds. African blackwood is the national tree of Tanzania.

THE TIMBER

The heartwood is dark purplish-black or brown with black streaks, which usually predominate so that the general effect is nearly black. The narrow sapwood is white and very clearly defined. The luster is dull. African blackwood is a very hard and heavy wood with an average reported specific gravity of 1.08 (ovendry weight/green volume), equivalent to an air-dried weight of 91 pcf. The grain is mostly straight with a fine texture. It has a slightly oily nature. Odor and taste are not distinct.

SEASONING

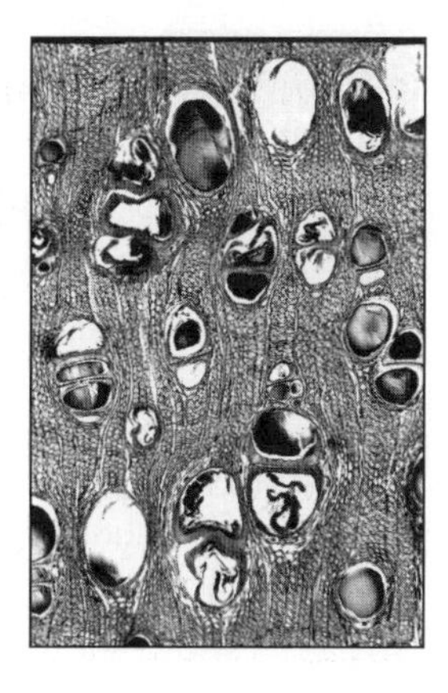

African blackwood dries very slowly and tends to split, especially in the log. It is advisable to coat the ends to minimize splitting. Average reported volumetric shrinkage (green to ovendry) is 7.6%. Once dried, the timber is slow to absorb moisture.

DURABILITY

The heartwood is reported to be very durable.

WORKABILITY

This wood has exceptionally good working qualities. It cuts very smoothly and evenly, taking an excellent finish directly from the drill or lathe. It can be tapped for screw threads almost like metal. It is considered to be the best wood available for ornamental turnery.

USES

A primary use is for woodwind instruments such as flutes, clarinets and bag pipes. It is superior to ebony when used in musical instruments because of its oily nature and resistance to climactic changes. The wood is also used for turnery in making items such as brush backs, knife handles, chessmen and pulley blocks.

SUPPLIES

The timber is exported from East African ports in the form of small logs 3 to 5 feet in length. It is then sometimes only sold by the importer in log form and priced by the pound. It may be considered expensive.

Dalbergia nigra

Brazilian rosewood

by Max Kline

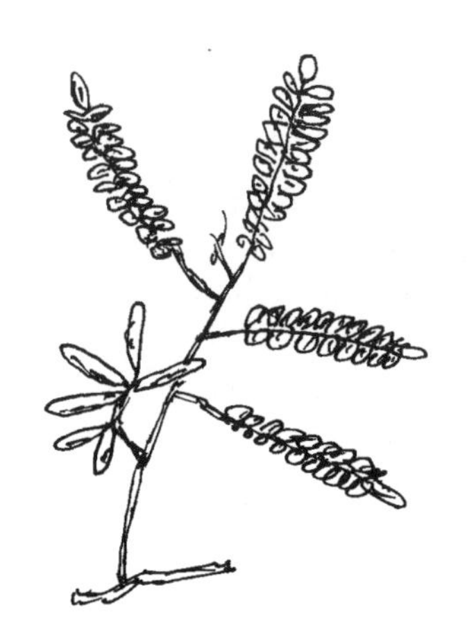

SCIENTIFIC NAME

Dalbergia nigra. Derivation: The genus name is in honor of Nicholas and C. G. Dalberg, Swedish botanists about 1730 to 1820. The specific epithet is Latin for black.

FAMILY

Fabaceae or Leguminosae, the legume family; (*Papilionaceae*) the pea or pulse group.

OTHER NAMES

jacaranda, pianowood, caviuna, pau preto, obuina.

DISTRIBUTION

In northern Espirito Santa, Bahia and eastern Minas sections of Brazil.

THE TREE

Brazilian rosewood grows to a height of 125 feet or more with a rather short irregular bole, often 3 or 4 feet in diameter. The trunk contains a large amount of sapwood, which is usually hewn off in the forest. Commercial logs of heartwood are seldom over 18 inches in diameter.

THE TIMBER

The color of rosewood is variegated with shades of brown and violet with irregular black streaks. The wood from young trees is brown and not as attractive as that from the old and often defective trees which produce the finest color and figure. The wood has a somewhat oily appearance, and the sapwood is nearly white and sharply defined. The luster is low to medium. Rosewood is very fragrant when cut and even when burned, and its name is derived from this odor. The smell is not as pronounced in wood from young trees. The grain is straight to wavy. The rather large pores of the timber are exceedingly irregular both in size and position. Average reported specific gravity ranges from 0.62 to 0.73 (ovendry weight/green volume), equivalent to an air-dried weight of 48 to 58 pcf. For the purposes Brazilian rosewood is utilized, its strength properties are more than adequate. In hardness, it exceeds by far any of the native hardwood species used to make furniture.

SEASONING

Brazilian rosewood timber must be dried slowly to avoid checking. Average reported shrinkage values (green to ovendry) are 2.9% radial, 4.6% tangential and 8.5% volumetric. Once dry, it absorbs moisture slowly and is dimensionally stable.

DURABILITY

Brazilian rosewood has an excellent reputation for durability with respect to fungus and insect attack, including termites.

WORKABILITY

The wood machines and veneers well. As with other woods in this density class, it can be glued satisfactorily providing the necessary precautions are taken to ensure good glue bonds. An exceedingly smooth and highly polished surface can be achieved. It seasons well and holds its shape.

USES

Rosewood has been used for 300 years in making furniture and cabinets and still maintains its long established reputation as one of the most esteemed woods of the world. It also has been commonly used for levels, knife handles, billiard tables, piano cases, brush backs, turned articles and marquetry. This wood has been the first choice as a tonewood for the finest of stringed musical instruments.

SUPPLIES

Dalbergia nigra is now considered an endangered species and restrictions for trade are in place. Related species are being substituted.

Dalbergia retusa

cocobolo

by Max Kline

SCIENTIFIC NAME

Dalbergia retusa. Derivation: The genus name is in honor of Nicholas and C. G. Dalberg, Swedish botanists about 1730 to 1820. The specific epithet is Latin for retuse meaning slightly notched at a rounded apex, which refers to the leaves.

FAMILY

Fabaceae or Leguminosae, the legume family; (*Papilionaceae*) the pea or pulse group.

OTHER NAMES

Nicaragua rosewood, granadillo, nambar, palisandro.

DISTRIBUTION

Western Costa Rica, Nicaragua, Panama, Mexico, Colombia, Salvador, Honduras and Guatemala.

THE TREE

Generally of poor form, cocobolo is a deciduous tree growing 45 to 60 feet in height with a diameter of 20 to 24 inches. Cocobolo varies in form and is sometimes low and spreading and sometimes high with a straight trunk and a more elongated crown. The compound leaves have about nine alternate leaflets which turn black upon drying. Usually, the flat legume fruit has only one seed but may contain as many as four.

THE TIMBER

The color of the heartwood varies. It might be described as rainbow-hued, but upon exposure to sunlight and air, the lighter colors lose their brilliance and merge into a deep red with a black striping or mottling. To some extent, the color can be removed when boiled in water. The odor is slightly pungent and mildly fragrant, and the taste is not distinctive. The luster is low, and the texture medium to fine. The grain is fairly straight to interwoven. Average reported specific gravity is 0.80 to 0.98 (ovendry weight/green volume), equivalent to an air-dried weight of 64 to 81 pcf.

SEASONING

Cocobolo requires care when seasoning as it is liable to caseharden, warp, and check when kiln dried from the green condition. A period of air drying prior to kiln drying reduces this tendency. Average reported shrinkage values (green to ovendry) are 2.7% radial, 4.3% tangential and 7.0% volumetric.

DURABILITY

Cocobolo is considered to be very durable. Prolonged or repeated immersion in soapy water has little effect on the wood, except to darken its color. This is due to the presence of an oily substance, which tends to waterproof the wood and keep it in shape after manufacture. If the smooth surface is rubbed with a cloth, it acquires a waxy finish without the use of any other finishing material.

WORKABILITY

This timber is not difficult to work. It turns readily and finishes very smoothly. It is unsuitable for gluing because of the waxy nature. The fine dust arising when working with this wood may produce a rash or dermatitis resembling poison ivy. Some people are immune to this poison while others are readily susceptible. It is believed that perspiration giving an acid reaction serves to offer resistance from the poison to workers while alkaline perspiration will allow susceptibility. This is based on the fact that alkali will dissolve the irritating agent, permitting its entrance into the skin pores.

USES

Cocobolo is one of the most important woods in the cutlery trade and is extensively used for knife handles because of its beautiful color, texture, grain and waterproof properties. It is also used for small tool handles, brush backs, musical and scientific instruments and steering wheels for boats. Some of the more figured wood is manufactured into jewelry boxes, canes, forks, spoons, buttons and chessmen.

SUPPLIES

Cocobolo supplies are becoming very limited and the price more and more costly. Occasionally veneers are marketed with highly figured grain patterns.

Dalbergia stevensonii

Honduras rosewood

by Max Kline

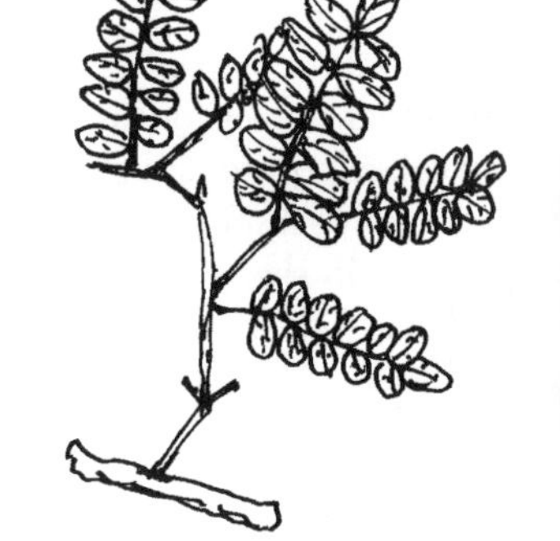

SCIENTIFIC NAME

Dalbergia stevensonii. Derivation: The genus name is in honor of Nicholas and C. G. Dalberg, Swedish botanists about 1730 to 1820. The specific epithet was assigned in 1927 by Paul C. Standley in honor of Neil S. Stevenson who collected the type herbarium specimen (in flower) in the Toledo District of former British Honduras (Belize) in the same year.

FAMILY

Fabaceae or Leguminosae, the legume family; (*Papilionaceae*) the pea or pulse group.

OTHER NAMES

palisandro de Honduras.

DISTRIBUTION

Restricted to the southern part of Belize between latitudes 16° and 17° north in the damp forests and along rivers.

THE TREE

Honduras rosewood grows to a height of 50 to 100 feet with diameters of about 3 feet. The trunk commonly forks at about 20 to 25 feet from the ground. The papery bark is about 0.25 inch thick.

THE TIMBER

Honduras rosewood timber is handsome in appearance, purplish-brown in color with irregular black markings. The grayish sapwood is sharply defined from the darker heartwood. It is mostly straight grained and of medium to rather fine texture. When freshly cut, the sapwood has a mild odor and is slightly bitter in taste. The odor and taste of seasoned heartwood are not distinct. Average reported specific gravity is 0.75 to 0.88 (ovendry weight/green volume), equivalent to an air-dried weight of 60 to 72 pcf. The strength properties of Honduras rosewood are not outstanding, and it is used in applications where other properties are of greater importance.

SEASONING

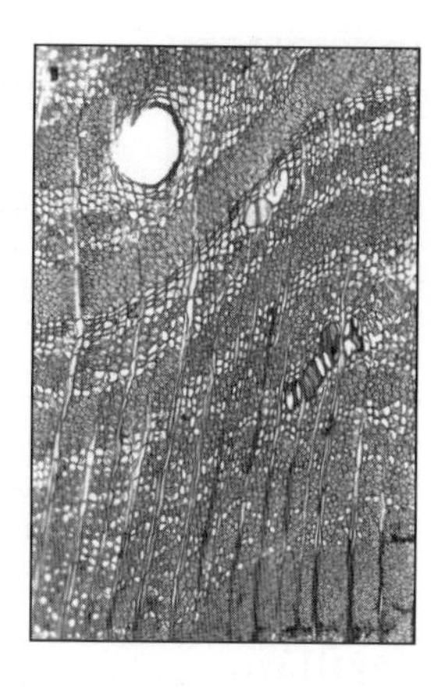

This species will air dry slowly but with a marked tendency to check. For certain uses a period of 2 to 7 years is employed during the seasoning process. Shrinkage values are reported to be similar to other American rosewoods which are unusually low.

DURABILITY

When used in contact with the ground, the sapwood decomposes rapidly but the heartwood is exceptionally stable. In one reported case, no changes occurred in the soundness of the wood after it had been in the ground for 37 years.

WORKABILITY

The timber is hard to work with hand tools, but is worked without difficulty with machine tools. Some dulling effect of the cutting edges does occur. It turns well and finishes smoothly with a good polish.

USES

The chief use is for the manufacture of bars for marimbas and xylophones in the United States. It is superior for this purpose over Brazilian rosewood (*Dalbergia nigra* WDS 097) due to its greater density, toughness and more highly resonant qualities. Only the finest straight grain logs are employed in making bars and waste may be as high as 70% to 80% after discarding all checked and inferior material. Honduras rosewood is also used to a limited extent for cabinetry and veneer.

SUPPLIES

Since the growth areas of Honduras rosewood are so small, the quantity available on the commercial market is very limited. The woodworker, however, can secure both veneer and lumber at high prices from importers.

Dialyanthera spp.

virola

by Jon Arno

SCIENTIFIC NAME

Dialyanthera spp. Derivation: The genus name is Greek for anthera, referring to the pollen bearing part of the stamen.

FAMILY

Myristicaceae, the nutmeg family.

OTHER NAMES

Within the Myristicaceae family, this Wood Data Sheet recognizes two genera, Dialyanthera and Virola. Note that the preferred common name virola applies to both genera and wood from both genera is commercially marketed under the common name virola. Other names include uangare, otoba, miguelario, coco, sebo and saba.

DISTRIBUTION

Virola is cut from several closely related species native to Central and South America from Costa Rica to Ecuador including Panama, Colombia and parts of Venezuela.

THE TREE

Virola can grow to a height of 100 feet with diameters of more than 4 feet. Currently, the main sources of supply are Colombia and Ecuador where the trees grow in almost pure stands in moist (swampy) areas and yield clear logs up to 50 feet long.

THE TIMBER

Both the sapwood and the heartwood are pale pinkish-tan. The grain is generally straight. Tangential surfaces show very little figure, and the surface has a fairly high luster. Texture ranges from medium to medium coarse depending upon the species. With an average reported specific gravity of only 0.36 (ovendry weight/green volume), equivalent to an air-dried weight of 27 pcf, virola is comparable to western white pine (*P. monticola*). Interestingly, despite being in the nutmeg family, virola has no strong odor or taste. It has a very low crushing strength. Virola is comparable to, but generally weaker than, western white pine in terms of most strength properties except elasticity where it is only slightly superior.

SEASONING

Virola air dries quickly. Average reported shrinkage values (green to ovendry) are 4.2% radial, 9.4% tangential and 12.0% volumetric. The heartwood can contain excessively moist streaks, which may collapse (honeycomb) as the wood seasons.

DURABILITY

Nondurable, the wood is especially prone to blue stain and insect attack.

WORKABILITY

Like most soft, easily crushed woods, virola requires sharp cutting tools, but is easy to work. It accepts glue, stain and other finishes and takes nails well. Although the wood has an attractive pinkish hue, it is otherwise almost without character, almost bland. On first examination, it is the kind of wood that might appear to make a good, secondary stock for interior panels, drawer sides and other utilitarian applications and it is in fact used for these purposes. Its high shrinkage, however, suggests it may not be any better suited for these applications than some less expensive U.S. domestic species such as American basswood (*Tilia americana* WDS 270), sweetgum (*Liriodendron tulipifera* WDS 162) and yellow-poplar (*Liquidambar styraciflua* WDS 161).

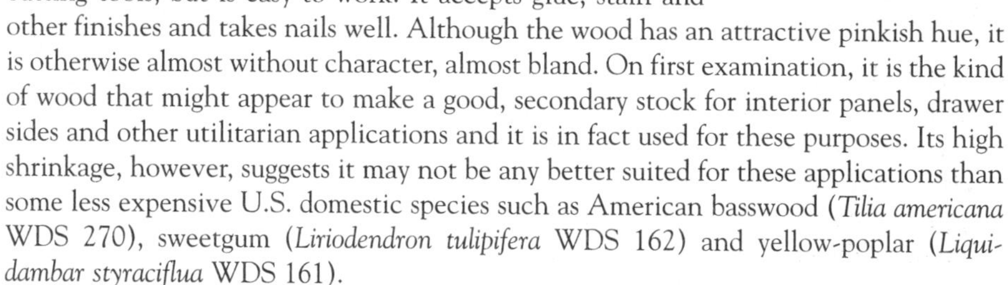

USES

Uses include paneling, interior trim, corestock particleboard, general carpentry and some furniture components.

SUPPLIES

Virola is fairly plentiful, but is much exploited to supply markets within its native range. Some is exported and is available both in the United States and in Europe, although it is not often featured by retailers catering to woodworkers. While moderately priced, it is not that interesting or attractive relative to U.S. domestic alternatives which are even less expensive and functionally superior.

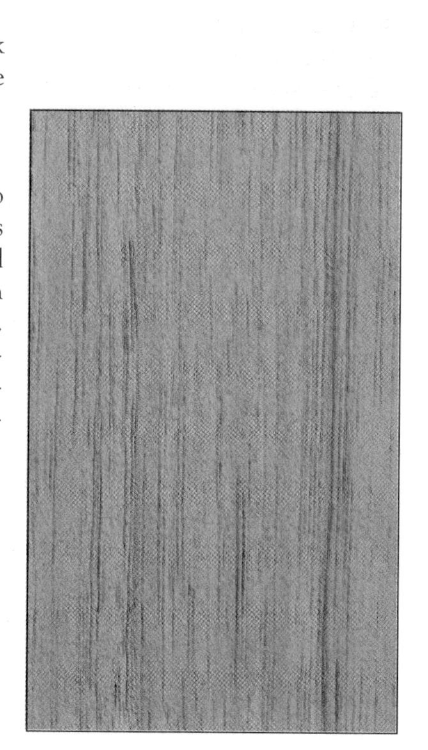

Diospyros celebica

Macassar ebony

by Max Kline

SCIENTIFIC NAME

Diospyros celebica. Derivation: The genus name is Greek meaning "of the God Zeus or Jupiter" and "grain" alluding to the edible fruit. The specific epithet is Latin meaning the plant is "of Celebes", which is now known as Sulawesi in the Indonesian Archipelago.

FAMILY

Ebenaceae, the ebony family.

OTHER NAMES

Indian ebony, camagon, golden ebony.

DISTRIBUTION

Southeast Asia and the Philippines.

THE TREE

Generally, the tree is small, averaging 50 feet in height with a clear bole of 8 to 15 feet. The average diameter is approximately 16 inches. The tree grows in well-drained rocky soil, sometimes near water, but never in swamps.

THE TIMBER

The color of Macassar ebony is black with reddish or reddish-brown streaks. The sapwood is uniform light red. It has a metallic luster. Odor and taste are not distinct. The wood has a fine texture and a straight to wavy grain. Average reported specific gravity varies from 0.60 to 0.80 (ovendry weight/green volume), equivalent to an air-dried weight of 46 to 64 pcf.

SEASONING

The heartwood is very difficult to season as the wood develops long, fine, deep checks, especially if cut in large dimensions. The best results are reported when the tree is girdled and allowed to stand for 2 years before felling followed by 6 months of seasoning. To prevent end and surface checks, the wood should be protected against drying too rapidly. Average reported shrinkage values (green to ovendry) are 5.4% radial and 8.8% tangential.

DURABILITY

Macassar ebony heartwood is reported to be very durable and resistant to decay. Preservatives cannot be used because of the wood's resistance to impregnation.

WORKABILITY

Working with Macassar ebony is slow, but in spite of the hardness little tool dulling is experienced if correct cutting angles are used. The wood takes a very high finish and turns well. It glues satisfactorily but the wood is too hard to use nails or screws without first pre-boring. The dust of this species can be an irritant.

USES

Uses include musical instrument parts, special fittings on furniture such as knobs and decorative items, inlay work and other applications where accent material is needed.

SUPPLIES

The timber is rare and is one of the most costly timbers on the commercial market. Veneer is more readily available but also costly.

Diospyros dendo

Gabon ebony

by Max Kline

SCIENTIFIC NAME

Diospyros dendo. Derivation: The genus name is Greek meaning "of the God Zeus or Jupiter" and "grain" alluding to the edible fruit. The specific epithet is probably a local name.

FAMILY

Ebenaceae, the ebony family.

OTHER NAMES

Nigerian ebony, black ebony, calabar, Lagos billetwood.

DISTRIBUTION

In Africa from Ghana to the Congo (formerly Zaire), especially in coastal areas.

THE TREE

Gabon ebony grows to a height of 50 to 60 feet with an average bole diameter of 2 feet. The evergreen leaves vary from 4 to 10 inches in length and 3 to 5 inches in width and are alternate and elliptic with acuminate tips. The fruit is globe-shaped, 1 to 2.5 inches in diameter and contains several flat seeds. When ripe it is orange in color.

THE TIMBER

The sapwood of Gabon ebony is light yellowish-white, and the heartwood is uniform jet black, sometimes streaked with greenish-black markings. The odor and taste are not distinct. It has a metallic luster, and the timbers are very fine textured. The grain is usually quite indistinct. Average reported specific gravity is 0.82 (ovendry weight/green volume), equivalent to an air-dried weight of 66 pcf. Gabon ebony is strong and hard with good resistance to compression and bending. The blackest grades tend to be the most brittle.

SEASONING

In small dimensions, Gabon ebony kiln dries well with little tendency to split or distort. Average reported shrinkage values (green to ovendry) are 5.5% radial and 6.5% tangential. After seasoning, the wood is dimensionally stable.

DURABILITY

Some damage from pin hole borers is occasionally found in Gabon ebony timbers. Because it is very resistant to decay and termites, it is not given a preservative treatment.

WORKABILITY

The abrasive effect on cutting edges of tools caused by machining Gabon ebony is well known. Very sharp tools are required in all processes. It turns well and can be carved with good results. Worked edges remain sharp, and the timber will finish to a high polish. Because the wood is hard, using nails or screws requires pre-boring.

USES

Gabon ebony is used for turnery, carving, inlaid work, novelties, brush backs, handles, piano keys, musical instruments and billiard cues. The light-colored sapwood is sometimes used for less expensive tool handles.

SUPPLIES

Demand is small but supplies are available commercially at a very high cost. It is usually produced in the form of billets or short logs of 4 to 8 feet.

Diospyros virginiana

common persimmon

by Max Kline

SCIENTIFIC NAME

Diospyros virginiana. Derivation: The genus name is Greek meaning "of the God Zeus or Jupiter" and "grain" alluding to the edible fruit. The specific epithet means the plant is "of Virginia".

FAMILY

Ebenaceae, the ebony family.

OTHER NAMES

date plum, possumwood, simmon, plaqueminier, persimmon.

DISTRIBUTION

Southeastern United States except the lower portion of Florida. Northward to the southern tip of Connecticut and New York, westward to Iowa and Kansas, south to Texas.

THE TREE

Although usually smaller, common persimmon can reach heights of 80 to 120 feet with trunk diameters of 18 to 24 inches. The fruit is a large berry, which before it is completely ripened is puckery to the taste. When ripe the blackish-purple fruit is sweet and edible and provides food for wildlife. [Ed. note: I like to eat them after the first frost when they are a bright orange color. jhf] *Diospyros virginiana* is the northern most member of the ebony family and one of only two species that grow in the United States.

THE TIMBER

Found only in older trees, the heartwood of common persimmon is very narrow and dark brown to black. The wide sapwood is creamy white mottled with dark spots turning a grayish-brown when exposed to air. The wood is odorless and tasteless, close grained, hard, tough and strong. Average specific gravity is 0.68 (ovendry weight/green volume), equivalent to an air-dried weight of 53 pcf. The texture is medium to fine. Very little figure is found in the timber.

SEASONING

This timber is difficult to season and shrinks when drying. Average reported shrinkage values (green to ovendry) are 7.9% radial and 11.2% tangential.

DURABILITY

The black heartwood is highly resistant to decay. Insects seldom seriously attack the tree but persimmon wilt, a fungal disease, is causing severe losses in the southeastern United States.

WORKABILITY

Common persimmon is difficult to work with tools, but will finish to a high polish. It does not glue well but is high in shock resistance and nail holding properties. The fibers do not fray and like flowering dogwood (*Cornus florida* WDS 088) the surface will stay smooth even under hard usage.

USES

The sapwood of common persimmon is used for shuttle blocks, bobbins, shoe lasts and, at one time, was considered the standard for golf club heads. It is also used for handles, spools and miscellaneous articles of turnery because of its toughness, hardness and ability to retain a smooth surface even after prolonged use.

SUPPLIES

Common persimmon is seldom found in veneer form and what little lumber is available is quite costly. [Ed. note: Dedicated wood collectors can usually find supplies in out-of-the-way places, especially small local sawmills.]

Distemonanthus benthamianus

ayan

by Max Kline

SCIENTIFIC NAME

Distemonanthus benthamianus. Derivation: The genus name is Greek for a flower with two stamens. The specific epithet is in honor of English botanist George Bentham (1800 to 1884), the author of numerous scholarly and fundamental works including *Flora Australiensis*.

FAMILY

Fabaceae or Leguminosae, the legume family; (*Caesalpiniaceae*) the cassia group.

OTHER NAMES

movingui, Nigerian satinwood, barre, ayanran, bonsamdua, ejen.

DISTRIBUTION

Widely but sparsely distributed throughout the forests of West Africa, mainly in Cameroon, Ghana and Nigeria.

THE TREE

Ayan attains an average height of 90 feet with a diameter of 2.5 feet. The bole is clear, reasonably straight and cylindrical. It has rather thin, weakly developed buttresses.

THE TIMBER

Ayan sapwood is narrow and varies in color from lemon-yellow to yellowish-brown. It is fairly distinct from the uniformly yellowish-cream or light golden-yellow heartwood. The odor and taste are not distinct. The luster is high, and the texture is fine to medium. The grain is straight to interlocked which produces a ribbon figure. The heartwood may contain up to 1.3% silica and a yellow extractive dye, which under moist conditions acts as a direct dye on textiles. Average reported specific gravity is 0.58 (ovendry weight/green volume), equivalent to an air-dried weight of 45 pcf. Ayan is comparable to oak in strength. It also has moderately good bending properties after being steamed.

SEASONING

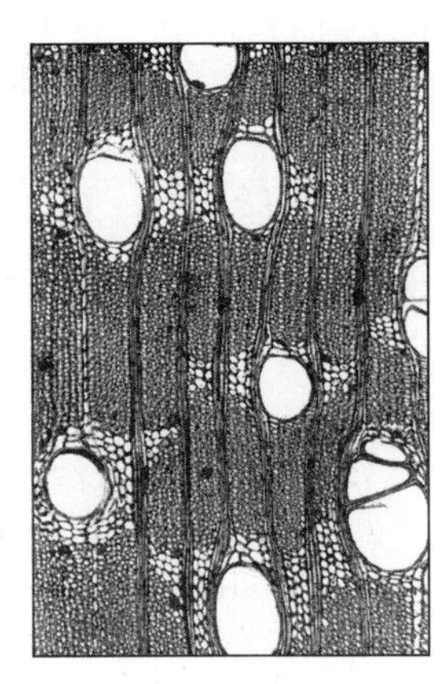

The timber can be seasoned with little degrade and is quite dimensionally stable. Average reported shrinkage values (green to ovendry) are 3.1% radial, 5.2% tangential and 10.7% volumetric.

DURABILITY

The timber is moderately resistant to all types of fungal attack and to wood-borer pests. It is also resistant to penetration by preservatives.

WORKABILITY

In general, ayan works fairly well with hand and machine tools. The dulling effect on cutters varies with the silica content. When the silica content is high, carbide-tipped saws must be used. Gum build up on saws can cause overheating. When bored, the wood tends to char. It takes stain and polish well requiring only a moderate amount of filler. It has a slight tendency to split when nailed but can be glued satisfactorily. It can be easily peeled for veneer.

USES

The primary uses are for cabinetry and interior joinery for window frames, door frames and sills. It should not be used in the construction of any product liable to come in contact with wet fabric because of the yellow extractive dye it contains. It is suitable for most normal flooring and is of interest for its decorative appearance in the form of veneer.

SUPPLIES

The wood can be found on the commercial market.

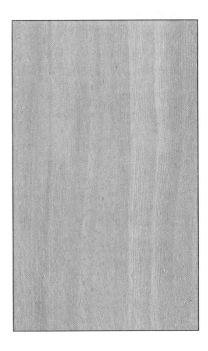

Dracontomelon dao

paldao

by Jim Flynn

SCIENTIFIC NAME

Dracontomelon dao. Derivation: The genus name is derived from the Greek *drakes* meaning dragon. The specific epithet is from a local venacular. Synonyms are *D. cumingiamum* and *D. edule*

FAMILY

Anacardiaceae, the cashew family.

OTHER NAMES

While paldao is the preferred common name for these species, the wood is often marketed as New Guinea walnut. It is not, however, a true walnut of the genus *Juglans*. Common names include lamio, dao, damoni, dorea, loup, and New Guineawood.

DISTRIBUTION

The growth range of paldao is widely distributed throughout Southeast Asia and the islands of the southwest Pacific.

THE TREE

A large tree towering to 120 feet in height, its bole is often clear to 80 feet and is supported by very high buttresses that may extend up to 20 feet.

THE TIMBER

Overall, paldao wood is a grayish-brown color with a greenish cast. It contains irregular bands of different color intensity as the width varies. These bands are often dark brown to almost black. The sapwood is very wide and pale in color with no distinguishing features. The grain is interlocked and sometimes wavy. It has a medium density. There is neither taste nor odor. The average reported specific gravity is about 0.50 (ovendry weight/green volume), equivalent to an air-dried weight of 38 pcf. Paldao is slightly lighter than American black walnut (*Juglans nigra* WDS 150) but fairly close to the European walnut, *J. regia*. The wood is considered fairly strong and tough.

SEASONING

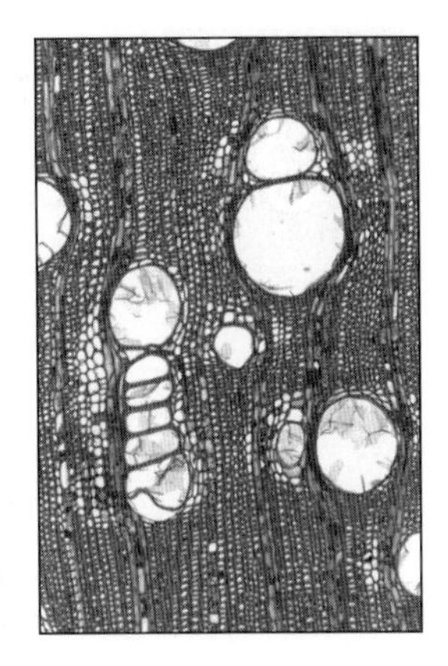

Moderate shrinkage occurs when drying, and the wood has a tendency to warp. Average reported shrinkage values (green to ovendry) are 3.9% radial and 7.5% tangential. Preliminary air-drying stacks should be stabilized with weights to minimize twisting and cupping. Once dry, the wood is dimensionally stable.

DURABILITY

The wood is not classified as resistant to termite and marine borer attacks. Depending upon the origin of the timber, durability characteristics may vary. Stock from Papua New Guinea seems to be more resistant to fungal attack. Preservatives do not satisfactorily penetrate the heartwood.

WORKABILITY

The timber is easy to work with both machine and hand tools. It is easy to bend. Peeling for veneer can be done successfully, although slicing is preferred in order to best display the figured grain. It glues easily and takes a good finish and polish.

USES

Primarily used as a veneer, paldao is also used for furniture and cabinetry, paneling, gunstocks, flooring and decorative work. It makes good moldings.

SUPPLIES

Paldao is available on the world market at a fairly high price.

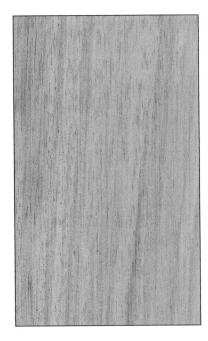

Dryobalanops aromatica

kapur

by Jon Arno

SCIENTIFIC NAME

Dryobalanops aromatica. Derivation: The genus name *dryo* is Greek for oak, and the balance of the name is probably a geographic location. The specific epithet indicates that the wood has a fragrant or aromatic odor when freshly cut. A synonym is *D. sumatrensis.*

FAMILY

Dipterocarpaceae, the dipterocarp family.

OTHER NAMES

Borneo camphorwood, keladan.

DISTRIBUTION

The half dozen or so species of *Dryobalanops*, the wood of which is marketed as kapur, range from Malaysia to Indonesia, including the island of Sumatera (formerly Sumatra).

THE TREE

Kapur is a huge, heavily buttressed tree capable of yielding cylindrical, blemish-free logs which when cut from the bole between the buttress and the lower branches of the crown can be up to 100 feet in length. The maximum overall tree height is about 250 feet. Although trunk diameters of over 10 feet are known, the diameter normally ranges between 4 and 6 feet as measured directly above the buttress.

THE TIMBER

Kapur is very similar to lauan (*Shorea* spp. WDS 249) in terms of the pore pattern and figure but is of a slightly finer texture than typical lauan. The narrow sapwood is a light yellowish-brown and is clearly demarcated from the reddish-brown heartwood. Kapur is more lustrous than the softer lauans and has a distinct camphor-like odor when freshly cut. This pleasant scent, however, is not as permanent as in true camphorwood. Kapur is easily confused with keruing, another closely related timber in the genus *Dipterocarpus*, but is generally a little less resinous than the latter which is known to literally bleed resin, sometimes even after it is seasoned and surfaced. Kapur's density varies by species and growing conditions. The average reported specific gravity can range between 0.57 and 0.65 (ovendry weight/green volume), equivalent to an air-dried weight of 44 to 51 pcf. In other words, compared to the more familiar species, kapur's density varies between that of sugar maple (*Acer saccharum* WDS 010), which has a specific gravity of 0.56, and pignut, the heaviest hickory, which has a specific gravity of 0.66. Kapur imported from Malaysia tends to be heavier than that coming from Borneo. Kapur compares closely to teak (*Tectona grandis* WDS 261) in most strength properties but is slightly more elastic.

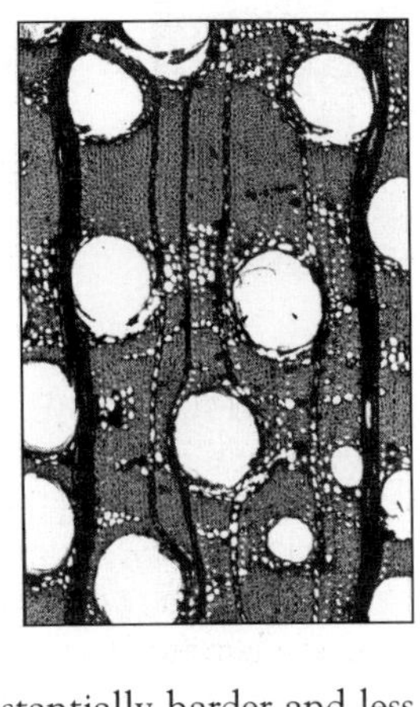

SEASONING

This somewhat resinous and dense wood dries slowly and has a pronounced tendency to cup. The ratio between its average reported tangential shrinkage of 10.2% (green to ovendry) and its radial shrinkage of 4.6% exceeds 2:1, and this relationship generally results in high drying stress.

DURABILITY

Kapur has good weathering properties and is resistant to fungi but vulnerable to termites.

WORKABILITY

Although kapur's pleasant scent offers some rationale for suggesting it as a substitute for true camphorwood, in my opinion, it is a poor stand-in (see Special Note). Being a substantially harder and less stable wood, it is both more difficult to work and less reliable once in use. While not quite as resinous as keruing, it has a tendency to gum up saw blades and, like keruing, the silica content is often high enough to dull cutters.

USES

Uses include heavy construction work, furniture, flooring, boat framing and plywood cores.

SUPPLIES

Supplies are plentiful and, for the moment, it is available in wide, clear boards. While it is comparable to the lauans in terms of price, it is a little more difficult to locate on the American market.

SPECIAL NOTE

The use of the trade name Borneo camphorwood for this wood is very misleading. Kapur is not related to true camphor wood (*Cinnamomum camphora* WDS 082), which belongs to the laurel family, Lauraceae. Since true camphorwood is becoming scarce and expensive, finding a substitute makes good sense, but clearly the better choice is the so-called African camphorwood, *Ocotea usambarensis*, which also belongs to the laurel family. It is considerably less dense than kapur.

Dyera costulata

jelutong

by Max Kline

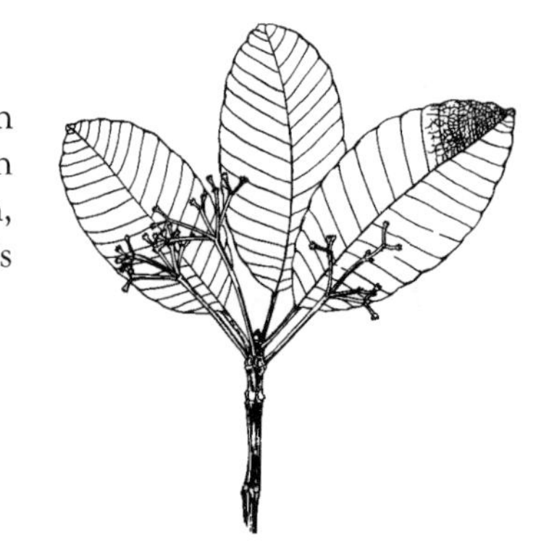

SCIENTIFIC NAME

Dyera costulata. Derivation: The genus name is in honor of Sir W. T. Thiselton (1843 to 1928), a renown botanist and the first President of the Botany Section, British Association in 1895. The specific epithet is Latin for ribbed, referring to the ribbed seedpod.

FAMILY

Apocynaceae, the dogbane family.

OTHER NAMES

jelutong bukit, jelutong paya.

DISTRIBUTION

Found in Malaysia and Indonesia.

THE TREE

Growing at elevations from sea level up to 1,400 feet, jelutong can reach a height of 200 feet with a diameter up to 8 feet. The bole is straight and cylindrical and often 90 feet long. Widely distributed in Malaysia, it occurs less frequently in Borneo.

THE TIMBER

The sapwood and heartwood are very similar in color, almost white when freshly cut, turning a pale straw-yellow color on exposure to air. Sometimes the wood is discolored due to attack by staining fungi. The texture is fine and even, and the grain is straight. Jelutong is light weight and is correspondingly low in strength. Average reported specific gravity is 0.36 (ovendry weight/green volume), equivalent to an air-dried weight of 27 pcf. The surface is slightly lustrous. It has no taste but does have a distinct slightly sour odor. The longitudinal surface of this wood has latex passages that occur at remarkably regular intervals of approximately 3 feet or so; therefore, the wood has to be used in short lengths.

SEASONING

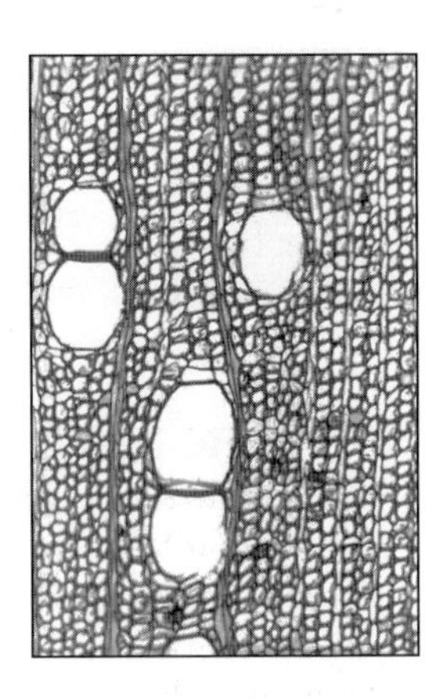

Jelutong can be air and kiln dried very easily with little tendency to warp or split, although staining is apt to be problematic. Average reported shrinkage values (green to ovendry) are 2.3% radial, 5.5% tangential and 7.8% volumetric. During air drying ample air circulation is essential, but once dried there is little dimensional change.

DURABILITY

The timber is not durable when used for exterior purposes. It is not resistant to termites or powder post beetles, but it can be easily treated with preservatives.

WORKABILITY

Since the grain is straight, jetulong will plane to a good clean surface. It has little dulling effect on sharp tools. It has good nail and screw holding properties and can be glued satisfactorily. The timber is best finished with paint or varnish, as it does not respond well to polish. It can be stained with satisfactory results.

USES

Because of its fine even texture, good working properties and stability, jelutong is acceptable as an alternative for other species for pattern making and drawing boards. It is popular for model making, carving, wood shoes and battery separators. The latex is used in chewing gum.

SUPPLIES

Recently, jelutong has become popular as a carving wood and is readily available worldwide in thick sizes.

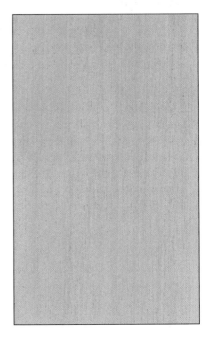

Ebenopsis ebano

ebony blackbead

by Max Kline

SCIENTIFIC NAME

Ebenopsis ebano. Derivation: The genus name is Greek for black. The specific epithet means ebony-like. A synonym is *Pithecellobium flexicaule*.

FAMILY

Fabaceae or Leguminosae, the legume family; (*Mimosaceae*) the mimosa group.

OTHER NAMES

Texas ebony, apes-earring, ebony ebano, palo fierro.

DISTRIBUTION

Southwestern Texas and southward in Mexico to Neuvo Leon and the Yucatan Peninsula.

THE TREE

Ebony blackbead is an evergreen shrub, which under the best growing conditions can reach a height of 50 feet with a trunk diameter of 2 to 4 feet. The tree is similar to the orange tree (*Citrus sinensis*) in appearance. The leaflets are shiny on the upper side, round and leathery. The light yellow flowers are very fragrant. The fruit is a pod; the seeds are edible and nutritious and when roasted are used by the Mexicans as a substitute for coffee.

THE TIMBER

Ebony blackbead sapwood is a clear yellow color while the heartwood is dark reddish-brown tinged with purple. It is heavy, compact, close grained and hard. It has an oily or waxy appearance and feel. Average specific gravity is 0.79 (ovendry weight/green volume), equivalent to an air-dried weight of 63 pcf. There is no distinctive odor or taste. The texture is fine, and the grain is fairly straight. The luster is outstanding, and it has a high natural polish. Ebony blackbead is noted for its strength.

SEASONING

Very little information has been recorded on this characteristic, and shrinkage values are not available. Ebony blackbead timbers are best cut and the ends sealed for later use. In the meantime, the timbers should be protected from excessive and uncontrolled weathering. When cut, the slabs are best stickered and air dried.

DURABILITY

The durability of ebony blackbead is excellent.

WORKABILITY

Because of its hardness, ebony blackbead must be carefully worked with hand tools, but it can be machined and turned with ease. It finishes with a high natural polish and a golden luster. To obtain maximum beauty, it should only be finished in its natural color.

USES

In Mexico this wood is used for fence posts, wheelwright work and fuel. Because of its strength, it is used for various farm purposes. It is a beautiful timber for craftwork and is well suited for use as knife handles and brush backs. If it were more easily obtained, it would find widely expanded uses by cabinetmakers.

SUPPLIES

Because of its limited growth range, ebony blackbead is not cut commercially for timber. Small quantities, however, are available from wood collectors who find it a beautiful species for novelties.

Elaeagnus angustifolia

Russian-olive

by Jon Arno

SCIENTIFIC NAME

Elaeagnus angustifolia. Derivation: The genus name is Greek for olive and the classical name for the chaste-tree. The specific epithet means narrow leaf.

FAMILY

Elaeaganaceae, the elaeaganus or oleaster family.

OTHER NAMES

oleaster.

DISTRIBUTION

Russian-olive is neither a true olive nor uniquely Russian. It is native to southern Europe and central Asia and was brought to America during colonial times. Because of its ability to resist drought and its extreme tolerance to cold, it has been widely used as a windbreak species throughout the plains states and central Canada.

THE TREE

Seldom exceeding 25 feet in height, Russian-olive is a short tree with a round crown. Its silhouette is similar to orchard-grown species such as apple or plum. The bole is usually very short and almost never straight. The underside of the leaves are a beautiful silver-gray color and contrast sharply with its dark brown, almost black, bark. Because of its striking color, small size and remarkable hardiness, it is a popular ornamental in urban areas. The small, olive-shaped fruit contains a single seed, and the outer flesh becomes dry and scaly when mature. While not palatable by human standards, many species of birds are attracted to its fruit.

THE TIMBER

Russian-olive is a ring-porous wood with a very attractive tangential figure. The heartwood is light cinnamon brown and contrasts sharply with the light cream-colored sapwood. The latewood is lustrous and lighter in color than the soft, dull and very porous earlywood, which highlights the striking ribbon grain in radially cut boards. Since Russian-olive is not recognized as a commercial timber species, there is virtually no data in the literature on its mechanical properties. A standard 3- by 6- by 0.5-inch sample had a specific gravity of 0.55 (ovendry weight/green volume), equivalent to an air-dried weight of 42 pcf.

SEASONING

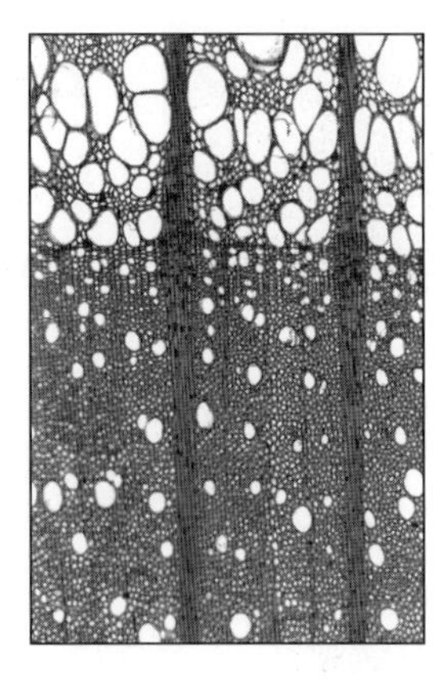

This species is difficult to season. The wood has a high tendency to ring check (separate along the annual rings) as it shrinks. Great care must be taken to thoroughly coat the end grain and to allow the wood to season slowly. Shrinkage values are not available.

DURABILITY

Again, there is no published data, but the wood's high susceptibility to checking suggests that it would perform poorly when exposed to extreme changes in humidity.

WORKABILITY

Russian-olive has working qualities somewhat comparable to black ash (*Fraxinus nigra* WDS 123). Because it is naturally lustrous, it can easily and quickly be rubbed to a fine polished surface. It takes glues and finishes well, but is virtually impossible to fasten with screws or nails without splitting. Even thin-gauged brads require carefully matched pilot holes. The most appealing features of Russian-olive are its natural warm brown color and its very attractive open-grained figure. The end result is rewarding, but getting there is not always pleasant.

USES

Russian-olive is not a commercial timber species. Carvers, turners and hobbyists with powerful curiosities and great patience use it on a limited basis.

SUPPLIES

There are no known commercial sources of supply, but it is readily available throughout most of temperate North America and Europe for those who forage wood piles and windfalls.

Endiandra palmerstonii

Queensland walnut

by Max Kline

SCIENTIFIC NAME

Endiandra palmerstonii. Derivation: The genus name is Latin for two stamens. The specific epithet is in honor of Australian H. J. Palmerston (1784 to 1865).

FAMILY

Lauraceae, the laurel family.

OTHER NAMES

oriental wood, walnut bean, Australian laurel, Australian walnut.

DISTRIBUTION

This tree is confined to northern Queensland, Australia chiefly in the coastal districts.

THE TREE

Queensland walnut reaches a height of 120 to 140 feet with a diameter up to 6 feet. The base is buttressed, but the bole above the buttress is usually well shaped and unbranched for 80 to 90 feet. It thrives well in areas of heavy rainfall and is one of the most common trees of the Queensland area.

THE TIMBER

The color of Queensland walnut varies from light pinkish-brown to dark brown, often with pinkish, grayish-green or almost black streaks. It has a peculiar, disagreeable odor when freshly cut that disappears when the wood is dried. The taste is not distinct. Although not a true walnut, the wood has a fine, even texture similar to so many walnut species but is more lustrous and heavier. Average reported specific gravity is 0.55 (ovendry weight/green volume), equivalent to an air-dried weight of 42 pcf. The grain is variable from straight to wavy which in veneer is beautiful and striking. The timber contains deposits of silica, which occur in the form of crystalline aggregates in the ray cells. This timber has good strength properties, considerably better than most of the walnut species.

SEASONING

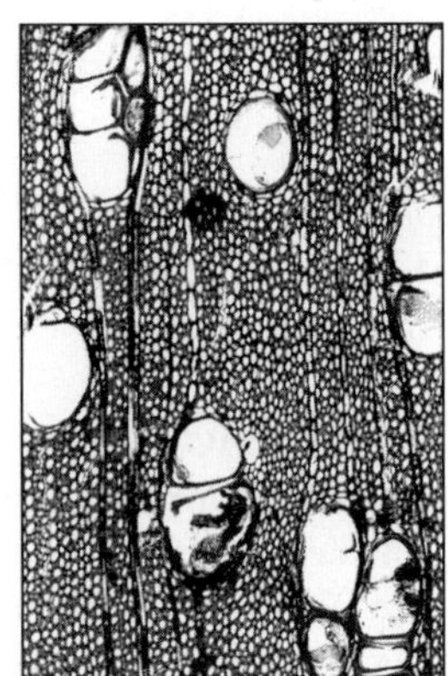

If degrade is to be avoided, Queensland walnut needs care during the seasoning process. When air drying, splits may occur in thick timbers unless end coatings are used, and in thin dimensions there is some tendency to warp that may be overcome by suitable weighting of the stack. Kiln drying takes place fairly rapidly in thinner stock with some tendency to warp. Average reported shrinkage values (green to ovendry) are 4.5% radial and 8.6% tangential. Once seasoned, the timber holds its shape well.

DURABILITY

Although occasionally damaged by wood-boring pests, Queensland walnut is fairly resistant to all types of fungal attack. For the purposes for which it is used, preservative treatment is unlikely.

WORKABILITY

When working this species, sawing is likely to prove troublesome and due to its high silica content tungsten-carbide tipped saws are essential. When very sharp tools are used, a good finish can be obtained. In other operations, the dulling effect is severe and frequent sharpening is an absolute necessity. The wood can be glued satisfactorily and nail and screw holding properties are good. It polishes excellently and peels well for veneer. Stains can be evenly applied. This species is not recommended for bending purposes.

USES

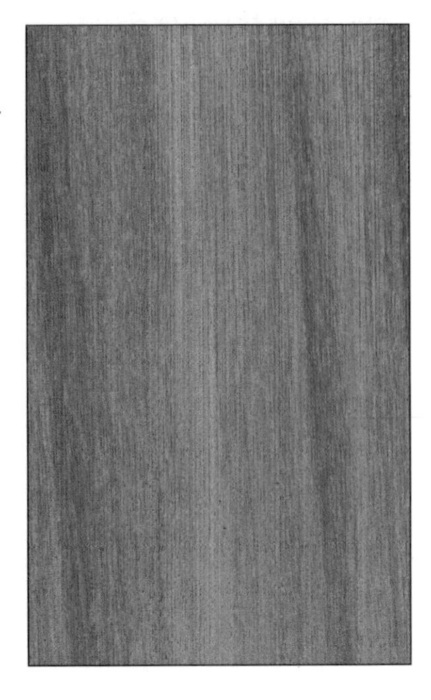

This wood is one of the most popular species in Australia for cabinetry, interior decoration, paneling and to a limited extent as a moderately wear-resistant flooring timber. An unusual use of Queensland walnut is for electrical insulation purposes where it is reported that its insulation value is 50 times greater than most woods.

SUPPLIES

Queensland walnut is scarce on the American market and considered costly.

Endospermum macrophyllum

kauvula

by Jim Flynn

SCIENTIFIC NAME

Endospermum macrophyllum. Derivation: The genus name is Greek meaning nutritive matter in seed-plant ovules derived from an embryo sac. The specific epithet means large leaves.

FAMILY

Euphorbiaceae, the spurge family.

OTHER NAMES

gubas, sesendok, sendok sendok, terbulan, ekor belangkas.

DISTRIBUTION

Western Pacific islands, including large growths in the Philippines, New Guinea and Fiji. In Malaysia, it is common in lowland forests.

THE TREE

Kauvula grows to heights of up to 100 feet and has heavily buttressed, clear boles that are up to 3 feet in diameter.

THE TIMBER

There is no difference in color between the sapwood and heartwood which are pale yellow with a buff hue. The texture is medium to coarse with shallowly interlocked grain. The wood is lustrous when dry and lightly sanded. Neither distinctive odor nor taste can be discerned. The wood is non-siliceous and very light. Average reported specific gravity is 0.38 (ovendry weight/green volume), equivalent to an air-dried weight of 28 pcf. Because of its similarity with Parana-pine (*Araucaria angustifolia* WDS 023), alerce (*Fitzroya cupressoides*), obeche (*Triplochiton scleroxylon* WDS 272) and other woods, the Forestry Department of the Republic of Fiji recommends kauvula as an acceptable substitute for these species.

SEASONING

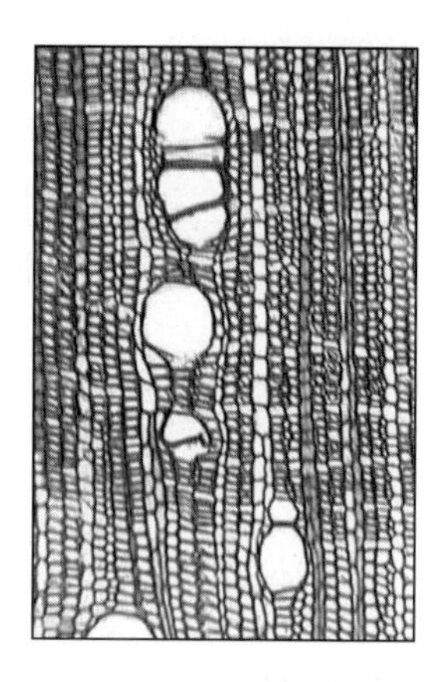

Green kauvula is readily kiln dried with little difficulty. Longitudinal shrinkage in tension wood is high causing warped boards and splitting. Selection of straight logs and grading of defective lumber after milling is necessary to ensure the high quality of finished stock. The average reported shrinkage values (green to 12% moisture content) are 2.0% radial and 3.6% tangential.

DURABILITY

Kauvula is nondurable and particularly prone to blue stain and powder post beetle attack. The wood can be treated with preservatives.

WORKABILITY

The workability of the timber is exceptionally good in both the green and dry states. Responsive to hand tools, it can be compared to eastern white pine (*Pinus strobus* WDS 212) in appearance and feel. It machines well producing crisp edges and smooth surfaces. Cutting against the grain may result in fibrous surfaces and some chipping, but this is easily handled by sanding. The stock glues and nails well. It is an excellent wood for staining and may require some light filling. Kauvula is easily turned.

USES

The wood is ideal for manufacturing matches. Traditional uses also include moldings, cabinets, furniture, paneling, pattern making, boxes and crates (e.g., banana), carvings and wood shoes. There is considerable potential for the wood to be used in the manufacture of plywood. With appropriate preservatives, it makes excellent shingles.

SUPPLIES

Commercial operations in the area where the species is available result in the marketing of significant stock. Kauvula is marketed under its own name, but does not seem to reach world markets on a large scale. A significant amount from the Republic of Fiji is exported to New Zealand at a fairly low cost.

Entandrophragma cylindricum

sapele

by Max Kline

SCIENTIFIC NAME

Entandrophragma cylindricum. Derivation: The genus name is Greek meaning within the male membrane. The specific epithet is Greek for cylindrical referring to the fusiform, cylindrical fruits.

FAMILY

Meliaceae, the mahogany family.

OTHER NAMES

scented mahogany, aboudikro, penkwa, muyovu, libuyu, sapele mahogany.

DISTRIBUTION

In Africa from Ghana to the Congo (formerly Zaire) and Uganda.

THE TREE

Sapele grows to a height of 150 feet with diameters at breast height of 4 to 5 feet. The grayish-brown bark flakes but often may be smooth. The size of the buttresses vary. The straight cylindrical bole is clear for 80 to 100 feet. The deciduous leaves are pinnate with five to nine pairs of leaflets. Each narrow leaflet is 5 inches long and pointed at the tip. The fruits are pendulous capsules about 4 inches long. These then split into five valves to reveal 15 to 20 seeds.

THE TIMBER

The timber is light red to dark reddish-brown usually with a purplish cast. The distinct sapwood is white or pale yellow. The grain is interlocked, sometimes wavy, producing a narrow, uniform figure when quartersawn. It has a cedar-like aromatic odor. The taste is not distinct. The texture is medium, and the luster high and golden. Average reported specific gravity is 0.55 (ovendry weight/green volume), equivalent to an air-dried weight of 42 pcf. The timber is in the same strength class as oak, being considerably stronger than either African or American mahogany.

SEASONING

Sapele seasons fairly rapidly with a marked tendency to distort. Average reported shrinkage values (green to ovendry) are 4.6% radial, 7.4% tangential and 14.0% volumetric. Excessive temperatures at the start of kiln drying should be avoided. Movement in service is rated as medium.

DURABILITY

Moderately resistant to termites, sapele is susceptible to attack by pinhole borers. Powder post beetles often damage the sapwood. It is resistant to preservative treatment.

WORKABILITY

Sapele is not difficult to work but will take the edge off tools more quickly than African mahogany. In planing and molding, the surface is likely to tear due to the interlocked grain. Glued joints are sound and nails and screws hold firmly but thin stock may split during nailing. The timber responds excellently to stain and polish treatment, but the finish may be non-uniform if the wood has not been properly surfaced.

USES

Since sapele belongs to the same botanical family as true mahogany, its uses are very similar. Uses include furniture and cabinetry, decorative veneers, paneling, flooring and plywood. It is widely used in joinery items such as staircases and window frames.

SUPPLIES

Sapele veneer and lumber are available at moderate prices.

Enterolobium cyclocarpum

guanacaste

by Max Kline

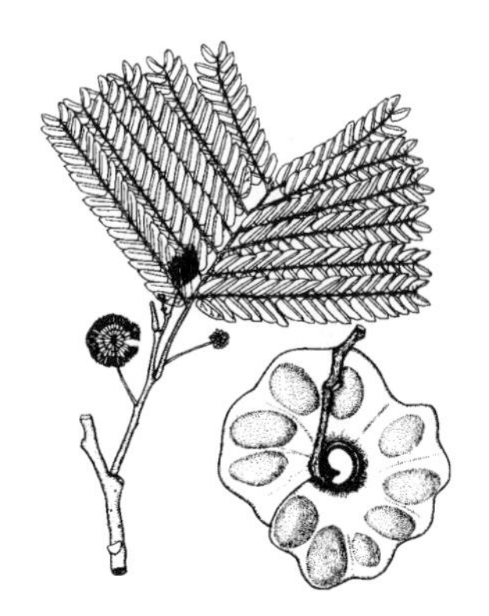

SCIENTIFIC NAME

Enterolobium cyclocarpum. Derivation: The genus name is Greek meaning intestine-like pod. The specific epithet is Latin referring to circularly rolled up fruit.

FAMILY

Fabaceae or Leguminosae, the legume family; (*Mimosaceae*) the mimosa group.

OTHER NAMES

kelobra, perota, jenisero, kolobra, eartree, jarina, ore-jero, carocaro.

DISTRIBUTION

Mexico and southward through Central America to Venezuela, Trinidad, Guyana and Brazil.

THE TREE

Growing to a height of 60 to 100 feet, guanacaste is an excellent shade tree with a broad top and a gray buttressed trunk that is 3 to 6 feet in diameter. The ear-shaped pods are called *orejas de negro* in Spanish. The bipinnate leaves are fern-like with 20 to 30 pairs of leaflets which combine to form a wide-spreading crown.

THE TIMBER

Frequently containing darker streaks, guanacaste heartwood is pale brown to dark walnut brown and is sharply demarcated from the whitish sapwood. The luster is high. Crotches are often highly figured, and the grain is typically interlocked. The texture is coarse. The wood is without distinctive odor or taste. Average reported specific gravity is 0.34 (ovendry weight/green volume), equivalent to an air-dried weight of 25 pcf.

SEASONING

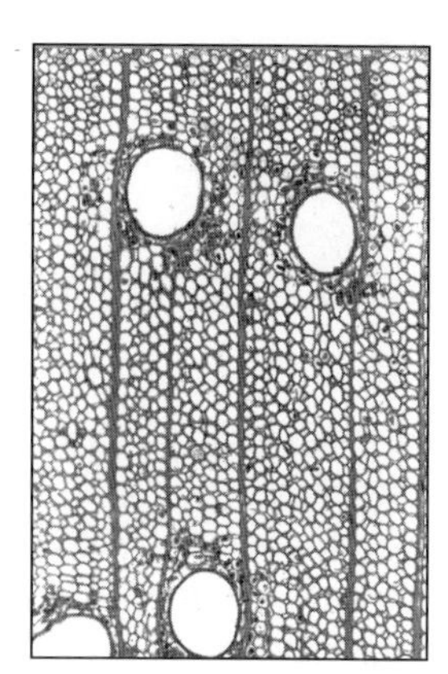

Guanacaste seasons with little difficulty and does not warp or check. Average reported shrinkage values (green to ovendry) are 2.0% radial, 5.2% tangential and 7.2% volumetric. It holds its place well after manufacturing.

DURABILITY

The wood is very durable in water. The large trunks can be used for canoes or water troughs. It is also resistant to drywood termite attack.

WORKABILITY

The timber is easy to work and finishes smoothly to a high luster. Tension wood often results in fuzzy grain. Raised and chipped grain is common in planing, and rough end grain can be experienced in shaping. Dust from working it has a disagreeable pungent odor and may be an allergen to some woodworkers.

USES

Guanacaste is useful and fairly attractive, but cannot be classified as a high-grade furniture wood even though it is used when making furniture. The wood is also used in cabinetry, interior finish, veneer, gunstocks, corestock and pattern woods. The pods are an excellent cattle food, and the seeds are sometimes cooked for human consumption. The fruit and bark are rich in tannin and used as a substitute for soap. Syrup made from the bark is used as a cold remedy.

SUPPLIES

Quartered veneer, crotches and swirls are scarce. The lumber is available at a medium price.

Erythrina crista-galli

cockspur coralbean

by Jim Flynn

SCIENTIFIC NAME

Erythrina crista-galli. Derivation: The genus name is derived from the Greek for red, referring to the large, bright red flowers. The specific epithet is Latin for cock spur.

FAMILY

Fabaceae or Leguminosae, the legume family; (*Papilionaceae*) the pea or pulse group.

OTHER NAMES

common coral tree, fireman's cap, cockspur coral bean, cresta de gallo, cry baby tree, dragon's tooth, immortelle, el siebo, amasisa, boro and ahuejote. In addition, there are at least five or more names in the vernacular of each county in which the tree grows.

DISTRIBUTION

There are over 130 species of *Erythrina* growing in various parts of the warm climates. This particular species is native to Brazil, Paraguay, Uruguay and northern Argentina. Because of its attractive flowers, it is widely cultivated in other parts of the world including the coastal plain of the Gulf States in the United States; it does not appear to have become naturalized.

THE TREE

Cockspur coralbean is the national tree of Uruguay. Its flower is the national flower of Uruguay and Argentina. In its native range, its growth is considered a part of the typical landscape of the country. It grows best on the banks of rivers and streams where its large racemes and terminals of bright red flowers brighten the area. The tree is not large, growing, at best, to about 25 feet in height. The alternate, trifoliate leaves are large, often 12 inches long including the petiole. The leaflets are 2 to 6 inches long and vary in shape from elliptic to oval or ovate. The petioles are slender, angular, finely grooved and somewhat spiny. The fruit is a cylindrical legume 3 to 10 inches long with seeds that vary in color from yellow to orange or red. The seeds produce a strong alkaloid that is used for making medicines and insecticides. The tree is widely used as an ornamental and readily takes root from stem cuttings.

THE TIMBER

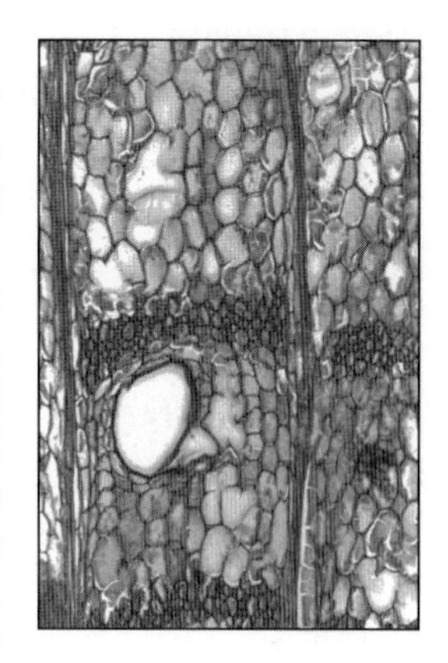

The wood is light weight and has an average specific gravity of only 0.25 (ovendry weight/green volume), equivalent to an air-dried weight of 18 pcf. The wood has a yellowish cast and may at times be streaked with lighter or darker colors. There is neither odor nor taste to the timber. There does not appear to be any distinguishable features to the heartwood, although some species of *Erythrina* may have chambered piths. Rays observed in the end grain are prominent and uniform and can be viewed with the naked eye. While the grain is usually straight, depending upon the growing conditions, it can acquire a twisted configuration. It often seems fuzzy to the touch as if it needs to be sanded. For its weight, the timber is considered tough and strong.

SEASONING

As there is no widespread commercial value, there is no data on this characteristic and shrinkage values are not available.

DURABILITY

Cockspur coralbean is not a durable wood and is subject to blue stain fungi. An application of household bleach usually arrests this condition. The wood is perishable when in contact with the ground unless it has been freshly cut, and in this situation it will grow.

WORKABILITY

Insufficient supplies of this wood prevent a deeper discussion of this characteristic. Machining sample patches of the timber revealed no problems in working because of its light weight and even grain. It sands well when dry and glues without any problem. Even though the grain is rather tight, it is doubtful that a decent finish can be obtained because of the softness of the wood. The wood can be carved with ease.

USES

Cockspur coralbean is often used to make woodenware, floats and corks.

SUPPLIES

Because the tree is widely planted in many parts of the world as an ornamental, there should be opportunities to harvest some of the wood for crafts and wood samples. It is probably not available in commercial markets except in areas where the trees abound.

Eucalyptus marginata

jarrah

by Max Kline

SCIENTIFIC NAME

Eucalyptus marginata. Derivation: The genus name is from the Greek *eu* meaning well and *kalyptos* meaning covered. The specific epithet is Latin and refers to the veins in the leaf margins.

FAMILY

Myrtaceae, the myrtle family.

OTHER NAMES

Western Australian mahogany.

DISTRIBUTION

Jarrah occurs in a 20-mile-wide belt along the coastal area of southwestern Australia from the latitude of Perth extending 200 miles southward.

THE TREE

Jarrah reaches a height of 100 to 150 feet with a diameter of 3 to 5 feet. Jarrah is one member of the great Australian family of Eucalypts and is one of the most important timber species of Australia.

THE TIMBER

The heartwood is reddish-brown but darkens with exposure. The texture is even, though fairly coarse, and the grain is commonly interlocked or wavy. The luster is medium. Average reported specific gravity is 0.68 (ovendry weight/green volume), equivalent to an air-dried weight of 53 pcf. The wood has no distinctive odor or taste. Gum veins or pockets are a common defect.

SEASONING

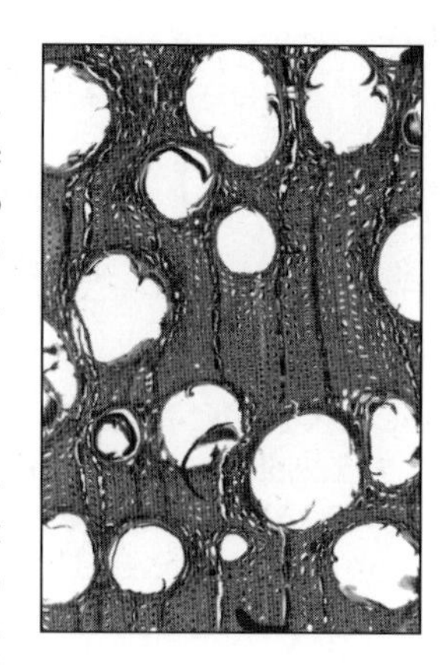

Warping is the most serious problem when kiln drying jarrah. It is recommended that partial air drying be done before placing the wood in the kiln. Wide stock has a tendency to check. Average reported shrinkage values (green to ovendry) are 7.7% radial, 11.0% tangential and 18.7% volumetric. Movement in use is rated as medium.

DURABILITY

The timber is classified as durable in Australia. It is long-lasting in the ground and is resistant to termites and not subject to marine borers. It is also fire-resistant. Jarrah is 10% to 20% superior to oak in most strength properties but 40% to 50% harder. It is moderately good for steam bending, and its resistance to abrasion is good.

WORKABILITY

Jarrah is not an easy timber to work with dull tools. With sharp machine tools, it can be worked cleanly. Pre-boring is advisable before nailing but nail and screw holding properties are good. It is rarely stained because of its natural color, but is often finished with wax or varnish. The wood can be glued satisfactorily. It can be sliced and peeled into veneer without difficulty.

USES

Uses include fencing, domestic and industrial flooring, general construction, railroad crossties, furniture components, paving blocks, piles, sleepers, framing of piers and jetties and interior joinery.

SUPPLIES

Jarrah is exported throughout the world, and the price is moderate. It can be obtained at a low price in Australia where it is readily available.

Eucalyptus regnans

mountain ash

by Max Kline

SCIENTIFIC NAME

Eucalyptus regnans. Derivation: The genus name is from the Greek *eu* meaning well and *kalyptos* meaning covered. The specific epithet means kingly.

FAMILY

Myrtaceae, the myrtle family.

OTHER NAMES

Australian oak, canary ash, giant gum, swamp gum, stringy gum, argento. [Ed. Note: Tasmanian oak is a trade name for a mixture of three species of eucalypts. Mountain ash is one of them. The other two are *E. delegatensis* (Alpine ash) and *E. oblique* (messmate stringybark). These three species may or may not be marketed separately. It appears that they are marketed separately in Australia but for export trade they are combined and marketed as Tasmanian oak.]

DISTRIBUTION

Northeastern Victoria in Australia and northeastern Tasmania.

THE TREE

Mountain ash is one of the tallest trees in the world, occasionally reaching over 300 feet in height with diameters of 7 feet or more. The boles are clean and only slightly tapered. The smooth bark is white. The alternate, petiolate, thin, lanceloate, adult leaves are 4 to 7 inches long by 0.75 to 1.25 inches wide. Mountain ash is a fast-growing tree so future supplies are assured.

THE TIMBER

The wood is pale yellowish-brown, sometimes with a pinkish cast and often with darker brown streaks. The luster is high and satiny. Odor and taste are not distinct. It is moderately hard. Average specific gravity is 0.52 to 0.53 (ovendry weight/green volume), equivalent to an air-dried weight of 40 pcf. The grain is straight to wavy, and the texture is medium. Flatsawn timber bears a general resemblance to European and American oak.

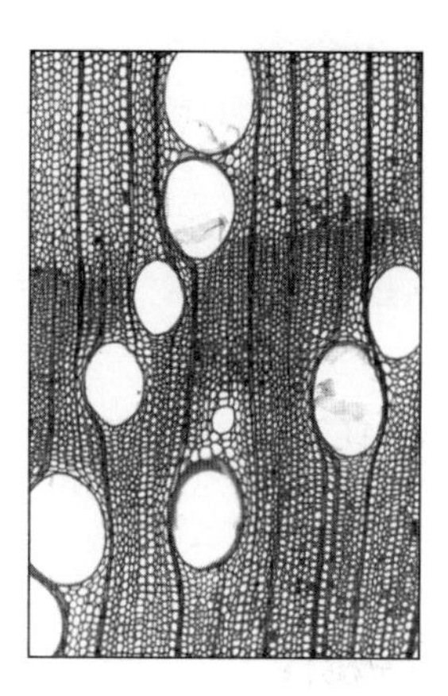

SEASONING

Care must be exercised when seasoning due to the tendency to warp, check and collapse. Shrinkage values are not available. Mountain ash does not have good dimensional stability.

DURABILITY

Mountain ash is about 50% stiffer, twice as tough and 20% stronger in bending than European oaks. It is classified as non-durable and is resistant to preservative treatment.

WORKABILITY

The wood works with moderate ease with hand and machine tools. It finishes well, takes stain and can be glued satisfactorily. Treatment with ammonia fumes produces a pale walnut color. The timber has a slight tendency to split when nailing but takes screws without difficulty.

USES

Mountain ash is used in the furniture and joinery industries. It is also used for boxes and crates, flooring, cooperage, handles, brooms, rakes, oars, veneer and plywood. It is also pulped to make paper. Beautiful turnings can be made with wood that has blister or fiddleback figure.

SUPPLIES

The timber is readily available in large sizes in Australia at medium prices and is exported throughout the world. See Other Names.

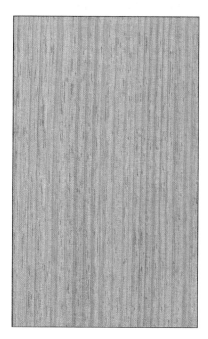

Eucryphia cordifolia

ulmo

by Jim Flynn

SCIENTIFIC NAME

Eucryphia cordifolia. Derivation: The genus name is Greek for good or well. The specific epithet is Latin for heart-shaped leaf.

FAMILY

Eucryphiaceae, the eucryphia family.

OTHER NAMES

gnulgu, muermo, roble de Chile.

DISTRIBUTION

Argentina and Chile. Ornamental plantings may be found outside of its native range.

THE TREE

There are only five species in the genus *Eucryphia*, some of which survive only as shrubs. Two of the species are native to Australia and Tasmania and include Tasmanian leatherwood, *E. lucida.* Ulmo is an evergreen tree with simple, opposite leaves. Large flowers are borne singly in leaf axils. The fruit is a leathery or woody capsule with several boat-shaped valves and imbricated winged seeds. The tree is reported to grow as high as 130 feet and 24 inches in diameter. The well formed boles are often fluted or slightly buttressed and 40 to 60 feet long. Ulmo is known as the dogwood of Chile because of its beautiful flowers in spring. These flowers, similar in appearance to the white Rose of Sharon, are rich in nectar and an excellent source for honey.

THE TIMBER

The heartwood is reddish or grayish-brown and sometimes variegated. It is not well demarcated from the lighter colored sapwood. The luster is rather high. The texture is fine and uniform and the grain generally straight. There is no distinctive odor or taste. This moderately strong and hard wood has an average specific gravity of 0.48, equivalent to an air-dried weight of 36 pcf. The growth rings are distinct thus producing an attractive figure on plainsawn surfaces.

SEASONING

Ulmo is rather difficult to season and is prone to severe surface and end checking. As with most woods, it warps badly when tension wood is present. It is stable when dry. Average shrinkage values (green to ovendry) are 4.5% radial, 8.2% tangential and 13.2% volumetric.

DURABILITY

Ulmo is a non-durable wood and is prone to insect attack. Both the sapwood and the hardwood respond well to preservative treatments with good lateral penetration.

WORKABILITY

There are no reports of adverse conditions related to working with this species. It saws and planes cleanly with both machine and hand tools, takes a fine finish and glues well.

USES

In Chile, ulmo is predominantly used for flooring. Other uses include light structural material, furniture, veneer, interior trim, railway sleepers, piling and turnery. The bark is used for the extraction of tannin, and the root wood is used for tobacco pipes.

SUPPLIES

Although supplies may be limited, the wood is sold commercially at a medium high price.

Excoecaria africana

tamboti

by Jim Flynn

SCIENTIFIC NAME

Excoecaria africana. Derivation: The genus name is derived from Latin and Greek meaning spiral/coiled spike-like flowers. The specific epithet means the plant is "of Africa". A synonym is *Spirostachys africana*.

FAMILY

Euphorbiaceae, the spurge family.

OTHER NAMES

There are many spelling variations of the common name including tambooti, tambouti and tambotie. The trade name African sandalwood is also recognized. Other vernacular names are masarakana, mtolo, munhiti, mutomboti, mutivoti, tsomvori, ubande and umthombothi.

DISTRIBUTION

This species is indigenous to southern Africa but extends along the eastern coast as far north as Kenya. It thrives best at low altitudes along river and stream banks.

THE TREE

This rare medium-sized African tree reaches heights of only 50 feet in the best growing situations. The stem diameter is about 20 inches. It has a rounded crown. The alternate, simple leaves are ovate to elliptic, 1 to 3 inches long and 0.5 to 1.5 inches wide. They are light green and turn brilliant yellowish-red in the autumn. The very small flowers are shaped in slender catkin-like spikes (hence the name of the genus) resembling a rat's tail. The fruit is a two- to three-lobed capsule that splits when mature. The seeds jump like Mexican jumping beans when infested with moth larvae. The dark gray to blackish bark is quite rough and forms chunky flakes. The milky latex exuding from cut bark is extremely toxic and is used as a fish poison and to tip arrowheads. It causes extreme irritation to sensitive skin and severe pain and damage to the eyes. Africans are warned of this from early childhood and their fear has resulted in a taboo-like avoidance of the tree.

THE TIMBER

Tamboti is a rare and beautiful wood. The narrow, light yellow sapwood contrasts with the light and dark chocolate-colored heartwood, which is formed by the alternating bands of early and latewood. There is a persistent sweet odor, but I did not attempt the taste test. The texture is fine and even. The grain is generally straight and well figured on the tangential and radial surfaces. It is heavy with a locally measured specific gravity of 0.83 (ovendry weight/green volume), equivalent to an air-dried weight of 67 pcf. The drawback is that the wood is apt to cause serious problems for those who are sensitive to its toxicity. Keith Coates Palgrave's *Trees of Southern Africa* reports that the wood is not used as yokes for oxen because it causes burn-like sores on the animals' necks.

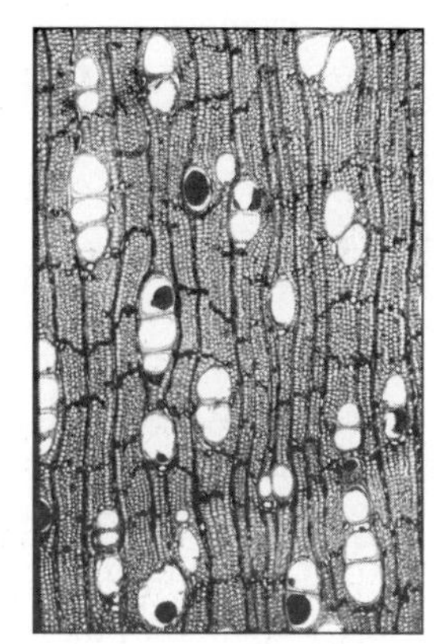

SEASONING

Very little data is available because the wood is not plentiful. According to a research report by the Zimbabwe Forestry Commission, it seasons well with minimal end checking if dried slowly and carefully stacked under cover. Average reported shrinkage values (green to 12% moisture content) are 2.55% radial and 4.8% tangential.

DURABILITY

The timber has an excellent durability rating. It is highly resistant to borers, termites and fungi. It also resists penetration by preservatives. It is said that insects avoid this wood and pieces are sometimes carried on a person's clothing as a repellent.

WORKABILITY

The struggle to acquire this species starts at the very beginning. The logs are very difficult to cut and tungsten-tipped saw teeth are a must. Dry stock is easier to saw due to the oil content. Nevertheless, it planes well and has a highly lustrous and smooth finish. Pre-drilling for nails and screws is required. It takes a high-gloss finish after the difficult sanding operations have been completed. It does not pose any problems when gluing. Care must be taken to avoid exposure to dust, both when inhaled and when in contact with the skin as it can be an irritant.

USES

Tamboti is one of the great woods for furniture, turnery, inlay work, carving and craft items of high value.

SUPPLIES

Tamboti is available on the commercial market but at a rather high price due to its scarcity and high demand.

Fagus grandifolia

American beech

by Max Kline

SCIENTIFIC NAME

Fagus grandifolia. Derivation: The genus name is the classical Latin name, from the Greek word meaning to eat, referring to the edible beechnuts. The specific epithet means large leaf. A synonym is *F. americana*.

FAMILY

Fagaceae, the beech family.

OTHER NAMES

red beech, white beech, stone beech, winter beech.

DISTRIBUTION

From Nova Scotia and New Brunswick to Minnesota, south to Florida, eastern Oklahoma and Texas.

THE TREE

The tree grows best in the bottomlands of the Ohio-Mississippi River valleys and along the western slopes of the southern Appalachians where it can reach a height of 100 feet with a diameter of 30 to 50 inches. The bluish-gray bark is smooth. The small, edible nuts furnish food for wildlife.

THE TIMBER

The heartwood of American beech is dark to light reddish-brown with very narrow, nearly white sapwood. The growth rings are thin and tiny, and the rays are very numerous. It is hard, heavy, strong and uniform in texture. The grain is straight or sometimes interlocked. Quartersawn timber is quite attractive with rays conspicuous as darker flakes. The sheen is somewhat silvery. There is no taste or odor. Average reported specific gravity is 0.56 (ovendry weight/green volume), equivalent to an air-dried weight of 43 pcf.

SEASONING

Considerable care is required when seasoning as the wood has a high rate of shrinkage which can cause distortion and splitting. Average reported shrinkage values (green to oven-dry) are 5.5% radial, 11.9% tangential and 17.2% volumetric. In use, the wood is not considered dimensionally stable.

DURABILITY

American beech is not durable when in contact with the soil, but readily absorbs preservatives. It is rated high in strength properties and shock resistance.

WORKABILITY

The wood is somewhat difficult to work with hand tools, but can easily be handled with power equipment. It is hard to nail, has a tendency to split, but holds nails and screws well. A good finish can be obtained, and it can be stained to match other hardwoods. When steamed, the wood is readily bent.

USES

American beech has a variety of uses. Because it does not impart taste or odor to food, it is used extensively for food containers, baskets and butcher blocks. It also finds use for chairs, flooring, woodenware, handles, novelties, turnery, railroad ties when preserved, veneer and playground equipment. It wears well when subjected to friction under water, which creates some uses in the distillation field. [Ed. note: American beech is a prime wood in the manufacture of clothes pins.]

SUPPLIES

American beech veneer and lumber are plentiful at a moderate price.

Flindersia brayleana

Queensland maple

by Max Kline

SCIENTIFIC NAME

Flindersia brayleana. Derivation: The genus name is in honor of Captain M. Flinders (1774 to 1854), an explorer in the British Naval Service. The specific epithet is in honor of Professor E. W. Brayley (1773 to 1854), a member of the Royal Society of London.

FAMILY

Rutaceae, the rue or citrus family.

OTHER NAMES

maple silkwood, Australian maple, silkwood.

DISTRIBUTION

Northern Queensland in Australia, extending into Papua New Guinea.

THE TREE

Queensland maple is not a very tall tree. It seldom exceeds 100 feet in height, but its diameter can grow to 4 feet.

THE TIMBER

This timber is light brown or pinkish but the color fades with age to light brown. Planed surfaces are lustrous. The texture is close and even. The grain may be wavy or curly with a wide range of figure. Interlocked grain produces a stripe figure when quartersawn. When freshly cut, the wood is slightly scented. Average reported specific gravity is 0.45 (ovendry weight/green volume), equivalent to an air-dried weight of 34 pcf. Queensland maple is strong for its weight, similar to oak in most respects, but poor for steam bending. It has very good resistance to shock loads.

SEASONING

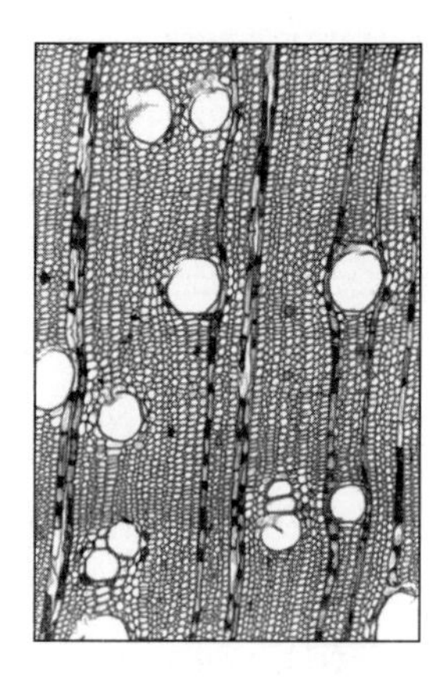

Due to the interlocked grain, some warping and cupping can occur when seasoning wide boards. Collapse can also be a problem, but there is no trouble with checking. Average reported shrinkage values (green to 12% moisture content) are 3.5% radial and 5.0% tangential.

DURABILITY

Queensland maple is rated as non-durable under conditions which promote decay. It can be considerably damaged by pinhole borers, but is moderately resistant to powder post beetle attack.

WORKABILITY

The timber works fairly satisfactorily with hand and machine tools with little dulling effect on the tool edges. It is not recommended for turnery, but is excellent for veneer and plywood. Care is needed in planing operations when the grain is highly interlocked to prevent the tendency to pick up. Nail and screw holding properties are good. It can be glued satisfactorily and will stain and polish evenly.

USES

Queensland maple is the premier cabinetwood of Australia, where it is widely used for decorative paneling, joinery and interior furniture, both in veneer and solid forms. Other uses include printing blocks, motor and railway vehicle construction, rifle stocks, frames, molding and airplane propellers.

SUPPLIES

The timber is available in Australia, but the price is high for well figured material. It is largely used for local consumption and exported wood is usually found only in veneer form at a high price.

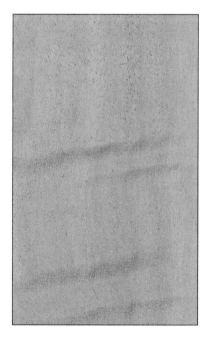

Fraxinus americana

white ash

by Max Kline

SCIENTIFIC NAME

Fraxinus americana. Derivation: The genus name is the classical Latin name for ash. The specific epithet means the plant is "of America".

FAMILY

Oleaceae, the olive family.

OTHER NAMES

American ash, Biltmore ash, smallseed white ash, cane ash.

DISTRIBUTION

From Nova Scotia and Maine, west to Minnesota and south to Texas and Florida. White ash is found in all states east of the Mississippi River and as far west as Kansas and Oklahoma.

THE TREE

White ash is the largest and most important of the 18 ash species native to the United States. It is a well shaped tree 70 to 80 feet in height with a diameter of 2 to 3 feet. The straight trunk is free of limbs for 30 to 50 feet. The dark brown to gray bark is deeply fissured and has broad, flat, scaly ridges. The compound, opposite, pinnate leaves are 8 to 12 inches long with sets of paired leaflets. The fruits are in dense, long clusters each with a dry, flat winged seed.

THE TIMBER

White ash wood is strong and hard with straight, close grain. The heartwood is brown to dark brown, sometimes with a reddish tint, while the narrow sapwood is almost clear white. White ash is free from taste and odor. It is tough and strong. Average reported specific gravity is 0.55 (ovendry weight/green volume), equivalent to an air-dried weight of 42 pcf. It bends well and has excellent elastic properties.

SEASONING

The timber air dries at a faster rate than average. Average reported shrinkage values (green to ovendry) are 4.9% radial, 7.8% tangential and 13.3% volumetric. White ash can be kiln dried with very satisfactory results provided initial temperatures are kept low. It holds its shape well after seasoning.

DURABILITY

White ash timber is prone to fungal decay. The sapwood is susceptible to attack by powder post beetles and furniture beetles.

WORKABILITY

White ash is fairly easy to work with only a slight dulling on cutting edges. The wood does not split easily when nailing. Nail and screw holding properties are satisfactory. Glue adheres well. It is an excellent species for veneering purposes. It responds well to most types of finishes.

USES

White ash is principally used for furniture, handles, cooperage, boxes, baseball bats and other sporting and athletic equipment such as boat oars. It is also extensively used for ladders, chair manufacture, agricultural implements, fork and shovel handles and boat building.

SUPPLIES

Supplies are plentiful at a reasonable cost.

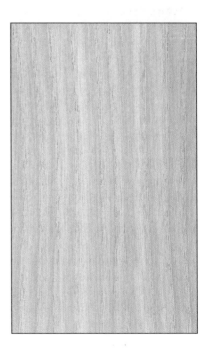

Fraxinus excelsior

European ash

by Jim Flynn

SCIENTIFIC NAME

Fraxinus excelsior. Derivation: The genus name is the classical Latin for ash. The specific epithet means taller.

FAMILY

Oleaceae, the olive family.

OTHER NAMES

common ash, ash, Italian olive ash. Other names are associated with the country where the tree grows such as French ash, Turkey ash, Hungarian ash, etc.

DISTRIBUTION

Native to Europe and Asia Minor, European ash is cultivated worldwide largely in temperate zones.

THE TREE

While there are not a large variety of trees native to "Old Europe", this species of ash is one of them and claims nobility with a long historical lineage. Tales and myths abound in the folklore of many countries concerning its power and use. It is said that the spears of Homer's heroes were made of this ash. In England, it is known as the "Venus of the Forest" because of its graceful form and elegant foliage. Depending on conditions where it grows, European ash can reach heights of 80 to 100 feet with diameters of 2 to 5 feet. The opposite, compound, pinnate leaves are 9 to 10 inches long with 9 to 13 leaflets; the terminal ones are the largest. In young trees, the bark is pale gray and smooth but as the tree matures it becomes thick with interwoven ridges.

THE TIMBER

When growing in a prime location, European ash develops a clear bole 30 to 50 feet long. The wood is white with shades of brown. When freshly cut, it often is a light pink color. Occasionally, the wood will contain heartwood with irregular dark brown or black colorations that are not necessarily associated with decay. The sapwood is not visually distinguishable from the heartwood. The grain is straight but may produce a decorative figure in plainsawn timber or rotary-cut veneer. Specific gravity varies from 0.45 to 0.55 (ovendry weight/green volume), equivalent to an air-dried weight of 34 to 42 pcf. The wood is very tough and flexible. Because of the very wide distribution of this species, from Turkey to the British Isles, a wide variance in the properties of the timber can be expected.

SEASONING

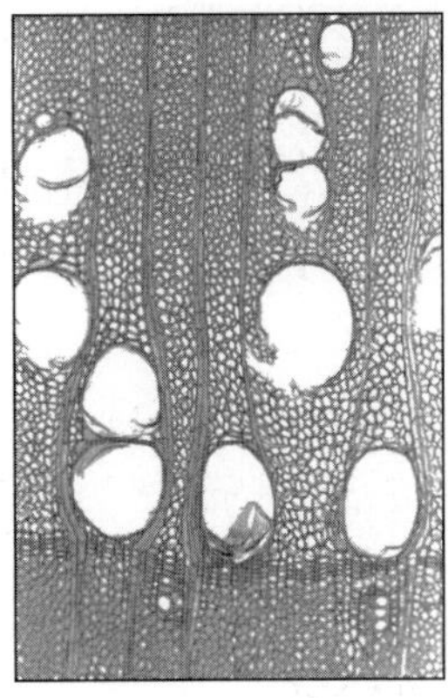

European ash seasons well providing kiln temperatures are kept low. It dries rapidly with very little splitting or checking. Average shrinkage values (green to 12% moisture content) are 5% radial and 12% tangential.

DURABILITY

The tree and the logs are subject to insect attack. The wood is perishable when in contact with the ground and in moist conditions. It is moderately resistant to preservative treatment.

WORKABILITY

All aspects of this characteristic are satisfactory indicating that European ash, while a dense wood, can be satisfactorily machined. Sanding, gluing and finishing all are performed without special effort.

USES

One of the more demanding uses for European ash was in the construction of aircraft during WWI. The wood was ideal for structural members of the aircraft because of its strength and flexibility. This heavy use nearly exhausted the timber supply and all kinds of incentives were used to find trees in isolated and private estates. Currently, the wood has a long list of uses such as sporting goods (hockey sticks, baseball bats, cricket stumps), tool handles, walking sticks, furniture and cabinets, fancy turnery, veneer and many more applications both simple and sophisticated.

SUPPLIES

There are adequate supplies of European ash in the marketplace at a reasonable price.

Fraxinus nigra

black ash

by Jon Arno

SCIENTIFIC NAME

Fraxinus nigra. Derivation: The genus name is the classical Latin name for ash. The specific epithet is Latin for black.

FAMILY

Oleaceae, the olive family.

OTHER NAMES

brown ash, swamp ash, basket ash, hoop ash.

DISTRIBUTION

Black ash is common in the Great Lakes region and ranges from central Minnesota to the Atlantic seaboard and from southern Ohio to central Ontario. Favoring low, moist soil, it is often found along streams and around bogs with white birch (*Betula papyrifera* WDS 039) and black spruce (*Picea mariana* WDS 204) in the northern portion of its range and with western larch (*Larix occidentalis* WDS 159) and yellow birch (*Betula alleghaniensis* WDS 037) in the midwestern United States.

THE TREE

Black ash is a slow grower. In favorable growing conditions, it is one of the tallest of the ashes reaching a height of over 90 feet, but it seldom exceeds about 2 feet in diameter. The compound leaves have nine stemless leaflets, which distinguishes it from white and green ash that have seven stemmed leaflets. The short, dart-shaped seed also differs from most other ashes in that it is wider and the wing extends down the side of the seed almost to the tip.

THE TIMBER

The wood is ring porous, and the rays are inconspicuous. The grayish-brown heartwood is substantially darker than other ashes. Due to its slow rate of growth, the porous earlywood is usually more closely spaced making the wood substantially lighter in weight. With an average reported specific gravity of 0.45 (ovendry weight/green volume), equivalent to an air-dried weight of 34 pcf, black ash is 10% to 12% less dense than white ash (*Fraxinus americana* WDS 121). Where strength is important, white ash is a better choice. Second-growth white ash, which is especially fast growing, is substantially superior in strength to black ash. Also, black ash has relatively low resistance to abrasion since the porous earlywood tends to break down or wear away quickly.

SEASONING

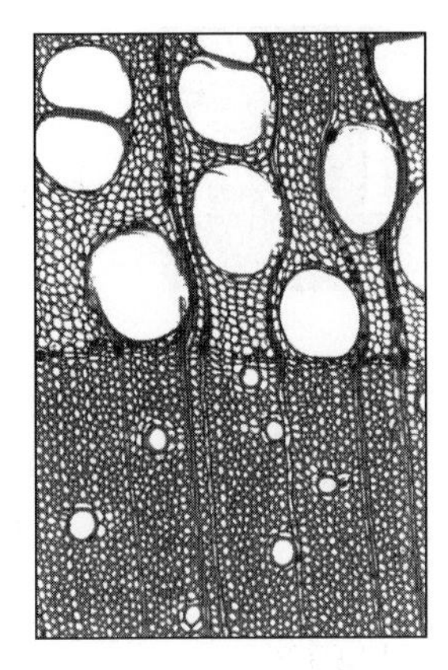

Black ash is easy to air dry with a moderately low tendency to warp. Average reported shrinkage values (green to ovendry) are 5.0% radial, 7.8% tangential and 15.2% volumetric.

DURABILITY

The ashes are among the least durable of common North American hardwoods. Being extremely porous, the wood takes up moisture quickly and is susceptible to microbial attack. Borers are a serious problem. Even kiln-dried ash can be invaded by powder post beetles.

WORKABILITY

Black ash is exceptionally good for cabinetry. It is softer and easier to work than the other ashes and its natural grayish-brown color is attractive with only several coats of clear varnish. Because of its close-spaced annual rings, tangentially cut boards exhibit a very showy figure. Black ash is one of the best woods for steam bending. Although it is softer than other ashes, it has adequate strength and spring for use in chairs, tables and other heavily used furniture. Like most ring-porous woods, filling is necessary when the project requires a perfectly smooth surface. Black ash polishes nicely, accepts glue well and is stable in use but splits easily along the annual rings.

USES

Uses include interior trim and furniture, especially parts requiring steam bending. Because the wood separates easily along the annual rings, producing thin and highly elastic strips of latewood, it has long been a favorite species for basket weaving.

SUPPLIES

Black ash does not have as extensive a range as either white or green ash; however, supplies are adequate. It is a little harder to find through retail channels of distribution and is often mixed with other species and marketed simply as ash. Virtually all species of ash are moderate to low in price. Because of its color, figure and friendly working characteristics, black ash is worth a slight premium over other U.S. domestic species.

Fraxinus sieboldiana

tamo

by Max Kline

SCIENTIFIC NAME

Fraxinus sieboldiana. Derivation: The genus name is the classical Latin name for ash. The specific epithet is in honor of Philip Franz Siebold (1796 to 1866) who, beginning in 1823, spent 6 years in Japan as a doctor at the Dutch embassy and became an authority on the Japanese language, literature and natural history. A synonym is *F. mariesii.*

FAMILY

Oleaceae, the olive family.

OTHER NAMES

Japanese ash, shioji, damo, yachidamo.

DISTRIBUTION

Japan, China, Korea, Manchuria, Siberia.

THE TREE

Tamo thrives best in the deep, rich and well watered alluvial soils of bottomlands. Frequently, the tree reaches a height of 100 feet with a diameter of 35 to 50 inches. The highly figured wood is confined to the lower part of the tree and varies from 4 to 12 feet in length.

THE TIMBER

Tamo is white to pale yellow or pale brown. The odor and taste are not distinct. The texture is coarse, and the luster is medium. The grain is straight, and figures ranging from swirls, fiddleback and mottle to the famous peanut shell that are found in tamo are in a class by themselves. The split wood sample photograph illustrates this characteristic. Each piece of wood has a distinctive pattern. Tamo is stronger than the American or European ash varieties. Average specific gravity is 0.41 to 0.42 (ovendry weight/green volume), equivalent to an air-dried weight of 31 pcf. It is shock resistant and not susceptible to marring or denting. Veneer can be bent to practically any shape without injury to the wood fibers.

SEASONING

Tamo seasons well, remains flat if properly dried and is dimensionally stable in use. Average reported shrinkage values (green to ovendry) are 3.5% radial and 6.0% tangential.

DURABILITY

There is scant information on this particular species of ash but one can assume that it has poor durability like the other ash species.

WORKABILITY

The lumber is easy to plane and joint. Veneer sheets are smooth and free from checks. It glues exceptionally well, but an excess of glue should not be applied as too much will penetrate the large, open pores causing darkening in these areas. The wood takes a beautiful finish with a high degree of sheen. The color does not fade with age in the finished product.

USES

The non-figured wood is used in its growth region for tool handles, agricultural implements, sporting and athletic goods, plywood and cabinetry. The veneer is outstanding and used for tabletops, other furniture and paneling. For marquetry, it is unsurpassed for providing bizarre effects.

SUPPLIES

Tamo lumber is not available in the United States, but the veneer and specially figured wood is obtainable from specialized craft suppliers for a high price.

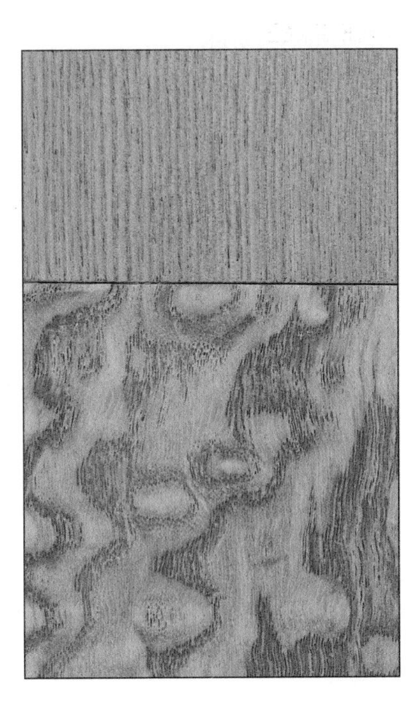

Ginkgo biloba

ginkgo

by Jim Flynn

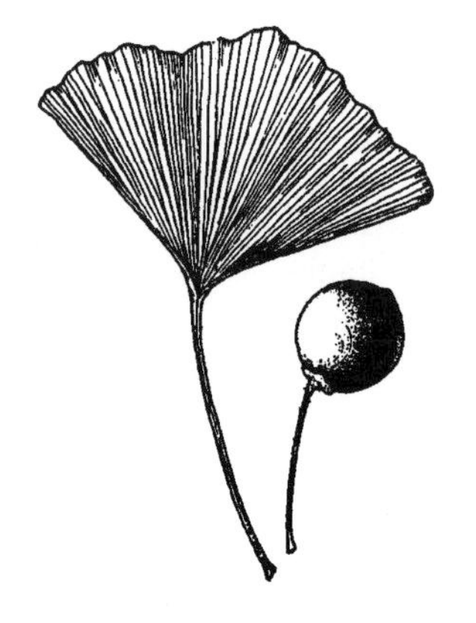

SCIENTIFIC NAME

Ginkgo biloba. Derivation: The name *ginkgo* is a corruption of a Japanese pronunciation of the Chinese ideograph for the tree, yin-hsing (silver apricot). A much earlier Chinese name was Ya Chio meaning "leaf of a duck's foot". Linnaeus assigned the specific epithet *biloba* for the two-lobed leaf on the specimens he had.

FAMILY

Ginkgoaceae, the ginkgo family, of which *Ginkgo biloba* is the only living member. At one time the genus *Ginkgo* was included under the family *Salisbureae* and later in the family *Coniferae.* In Japan in 1896, Hirase, a servant in a Japanese laboratory, discovered the reproductive mechanism of the species, the motile sperm cells in the pollen tube (the most remarkable botanical event in the 19th century). This established ginkgo's lineage, and its current classification.

OTHER NAMES

maidenhair tree, yin-shing (Chinese).

DISTRIBUTION

Throughout the temperate regions ginkgo is a widely grown ornamental. It is found in its native state in eastern China at the borders of the Chekiang and Anhwei Provinces.

THE TREE

Gingko has survived for over 200 million years, and Charles Darwin called it "a living fossil". Because of the likeness of its leaves to maidenhair fern, ginkgo is also called the maidenhair tree. Instantly identifiable, the leaves are fan-shaped with or without one or more sinuses. The species is dioecious (having male and female structures not on the same tree). The round and plum-like fruit of the pistillate trees produce an extremely obnoxious odor when ripe. Because the tree is relatively disease resistant and tolerates noxious fumes and pollutants, it is used as a street planting. Ancient trees in the Far East estimated to be 1,200 years old grow to heights approaching 100 feet with circumferences of 25 feet or more. The oldest ginkgo in the western world is in the Botanical Garden at Utrecht, Holland, planted in the early 1700s. The oldest tree in North America was planted in 1784 at the Woodlands Cemetery in West Philadelphia and died a few years ago, but one at John Bartram's Garden in Philadelphia is of comparable age and thriving.

THE TIMBER

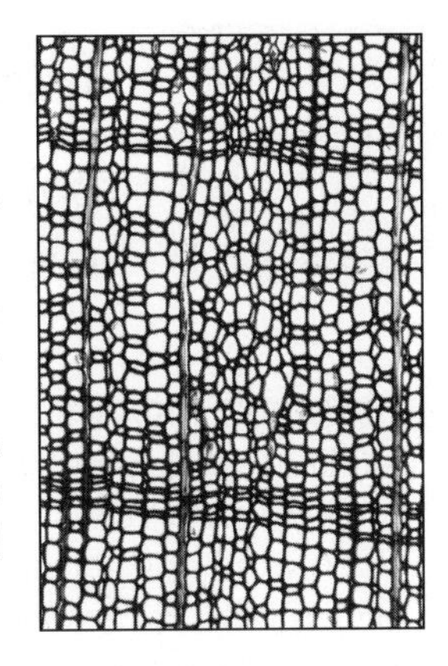

The Pulp and Paper Research Institute of Canada and the Department of Chemistry, McGill University, Montreal, conducted an interesting and revealing study. "The wood and bark of *Ginkgo biloba* are similar to those of most conifers, both as regards their structure and as to their general chemical composition." As a result of my experience, I compare it to sugar pine (*Pinus lambertiana* WDS 210) because of its lightness and feel, straight grain and straw-like color. I calculated the specific gravity of a sample as 0.40 (ovendry weight/green volume), equivalent to an air-dried weight of 30 pcf.

SEASONING

Gingko appears to season well when protected from excessive moisture, which will cause staining by fungus. Shrinkage values are not available.

DURABILITY

Because the wood is not used extensively in commercial applications technical data is absent, but it probably is similar to soft pine.

WORKABILITY

Gingko is an excellent wood to work and hand tools can be used with relative ease. It glues easily and can be stained and finished readily.

USES

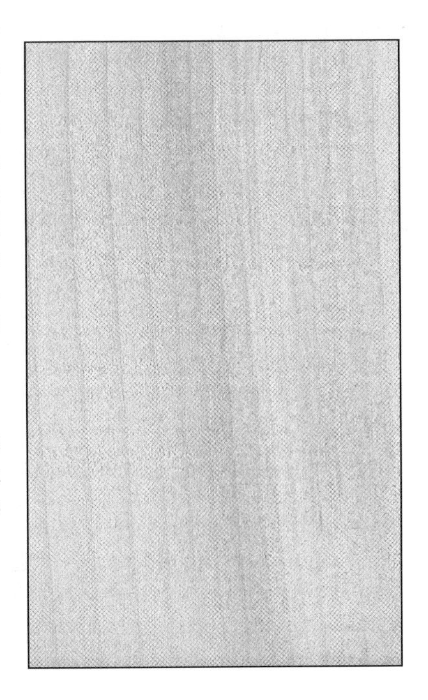

There are many recorded uses for the tree but few references apply to the wood. According to an article in *The Virginia Journal of Science* of July 1959 titled "*Gingko Biloba L*: Historical Summary and Bibliography", the most specific use in China was for abacus beads, chessmen and as the base for fine oriental lacquerware. It also stated that Japanese school children placed gingko leaves in their books to "scare the worms away". The wood can be used for many applications where white pine is used.

SUPPLIES

The wood is not marketed to any great extent but, because the tree is widely planted as an ornamental, woodworkers and sample collectors should have no trouble locating small supplies.

Gleditsia triacanthos

honey locust

by Jim Flynn

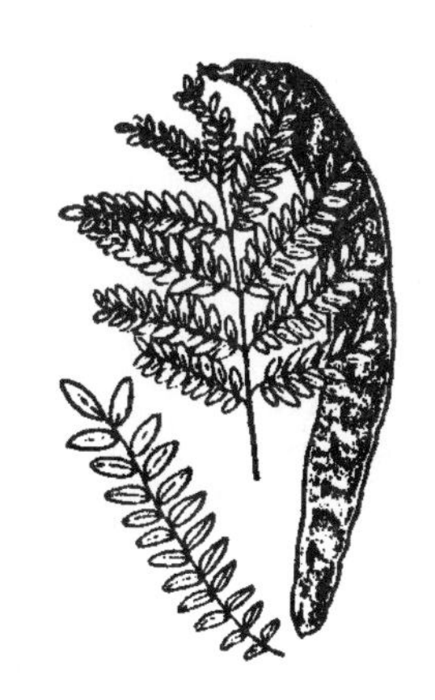

SCIENTIFIC NAME

Gleditsia triacanthos. Derivation: The genus name is in honor of Johann Gottlieb Gleditsch (1714 to 1786), the director of the Berlin botanical garden. The specific epithet means three-thorn, which refers to the large branched thorns. There is also a thornless honey locust, G. *triacanthos* var. *inermis*.

FAMILY

Fabaceae or Leguminosae, the legume family; (*Caesalpiniaceae*) the cassia group.

OTHER NAMES

sweet locust, thorny acacia, three-thorned acacia, thorny locust, honeyshucks. In some areas of the southern United States it is called Confederate pintree because its large spines were used to pin together the tattered uniforms of Confederate soldiers.

DISTRIBUTION

Honey locust is found scattered in the east-central United States from central Pennsylvania west to southeastern South Dakota, south to central and southeastern Texas, east to southern Alabama, then northeasterly through Alabama to western Maryland. Outlying populations may be found in northwestern Florida, west Texas and west-central Oklahoma. It is naturalized east to the Appalachian Mountains from South Carolina north to Pennsylvania, New York and New England. In the early 1700's, it was introduced in England and widely distributed throughout Europe where it still thrives.

THE TREE

Gleditsia is a genus with about 12 different species of deciduous, mostly spiny trees native to North America and Asia; one species is found in Argentina. Honey locust is a moderately fast-growing tree found on moist bottomlands or limestone soils. It has been widely planted for windbreaks and soil erosion control. In urban areas, it was planted to replace elm. The iron-gray, thick bark has deep fissures. The pinnately compound leaves are dark shiny green above and a duller yellowish-green below with 14 to 16 leaflets that are 7 to 8 inches long. Many of these trees reach 100 feet or more in height with circumferences of 4 feet or more. Clusters of brown flat pods that are 12 to 15 inches long contain hard, shiny brown seeds which are about the size of navy beans and are excellent forage for cattle and wildlife.

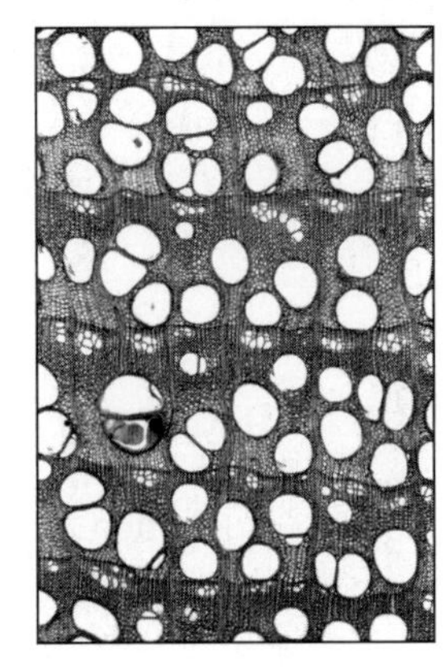

THE TIMBER

This is a remarkably strong and durable timber. The wide yellow sapwood is clearly distinguished from the light red to reddish-brown heartwood. It is ring porous with large and plainly visible springwood pores. Because of its color and grain structure, it is often mistaken for Kentucky coffeetree (*Gymnocladus dioicus* WDS 137). The wood lacks characteristic odor and taste. Overall, it has a very attractive figure and striking grain pattern. Average reported specific gravity is 0.60 (ovendry weight/green volume), equivalent to an air-dried weight of 46 pcf.

SEASONING

The usual precautions should be taken when seasoning. Average reported shrinkage values (green to ovendry) are 4.2% radial, 6.6% tangential and 10.8% volumetric.

DURABILITY

This timber is moderately durable when in contact with the soil and in moist environments.

WORKABILITY

Honey locust is hard to split which suggests a close-knit structure that will resist machining with dull tools. Nevertheless, good clean cuts with sharp hand tools have been achieved as evidenced by the long, curly slices produced by a hand plane. The wood has an attractive luster when sanded, takes glue very well and finishes very attractively. With the exception of the problems caused by interlocked grain, it is quite similar to oak in workability.

USES

A 150-year-old reference states it is only useful for hedges and as an ornamental. Today even though the wood possesses many desirable qualities, because of its scarcity it is little used. In its growth area, the wood is extensively used for fence posts, pallets, crating and general construction.

SUPPLIES

The timber is not commercially marketed except where available at local levels, but widespread use as an ornamental suggests that woodworkers and collectors can obtain samples.

Gmelina arborea

gmelina

by Jon Arno

SCIENTIFIC NAME

Gmelina arborea. Derivation: The genus name is in honor of Johann Gottlieb Gmelin (1709 to 1755), a German traveler and naturalist. The specific epithet is Latin meaning tree-like.

FAMILY

Lamiaceae (Verbenaceae), the mint family.

OTHER NAMES

gumhar, yemane.

DISTRIBUTION

Gmelina is native to Southeast Asia from India to Vietnam. Although this species prefers low-lying moist soils, it is not abundant in mature rain forests where slower growing but ultimately more massive species tend to displace it. In managed plantation environments, however, gmelina is incredibly productive (see Supplies).

THE TREE

While small by rain forest standards, this species has good form for lumber production and will very rapidly reach heights of 100 feet with diameters of 2 feet or more.

THE TIMBER

Gmelina is a coarse-textured, yellowish wood with little contrast between the heartwood and the sapwood. Typically rather bland in appearance, sometimes this species has rose-colored streaks in the heartwood and interlocked fiddleback grain. Because the wood is also somewhat lustrous, samples which have these special features can be quite attractive, but are not representative of typical gmelina. Gmelina is slightly stronger than white pine in most respects. Lacking in both elasticity and crushing strength, it is a poor choice in applications where it would be put under repetitive stress, such as chair legs and spindles. Average reported specific gravity is 0.41 (ovendry weight/green volume), equivalent to an air-dried weight of 31 pcf.

SEASONING

With average reported shrinkage values (green to ovendry) of 2.4% radial, 4.9% tangential and 8.8% volumetric, gmelina experiences very low drying stress and is easy to season.

DURABILITY

Unlike its relative teak (*Tectona grandis* WDS 261), which is renown for its durability, gmelina is non-durable in exterior or marine applications.

WORKABILITY

Even though this species lacks strength and durability, it is an exceptionally pleasant and predictable wood to work. Although the adjective limp is inappropriate for describing any woody tissue, it somehow seems appropriate for gmelina. For such a coarse-textured wood, it shapes easily, accepts nails without splitting and can be brought to a smooth finish with much less effort than is the case with either butternut (*Juglans cinerea* WDS 149) or trembling aspen (*Populus tremuloides* WDS 219), two U.S. domestic woods with comparable density and texture.

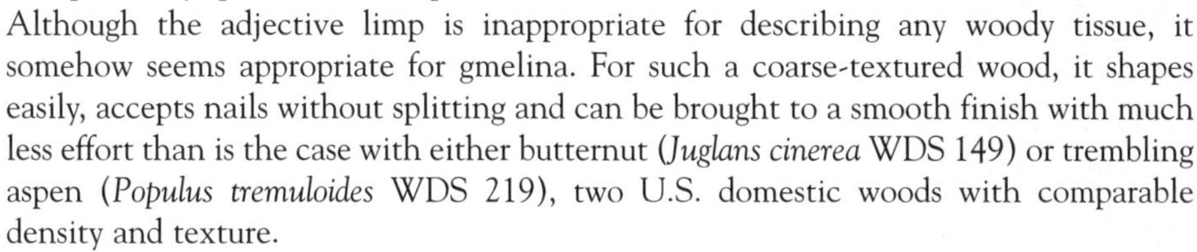

USES

Uses include light-duty carpentry, furniture components, plywood, composition boards, pulp for paper, decorative carvings and craft items. Gmelina is a very utilitarian wood and one of the easiest hardwoods in the world to process into rotary-cut veneer. Given its exceptional stability, it makes an excellent secondary wood for drawer sides and interior panels in furniture.

SUPPLIES

Although originally far less abundant than other commercially important species from Southeast Asia, such as the lauans (*Shorea* spp. WDS 249), gmelina is destined to become a major international timber as the world's rain forests diminish. Because it grows rapidly and is so easy to season and process into a wide range of commercial products, gmelina is one of the half dozen or so most popular species for reforestation projects and plantation ventures in the tropics. While the wood may not be flashy, the plant kingdom has precious few candidates that can match gmelina in terms of the efficient conversion of sunlight and abundant moisture into lumber.

Gonystylus bancanus

ramin

by Jon Arno

SCIENTIFIC NAME

Gonystylus bancanus. Derivation: The genus name *gono* is Greek meaning reproductive and *stylus* meaning style. The specific epithet most likely refers to the Indonesian island of Banca, now called Pulau Bangka.

FAMILY

Thymelaeaceae, the mezereon family.

OTHER NAMES

melawis, garu buaja, buaja, lanutan-bagio.

DISTRIBUTION

Ramin's native range extends from Malaysia eastward through some of the islands of Indonesia and northward into the Philippines.

THE TREE

While this species seldom develops trunks much over 2 feet in diameter, it produces a straight, cylindrical bole that yields clear logs up to 60 feet long. Preferring wet soils, it is found along rivers and in dense swampy forests.

THE TIMBER

Ramin is a light yellowish-tan timber with little color variation between the heartwood and the sapwood. Although the wood is generally fine textured, it contains widely spaced, yet fairly numerous, large vessels which pepper the surface with fine brown lines that give it a subtly attractive figure. Ramin has low surface luster, and the grain is often wavy. When wavy grain is present, it refracts light giving the wood a rich translucent quality. A musty odor, akin to freshly cut elm, provides a fairly reliable clue to identifying ramin. One reference suggests this odor results from the logs being stored in ponds prior to milling. Average reported specific gravity is 0.52 (ovendry weight/green volume), equivalent to an air-dried weight of 40 pcf. Strength is adequate for most cabinetry purposes.

SEASONING

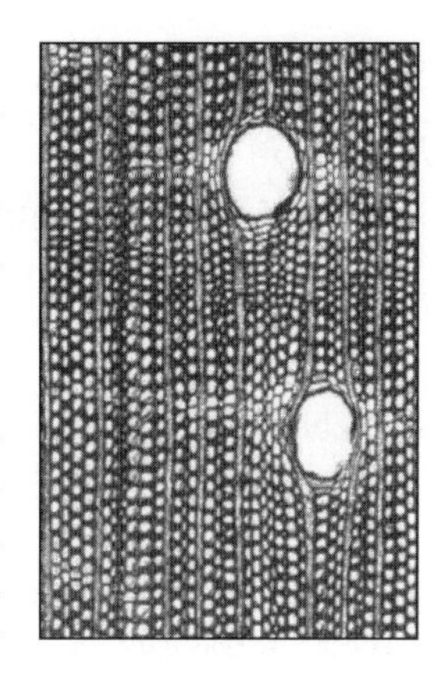

Ramin must be seasoned carefully to avoid both staining and checking; but, it is not highly susceptible to warping. Average reported shrinkage values (green to ovendry) are 4.3% radial, 8.7% tangential and 13.4% volumetric. Ramin has a reputation for excessive in-use movement.

DURABILITY

Durability is poor. This species has very low resistance to fungi and quickly blue stains. It is also highly susceptible to attack by termites and beetles. These negative aspects, in combination with its tendency to end check, may suggest a motive for keeping the logs saturated in storage ponds prior to milling.

WORKABILITY

With an average specific gravity of 0.52, ramin is comparable in density to black walnut (*Juglans nigra* WDS 150) whose specific gravity is 0.51. Ramin is a relatively pleasant wood to work. The most attractive ramin is found in logs with wavy grain, so the woodworker must deal with the wood's tendency to chip when planing. But, this is no more difficult to handle than is curly maple. In fact, ramin's working characteristics are quite similar to those of hard maple, except it is about 5% to 10% less dense and slightly more porous. Like maple, it takes glues and finishes well. Because it is susceptible to fungi attack, it is capable of producing attractively spalted material for turning. Its tendency to split and its high in-use movement are serious drawbacks, but maple has essentially the same limitations.

USES

Ramin is used in a broad range of cabinetry applications, including furniture, paneling, moldings, flooring, plywood, handles and other turned items.

SUPPLIES

Plentiful within its native range, ramin functions as a general purpose utility timber for interior projects. Because its uses are redundant with those of some of the plentiful North American hardwoods (such as maple) and shipping costs impact its ability to be competitively priced, ramin is seldom available in the United States in lumber form. It is, however, frequently encountered in the United States and in Europe in the form of processed materials such as plywood corestock, dowels and moldings.

Gordonia lasianthus

loblolly-bay

by Jim Flynn

SCIENTIFIC NAME

Gordonia lasianthus. Derivation: The genus name is in honor of James Gordon (1728 to 1791), a British nurseryman. The specific epithet *lasianthus*, which was an earlier name for the genus, means hairy flowered.

FAMILY

Theaceae, the tea family.

OTHER NAMES

gordonia, bay, holly-bay, black laurel, swamp laurel, tan bay.

DISTRIBUTION

Loblolly-bay grows along the Atlantic and Gulf coastal plains of the eastern seaboard of the United States. Starting in North Carolina it can be found as far south as central Florida and as far west as the Appalachicola River in the Florida panhandle. Small, disconnected populations can be found in southern Mississippi and Louisiana. At one time, tracts of 50 to 100 acres in continuously moist pine barrens that were at low elevations contained pure stands of loblolly-bay and were thus called bay swamps.

THE TREE

This is one of the puzzling geographic distribution cases and points out the amazing plant migration process. While there are at least 30 other species of the genus *Gordonia*, loblolly-bay stands out as the only species that is native to the United States. The other species are all Asiatic. Sometimes included in this genus is G. *franklinia*. Extinct except in cultivation, it is now properly classified as *Franklinia alatamaha*. Loblolly-bay is an evergreen which grows to a height of 80 feet or more with a diameter of 20 inches or more. The U.S. National Register of Big Trees rates one in the Ocala National Forest in Florida as the champion. This specimen is 95 feet in height and has a diameter of 164 inches. The thick bark is reddish-brown and becomes deeply furrowed with age. Shiny on the upper surface, the evergreen leaves are 3 to 6 inches long, alternate, oval-acuminate, slightly toothed and smooth. The white, sweet smelling flowers are 1 inch wide and blossom from July through September. The tree is considered one of the most beautiful and unique trees of the southern states, and its use as an ornamental has long been encouraged.

THE TIMBER

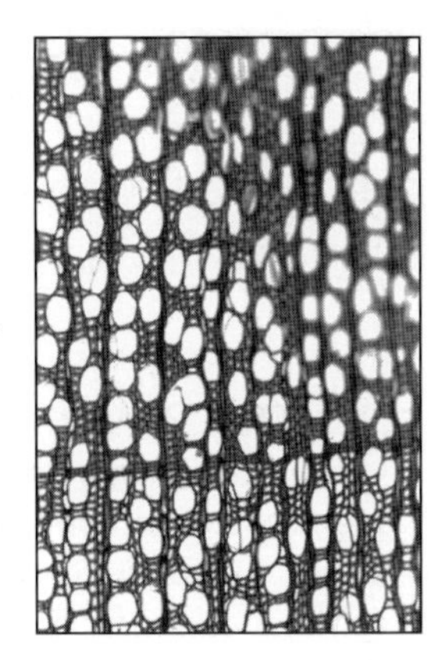

Loblolly-bay has pinkish heartwood which is not sharply demarcated from the pale brown sapwood. It has a fine silky texture, is diffuse porous and has no odor or taste. The grain is straight and has the appearance of birch although it is not as strong or dense. Average specific gravity is 0.42 (ovendry weight/green volume), equivalent to an air-dried weight of 31 pcf.

SEASONING

While shrinkage values are not available, it does not seem that this species would pose any unusual seasoning problems.

DURABILITY

This is not a durable wood and should not be used in places subject to moisture and the associated fungi, bacteria and insects. Because the wood is not available in sufficient quantities to warrant the effort, it is not known if preservatives can be successfully applied.

WORKABILITY

When loblolly-bay dries properly and has reached its equilibrium moisture content, it becomes easy to work. There are relatively few problems in milling and machining. It sands, glues, stains and finishes exceptionally well.

USES

Early records indicate that the wood was used for interior cabinetry because of its pale mahogany-like appearance. There is information from the early part of this century that 27-inch blocks of loblolly bay were processed for veneer. Today, unfortunately, it seems that the majority of the growth in U.S. southern forests is being harvested and pulped for paper. While this is the only recorded use, it can be assumed that other small-scale uses include furniture and craftwork.

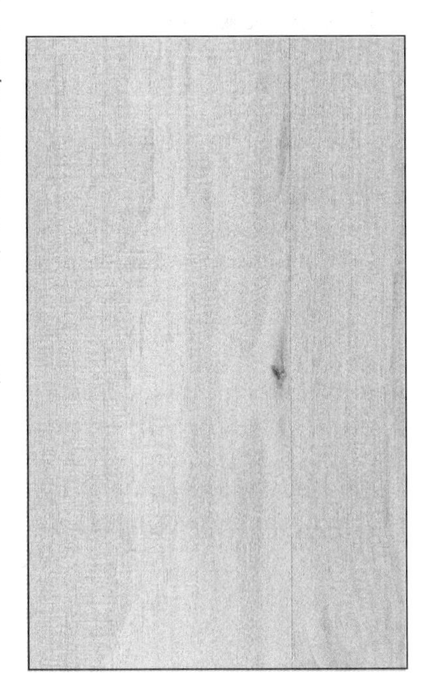

SUPPLIES

Within its growth range, quantities can be found at a relatively low cost.

Gossweilerodendron balsamiferum

agba

by Jon Arno

SCIENTIFIC NAME

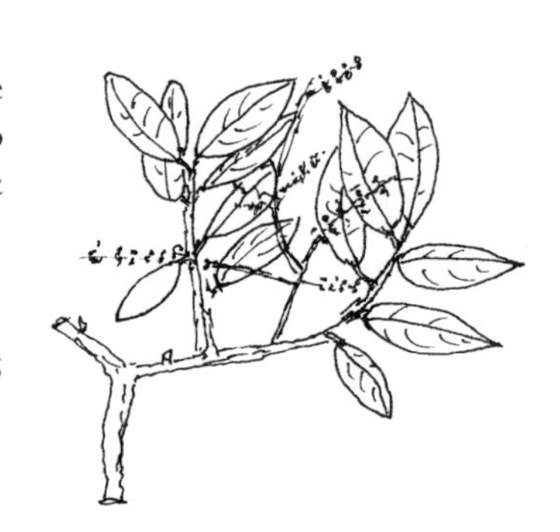

Gosssweilerodendron balsamiferum. Derivation: The genus name is in honor of John Gossweiler (1873 to 1952), a plant collector in West Africa. The specific epithet is Latin for balsam yielding.

FAMILY

Fabaceae or Leguminosae, the legume family; (*Caesalpiniaceae*) the cassia group.

OTHER NAMES

tola, tola branca, achi.

DISTRIBUTION

Agba is native to the rain forests of west central Africa, from Nigeria southward into the Congo basin. It prefers deep, moist soils and possesses the genetic potential to dominate other flora on ideal sites.

THE TREE

Agba is a giant among African timbers, reaching heights of over 200 feet with diameters up to 8 feet. Because it lacks the pronounced buttress typical of most tropical timbers, especially those favoring moist soils, agba has an excellent form for the production of timber. Clear logs of 100 feet or more in length are not uncommon.

THE TIMBER

The heartwood is generally light yellow or straw-colored but may occasionally have a pinkish hue. The transition between the heartwood and sapwood tends to be gradual with the sapwood band being 4 inches or more in width. Interlocked grain is quite common and radially cut boards that accent this feature are very attractive since this species also has a rather high luster. Gum pockets frequently mar the quality of the wood. Its distinct resinous odor is a strong clue to identifying agba. Average reported specific gravity is 0.40 (ovendry weight/green volume), equivalent to an air-dried weight of 30 pcf. Strength is comparable, but generally slightly inferior to, the true mahoganies (*Swietenia* spp. WDS 254 and *Khaya* spp. WDS 155). It is not a good choice for furniture that will experience hard use.

SEASONING

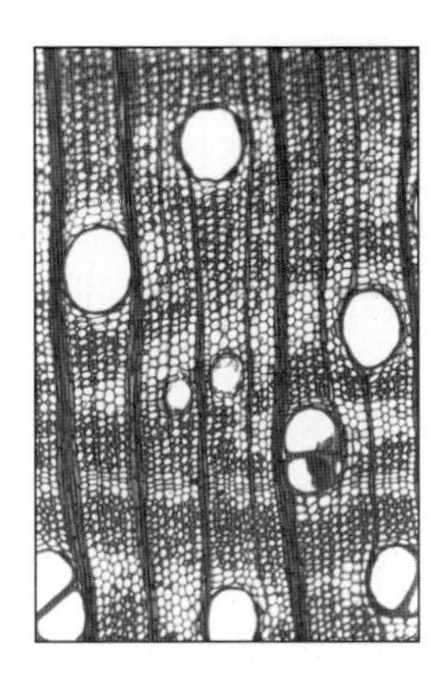

Agba seasons quickly without difficulty and remains stable in use. Blue stain is uncommon but its high gum content may bleed to the surface when the wood is exposed to harsh kiln schedules. Average reported shrinkage values (green to 12% moisture content) are 1.5% radial and 3.0% tangential.

DURABILITY

The heartwood is rated as durable and resistant to fungi. While the sapwood is susceptible to attack by beetle larvae, the heartwood is an acceptable choice for exterior or marine applications where strength is not a vital concern.

WORKABILITY

With an average specific gravity of 0.40, agba is comparable in density to yellow-poplar (*Liriodendron tulipifera* WDS 162) but somewhat coarser in texture. It is an easy wood to work and can be counted on for in-use stability. Its exceptionally high gum content will quickly clog sandpaper and foul up blades. The gum does not seem to affect the holding power of adhesives, but may bubble up under finishes especially on the end grain. Agba is a good, general purpose timber and the ribbon-grain figure on quartersawn boards provides interesting visual impact. Unfortunately, its high gum content and relative softness make it a questionable choice as the primary wood in fine furniture. On the other hand, its in-use stability and availability in wide, clear boards make it a very acceptable secondary wood for interior panels, drawer sides and veneer corestock. Agba's resinous scent is pleasing and after the project is completed you can enjoy the aroma while cleaning the gum off of the saw blades.

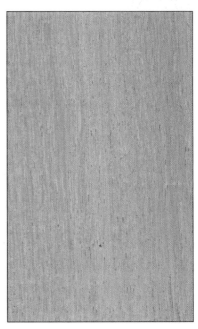

USES

Uses include general purpose construction, boat building, plywood corestock, decorative veneer (when radially cut), furniture and millwork.

SUPPLIES

Agba is plentiful and relatively inexpensive, but it is not often available through retailers of furniture quality hardwoods in North America. It is more readily available in Europe, but is also often encountered in the United States as the corestock in plywood or the secondary wood in furniture of European manufacture. The plentiful domestic supplies of equally functional secondary woods make it less competitive in the United States.

Grevillea robusta

silky oak

by Jon Arno

SCIENTIFIC NAME

Grevillea robusta. Derivation: The genus name is in honor of Charles Francis Greville (1749 to 1809), a British introducer of exotic plants. The specific epithet is Latin meaning robust.

FAMILY

Proteaceae, the protea family.

OTHER NAMES

grevillea, lacewood, southern silky oak.

DISTRIBUTION

Native to eastern Australia, silky oak's ability to withstand drought has made it a widely planted cultivar in warm, dry regions, including the southwestern United States. Because its fern-like foliage represses sunlight, it is used to protect sun-sensitive commercial crops such as tea and coffee.

THE TREE

Silky oak reaches a height of 80 to 100 feet with a diameter of 2 to 3 feet. On ideal sites, specimens can reach 150 feet in height with diameters in excess of 4 feet. By Australian standards, it is not a particularly large tree and is displaced on more favorable growing sites by numerous species of eucalypts.

THE TIMBER

Silky oak is a moderately coarse-textured, diffuse-porous wood. The transition between the light-colored pinkish-yellow sapwood and the darker, pinkish buff-brown heartwood is not sharply demarcated. The most striking feature is the woven pattern of its figure resulting from its large and plentiful rays. While these conspicuous rays may explain the use of oak in its common name, silky oak is not closely related to the true oaks (*Quercus*) which belong to the beech family (Fagaceae). With an average reported specific gravity of 0.51 (ovendry weight/green volume), equivalent to an air-dried weight of 39 pcf, it is much softer than most of the true oaks. Because its ray tissue has extremely high luster in comparison to the rather dull background tissue, carefully sliced radial veneer reflects light to make this one of the showiest inlay woods in the world. Silky oak compares closely to bigleaf mahogany (*Swietenia macrophylla* WDS 254) in terms of most strength properties. It is much weaker than virtually all of the true oaks, but is strong enough for use in fine furniture.

SEASONING

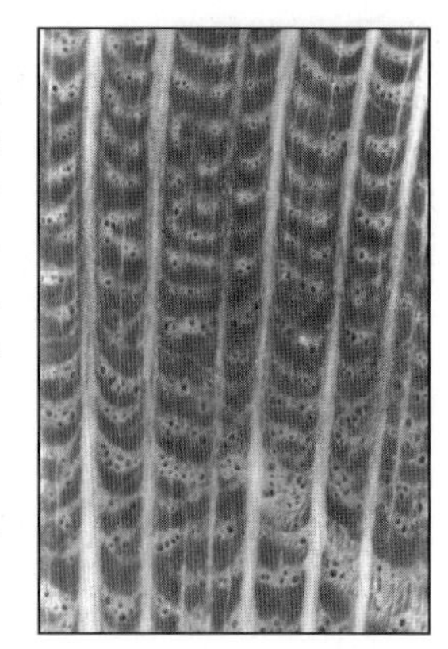

While silky oak has moderately low volumetric shrinkage, the relatively high contrast between its tangential shrinkage of 7.7% (green to ovendry) and its radial shrinkage of only 2.7% suggests a potential for considerable drying stress. Honeycomb and collapse sometimes occur, especially when thicker stock is exposed to harsh, overly accelerated, kiln schedules.

DURABILITY

The heartwood is moderately durable, but this is not an outstanding wood for exterior marine applications.

WORKABILITY

Due to its dominant rays, silky oak tends to chip when planed. Its moderately low density makes it somewhat more pleasant to work than most of the true oaks. For such a highly figured wood, it turns well. Most highly figured woods tend to be ring porous and these open-grained woods often do not perform well on the lathe. Silky oak, on the other hand, achieves its flamboyant figure by virtue of its unusually vivid rays, but otherwise has the working qualities of a diffuse-porous wood. It steam bends well, accepts virtually all glues and finishes and will accommodate large diameter fasteners without splitting. A major drawback is that it contains some toxic, phenolic compounds that can cause serious skin rash. As with poison ivy, not everyone is susceptible, but if exposure causes itchiness, leave it alone.

USES

Uses include furniture, inlay and plywood veneer, turnery and decorative but not durable parquet flooring.

SUPPLIES

Although silky oak is not rare in Australia, the costs of transportation and maximizing its figure make it rather pricey on the American market.

SPECIAL NOTE

In some references, especially those of British origin, *Cardwellia sublimis* (WDS 060) is defined as lacewood or silky oak. This less drought resistant and more northerly species is native to the coastal rain forests of Queensland and is a closely related member of the Proteaceae family.

Guaiacum officinale

lignumvitae

by Max Kline

SCIENTIFIC NAME

Guaiacum officinale. Derivation: The genus name is derived from the Carib Indian name guayacan. The specific epithet means medicinal.

FAMILY

Zygophyllaceae, the caltrop family.

OTHER NAMES

guaiacum wood, guayacan, ironwood, palo santo.

DISTRIBUTION

Florida, West Indies, Central America.

THE TREE

Lignumvitae grows to a height of 30 feet with an average diameter of 10 to12 inches but occasionally 18 to 30 inches. It has a dense, rounded crown and dark green foliage. The smooth bark is light brown and mottled and peels off in thin scales. The opposite, pinnate leaves have four or more stalkless, oblique, broadly elliptic or ovate leathery leaflets. The pale blue flowers with five petals provide a means of positive tree recognition when in bloom.

THE TIMBER

The heartwood of lignumvitae is a peculiar dark greenish-brown or nearly black color, which is sharply demarcated from the narrow, yellowish sapwood. The wood is especially oily or waxy due to a natural gum resin, which is responsible for about 30% of its weight. It has a mildly scented and pleasant odor and is slightly acid to the taste. The wood has a fine and uniform texture with a low luster. It is extremely heavy, in fact, one of the heaviest timbers known to commerce. Average reported specific gravity is 1.05 (ovendry weight/green volume), equivalent to an air-dried weight of 88 pcf. The grain is interlocked, very irregular and wavy. The strength properties of lignumvitae are very high. Of 405 woods tested, its resistance to indentation was ranked first.

SEASONING

Care is needed when seasoning ligumvitaes or serious splitting may occur. If carefully seasoned, the wood will dry slowly with only very slight distortion. Average reported shrinkage values (green to ovendry) are 2.5% radial and 3.8% tangential. After drying, the stability factors are very satisfactory.

DURABILITY

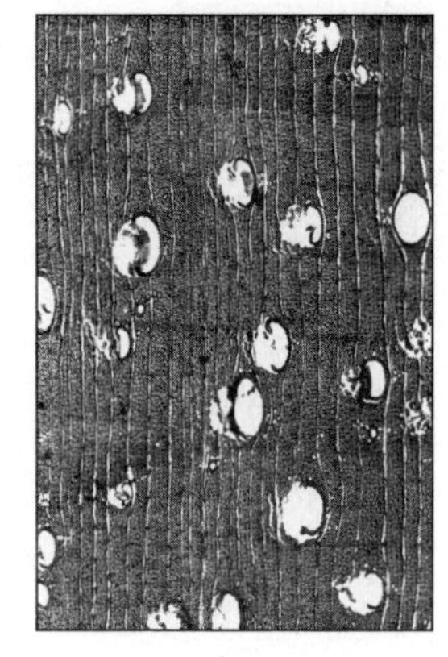

Unseasoned logs are subject to longhorn woodborer beetle damage, but once seasoned the timber has very high standards of resistance to both insects and fungi. Even when used under the most exacting requirements, it needs no preservative treatment.

WORKABILITY

The heartwood is extremely difficult to work with hand tools and very hard to saw and machine. The sapwood is much less hard and brittle. When planing, a lowered cutting angle is useful but even then the timber will attempt to ride over the cutters. It is excellent for turnery but sand paper will quickly clog from the resin contained in the wood. Pre-boring for screws is essential and they will hold well. Drilling operations should be carried out at low speeds. Some glues do not adhere well, and the best results are obtained when using synthetic glues. Polishing is the only form of finishing treatment normally given lignumvitae as there is no advantage to be gained from further treatment.

USES

No wood has been found equal to lignumvitae for ship's propeller shaft bushings and bearings. Because of its self-lubricating properties, it normally lasts about three times as long as steel or bronze. In this application, its ability to withstand working pressures of 2,000 psi is also essential. Small lignumvitae bearings are widely used for clocks, fans and air conditioning units. Miscellaneous uses include mallets, rollers, casters, small wheels and band saw blocks. Formerly ligumvitae was used for bowling balls. It has been an article of trade since 1508 when it was introduced in Europe for medical purposes. The resin is sometimes used for treating gout and for rheumatism.

SUPPLIES

It is imported in the form of small logs and sold by the pound. Lignumvitae is not usually sold in lumber form but rather in the fabricated shape for specific uses. [Ed. note: This species of lignumvitae and G. *sanctum* have been designated endangered species and restrictions on international trade have been imposed.]

Guarea cedrata

guarea

by Jim Flynn

SCIENTIFIC NAME

Guarea cedrata and G. *thompsonii*. Derivation: The genus name is from the Cuban name guara. The specific epithet *cedrata* refers to the cedar-like scent of its timber. The specific epithet *thompsonii* is in honor of Sir E. Maude Thompson (1859 to 1907), a librarian at the British Museum. These two species are frequently marketed under a single trade name; for many applications, differences between them are unimportant.

FAMILY

Meliaceae, the mahogany family.

OTHER NAMES

bosse, kwabohoro, obobo, edoucie, pink mahogany, pink African cedar, Benin mahogany, piqua, African cedar, cedron, scented guarea.

DISTRIBUTION

The range of both species overlaps in the Ivory Coast, Ghana and southern Nigeria. G. *cedrata* extends into Cameroon while G. *thompsonii* reaches into Liberia.

THE TREE

Guarea is a large tree reaching 160 feet in height with heavy buttresses at the base of a clear trunk. Above the buttresses, it has a diameter of 3 to 4 feet. The boles are long, straight, cylindrical and produce ideal sawlogs. The bark is light grayish-yellow and exudes a fragrant odor when cut. The leaves are pinnate with a winged leaf stalk with up to six pairs of leaflets and a terminal leaflet. While the tree is evergreen, it produces new leaves at irregular intervals. These new leaves are bronzed and conspicuous. The fragrant greenish-yellow flowers form between April and August. The fruit is a leathery, brownish capsule about 1.5 inches long that is relished by monkeys.

THE TIMBER

Guarea is an attractive wood closely resembling New World mahogany. Its heartwood is pinkish-brown and darkens considerably on exposure. The sapwood, variable in width, is pale in color and, at times, well demarcated from the heartwood. The texture is medium to fine. The grain is straight, wavy or interlocked. G. *cedrata* sometimes produces an attractively mottled or curly figure. G. *thompsonii* tends to have straighter grain and a plainer appearance. The wood exudes a pleasant cedar-like aroma when freshly cut. Both species contain resin in canals that are often under pressure in the living tree. When the timber is cut, the resin exudes onto the surface, some of which forms flecks of different sizes, often causing problems in use and gumming cutting tools and machinery. Average reported specific gravity is 0.48 (ovendry weight/green volume), equivalent to an air-dried weight of 36 pcf.

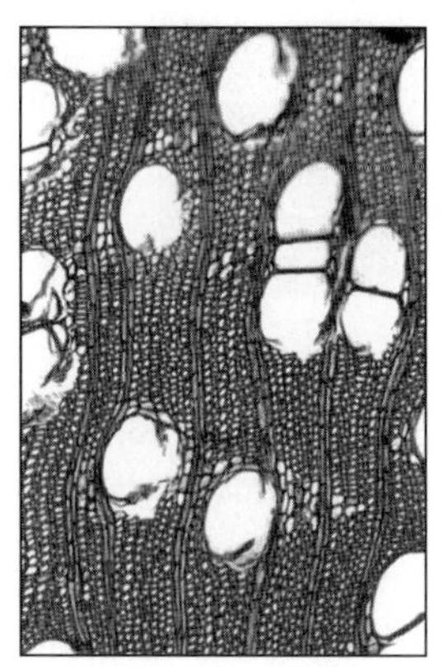

SEASONING

The wood dries rapidly with little tendency to warp, but care must be exercised to prevent checking. Average reported shrinkage values (green to 12% moisture content) are 2.0% to 2.5% radial and 3.5% to 4.0% tangential. As noted, resin exudation may adversely affect the appearance of the seasoned timber. When dry, the wood is stable in use.

DURABILITY

The sapwood is rarely attacked by powder post beetles and has been reported as moderately resistant to termites in Africa. The heartwood is rated very durable. Except for the sapwood, the wood is extremely resistant to preservative treatments.

WORKABILITY

There are slight differences in the working characteristics of the two species. Both are rated as easy to work with machine and hand tools. There is some chipping when planing if the wood has interlocked grain. It has good gluing attributes. It stains and polishes well if using a wood filler, but care must be taken to sand off the resin spots. This is more troublesome with G. *cedrata*. Steam bending can be done effectively. Precautions should be taken to avoid exposure to the dust which can be an irritant.

USES

Traditional uses for guarea include furniture making, boats, canoes, cabinets, paneling, decorative veneer, turnery, flooring and other up-scale joinery.

SUPPLIES

Since African forest products are generally distributed to the European market, this timber does not appear in stock lists of lumber suppliers in United States. It does, however, appear on sample lists for collectors.

Guibourtia arnoldiana

benge

by Jon Arno

SCIENTIFIC NAME

Guibourtia arnoldiana. Derivation: The genus name is in honor of N. J. B. Guibourt (1790 to 1861), a French pharmacologist who wrote a history of plants used for medicine (economic plants). The specific epithet is in recognition of the Arnold Arboretum in Boston, Massachusetts.

FAMILY

Fabaceae or Leguminosae, the legume family; (*Caesalpiniaceae*) the cassia group.

OTHER NAMES

mutenye, mbenge.

DISTRIBUTION

The range of benge is more restricted than the other species of *Guibourtia.* Benge is found sparingly from Cameroon southward into extreme western Congo (formerly Zaire).

THE TREE

Smaller than most of its close relatives, benge reaches a height of about 90 feet with diameters of 20 to 30 inches. Although logs up to 60 feet long are obtainable, they are not normally the straight, evenly cylindrical-shaped logs that are produced by the more ideal timber species.

THE TIMBER

The light chocolate-brown heartwood, similar in color to European walnut, generally has a yellowish or sometimes pinkish tinge. The sapwood is yellowish-gray and sharply demarcated from the heartwood. The figure is highlighted by attractive black striping and this, in combination with its interlocked and often wavy grain, gives the wood a great deal of character even though it is fine textured and very dense. Average reported specific gravity is 0.64 (ovendry weight/green volume), equivalent to an air-dried weight of 50 pcf. Benge is strong, superior to comparably dense Brazilian rosewood (*Dalbergia nigra* WDS 097) in bending strength, elasticity and crushing strength.

SEASONING

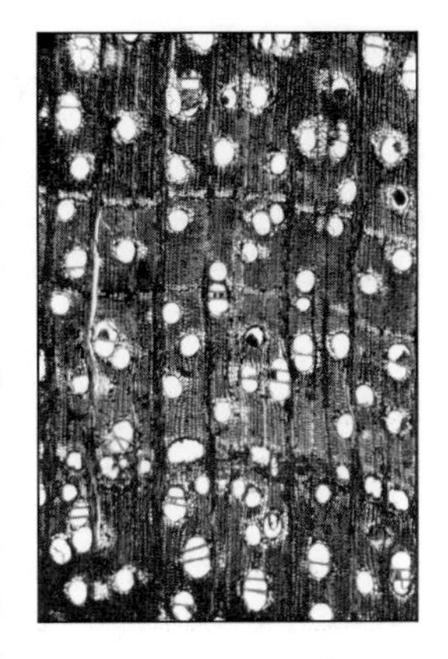

With adequate care, benge is reported to season well. Average reported shrinkage values (green to ovendry) range from 4.6% to 5.8% radial, 8.0% to 9.2% tangential and 10.5% to 14.7% volumetric. (See Special Note.)

DURABILITY

Benge is moderately durable and resistant to termites. It is not as durable as those species marketed as bubinga.

WORKABILITY

European references cite this species as a substitute for walnut, but in my opinion it would be better to think of it as a "poor man's rosewood". Its average specific gravity of 0.64 puts it in a density range with the softer Brazilian rosewoods, while its finer texture is somewhat comparable to Honduras rosewood (*Dalbergia stevensonii* WDS 099). Benge lacks the luster and fragrant scent of rosewood, but under a varnished surface this point is academic. Benge's interlocked wavy grain and rosewood-like black striping make it very eye pleasing. Given its density, however, there is no charitable way to describe benge as a pleasant wood to work. While its fine texture suggests it would be a good candidate for turning, this is offset by the somewhat high risk of chipping caused by the interlocked grain and the fact that it is far less stable than its rosewood counterparts.

USES

Uses include decorative veneer, flooring, furniture and turnery.

SUPPLIES

Benge is not plentiful in comparison to other species of *Guibourtia*. The price is moderate for an imported species and substantially lower than for any of the true rosewoods.

SPECIAL NOTE

Wood Data Sheet 135 describes bubinga (G. *tessmannii*), a close relative of benge. The genus *Guibourtia* supplies four fairly well-known timbers, which are remarkably different in color and appearance considering their close botanical kinship. All are native to Africa and might best be summed up as follows: **Bubinga**, most plentiful, is cut from G. *tessmannii*, G. *pellegriniana*; **Benge**, more commonly known as mutenye in Europe, comes from G. *arnoldiana*; **Ovangkol**, also known as ehie, is supplied by G. *ehie*; **Rhodesian copalwood**, less plentiful in international trade, comes from G. *coleosperma*.

Guibourtia tessmannii

bubinga

by Max Kline

SCIENTIFIC NAME

Guibourtia tessmannii. Derivation: The genus name is in honor of N. J. B. Guibourt (1790 to 1861), a French pharmacologist who wrote a history of plants used for medicine (economic plants). The specific epithet is in honor of Gunther Tessmann (1904 to 1926), an ethnologist and South American plant collector.

FAMILY

Fabaceae or Leguminosae, the legume family; (*Caesalpiniaceae*) the cassia group.

OTHER NAMES

African rosewood, eban, kevazingo, kssingang, amazakone, akume.

DISTRIBUTION

Cameroon, Gabon, Ivory Coast.

THE TREE

Bubinga is a large tree that commonly exceeds 100 feet in height with a clear bole that is 30 to 60 feet long and 3 feet in diameter. The base of the tree is heavily buttressed. The alternate compound leaves are 4 inches long and consist of two asymmetrical leaflets that are about 2 inches long and are similar in appearance to butterfly wings.

THE TIMBER

The heartwood of bubinga is light reddish-brown attractively veined with pink or red stripes; the sapwood is paler in color. The wood is very hard and heavy. Average reported specific gravity ranges from 0.65 to 0.78 (ovendry weight/green volume), equivalent to an air-dried weight of 51 to 62 pcf. It is fine in texture, and the luster is high. Odor and taste are not distinct. The grain is wavy.

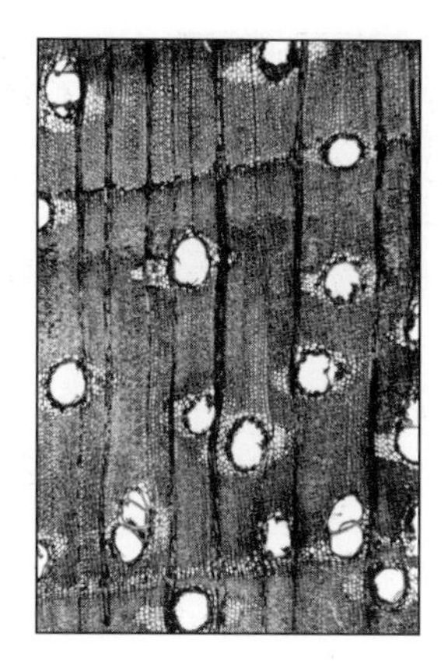

SEASONING

The key to effective seasoning is to dry slowly. Average reported shrinkage values (green to ovendry) are 4.0% to 7.6% radial, 6.6% to 10.2% tangential and 9.4% to 16.6% volumetric.

DURABILITY

Bubinga is highly resistant to the termites in West Africa and moderately resistant to marine borers.

WORKABILITY

Although hard and heavy, bubinga can be sawn without difficulty, and it takes a fine finish. When rotary cut for veneer, it is sometimes called kevazingo, which most often comes from Gabon and is highly figured.

USES

Bubinga is used mostly in veneer form for decorative paneling and inlay work but also finds some use for up-scale furniture and fancy turnery work.

SUPPLIES

Bubinga is imported into the United States and is available in lumber form. The imported logs are of tremendous size, some weighing as much as 10 tons. Another timber belonging to the same botanical genus as bubinga is benge (G. *arnoldiana* WDS 134), but it is finer textured and differs markedly with respect to color. It too is available on the American market. These timbers are in the medium to high price range.

Gymnanthes lucida

oysterwood

by Max Kline

SCIENTIFIC NAME

Gymnanthes lucida. Derivation: The genus name is Greek indicating a naked flower (the flowers with perianth reduced to bract-like scales or absent). The specific epithet means shiny which refers to the leaves.

FAMILY

Euphorbiaceae, the spurge family.

OTHER NAMES

crabwood, poisonwood, acetillo, grandillo, false-lignumvitae, branquilho, Cuban oysterwood.

DISTRIBUTION

Southern Florida, southern Mexico, West Indies, Belize, Brazil.

THE TREE

Oysterwood is a small tree with a narrow crown whose height rarely exceeds 30 feet. Its slender trunk, often irregularly ridged and scaly, is commonly less than 10 inches in diameter. The alternate, simple leaves are 2 to 3 inches long, evergreen and leathery. The fruit is a dark brown, long-stalked, three-lobed capsule which contains two or three round, shiny brown seeds.

THE TIMBER

The variegated heartwood is olive brown with alternating light and dark zones. It is sharply demarcated from the white sapwood. The luster is low, and the odor and taste are not distinct. Oysterwood is hard and strong. Average specific gravity ranges from 0.84 to 0.92 (ovendry weight/green volume), equivalent to an air-dried weight of 68 to 75 pcf. The texture is very fine and uniform. The grain is straight to slightly irregular or wavy. This timber has outstanding strength properties.

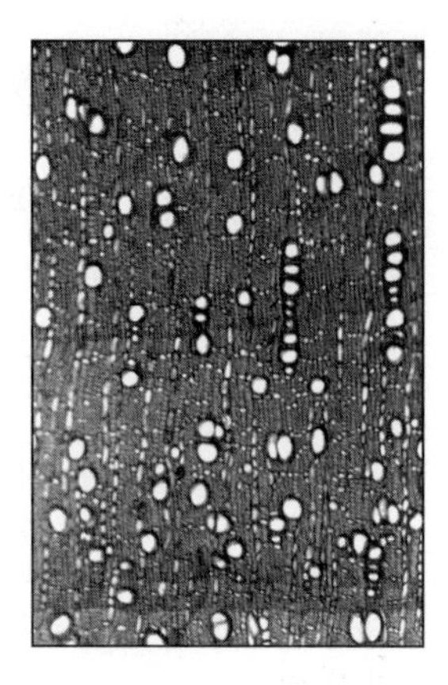

SEASONING

Oysterwood can be seasoned but is almost always found with cracks in the center of the heartwood, even before the seasoning process is started. Further cracking beyond this original shake does not usually take place. Shrinkage values are not available.

DURABILITY

Oysterwood is classified as a durable timber.

WORKABILITY

Oysterwood is not difficult to work and finishes well. It is particularly good for turning and carving projects.

USES

In its native range, the timber is used for poles, posts, stakes, tool handles and small articles of turnery. Occasional shipments have been exported and used for brush backs, walking sticks, umbrella handles and inlay veneer for marquetry. The beauty of the wood can be highlighted when made into various novelties.

SUPPLIES

There is almost no oysterwood on the market, and understandably the demand is very high. It can only be obtained from a wood collector living in the growth area or from someone who has visited and collected from these localities.

Gymnocladus dioicus

Kentucky coffeetree

by Max Kline

SCIENTIFIC NAME

Gymnocladus dioicus. Derivation: The genus name is Greek meaning naked branch. The specific epithet refers to the Latin name *dioecious* implying that the staminate and pistillate flowers generally are on different trees.

FAMILY

Fabaceae or Leguminosae, the legume family; (*Caesalpiniaceae*) the cassia group.

OTHER NAMES

American coffee bean, coffeenut, stump tree, American mahogany, chicot, dead-tree.

DISTRIBUTION

From southern Ontario and western New York to Minnesota and eastern Nebraska, southward through eastern Oklahoma to northern Louisiana and on rich bottomlands within the Appalachian Mountains to middle Tennessee.

THE TREE

Kentucky coffeetree reaches a maximum height of 120 feet with a diameter of about 4 feet. At a distance of 10 to 15 feet from the ground, the main trunk commonly divides into three or four stems. The doubly compound leaf with 40 to 60 leaflets is 1 to 3 feet long and often 2 feet wide. For about 6 months of the year, the tree shows no signs of life which led to its common names dead-tree and stump tree. Kentucky coffeetree is the state tree of Kentucky.

THE TIMBER

Early settlers roasted the seeds for a coffee substitute, which explains its name. The heartwood is light cherry-red or reddish-brown that is sharply demarcated from the greenish-white sapwood. The wood is moderately hard, tough and strong. Average reported specific gravity is 0.53 (ovendry weight/green volume), equivalent to an air-dried weight of 40 pcf. The texture is coarse, and the luster is medium. This straight-grained, ring-porous wood is often mistaken for honey locust (*Gleditsia triacanthos* WDS 126).

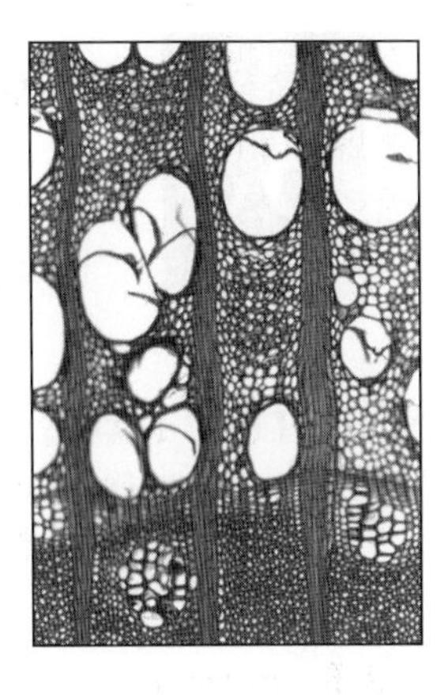

SEASONING

Kentucky coffeetree is difficult to season without splitting. Average reported shrinkage values (green to ovendry) are 4.1% radial, 7.6% tangential and 11.9% volumetric.

DURABILITY

The timber is durable when in contact with the soil.

WORKABILITY

This wood is easily worked with tools, glues well, has good nail and screw holding properties but has a tendency to split if not carefully handled. It finishes well and may be stained, waxed or painted.

USES

A great deal of Kentucky coffeetree is used for fence posts, railroad ties, poles and construction material. Moderate amounts are used for furniture.

SUPPLIES

The tree is too sparsely distributed to be of extensive commercial importance. It is sometimes sold in combination with miscellaneous hardwoods but is very scarce except in its growth areas. When available, the price is medium.

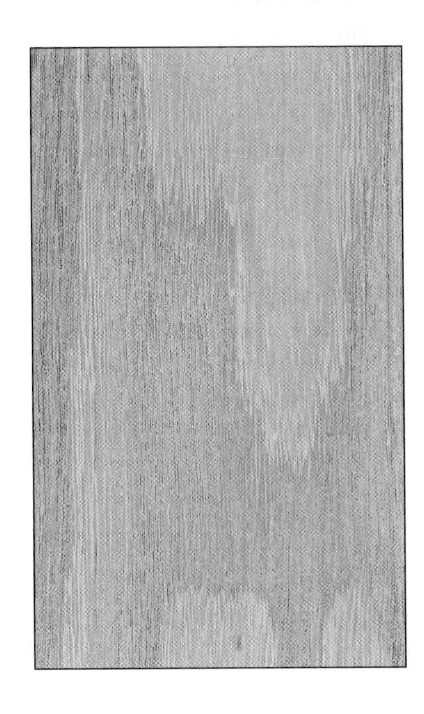

Haematoxylum campechianum

logwood

by Alan Curtis

SCIENTIFIC NAME

Haematoxylum campechianum. Derivation: The genus name *haemo* is Latin meaning blood and xylum meaning wood. The specific epithet refers to Campeche, a geographical area on the west coast of the Yucatan Peninsula of Mexico where the plant flourished.

FAMILY

Fabaceae or Leguminosae, the legume family; (*Caesalpiniaceae*) the cassia group.

OTHER NAMES

palo de tinta, tinto, campeche.

DISTRIBUTION

Native to the Yucatan Peninsula of Mexico, Belize and Guatemala. Logwood has naturalized on many of the Caribbean islands and in portions of central and northern South America.

THE TREE

Logwood is a small to medium-sized tree that occasionally reaches 40 feet in height with a diameter of 2 feet. Typically there are several short, very fluted, crooked trunks growing together. Logwood is often the largest tree found in swamp forests that are inundated for 6 months or more. It also grows well on drier sites. When in flower, the tree is very showy and honey bees are attracted to it.

THE TIMBER

The heartwood is bright orange to red and becomes dark red after exposure. The narrow sapwood is yellowish or white. The luster of the heartwood is high and golden. The grain is interlocked, and the texture is fine. Fresh cut wood has a slight odor and sweet taste. Average specific gravity is 0.76 to 0.79 (ovendry weight/green volume), equivalent to an air-dried weight of 60 to 63 pcf. The wood is strong but brittle.

SEASONING

Logwood is very difficult to dry without deep longitudinal checks developing. Average shrinkage values (green to 12% moisture content) are 7% tangential and 5% radial.

DURABILITY

Logwood is highly resistant to decay. When in contact with the ground, it should last 15 to 20 years.

WORKABILITY

Logwood is rather difficult to saw and plane but turns and carves easily. A lustrous finish can be easily obtained. Preboring is advised before using nails and screws.

USES

Logwood became an important article of commerce in the 16th century after it was discovered that dye could be obtained from the heartwood. Belize (formerly British Honduras) was founded as a colony in 1638 mainly for the export of mahogany and logwood. Today the dye has largely been replaced by synthetics but is used locally for coloring wool textiless. Hematoxylin is used as a stain for biological microscope slides and inks, and as a drug is used as an astringent, such as for dysentery and diarrhea. Use for purposes other than dye extraction is restricted by the irregular form of the trunks. Minor uses have included posts, tool handles, interior trim, carvings, toys and turnery. It is a good wood for cooking fire fuel.

SUPPLIES

This wood is not marketed commercially. If the wood could be stabilized during drying, logwood would become a popular wood with turners. After Hurricane Andrew hit Florida in 1992, wood from logwood trees was salvaged and distributed to woodworkers. Everyone had trouble with excessive checking while the wood was drying, even when vases were roughed out of green wood and left in plastic bags to dry slowly. I cut material into boards, sealed the ends and watched them crack.

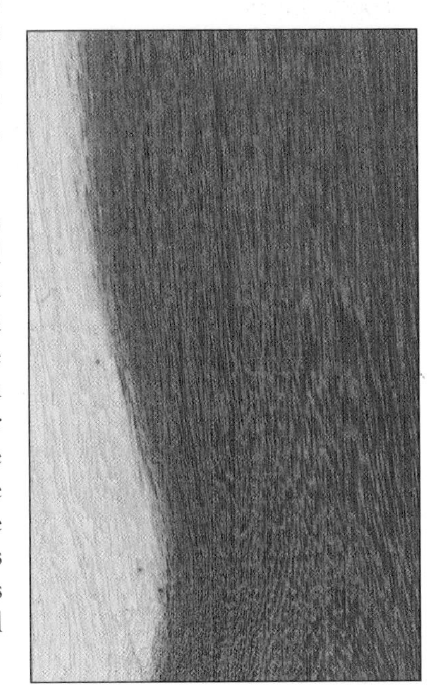

SPECIAL NOTE

A quick way to determine if an unknown wood is logwood is to whittle a piece of the heartwood and place the shavings in a glass of water. In a few minutes the dye will begin to dissolve from the wood and purple streaks will appear in the water. A personal anecdote is that I was using a chainsaw to cut logwood during the cleanup after Hurricane Andrew. The day was windy and the sawdust blew in my face and on my hands. The sweat on my face reacted with the dye in the wood and I 'turned purple'. I couldn't see this hilarious situation occurring, but my co-workers finally stopped their laughing long enough to tell me what had happened.

Halesia carolina

Carolina silverbell

by Jim Flynn

SCIENTIFIC NAME

Halesia carolina. Derivation: The genus name is in honor of Stephen Hales (1677 to 1761), a British clergyman and author of *Vegetable Staticks* (1722). The specific epithet means the plant is "of the Carolinas".

FAMILY

Styracaceae, the storax family.

OTHER NAMES

mountain silverbell, snowdrop tree, opossum-wood, belltree, calicowood, tiss-wood, wild olive, rattlebox. The large type which grows in the southern Appalachian Mountains was once considered a separate species, *H. monticola*, but is now synonymous with *H. carolina*.

DISTRIBUTION

Found mostly in the mountain areas from southwest Virginia, south West Virginia and south Ohio, west to extreme south Illinois, south to west Tennessee, Alabama, north Florida and northeast to central North Carolina. It is also found in the mountains of Arkansas and southeast Oklahoma. It has been cultivated as far north as southern New England and as far west as California. It was introduced in England in 1756 and can be found in many parts of Europe as an ornamental.

THE TREE

Carolina silverbell is an attractive shrub or small tree. It grows best in moist soils along streams in hardwood forests. The rate of growth and longevity is variable but on average lives for 100 years or more and most trees reach heights of about 40 feet with diameters of 18 to 24 inches. In the Great Smoky Mountains, specimens exceed 100 feet in height, and the U.S. National Register of Big Trees records three co-champions that exceed this height and have large circumferences of more than 10 feet. The bark is dark brown with many irregular fissures. The ovate-acuminate leaves are serrate, with the middle depressed, and 4 to 5 inches long. The pure white bell-shaped flowers and small size make it desirable for landscaping. Squirrels eat the seeds, and honeybees find the flowers highly desirable. The ripe fruit has been eaten by humans and, when green, sometimes used as a pickle, but I do not recommend it.

THE TIMBER

The wide sapwood is white and sharply demarcated from the pale brown heartwood. The luster is medium. The texture is fine and uniform with occasional pin-knots that add a pleasing pattern to the wood. The pores are exceedingly fine and rays are thin and ill-defined making it a close-textured, smooth wood. There is no detectable odor or taste. Average reported specific gravity is 0.42 (ovendry weight/green volume), equivalent to an air-dried weight of 31 pcf.

SEASONING

The ends of freshly cut billets should be immediately coated with wax and the wood allowed to season slowly. Cutting longitudinally through the center with a band saw will help immeasurably to prevent the wood from splitting. Average reported shrinkage values (green to ovendry) are 3.8% radial, 7.6% tangential and 12.6% volumetric.

DURABILITY

Carolina silverbell, scarce and in short supply, is not considered an exterior-use wood. Reports from the early 19th century suggest the wood was not considered durable.

WORKABILITY

The wood is a pleasure to work. Because of its close and even grain, it cuts and machines nicely. The edges of milled stock remain sharp and crisp. It sands, scrapes, stains and finishes well and glues to perfection.

USES

Early in the 19th century, D. J. Brown reported in *Trees of America: Native and Foreign* that "owing to its small size and comparative scarcity, (this species) is appropriated to no particular use in the arts". Since then records show use as veneer with interesting patterns if cut with rotary knives. The wood makes attractive cabinets, small furniture, is recommended for carvers and turners and is a favorite for tourist trade crafters. It has been known to sell in local markets as cherry or birch.

SUPPLIES

Carolina silverbell is not found on the world market. The best source is ornamental trees that are being cut or pruned. In its native growth area, it may be found by a dedicated search.

Hardwickia mopane
Colophospermum mopane
mopane
by Jim Flynn

SCIENTIFIC NAME

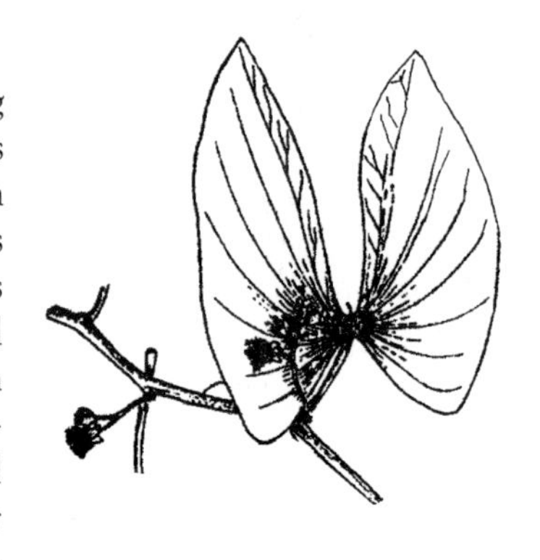

Taxonomists have been engaged in a continuing debate over whether to place this species in the genus *Hardwickia* or *Colophospermum*. While there is a slight margin in favor of *Hardwickia*, the authors decided to list both genera. Derivation: The genus name *Hardwickia* is in honor of Major General Thomas Hardwicke (1757 to 1835) of the East India Company's Artillery. The genus name *Colophospermum* is from the Greek meaning a resin-coated seed. The specific epithet is an African vernacular meaning butterfly, which refers to the shape of the leaves.

FAMILY

Fabaceae or Leguminosae, the legume family; (*Caesapliniaceae*) the cassia group.

OTHER NAMES

iphane, musaru, mwani, shantsi, mopanie. Because of the different tribal languages in tropical Africa, there are many more names.

DISTRIBUTION

Mopane is a dominant species thriving in hot, tropical Africa. From its southernmost limit in Transvaal's Kruger National Park, it is scattered throughout Namibia, Botswana, Zimbabwe, Zambia and Angola. Soil conditions often dictate where the trees will thrive. In some areas, the trees grow in pure stands while in others none can be found over large expanses.

THE TREE

This medium to large tree reaches heights of 60 feet or more with diameters of 20 inches. The dark gray bark is deeply fissured and flakes in narrow strips. Resembling a butterfly, the compound alternate leaves have two leaflets that grow closely together and droop. The small, green, inconspicuous flowers are erratic in their growing season. The fruit is a flat pod with one seed. Together the trees form one of the most distinctive vegetation groups in southern Africa earning the name mopane woodland. When conditions are not ideal, the plants remain stunted and scrub-like. The leaves are an important food for many animals, and the turpentine-smelling seedpods are eaten off the ground with no ill effect. Mopane is resinous, and the tree will burn readily when the bark is removed. Once the canopy is opened by elephants and coarse grasses invade the area, a dangerous fire hazard exists. A friend who spent many years in the African bush says the wood burns so thoroughly it produces a white ash finer than flour.

THE TIMBER

The narrow sapwood is straw-colored and sharply demarcated from the dark reddish-brown to almost black heartwood. The texture is fine and even. The wood exhibits narrowly streaked features on both the tangential and radial surfaces. The growth rings are well defined but the many rays are extremely narrow and difficult to see. Average specific gravity is 0.92 (ovendry weight/green volume), equivalent to an air-dried weight of 75 pcf.

SEASONING

The wood is prone to distortion and checking, so seasoning must be carefully controlled. The Zimbabwe Forestry Commission recommends keeping the logs in the round for long periods before cutting. Average reported shrinkage values (green to ovendry) are 3.1% radial and 5.5% tangential.

DURABILITY

The timber is extremely durable. It is highly resistant to borers, termites and fungi.

WORKABILITY

Because of its hardness, mopane is extremely difficult to work especially with hand tools. The ends often chip when cross cut. IWCS member Alan Bugbee, however, has turned mopane with an ornamental lathe and expressed enthusiasm for working with it. It is reported to take a high-gloss varnish finish and glue firmly.

USES

Uses include furniture, mine props, railway sleepers and poles. It has also been used in carvings and turnery. Mopane has a high calorific value for firewood. African natives often cook raw meat on the live coals and consume the feast with the ash coverings.

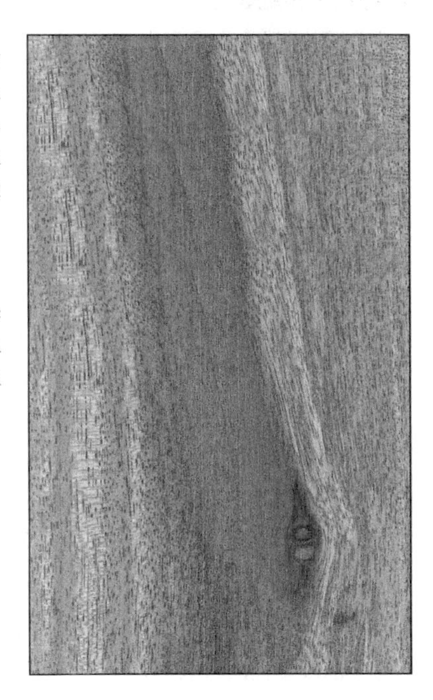

SUPPLIES

Many wood dealers specializing in the craft trade carry this wood. Some wood samples have been available in Florida from trees salvaged from Hurricane Andrew.

Heritiera simplicifolia

mengkulang

by Jim Flynn

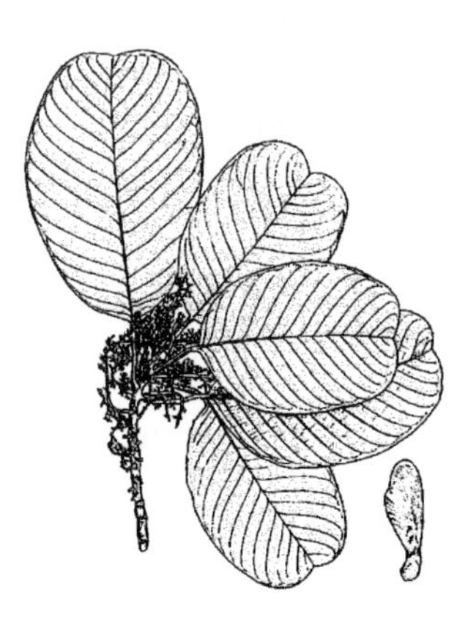

SCIENTIFIC NAME

Heritiera simplicifolia. Derivation: The genus name is in honor of Charles Louis l'Heritier de Brutelle (1746 to 1800), a French botanist and author. The specific epithet is Latin for simple foliage.

FAMILY

Sterculiaceae, the stericulia or cacao family.

OTHER NAMES

kembang, lumbayan, kanze, kanzo, chumprak, huynh, teraling, melima, balong ayam. Prior to 1959, mengkulang was classified in the genus *Tarrietia.*

DISTRIBUTION

This species can be found in Malaysia, Indonesia, the Philippines and other Pacific islands.

THE TREE

Mengkulang is a large-sized tree with steep buttresses and a reddish-brown bark that is slightly fissured. The tree reaches heights of 100 to 150 feet. The well formed bole can be 80 feet long with diameters above the buttresses measuring 2 to 4 feet. The wood will not float when freshly cut.

THE TIMBER

The multi-colored wood varies from medium pink to reddish-brown or dark reddish-brown. Sometimes dark streaks on the longitudinal surfaces are present. It is similar in appearance and related to *Tarrietia utillis* which comes from western Africa and bears the common name niangon. This latter wood is used extensively in Europe as a general construction timber and in some shipbuilding applications. The sapwood of mengkulang is 2 to 5 inches wide and is not always clearly defined. Typically the wood is interlocked and often irregular in direction. Surfaces that are quartersawn have a broad stripe figure and, when precisely cut, have a pronounced ray figure as does niangon. The texture is moderately coarse. Its average reported specific gravity (oven-dry weight/green volume) is in the range of 0.52 to 0.59, equivalent to an air-dried weight of 40 to 46 pcf. The weight of this timber varies according to growth region. In strength, it is comparable to European beech (*Fagus sylvatica*). It has an unpleasant odor when freshly cut.

SEASONING

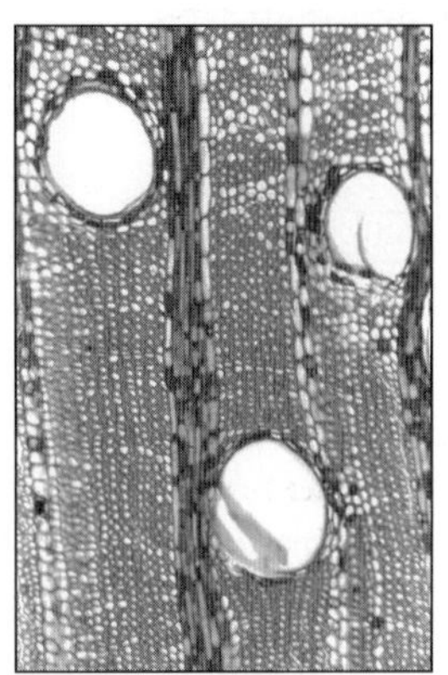

Mengkulang seasons rapidly with some tendency to warp and surface check. Average reported shrinkage values (green to 15% moisture content) are 1.6% radial and 3.1% tangential. Once dry, it is dimensionally stable in use.

DURABILITY

The heartwood is not durable and will be destroyed in less than 2 years if used in contact with the ground. Mengkulang is moderately resistant to preservative treatments and is not resistant to marine borers.

WORKABILITY

Because the wood contains silica, it is difficult to cut. But, knowing this, tooling can be adjusted accordingly. The wood takes a good finish, glues well and can be nailed satisfactorily. Staining, finishing and polishing can be achieved with good results providing fillers are used. It is a good timber for rotary peeling but is difficult to turn and chisel.

USES

The wood is used for face veneer, particleboard and hardboard manufacture, flooring, furniture, interior trim, boat building and as a general utility lumber.

SUPPLIES

Mengkulang is often sold with red meranti, red luan and other species of *Heritiera*. Records show that it is imported into Australia. Most supplies, except veneer and plywood, are distributed predominantly in the southwest Pacific areas. Mengkulang timber is inexpensive.

Hevea brasiliensis

rubbertree

by Alan B. Curtis

SCIENTIFIC NAME

Hevea brasiliensis. Derivation: The genus name is a Latinized form of the Brazilian name for the Para rubbertree. The specific epithet means the plant is "of Brazil". A synonym is *Siphonia brasiliensis*.

FAMILY

Euphorbiaceae, the spurge family.

OTHER NAMES

rubberwood, Para rubbertree.

DISTRIBUTION

Native to the Amazonian region of Brazil where there are 12 species of *Hevea*, only this species is cultivated for its latex. It has been widely planted in other tropical areas, including West Africa and Southeast Asia. Malaysia alone has 5 million acres of rubbertree plantations.

THE TREE

Rubbertrees are grown for the valuable white latex from which natural rubber is produced. In the wild, rubbertrees may grow from 100 to 125 feet in height with large cylindrical trunks. Plantation trees when cut are 50 to 80 feet in height with diameters from 9 to 16 inches. The time of cutting is not dictated by the tree's size but by its diminishing latex production, which occurs after 25 years of age. Plantation-grown trees may have clear boles up to 30 feet in length, although they taper rapidly and their crowns have many branches. A shortage of skilled laborers willing to work as tappers is causing concern in producing countries.

THE TIMBER

Timber production is secondary to the latex rubber industry. The wood is whitish-yellow when freshly cut and dries to a pale cream color, often with a pinkish tinge. The sapwood is not differentiated from the heartwood. The texture is moderately coarse and even, and the grain is straight to shallowly interlocked. The luster is low, and the wood has a characteristic sour smell. The wood's average reported specific gravity (ovendry weight/green volume) is 0.46 to 0.52, equivalent to an air-dried weight of 35 to 40 pcf.

SEASONING

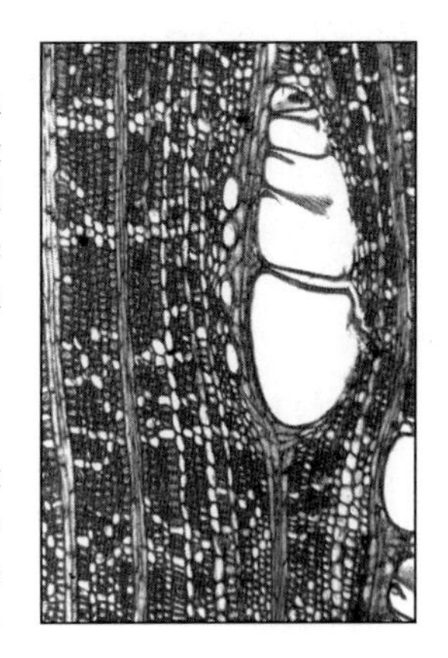

The wood air dries rapidly, but is usually kiln dried at high temperatures and humidity to prevent fungal and insect infestation and to reduce warping. Shrinkage is low and average reported shrinkage values (green to ovendry) are 2.3% radial, 5.1% tangential and 7.4% volumetric. When properly dried, the wood is dimensionally stable.

DURABILITY

Untreated rubbertree is very susceptible to rot and insect attack. In Malaysia, the rubberwood industry uses vacuum-pressure impregnation to treat all wood with a preservative (borax/boric acid) that is forced into the wood using hydraulic pressure. The treatment provides permanent protection.

WORKABILITY

Rubbertree is easy to saw and planes easily to a smooth surface. It is easy to turn and bore. It can be steam bent, and when set the bends are very stable. The wood may split when nailed. It is reportedly as strong as European Scotch pine.

USES

With its attractive appearance and easy workability, rubbertree is becoming a popular wood for furniture. Chairs, tables, dining and bedroom sets are made from solid wood or laminated components. The wood is also used for paneling and is sufficiently hard for parquet and strip flooring. Lamination provides larger pieces for stairs, railings and cabinet doors. The wood can be peeled to produce veneer for plywood and has been used to make one-piece molded chair frames. Other uses include paper production, particleboard and wood cement boards.

SUPPLIES

Malaysia's well-developed latex industry is dependent on the availability of rubbertree. Many plantations are re-established with higher latex-yielding trees, and it is anticipated that the supply is adequate for some time to come. Relatively little rubbertree lumber is imported into the United States, but occasionally it is available at a moderate price. Pre-finished rubbertree parquet flooring is also available. Because it is a sustainable tropical timber, we can expect to see more lumber and other forms of rubbertree in the future.

Hibiscus elatus

blue mahoe

by Max Kline

SCIENTIFIC NAME

Hibiscus elatus. Derivation: The genus name is from Ancient Greek and Latin meaning marshmallow. The specific epithet is Latin for tall.

FAMILY

Malvaceae, the mallow family.

OTHER NAMES

mahoe, mountain mahoe, seaside mahoe.

DISTRIBUTION

Occurs naturally in Jamaica and Cuba but has been widely planted in southern Florida, Trinidad, Tobago and from Mexico to Peru and Brazil.

THE TREE

Blue mahoe commonly grows to a height of 60 to 70 feet with a diameter of 12 to 18 inches. The boles are straight and of fairly good length. The soft-textured bark is dull gray and consists of many layers that can be separated after beating. The inner bark is used for making rope and cord that is reported to be very durable in salt water. The tree grows in a wide range of soil conditions and elevations from sea level to 4,000 feet, but not where rainfall is less than 60 inches annually.

THE TIMBER

The narrow sapwood is nearly white. The heartwood, usually with lighter and darker streaks, is often richly variegated shades of purple, metallic blue and olive brown or is plain olive. The luster is medium to high. Odor and taste in dry wood are not distinct. The texture is medium, and the grain is straight to slightly wavy. Average reported specific gravity is 0.62 (ovendry weight/green volume), equivalent to an air-dried weight of 48 pcf. Data is very limited on its strength properties, but it is considered a hard, tough and elastic wood of overall good quality.

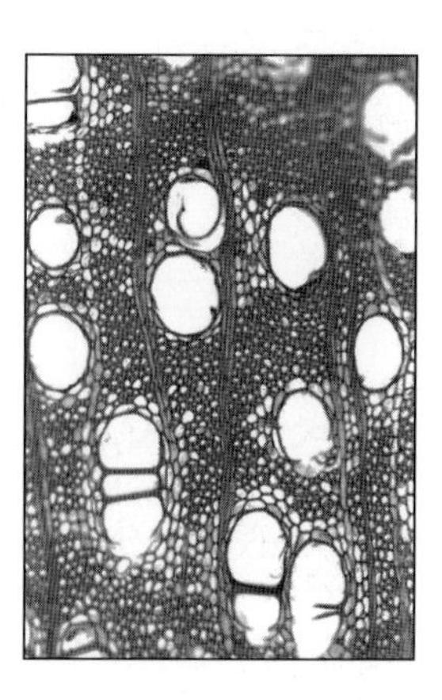

SEASONING

When dried slowly in stickered and weighted stacks, the timber can be seasoned with minimum degrade. No shrinkage values are available.

DURABILITY

Blue mahoe is generally reported to be durable to very durable, although not much data is available.

WORKABILITY

Blue mahoe works easily but requires care to provide a good polish, otherwise a dull appearance will result.

USES

In its growth region, especially in Jamaica, blue mahoe is considered one of the finest timbers and is in demand for furniture, interior trim, gun stocks, fishing rods, inlay work, cabinets, boats, railway sleepers and shafts and bodies of carts and wagons. It is also used in construction for sills, frames, flooring, shingles and footings. Since the wood is scarce and costly, most of it is consumed locally for high-grade applications.

SUPPLIES

The supply of blue mahoe in Jamaica and Cuba is not equal to the demand, and therefore none is available for export. Because of its unusual color, it is well worth acquiring if at all possible at the source.

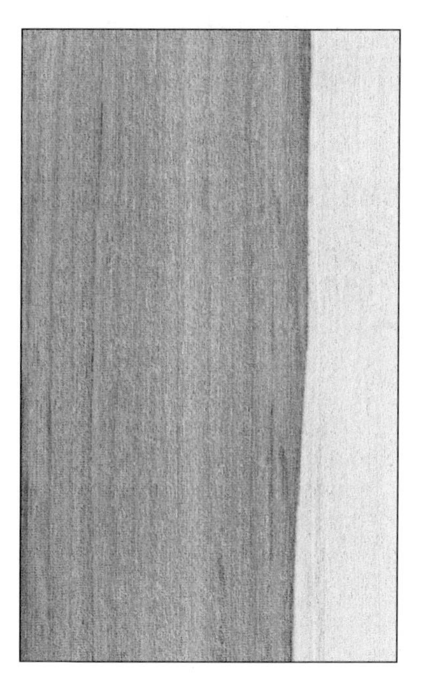

Hura crepitans

sandbox tree

by Jim Flynn

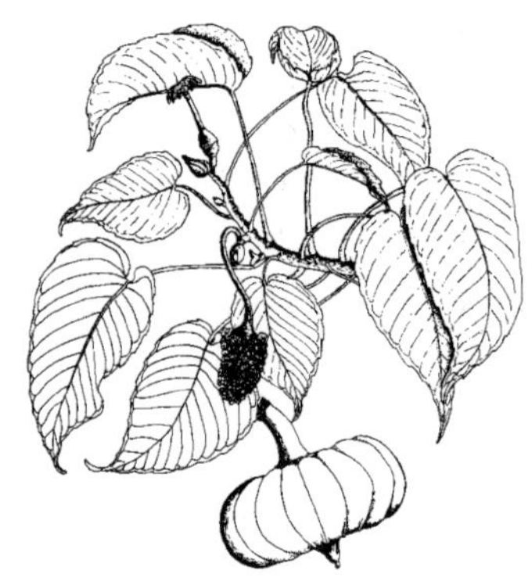

SCIENTIFIC NAME

Hura crepitans. Derivation: The genus name is a Latin version of the Spanish name for the tree. The specific epithet is Latin for crackling, which refers to the pumpkin-like fruit that violently explodes when dry scattering the wafer-like seeds in all directions.

FAMILY

Euphorbiaceae, the spurge family.

OTHER NAMES

At the very minimum, there are at least 50 common names associated with this species. Some of the more common ones are possumwood, hura wood, assacu, rakuda, javillo, acuapa, habillo and catahua. The common name sandbox tree was derived from the early practice of hollowing out the immature pods and using them as containers for ink blotting sand.

DISTRIBUTION

This species is widely distributed from the West Indies and southern Mexico to northern Brazil. A related species, *H. polyandra*, has a more limited distribution in Mexico. Except for some key morphological features, both species are relatively the same. The best development of these trees takes place on the low reefs of the coastal plains of Suriname.

THE TREE

Growing in moist sandy loam, sandbox tree reaches a maximum height of 90 to 130 feet. Clear boles of 40 to 75 feet in length and 3 to 5 feet in diameter are often harvested. Larger diameters of 6 to 9 feet have been encountered but are now rare. The branches and trunk often have sharp spines. The latex of the bark is caustic and may cause poison ivy-type blistering on human skin. The deciduous leaves are long-petioled, cordate-ovate and toothed. The stamens are arranged on a long column. The fruit is a 3- to 4-inch-wide capsule, which is shaped like a pumpkin and contains about 15 one-seeded woody cells which resemble the sections of an orange. When ripe the fruit explodes violently with a loud report, scattering the seeds far away from the tree. The tree is often planted as an ornamental and for shade.

THE TIMBER

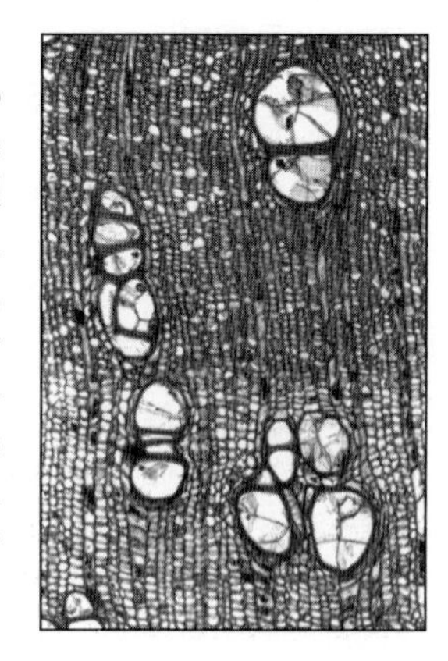

The wood varies in color from a lustrous creamy white to yellowish-brown or olive-gray. The heartwood and the sapwood are not sharply demarcated. The texture is fine and mostly uniform. The grain is straight and may be interlocked. There is neither distinct odor nor taste. It is light and soft, and the average reported specific gravity (ovendry weight/green volume) is 0.33 to 0.38, equivalent to an air-dried weith of 24 to 28 pcf. Some of the wood is apt to have a very attractive wavy grain which is more pronounced in crotches. The wood is known as an inexpensive substitute for Spanish-cedar (*Cedrela odorata* WDS 070).

SEASONING

Air drying sandbox tree is moderately difficult because of the tendency to dry too rapidly which causes severe warping and checking. Rapid air drying, however, prevents the development of mold and sap stain. Controlled kiln drying is recommended. Average reported shrinkage values (green to ovendry) are 2.7% radial, 4.5% tangential and 7.3% volumetric. Dimensional stability in use is considered moderate.

DURABILITY

There is a wide variance in the resistance of this wood to attack by decay fungi. It is highly susceptible to blue stain and very susceptible to dry-wood termites. It is easy to treat with a penetrating preservative.

WORKABILITY

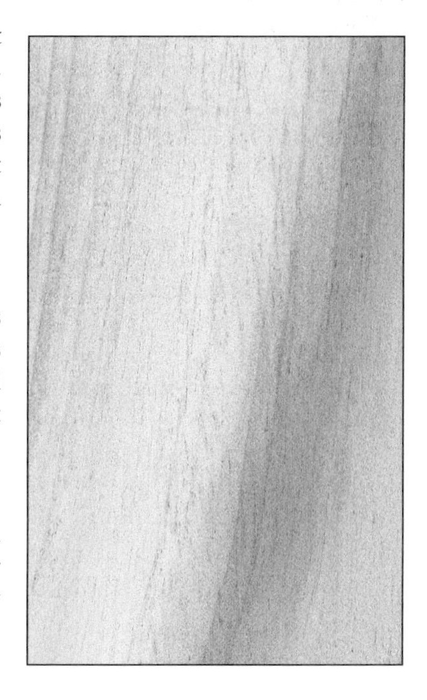

Because sandbox tree is not hard and dense, it saws and machines well with few exceptions. When green, it is apt to produce fuzzy surfaces because of tension wood. When dry, it planes well without undue dulling of cutting tools. It finishes and glues well and is easy to nail. It can be readily peeled and sliced.

USES

Sandbox tree is used primarily in applications that are not subject to severe wear such as general carpentry, boxes and crates, veneer and plywood, fiberboard and particleboard, light joinery and furniture.

SUPPLIES

Sandbox tree can be found on the commercial market at a reasonable price and is suitable for less-demanding applications than more costly exotic timbers.

Hyeronima alchorneoides

pilon

by Jon Arno

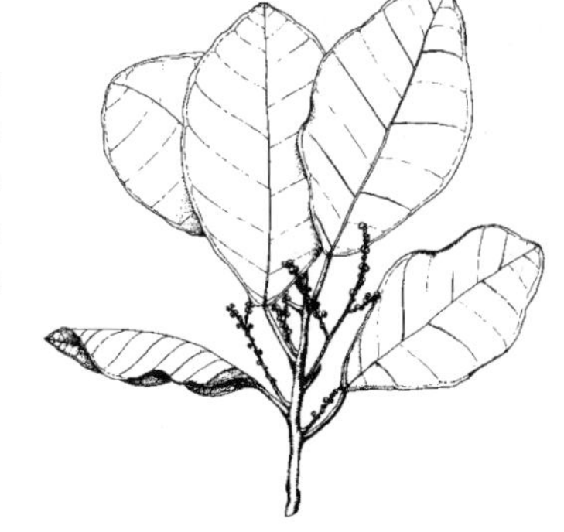

SCIENTIFIC NAME

Hyeronima alchorneoides. Derivation: The genus name is in honor of Jeronimo Serpa, a Brazilian botanist and colleague of Brother Allemao who described the tree. The specific epithet refers to the plant's similarity to plants in the genus *Alchornea*, a genus in the same family.

FAMILY

Euphorbiaceae, the spurge family.

OTHER NAMES

suradan, trompillo, tapana, curtidor, nancito, torito.

DISTRIBUTION

While *H. alchorneoides* is the most dominant source, several species of Hyeronima produce wood which is marketed as pilon. They range from southern Mexico through Central and South America to southern Brazil.

THE TREE

Pilon can reach heights of 130 feet with diameters in excess of 3 feet when growing in ideal forest conditions. Pilon grows straight and has excellent form for timber production. Although the tree can be free of branches for as much as 70 feet, the large buttress limits the length of the logs to about 50 feet.

THE TIMBER

Both the color and density of this wood can vary considerably. The heartwood is generally reddish-brown, although it is not uncommon to find pilon that is dark chocolate-brown, comparable in color to black walnut (*Juglans nigra* WDS 150). The sapwood is normally light pinkish-tan and is much softer than the heartwood. Pilon's somewhat coarse texture and often interlocked grain give it a mahogany-like appearance, but it is much heavier and less lustrous. Pilon's average reported specific gravity (ovendry weight/green volume) ranges between 0.60 and 0.67, equivalent to an air-dried weight of 46 to 52 pcf. The wood is very strong and elastic. It is only slightly less rugged than Brazilian rosewood (*Dalbergia nigra* WDS 097) in most respects.

SEASONING

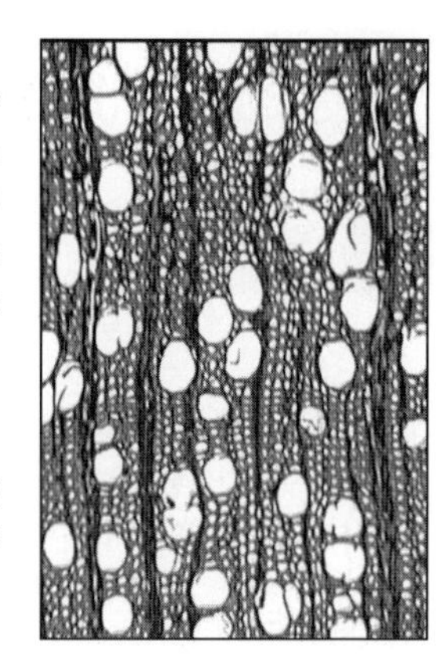

Pilon is very difficult to season even though it dries quickly relative to similarly dense woods. Pilon has a very high average volumetric shrinkage of 17.0% (green to ovendry) and the more than 2:1 ratio between its tangential shrinkage of 11.7% and its radial shrinkage of 5.4% encourages severe warping.

DURABILITY

Pilon is exceptionally durable, has excellent weathering properties and holds up well in marine applications. It is moderately resistant to insect attack.

WORKABILITY

On first examination, one would think pilon should be more popular than it is. It has adequate strength for most furniture applications, its color and figure rival both mahogany and walnut, and it machines quite well, except for its slight tendency to chip when pieces with extremely interlocked grain are planed. Like most dense woods, screws and heavy-gauge nails require pilot holes, but it is easy to finish. With reasonable care in selecting uniformly colored material for a given project, it is dark enough so as not to need staining and it takes varnish beautifully. On the down side, I suspect pilon has not caught on as an important furniture wood on an international basis because of its extreme instability. While it does not have an inordinate tendency to check, it is so prone to warping and volumetric change it is difficult to find boards that are flat enough for use in anything but small projects. Router bits and belt sanders cause the wood's resin to burn or darken quickly.

USES

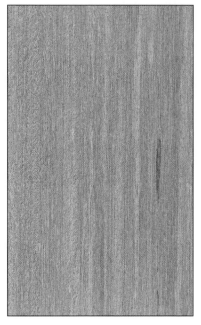

Within its native range, pilon is used for heavy construction such as beams and bridge timbers where its strength and elasticity are especially advantageous. Also, its durability makes it a good wood for both shipbuilding and railroad ties. Some is used for flooring, plywood, veneer, turnery and furniture.

SUPPLIES

When available, pilon is not expensive relative to other imported woods.

Hymenaea courbaril

courbaril

by Max Kline

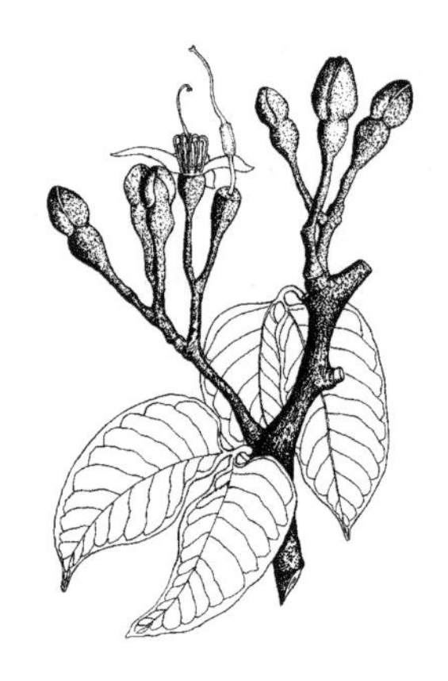

SCIENTIFIC NAME

Hymenaea courbaril. Derivation: The genus name is Greek for the goddess of marriage in allusion to the twin leaflets. The specific epithet is a Native American name.

FAMILY

Fabaceae or Leguminosae, the legume family; (*Caesalpiniaceae*) the cassia group.

OTHER NAMES

South American locust, algarrobo, guapinol, jutahy, jatoba.

DISTRIBUTION

West Indies, Mexico, Central America, Suriname, Guyana, Venezuela, Paraguay, Peru and Brazil.

THE TREE

Courbaril may reach a height of 60 to 130 feet with a diameter of 20 inches to 6 feet or more. Topped by a spreading crown of heavy branches, the bole is usually straight and free of branches for 40 to 70 feet. A yellow or orange-colored resin exudes from the bark and is known as South American copal. It is used in cements and varnishes.

THE TIMBER

The sometimes very wide sapwood is white to gray and is sharply demarcated from the heartwood which often has dark streaks. When freshly cut, the heartwood is bright red to orange brown, but after exposure becomes russet. Odor and taste are not distinct, and the luster is golden. It is hard and heavy with an average reported specific gravity of 0.71 to 0.82 (ovendry weight/green volume), equivalent to an air-dried weight of 56 to 66 pcf. The grain is commonly interlocked, and the texture is medium to rather coarse. This timber has good strength properties and may be used for all normal structural purposes for which oak would be suitable.

SEASONING

During seasoning, courbaril handles well with only slight degrade in the form of checking or warping. Air drying should be done slowly. Average reported shrinkage values (green to ovendry) are 4.5% radial, 8.5% tangential and 12.7% volumetric.

DURABILITY

Courbaril is only moderately resistant to fungal attack when used in contact with the ground or for exterior use. Durability is poor when subject to attack by marine pests.

WORKABILITY

Courbaril is not an easy wood to work, and it may have a marked dulling effect on the cutting edges of tools. It is a good species for turnery, and it bends well after steaming. Nail and screw holding properties are good, glue adheres satisfactorily and it can be finished smoothly.

USES

The timber is used for construction of all kinds, boatbuilding, general carpentry, turnery and bentwood products.

SUPPLIES

In the United States, the supply is very limited. In the past, some quantities have been available in Europe. Wood collectors can usually obtain samples.

Ilex opaca

American holly

by Max Kline

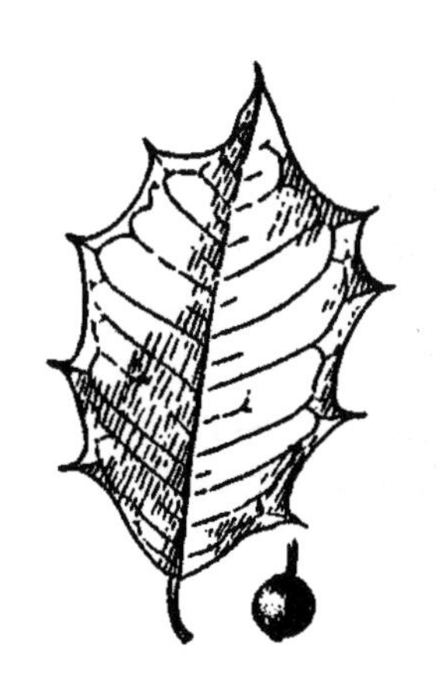

SCIENTIFIC NAME

Ilex opaca. Derivation: The genus name is the classical Latin name *Quercus ilex*, holly oak, of Europe, which has holly-like leaves. The specific epithet means opaque or dark, referring to the dull green leaves.

FAMILY

Aquifoliaceae, the holly family.

OTHER NAMES

Christmas holly, evergreen holly, white holly, prickly holly.

DISTRIBUTION

In the United States coastal region from Maine to Florida, north to the Mississippi Valley to Illinois, southwest through Missouri to Oklahoma.

THE TREE

American holly is the most important of the 19 species growing in the United States. It thrives on rich, deep, moist soil. Commonly 40 to 50 feet in height, the tree can reach heights of 80 feet. The trunk, which tapers rapidly, may be 1 or 2 feet in diameter. An evergreen famous for its red berries, it is used during the Christmas season. The tree is well known from its foliage, which are yellow-green spiny leaves that fall every 3 years during the spring instead of the autumn. American holly is the state tree of Delaware.

THE TIMBER

The wood is white or ivory with a low luster more closely resembling ivory than any other wood. The sapwood is quite wide and much whiter than the heartwood. It is moderately hard, close grained and tough. There is a total absence of any figure. American holly is odorless and tasteless. Average reported specific gravity is 0.50 (oven-dry weight/green volume), equivalent to an air-dried weight of 38 pcf.

SEASONING

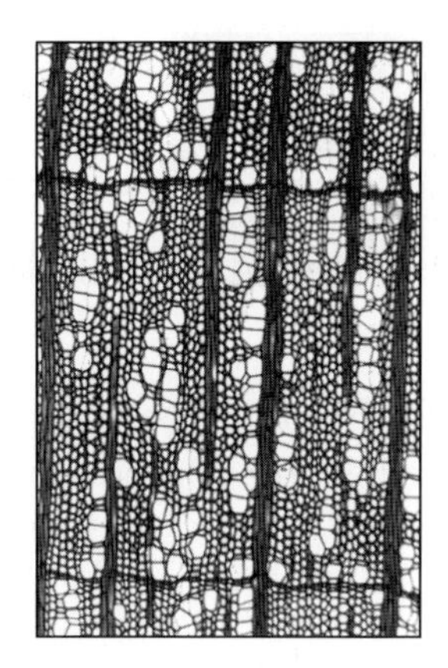

American holly may discolor during seasoning, so it is advisable to dry it fairly quickly. It should be cut in the winter and manufactured before hot weather to retain the original white color. Some distortion may take place during kiln drying but other types of degrade are not excessive. Average reported shrinkage values (green to ovendry) are 4.8% radial, 9.9% tangential and 16.9% volumetric.

DURABILITY

American holly is perishable when exposed to fungal attack. For the purposes for which it is used, preservative treatment is unlikely.

WORKABILITY

The wood is somewhat difficult to work due to its irregular grain, but it takes a high polish with the proper abrasives. In drilling operations, slow speeds should be used to prevent charring. Screws hold well and glues adhere without difficulty. Even in small sections, bending properties are poor. American holly finishes well and very uniformly to a high glaze. It is easy to stain to match other wood species. Sometimes it is dyed black and used as a substitute for ebony.

USES

American holly is used for piano, organ and accordion keys. Engravers find it desirable for certain types of work. It is used in inlays because of its white color. It is a good wood for turnery, and carvers find it desirable.

SUPPLIES

Because of the small size to which the trees usually grow, quantities are limited The large demand for leaves and berries during the holiday season restricts cutting the tree for lumber.

Intsia spp.

merbau

by Jim Flynn

SCIENTIFIC NAME

Intsia spp. The two closely related species *I. bijuga* and *I. palembanica* are generally marketed without species distinction. They will be discussed herein at the genus level except where noted. The genus name is a local vernacular in Madagascar.

FAMILY

Fabaceae or Leguminosae, the legume family; (*Caesalpiniaceae*) the cassia group.

OTHER NAMES

ipil, tat-talun, lumpa, lumpho, kwila, vesi.

DISTRIBUTION

Both species of *Intsia* are native to the tropical rain forests from Southeast Asia (especially the Philippines, Thailand, Malaysia and Indonesia) to the islands of southwest Pacific (including Papua New Guinea, the Solomons, Fiji and Samoa). They rarely grow together. *I. bijuga* grows in coastal areas bordering mangrove swamps, rivers or river floodplains. *I. palembanica* grows farther inland on low, hilly slopes or well-drained river flats. Both species tolerate a wide array of soil textures ranging from sandy and gravely to clay.

THE TREE

These species are large, deciduous, broad-crowned trees often growing to 100 feet or more in height with heavily buttressed boles. Circumferences of 5 feet have been reported. The trunk often is crooked, short and heavily fluted. The two species differ in the number of leaflets comprising their compound leaves: *I. bijuga* has 4 to 6 and *I. palembanica* has 14 to 18. The trees have pronounced taproots and many lateral roots which enable them to exploit minerals in large, deep volumes of soil. The trees have been so depleted that few remain in natural stands. While they do well on native sites, they do poorly on plantations in denuded tropical uplands where reforestation is needed.

THE TIMBER

Although saw logs may be short and somewhat out-of-round, they are usually sound to the heartwood and free of defects. Sharply demarcated from the pale yellow sapwood, the heartwood is yellow or orange-brown when freshly cut but deepens to bronze or dark red over time. When aged and weathered, the wood takes on a silver-gray cast not unlike teak (*Tectona grandis* WDS 261). Freshly cut wood shows a distinctive sulfur yellow deposit in the pores. Average reported specific gravity (ovendry weight/green volume) is 0.68, equivalent to an air-dried weight of 53 pcf. The interlocked grain is sometimes wavy resulting in ribbon grain patterns and an attractive fiddleback figure. The wood has a characteristic odor when worked and an astringent taste. The timber has qualities close to chanfuta (*Afzelia quanzensis* WDS 013) and was once classified in the same genus.

SEASONING

Logs dry well with little loss in quality. Average reported shrinkage values (green to ovendry) are 2.7% radial, 4.6% tangential and 7.8% volumetric. Dimensionally stable in use, the wood is readily kiln dried with little or no degrade except moderate twist in some boards.

DURABILITY

Intsia is among the most decay-resistant timbers. It is highly resistant to dry rot, subterranean termites and other insects and fungi. Yellow deposits (flavonol) in the pores help confer insect and fungal resistance to the wood but are water-soluble and leach out in wet conditions.

WORKABILITY

This dense non-siliceous wood is hard to work with hand tools. It has a blunting effect on saw blades and other cutting and milling tools. Saw teeth may gum on occasion. With sharp blades, it will plane to a smooth finish, but it is difficult to cut across the grain. It glues and finishes well. It is difficult to nail but has good holding power. As with most tropical timbers, protection from breathing saw dust is recommended.

USES

Handsome for flooring, its warm, nut-brown appearance is also valued for fine paneling, furniture, decorative turnery and high-grade joinery. When high quality is demanded, this wood should be used.

SUPPLIES

Other than manufactured flooring, merbau is not common in commercial use. Supplies in milled form can be obtained from some of the major timber importers.

Juglans cinerea

butternut

by Max Kline

SCIENTIFIC NAME

Juglans cinerea. Derivation: The genus name is the classic Latin name for walnut, meaning nut of Jupiter. The specific epithet is Latin for ash-color, referring to the bark.

FAMILY

Juglandaceae, the walnut family.

OTHER NAMES

oilnut, white walnut.

DISTRIBUTION

Extends from Quebec, south through the northeastern section of the United States, westward to North Dakota and as far south as Arkansas, Mississippi and Alabama.

THE TREE

Butternut is a small to medium-sized tree, which usually grows in open areas. The tree resembles its relative the black walnut (*Juglans nigra* WDS 150), but is shorter with a more spreading crown. It will grow on poorer soil and under greater temperature extremes than walnut. Butternut trees are not long-lived and seldom exceed 75 years of age. The sweet oily nuts, a favorite of squirrels, are gathered in quantity for home consumption and candymaking.

THE TIMBER

The heartwood is chestnut brown with darker zones, while the narrow sapwood is white. Butternut has a satiny, generally straight grain, and its texture resembles that of black walnut. It is without odor or taste. Average reported specific gravity is 0.36 (oven-dry weight/green volume), equivalent to an air-dried weight of 27 pcf. Lacking in stiffness, it is moderately weak in bending strength and in end-wise compression.

SEASONING

Butternut is relatively easy to dry and dries with minimal shrinkage. Average reported shrinkage values (green to ovendry) are 3.6% radial, 8.1% tangential and 12.5% volumetric. Once dry, it is dimensionally stable.

DURABILITY

The butternut tree is often attacked by insects or fungus growth before reaching maturity and when the timber is placed under conditions favorable to decay. Butternut's durability is rated below black walnut.

WORKABILITY

The wood is easy to work with tools, but sharp cutting edges are desirable due to the softness of the species. It is capable of taking a rich, lustrous finish.

USES

The major use for butternut lumber is for furniture with some going into interior trim and paneling. In the past, many church altars were made of butternut and carvers enjoy using it. Butternut tree sap is sweet and a syrup of fair quality can be made from it.

SUPPLIES

Supplies of butternut are limited and diminishing. Butternut is more valuable for its nuts than its lumber.

Juglans nigra

black walnut

by Max Kline

SCIENTIFIC NAME

Juglans nigra. Derivation: The genus name is the classic Latin name for walnut, meaning nut of Jupiter. The specific epithet means black, referring to the dye in the fruit husk or the dark brown wood.

FAMILY

Juglandaceae, the walnut family.

OTHER NAMES

American walnut, eastern black walnut, American black walnut, gunwood.

DISTRIBUTION

From Massachusetts to southern Ontario and Nebraska, southward throughout the eastern half of the United States, except the Atlantic coastal plain south of Virginia, the Gulf coast and the lower Mississippi valley.

THE TREE

Black walnut reaches 100 feet in height with a diameter of 3 to 4 feet. The clear bole can be 50 to 60 feet in length. Trees require deep, rich, moist but well-drained soils. Its nuts are found in a very hard and deeply grooved shell.

THE TIMBER

The heartwood is a rich purplish-brown shade. In most cases, wood from less mature trees has a chocolate-brown tint. The narrow sapwood is nearly white. The wood has a mild persistent odor. The texture is moderately coarse but uniform. Black walnut produces a greater variety of figure types than any other tree including crotches, swirls, stumpwood, stripe, ribbon, mottle and snail and occasional burls. Black walnut is a strong wood with more than adequate properties for its usual uses. For many purposes, it is stronger than white oak (*Quercus alba* WDS 230). Average reported specific gravity is 0.51 (ovendry weight/green volume), equivalent to an air-dried weight of 39 pcf.

SEASONING

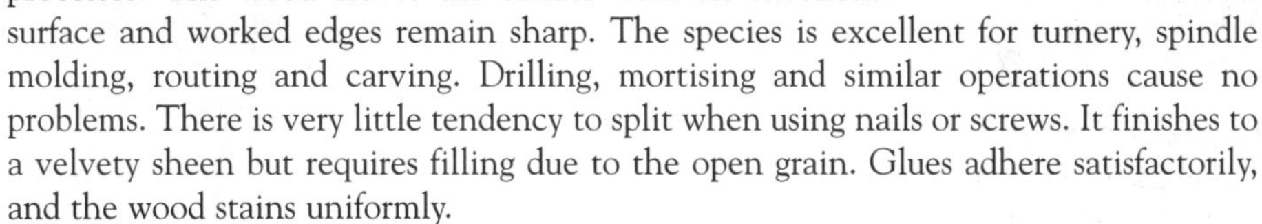

Black walnut should be dried slowly, and air or kiln drying produces satisfactory results. Average reported shrinkage values (green to ovendry) are 5.5% radial, 7.8% tangential and 12.8% volumetric. Once seasoned, it will remain exceptionally stable, with only negligible movement.

DURABILITY

The heartwood is highly resistant to decay. The tree is relatively resistant to fungus and insect attacks.

WORKABILITY

The timber works with ease in all hand and machine tool processes. The wood leaves the sander with an excellent surface and worked edges remain sharp. The species is excellent for turnery, spindle molding, routing and carving. Drilling, mortising and similar operations cause no problems. There is very little tendency to split when using nails or screws. It finishes to a velvety sheen but requires filling due to the open grain. Glues adhere satisfactorily, and the wood stains uniformly.

USES

Black walnut is the foremost American wood for cabinetwork. It is superior to all other woods for gunstocks because it keeps its shape, is light weight and absorbs recoil better than any other wood. It finds much use as fine figured veneers and cabinets, furniture, novelties and moldings.

SUPPLIES

Black walnut lumber is available but supplies are not as abundant as they once were which has increased the price.

Juniperus ashei

Ashe juniper

by Jim Flynn

SCIENTIFIC NAME

Juniperus ashei. Derivation: The genus name is the classical Latin name. The specific epithet is in honor of William Willard Ashe (1872 to 1932), a pioneer forester in the U.S. Forest Service. Ashe collected this species in Arkansas. A strange coincidence is that another species of juniper, *J. pinchottii*, named in honor of another U.S. forester, Gifford Pinchot, hybridizes with the Ashe juniper. A synonym is *J. mexicana*.

FAMILY

Cupressaceae, the cypress family.

OTHER NAMES

mountain cedar, rock-cedar, post-cedar, Mexican juniper, Ozark white cedar, yellow cedar, Arbuckle white cedar, Texas cedar, sabino, enebro, tascate, cedar brake, cedro.

DISTRIBUTION

Ozark Mountains of southern Missouri, northern Arkansas and northeastern Oklahoma, southern Oklahoma in the Arbuckle Mountains and central Texas on the Edwards Plateau. This species is also native to northeastern Mexico. It grows best on limestone hills. Reportedly resistant to cedar-apple rust, the tree is planted as an ornamental and may be found beyond of its natural range.

THE TREE

Ashe juniper is a small tree or shrub that in ideal conditions may grow 40 feet in height with a trunk diameter of 24 inches. It is often compared to the eastern redcedar (*Juniperus virginiana* WDS 152) in physical terms. It has a globular shape when grown in the open but, in dense pastures, drops its lowest branches so it has a more columnar profile. The small, opposite, scale-like, blunt pointed, dark bluish-green leaves are fringed with minute teeth and are more bunchy than eastern redcedar. The round seeds are larger and a deeper blue than eastern redcedar. The fibrous and shreddy bark is a light grayish-brown and is usually covered with a fungus giving the tree a blue-gray cast. The U.S. National Register of Big Trees lists the champion in Comal County, Texas. It is 38 feet in height with a main stem circumference of 115 inches.

THE TIMBER

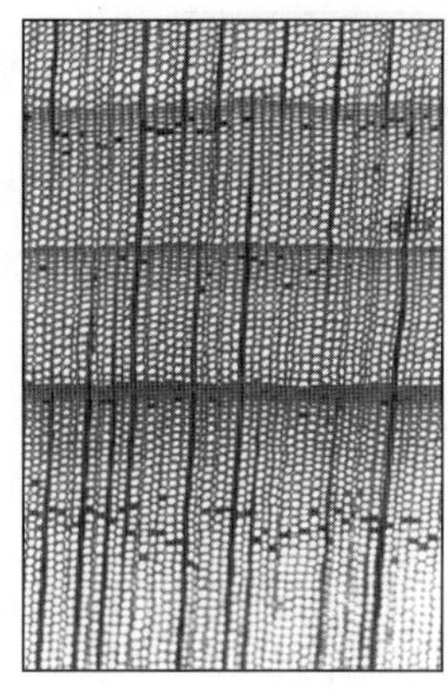

The wood is a pleasant brown color with a slight aromatic odor. The apparent slow growth of this species produces narrow growth rings, which can be shown to advantage when sawn on practically any plane. For a softwood, it is rather heavy with an average specific gravity of 0.65 (ovendry weight/green volume), equivalent to an air-dried weight of 51 pcf. The luster is medium. The grain is apt to wander in an attractive style over longer lengths of sawn boards unless the main stem was straight. Ashe juniper is not considered a strong wood.

SEASONING

Because of the small volumes of Ashe juniper used commercially, there seems to be little data available on this characteristic and shrinkage values are not available. The seasoning characteristics of other species in the genus, however, probably fit well. No unusual problems should be expected.

DURABILITY

As proof of its durability, the wood has commonly been used for fence posts and railroad crossties.

WORKABILITY

If the grain direction and patterns are considered when machining, Ashe juniper is fairly easy to work with both hand and machine tools. It is hard and dense enough to retain crisp edges and easy to hand scrape and sand. Gluing and finishing present no problems. It turns well but may occasionally splinter.

USES

In addition to the obvious external use for fence posts and fuel, Ashe juniper is used in its growth area for small items of furniture and craft work.

SUPPLIES

The best chances of obtaining Ashe juniper are through contacts in the growth area. Small sawmills tucked in and out of the byways should always be explored for this or other local species. Many IWCS members have this wood in their collections, and samples can be acquired through trading or through auctions.

Juniperus virginiana

eastern redcedar

by Max Kline

SCIENTIFIC NAME

Juniperus virginiana. Derivation: The genus name is the classical Latin name. The specific epithet means the plant is "of Virginia".

FAMILY

Cupressaceae, the cypress family.

OTHER NAMES

juniper, pencil cedar, Virginia juniper, eastern juniper.

DISTRIBUTION

Eastern redcedar grows in the eastern half of North America from Maine, Nova Scotia, New Brunswick, southern Quebec and Ontario to southeastern North Dakota, southward to eastern Texas and eastward to northern Florida. It is absent along the Gulf coast and from the high Appalachians. [Ed. note: Often confused with the eastern redcedar is southern redcedar (*Juniperus silicicola*) which is found on the coastal plain, chiefly near the coast, from northeastern North Carolina south to central Florida and from there west to southeastern Texas. There remains some taxonomic controversy, however, as to the classification of the southern redcedar as a distinct species.]

THE TREE

Eastern redcedar usually grows from 20 to 50 feet in height with a short trunk 1 to 2 feet in diameter. On good soil it may reach a height of 120 feet with a trunk diameter of 4 feet. It grows quite slowly, living up to 300 years of age. Dark purple-blue berries grow on the small leaf sprays and are relished by birds. The light reddish-brown bark is scarcely more than 0.125 inch thick and peels off in fibrous strips. Eastern redcedar is not a true cedar but is instead a juniper.

THE TIMBER

The very narrow sapwood is nearly white or a light cream color. The heartwood, often with many small knots, is dull to bright pinkish-red, sometimes with a purple-red tinge and often streaked a deep reddish-brown. The wood is highly aromatic. It is fine in texture, and the grain is usually straight. Average reported specific gravity is 0.44 (ovendry weight/green volume), equivalent to an air-dried weight of 33 pcf.

SEASONING

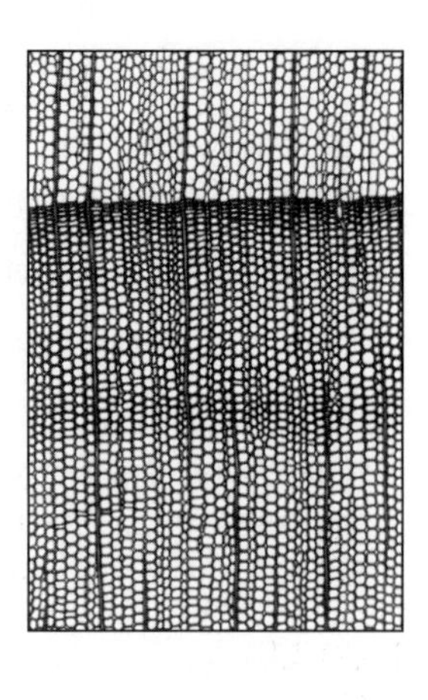

There is little warping or shrinkage when drying eastern redcedar. Average reported shrinkage values (green to ovendry) are 4.7% radial, 3.1% tangential and 7.8% volumetric. The wood is dimensionally stable in use.

DURABILITY

The heartwood is considered as durable as any of the North American woods. The chief enemy of the tree is fire, as the very thin bark does not offer much fire resistance. When growing in an area of apple orchards, the cedar-apple rust disease may spread from the cedars to the apple trees. Because of this, many cedars have been eradicated, especially in the Shenandoah Valley of Virginia.

WORKABILITY

Eastern redcedar is easily worked with all types of tools, has good carving and whittling properties and takes a beautiful natural finish.

USES

For many years, eastern redcedar was used exclusively as the wood in pencils, but with diminishing supplies the industry turned to incense-cedar (*Calocedrus decurrens* WDS 054) for this purpose. Eastern redcedar has long been used for chest linings, because its odor is believed to be disagreeable to the moth and buffalo bugs that are injurious to stored clothing. For similar reasons, it has found use in closet, wardrobe and cabinet building materials. It is a favorite for souvenir novelties of all kinds. Because of its durability, it is desirable for buckets, shingles, small boat construction, posts and poles. Cedar leaf oil distilled from the leaves is used in medicine. Oil distilled from the wood finds a place in perfume.

SUPPLIES

Because eastern redcedar grows slowly and has been destructively cut in the past, large trees are no longer available for quantities of timber. Trees currently available usually yield timber of small dimensions that are filled with numerous knots. The woodworker, however, can usually find ample supplies for wood-crafted articles at a moderate cost.

Kalmia latifolia

mountain-laurel

by Max Kline

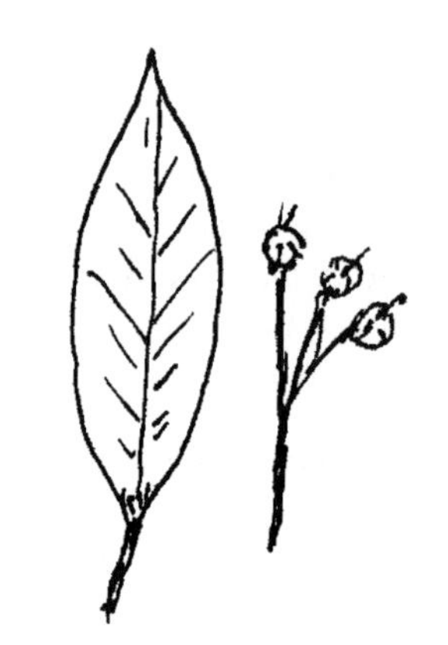

SCIENTIFIC NAME

Kalmia latifolia. Derivation: This genus name was dedicated by Linnaeus to his student Peter Kalm (1716 to 1779), a Swedish botanist who traveled and collected in Canada and the eastern United States. The specific epithet is Latin for broad-leafed.

FAMILY

Ericaceae, the heath family.

OTHER NAMES

American laurel, poison laurel, big-leaved ivy, calico bush, ivywood, kalmia, sheep laurel, spoonwood, ivy leaf laurel, ivy.

DISTRIBUTION

From New Brunswick to Lake Erie south to Virginia, Ohio, Indiana, Tennessee, the Appalachian Mountains, to western Florida, west through the Gulf states to western Louisiana.

THE TREE

Mountain-laurel is usually a small evergreen shrub, but in the Carolinas it grows to a tree height of 30 to 40 feet with a diameter of 18 to 20 inches. The leaves are leathery, and the pink to white flowers blooming in clusters are beautiful in the springtime. The leaves are poisonous if eaten causing convulsions and eventually paralysis of the limbs.

THE TIMBER

The wood is heavy, hard, rather brittle and strong with close, straight grain and little or no outstanding figure. The yellowish-brown heartwood is often tinged with red. The sapwood is slightly lighter in color. When quartersawn, mountain-laurel is quite attractive. Average reported specific gravity is 0.62 (ovendry weight/green volume), equivalent to an air-dried weight of 48 pcf.

SEASONING

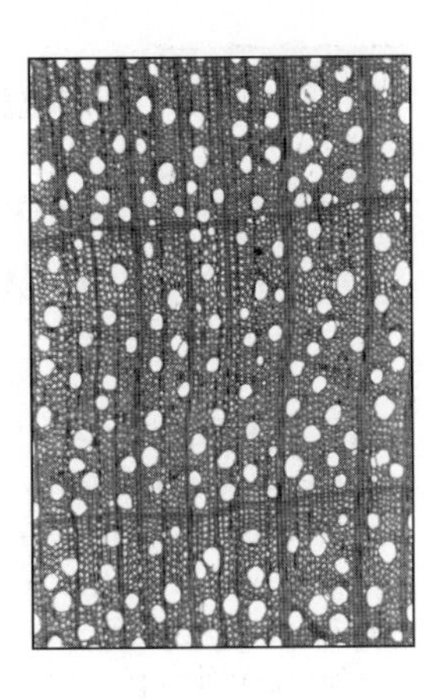

Mountain-laurel air dries quite well if the logs are debarked and split down the center. Average reported shrinkage values (green to ovendry) are 5.6% radial, 8.0% tangential and 14.4% volumetric.

DURABILITY

Mountain-laurel is not considered a durable wood.

WORKABILITY

Due to its close and straight grain, mountain-laurel is readily worked. It is especially good for use with carving tools or on the lathe.

USES

The wood is used for small objects and novelties and for carving and turnery. It has some commercial applications for tool handles, woodenware and fuel. The root burls make attractive smoking pipes.

SUPPLIES

Because of its widespread distribution in the eastern United States, the wood is readily obtainable by the wood collector, but it is not found in commercial lumber yards.

Kalopanax septemlobus

sen

by Max Kline

SCIENTIFIC NAME

Kalopanax septemlobus. Derivation: The genus name is derived from the Greek *kalos* meaning beautiful and *panax* is the name of a related genus. The specific epithet is a Latin name for seven lobed. A synonym is *K. pictus*.

FAMILY

Araliaceae, the ginseng family.

OTHER NAMES

hari-gara, castor aralia, nakora, Japanese ash.

DISTRIBUTION

Northern part of Honshu and in all parts of Hokkaido, Japan. Also occurs in China, Manchuria and Korea.

THE TREE

Sen can grow to an exceptionally large size of 100 feet or more in height but usually reaches a height of 80 feet with a diameter up to 40 inches. The thick branches have short, thick prickles. The alternate leaves have five to seven lobes and are 4 to 8 inches wide with petioles that are up to 12 inches long. The fruit is a globose, bluish-black drupe with a juicy pericarp about 0.25 inch in diameter.

THE TIMBER

The sapwood is white with no sharp line of demarcation between it and the heartwood. The heartwood can vary from cream to pale yellow to grayish-brown. The wood is straight grained and lustrous with no distinct odor or taste. The texture is coarse. Average specific gravity is 0.38 (ovendry weight/green volume), equivalent to an air-dried weight of 28 pcf. In general, sen resembles ash or American elm (*Ulmus americana* WDS 274) in texture and appearance. In strength, sen is inferior to both of these species and is more nearly comparable to red alder (*Alnus rubra* WDS 018) or yellow-poplar (*Liriodendron tulipifera* WDS 162).

SEASONING

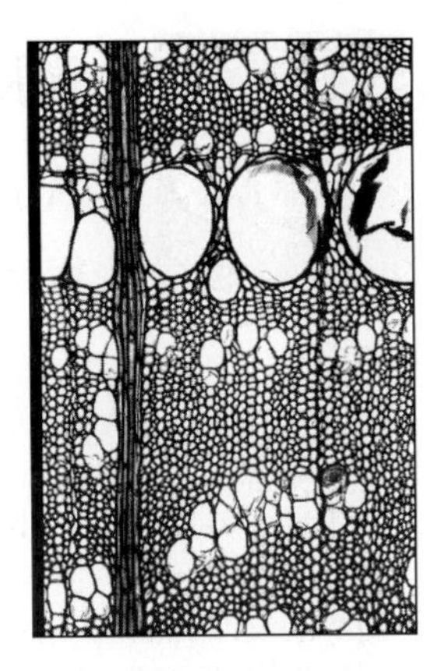

Sen is not difficult to season but has a high shrinkage rate. When dry, the wood is always subject to warping, swelling or shrinking with changes in humidity.

DURABILITY

The timber is non-durable and best suited to interior use.

WORKABILITY

Sen works well with all types of tools, but is inclined to be brittle. It is easily peeled and sliced into veneer. Fillers are required when finishing to obtain a smooth surface.

USES

In Japan, sen is used for a variety of purposes where strength or durability is not required. It is used in general construction, furniture, cabinets, sporting and athletic goods, chests, musical instruments and turnery. The exported timber is usually manufactured into plywood for paneling and doors.

SUPPLIES

Sen can be purchased for a medium price from importers. Japan exports sen to the United States, Europe and South Africa.

155 *Khaya* spp.

African mahogany

by Max Kline

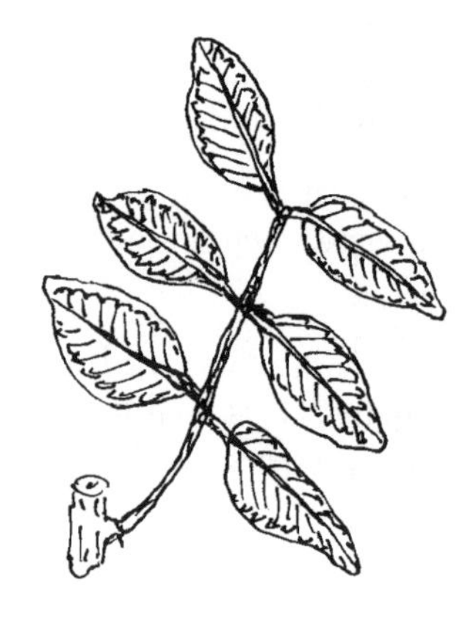

SCIENTIFIC NAME

Khaya spp. Derivation: The genus name is an African vernacular. Most of the commercial African mahogany exported is *Khaya ivorensis* but since other species, namely *K. anthotheca*, *K. grandifoliola* and *K. senegalensis*, are sometimes exported and have similar properties, they will be treated as a common group in this Wood Data Sheet.

FAMILY

Meliaceae, the mahogany family.

OTHER NAMES

Benin wood, Lagos wood, acajou, khaya, Ivory Coast mahogany, Nigerian mahogany, Gold Coast mahogany, degema, grand bassam.

DISTRIBUTION

African mahogany grows in all of the timber producing countries of West Africa with the greatest stands in the Ivory Coast, Ghana and Nigeria.

THE TREE

Khaya ivorensis grows to a height of 110 to 140 feet with a diameter up to 6 feet. It has a clean, cylindrical bole of 40 to 80 feet above the buttresses. It grows in rain forests on low-lying land. *K. anthotheca*, *K. grandifoliola* and *K. senegalensis* are found inland from the coast and do not require as much rainfall.

THE TIMBER

The color of African mahogany varies from light pinkish-brown to a deep reddish shade, often with a purplish cast. The luster is high and golden, and odor and taste are not distinct. The grain is straight but often has a ribbon figure. Crotch and swirl figures are also quite common. While the wood's characteristics are fairly close to those of the Central American species of *Swietenia*, African mahogany is more resistant to splitting and is unsuitable for bending. Average reported specific gravity for *K. ivorensis* and *K. anthotheca* is 0.44 (ovendry weight/green volume), equivalent to an air-dried weight of 32 pcf. Average reported specific gravity for *K. grandifoliola* and *K. senegalensis* is about 0.55 to 0.65 (ovendry weight/green volume), equivalent to an air-dried weight of 42 to 51 pcf.

SEASONING

These timbers season fairly well with little degrade. Some trees, however, have spongy heartwood, and this lumber will distort badly and shrink excessively and unevenly. Average reported shrinkage values for *K. ivorensis* and *K. anthotheca* (green to ovendry) are 3.2% radial and 5.6% tangential. Average reported shrinkage values for *K. grandifoliola* and *K. senegalensis* (green to 12% moisture content) are 2.5% radial and 4.5% tangential.

DURABILITY

In log form, African mahogany is susceptible to attack by beetles and borers. The timber is fairly resistant to wood-rotting fungi. It cannot be impregnated with preservatives.

WORKABILITY

In general, this timber works easily, but if the grain is exceptionally interlocked, it is difficult to surface without tearing. It holds glue well and will split when nailing only in thin dimensions. It will stain evenly and will take a very satisfactory polish. In damp conditions, it will react with iron resulting in dark stains on the wood, therefore, coated or non-ferrous fastenings should be used for assembly processes.

USES

Khaya is a standard timber for furniture, up-scale joinery, boat building, paneling and interior work. It has frequently replaced American mahogany due to its greater abundance and lower cost.

SUPPLIES

African mahogany is readily available in a wide range of sizes at a moderate cost. It is also easily obtained in plywood form from many lumber suppliers.

Koompassia malaccensis

kempas

by Jim Flynn

SCIENTIFIC NAME

Koompassia malaccensis. Derivation: The genus name is from *kempas*, the vernacular name in Singapore. The specific epithet means the plant is "of Malacca" in Southeast Asia.

FAMILY

Fabaceae or Leguminosae, the legume family; (*Caesalpiniaceae*) the cassia group.

OTHER NAMES

impas, mengris, tualang.

DISTRIBUTION

Malaysia and Indonesia. Growing mainly on leached sandstone and shale soils, in swampy forests and irregularly on hilly areas.

THE TREE

Kempas is a large tree growing 120 to 150 feet in height with widely spreading buttresses. The bole is generally straight up to 80 to 90 feet, and the diameter may be as large as 6 feet or more. The pockmarked main stem is brownish-gray with lenticels. The alternate, compound leaf has five to nine ovalate leaflets, one of which is more or less terminal.

THE TIMBER

Kempas is a heavy wood. Its average reported specific gravity is 0.72 (ovendry weight/green volume), equivalent to an air-dried density of 57 pcf. Green logs do not float. At first glance, the wood appears to be some sort of palm and the end grain will give the same impression. The white or pale yellow sapwood is wide and well demarcated from the heartwood. The heartwood is pinkish-brown but turns dark reddish-brown or yellowish-red upon exposure. In spite of its stringy-looking grain, the texture is even and fine. The grain is interlocked. The wood is slightly acidic and may cause a reaction when in contact with metal. A common defect in kempas is the presence of streaks of a hard, stone-like tissue that may be 0.25 inch wide radially, several inches wide tangentially and several feet along the grain. This may cause splitting during seasoning. Kempas has no odor.

SEASONING

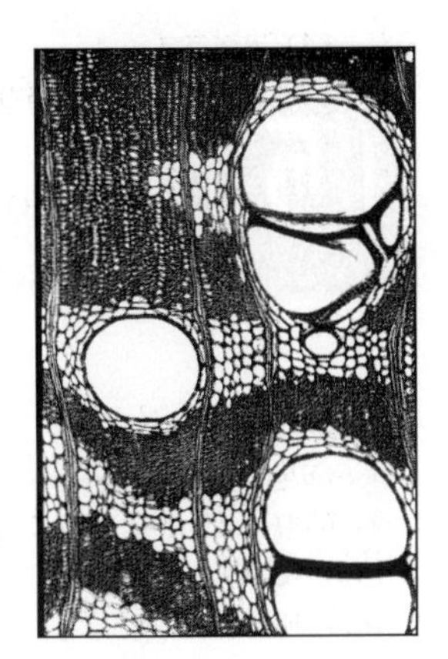

The wood is difficult to season. There is a tendency to cup and warp when drying 4/4 stock. Thicker pieces are more stable when dry. It is not considered practical to kiln dry from the green state, thus a period of air drying is suggested. Average reported shrinkage values (green to ovendry) are 6.0% radial, 7.4% tangential and 14.5% volumetric.

DURABILITY

Kempas is moderately durable but subject to termite damage. It readily accepts preservatives although some references state to the contrary. The wood has excellent wearing properties.

WORKABILITY

Because of its hardness, the wood is difficult to work with both machine and hand tools, but proper cutting angles on planers and other tooling will largely overcome this problem. Turning qualities are reported to be poor. It can be drilled and mortised effectively and smooth surfaces obtained. It nails and glues well and takes a good finish providing a filler is applied.

USES

Uses for kempas include structural timbers, heavy and light-duty flooring, furniture, agricultural implements and boxes and crates. After preservative treatment, it is used for shingles and railway sleepers.

SUPPLIES

Most probably, kempas can be found as a specialty flooring item in the north temperate zone with the more common usages found in its growth area. While this timber is not classified as endangered, it is noted that the Rainforest Action Network's booklet *Wood Users Guide* states that kempas requires conservation action to prevent its exploitation.

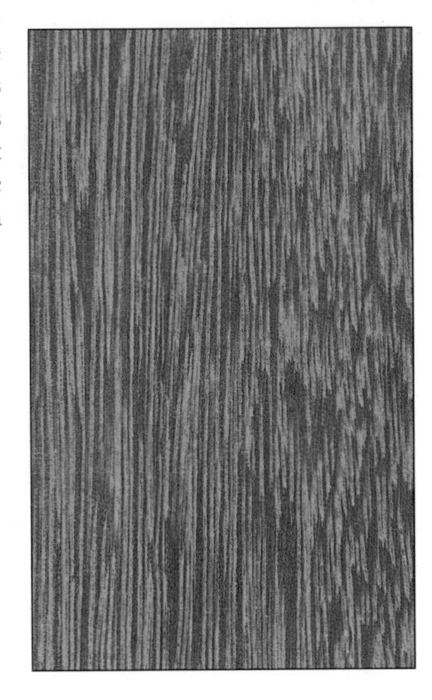

Krugiodendron ferreum

leadwood

by Max Kline

SCIENTIFIC NAME

Krugiodendron ferreum. Derivation: The genus name, meaning Krug's tree, is in honor of Carl Wilhelm Leopold Krug (1833 to 1898), a German businessman, botanist and patron of science who resided in Puerto Rico and studied the flora of the West Indies. The specific epithet means iron.

FAMILY

Rhamnaceae, the buckthorn family.

OTHER NAMES

acero, coronel, axemaster, chimtoc, bois de fer, bariaco, black ironwood.

DISTRIBUTION

Southern Florida, the West Indies from the Bahamas to St. Vincent, the Yucatan Peninsula in Mexico and northern Belize.

THE TREE

This often shrubby evergreen tree occasionally grows 30 to 40 feet in height with a diameter of 15 to 20 inches. Yellow-green flowers appear in clusters. The thin, light gray bark is rough. The simple, opposite leaves are about 1.5 inches long and 1 inch wide and are leathery in feel. The small, black fruit is egg-shaped to almost round and less than 0.5 inch in diameter. It is slightly fleshy and contains a single-seeded stone.

THE TIMBER

The heartwood is orange to dark brown, usually with dark streaks and a waxy appearance. It is sharply demarcated from the yellowish sapwood. The luster is medium to high. Odor and taste are absent or not distinctive. The grain is straight to slightly wavy, and the texture is very fine and uniform. It is exceedingly dense with an average specific gravity (ovendry weight/green volume) of 0.99 to 1.03, equivalent to an air-dried weight of 82 to 86 pcf. Leadwood has the highest specific gravity of any native wood in the continental United States. Leadwood has very high strength properties.

SEASONING

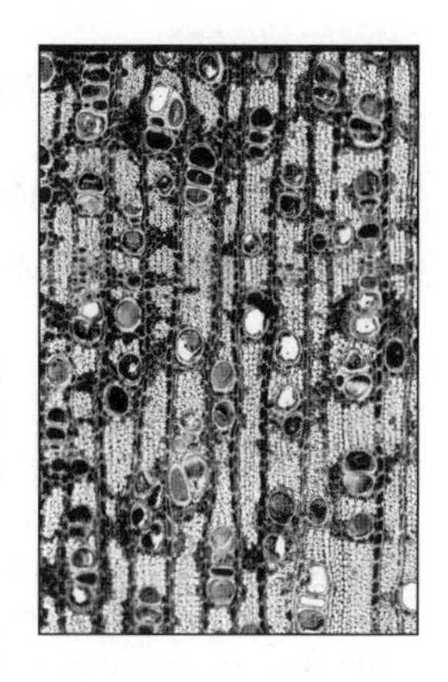

Drying data and shrinkage values are not available.

DURABILITY

Leadwood is very resistant to decay.

WORKABILITY

The timber is difficult to cut, but fairly easy to split. It turns and carves well and takes a very nice, high polish. Leadwood is of no commercial importance, but is of interest to the hobbyist.

USES

Leadwood is used in its growth area for turnery, carved articles and woodenware.

SUPPLIES

Leadwood is not available commercially, but small pieces can be obtained by personal collecting or from other IWCS members.

Lagarostrobos franklinii

Huon pine

by Max Kline

SCIENTIFIC NAME

Lagarostrobos franklinii. Derivation: The genus name is Greek for narrow cone. The specific epithet is in honor of Sir John Franklin (1786 to 1847), a Governor of Tasmania and a friend of Sir Joseph Hooker. A synonym is *Dacrydium franklinii.*

FAMILY

Podocarpaceae, the podocarp family.

OTHER NAMES

Huon pine has no other names.

DISTRIBUTION

This species is confined entirely to Tasmania. It grows on riverbanks in swampy locations. Its growth area extends from the upper Huon River around the southwest coast, reaching the most northerly point along the Stanley River. On the west coast, it extends up the Gordon River as far as the Serpentine River.

THE TREE

Huon pine reaches a medium height of about 80 feet with an average diameter at breast height of 3 to 3.5 feet. This is an extremely slow-growing tree which grows as little as 1 millimeter per year. It makes an excellent container plant.

THE TIMBER

Huon pine lumber is pale yellowish-brown. It is usually straight grained with very fine and close growth rings. Sometimes bird's eye figure is found in this species. The wood is smooth and oily to the touch, due to the presence of methyl eugenol, which gives a characteristic aroma. The sapwood is paler in color and not always easy to differentiate from the heartwood. The texture is fine. Average reported specific gravity varies from 0.42 to 0.52 (ovendry weight/green volume), equivalent to an air-dried weight of 31 to 40 pcf.

SEASONING

Huon pine may be air or kiln dried readily without degrade. Average reported shrinkage values (green to 12% moisture content) are 3.0% radial, 4.2% tangential and 6.2% volumetric.

DURABILITY

Huon pine is noted for its durability. Its resistance to decay is exceedingly high due to the presence of methyl eugenol, which acts as a preservative.

WORKABILITY

Huon pine is easy to work with hand or machine tools, it turns well and takes a good finish. It is moderately light, soft and fairly strong. Bending properties are good, nailing qualities are satisfactory and screws hold firmly.

USES

It is an ideal timber for boat building because it is easy to work and bend, has low shrinkage and is durable. Other uses include joinery, pattern making, drawing boards, ornaments and furniture.

SUPPLIES

Huon pine is considered rare because it requires 600 to 1,000 years for a tree to mature and its distribution in Tasmania is very limited. The total volume of production is small, and only small stocks are available at a high price.

Larix occidentalis

western larch

by Jim Flynn

SCIENTIFIC NAME

Larix occidentalis. Derivation: The genus name is the classical name of *Larix decidua,* European larch. The specific epithet means the plant is "of the Western Hemisphere".

FAMILY

Pinaceae, the pine family, a temperate Northern Hemisphere family of tress and shrubs with some nine genera and 210 species.

OTHER NAMES

larch, tamarack, western tamarack, hackmatack, mountain larch, Montana larch.

DISTRIBUTION

Western larch grows in the high mountains of the upper Columbia River basin in southeastern British Columbia, northwestern Montana, northern and central Idaho, Washington and north and northeastern Oregon.

THE TREE

Western larch is a long-lived tree, and specimens have been found that are over 900 years old. It is also one of the largest of the world's larches and normally is 100 to 180 feet in height when mature. For the first century of its life, it grows faster than any other conifer in the northern Rockies. It is not tolerant of shade so it needs to grow quickly to survive. Western larch has a narrow crown consisting of slender, irregularly whorled, and either spreading or drooping branches. The orange-brown twigs are brittle and stout. The bark is quite thick, usually from 3 to 6 inches, affording the tree protection from forest fires. The reddish-cinnamon colored tree is generally straight as well as tall and is scarred with massive ridges. Western larch trees can be easily distinguished from other cone-bearing trees by their leaves, which turn lemon-yellow in the autumn and fall off leaving the tree bare in the winter. The leaves are 1 to 1.75 inches long and grow in clusters of 14 to 30.

THE TIMBER

Western larch is coarse grained, strong and durable with reddish-brown heartwood and narrow, nearly white sapwood. It is the most important timber species of the *Larix* genus. The growth rings are very narrow (30 to 60 per inch). The wood is slightly resinous and has a distinct oily appearance but has neither odor nor taste. Average reported specific gravity is 0.48 (ovendry weight/green volume), equivalent to an air-dried weight of 36 pcf.

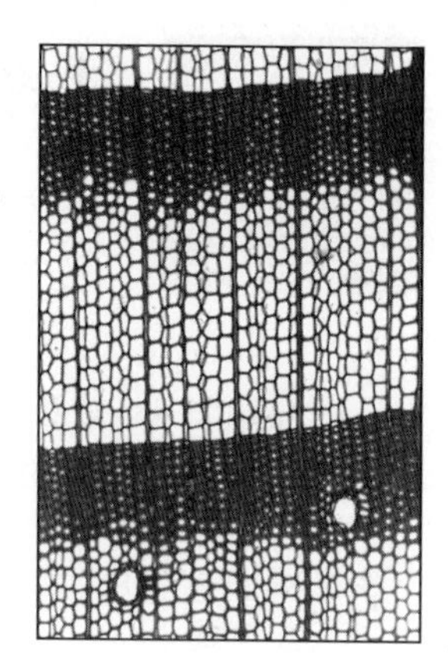

SEASONING

Western larch splits rather easily and is subject to ring shake, which is a condition where there is separation along the grain between the annual rings. It seasons well but warping and checking are problematic. Average reported shrinkage values (green to ovendry) are 4.5% radial, 9.1% tangential and 14.0% volumetric.

DURABILITY

Western larch is moderately resistant to heartwood decay. Where decay hazards exist, some form of preservative is required.

WORKABILITY

Western larch is somewhat difficult to work because of its stringy grain. It glues easily but must be properly primed to achieve a durable finish. It holds nails well but to minimize splitting blunt-pointed nails should be used.

USES

Construction lumber is the predominant use of this species. Timbers are used extensively for electrical transmission lines and telephone poles where long lengths and high strength are required. Applications are also found in interior finish, boxes and crates and veneer. One of the more interesting applications is the extraction of arabino galactan, which is a water-soluble gum used for offset lithography and in food, pharmaceutical, paint, ink and other industries.

SUPPLIES

Western larch, while plentiful, may be difficult to isolate commercially because it is used interchangeably with Douglas-fir (*Pseudotsuga menziesii* WDS 224) under the name Doug fir-larch. It is in the medium price range with other construction timbers.

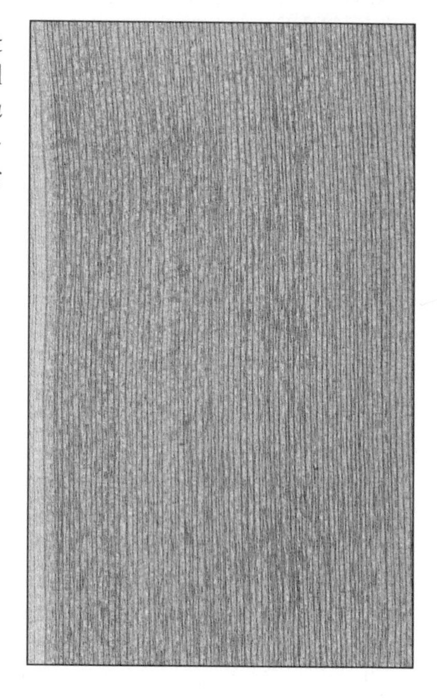

Leptospermum scoparium

tea tree

by Jim Flynn

SCIENTIFIC NAME

Leptospermum scoparium. Derivation: The genus name *lepta* is from the Greek meaning slender and *sperma* meaning seed. The specific epithet means broom-like.

FAMILY

Myrtaceae, the myrtle family.

OTHER NAMES

broom tea tree, red tea tree, Australian tea tree, New Zealand tea tree, manuka.

DISTRIBUTION

Australia, Tasmania and New Zealand. Tea trees, as well as other species of *Leptospermum*, have been planted as ornamentals worldwide in warm climates because of their beautiful white and pink flowers. It has become naturalized in many places including Hawaii.

THE TREE

The common name of this evergreen tree dates back to the time of Captain Cook when, during his first voyage to New Zealand in 1769, he discovered that a tasty tea could be brewed from the dried leaves. In ideal climates, tea tree grows to 30 feet in height. The sharp-pointed, glabrous leaves are rather small measuring less than 0.5 inch long. They are dotted with oil glands that exude a gingery smell when crushed. This slender tree usually grows in dense stands. The brown bark can be peeled in large strips and was used by the aborigines of New Zealand for roof coverings. Blossoming between March and May, the flowers are about 0.75 inch wide and have five petals. Because the tree will not tolerate heavy soil, it grows best in light sandy areas. Tea tree has been widely planted in California in attempts to stabilize shifting sands.

THE TIMBER

The wood of the tea tree is well figured and reddish in color. The sapwood is light tan. The grain is even and fine. The wood is heavy, and a piece I own has a specific gravity of 0.83 (ovendry weight/green volume), equivalent to an air-dried weight of 67 pcf. The pronounced growth rings tend to add a flash of contrast to quartersawn surfaces thus adding to the beauty of the wood. Unfortunately, it cannot be obtained in long lengths because of the limited growth of the tree. Although present in freshly sawn timber, there is neither detectable odor nor taste in old well-seasoned wood.

SEASONING

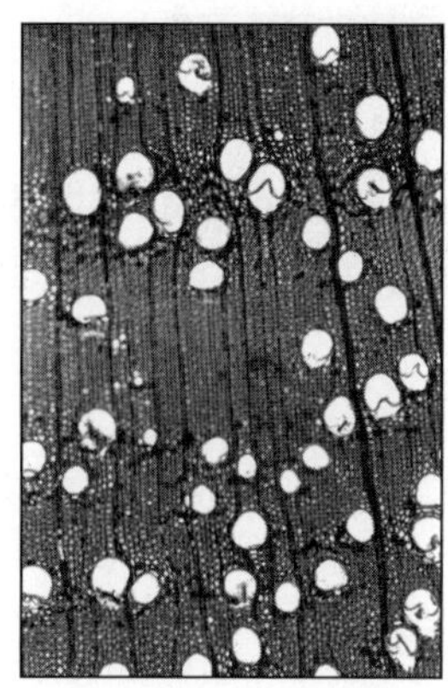

Shrinkage values are not available. This is not surprising because it is not available in large quantities. Because of the high density of the wood, slow seasoning under controlled conditions seems to be in order.

DURABILITY

While there is very little data available on the durability of tea tree, some species of *Leptospermum* have been reported to be extremely durable. For the purposes for which this wood is generally used, this characteristic is relatively unimportant.

WORKABILITY

Wood with a density as high as tea tree is always a tough product to tame with a saw and chisel. Needless to say the sharpness of the tools and the speed with which the cutters are applied will determine the success of the operation. The wood can be cut and shaped crisply, will glue and finish well, and if sanded with fine abrasives need not be finished at all.

USES

Tea tree is an outstanding timber for turning. It is an excellent choice for tool handles as well as the production of small craft items such as jewelry, boxes, carvings and the like. Maoris used the wood for paddles and spears. Tea tree is a popular firewood in New Zealand, and this use threatens the survival of the species.

SUPPLIES

Do not expect to find tea tree on the commercial market. But, because the tree has been widely marketed as an ornamental in many parts of the world, the chances of finding a supply are great if one keeps an eye open for it. Even in Great Britain, the tree is listed in the Hillier Nurseries catalog. A hint for wood-hunters in England: the Hillier Manual indicates that it is best grown near a protective wall.

Liquidambar styraciflua

sweetgum

by Max Kline

SCIENTIFIC NAME

Liquidambar styraciflua. Derivation: The genus name is Spanish from the common name in Mexico which refers to the fragrant resin. The specific epithet is the old name of the genus meaning styrax flowing, alluding to the medicinal storax. A synonym is *L. macrophylla.*

FAMILY

Hamamelidaceae, the witch-hazel family.

OTHER NAMES

alligator tree, alligator wood, liquidambar, hazel pine, sapgum, star-leafed gum, satin walnut, redgum, gum, bilsted.

DISTRIBUTION

From Connecticut and New York, west to Ohio, Illinois, Missouri and Oklahoma south to Florida and Texas.

THE TREE

Sweetgum is a beautifully shaped tree that grows from 80 to 120 feet in height with a diameter of 2 to 3 feet. It is superb for planting as an ornamental and has splendid colors in the autumn. The tree thrives best in rich moist bottomlands.

THE TIMBER

The heartwood of sweetgum is bright brown tinged with red, frequently with darker streaks. The wide sapwood is creamy white. The wood often has interlocked grain which forms a ribbon stripe. Few North American woods equal sweetgum in the beauty of its grain, and the wood with the best grain figure is marketed as "figured red gum". The wood has a satiny luster and is very uniform in texture. Average reported specific gravity is 0.46 (ovendry weight/green volume), equivalent to an air-dried weight of 35 pcf. The wood is moderately strong and stiff and can be obtained in wide dimensions of very high quality.

SEASONING

This timber can be satisfactorily air or kiln dried. In the early stages of drying, the wood, especially thin stock, may show a tendency to warp. Average reported shrinkage values (green to ovendry) are 5.3% radial, 10.2% tangential and 15.8% volumetric. After seasoning, it is dimensionally stable.

DURABILITY

Sweetgum is not highly resistant to either fungal or insect attack. For the purposes for which sweetgum is used, durability is not a concern.

WORKABILITY

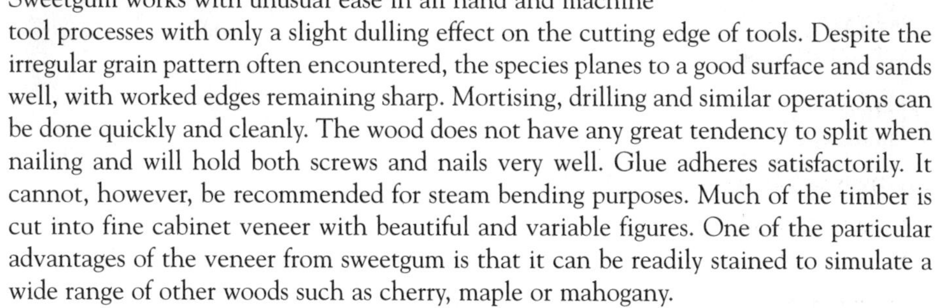

Sweetgum works with unusual ease in all hand and machine tool processes with only a slight dulling effect on the cutting edge of tools. Despite the irregular grain pattern often encountered, the species planes to a good surface and sands well, with worked edges remaining sharp. Mortising, drilling and similar operations can be done quickly and cleanly. The wood does not have any great tendency to split when nailing and will hold both screws and nails very well. Glue adheres satisfactorily. It cannot, however, be recommended for steam bending purposes. Much of the timber is cut into fine cabinet veneer with beautiful and variable figures. One of the particular advantages of the veneer from sweetgum is that it can be readily stained to simulate a wide range of other woods such as cherry, maple or mahogany.

USES

Sweetgum is one of the foremost furniture, interior finishing and cabinet woods. Its veneer was once a favorite wood for radio cabinets but has now been largely replaced by synthetics. Other uses include cigar boxes, woodenware, crating, inexpensive flooring and plywood. It is in great demand in England, France and Germany for furniture manufacture.

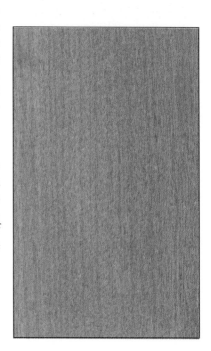

SUPPLIES

The largest quantity of sweetgum comes from the lower Mississippi Valley. The heartwood, which is sometimes 50% of the tree's volume, is marketed separately under the name redgum, while the sapwood is sold under the name of sapgum. Adequate supplies are available, and the price ranges from medium to inexpensive.

Liriodendron tulipifera

yellow-poplar

by Max Kline

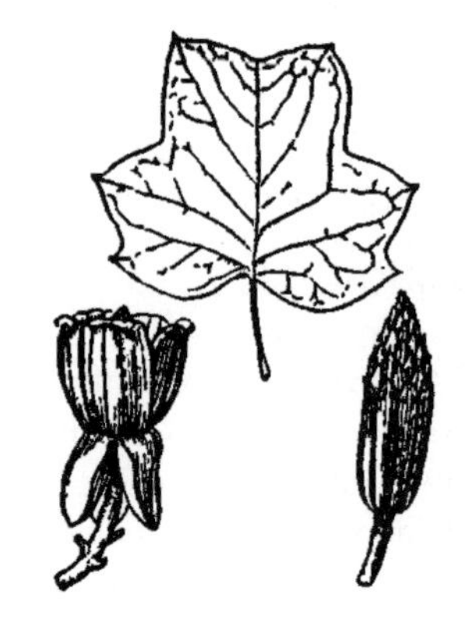

SCIENTIFIC NAME

Liriodendron tulipifera. Derivation: The genus name is from the Greek lily and tree, referring to the showy lily-like flowers. The specific epithet is an old generic name meaning tulip-bearing.

FAMILY

Magnoliaceae, the magnolia family.

OTHER NAMES

canary wood, canoe wood, tuliptree, tulip poplar, whitewood, hickory poplar, white-poplar.

DISTRIBUTION

From southern New England through New York to southern Michigan and south to west central Louisiana and northern Florida.

THE TREE

Yellow-poplar is a very tall stately tree which grows 100 to 150 feet in height with trunk diameters of 8 to 10 feet. The straight trunk is free of limbs for about 40 to 50 feet and occasionally up to 80 to 90 feet. The flowers and leaves make the tree particularly attractive. The flower resembles a tulip and is greenish-yellow and orange in color. Yellow-poplar is the state tree of Indiana, Kentucky and Tennessee.

THE TIMBER

Yellow-poplar is a lightweight hardwood of fine texture. Average reported specific gravity is 0.40 (ovendry weight/green volume), equivalent to an air-dried weight of 30 pcf. The light yellow to brown heartwood turns greenish upon exposure. The sapwood is creamy white. The grain is usually straight but sometimes an attractive blister figure is found. This timber is somewhat weak and brittle and only moderately strong. The wood has no taste or odor.

SEASONING

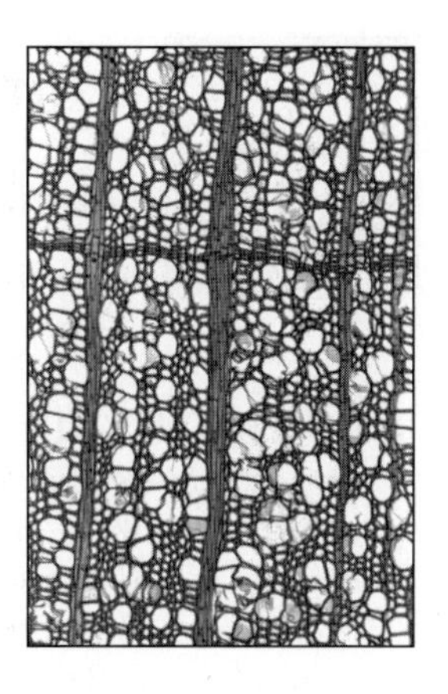

Yellow-poplar is easy to season and is dimensionally stable. Average reported shrinkage values (green to ovendry) are 4.6% radial, 8.2% tangential and 12.7% volumetric.

DURABILITY

The tree is usually free from pests, but frequently unsightly brown spots caused by gall insects cover the leaves.

WORKABILITY

Yellow-poplar is easily worked and takes paint very well. It nails easily but does not hold nails well. It glues easily and shrinks moderately.

USES

There are a wide variety of uses for yellow-poplar. Large quantities are used in furniture, and it has been used for core stock for pianos and television cabinets. It finds use for general cabinetwork such as sash, doors, shelving, boxes and crates, baskets and woodenware. Great amounts are used for shipping pallets. Small amounts are cut for pulpwood and made into paper. It is a fine wood for carving.

SUPPLIES

Yellow-poplar is readily available and inexpensive. Veneer is also easily obtainable.

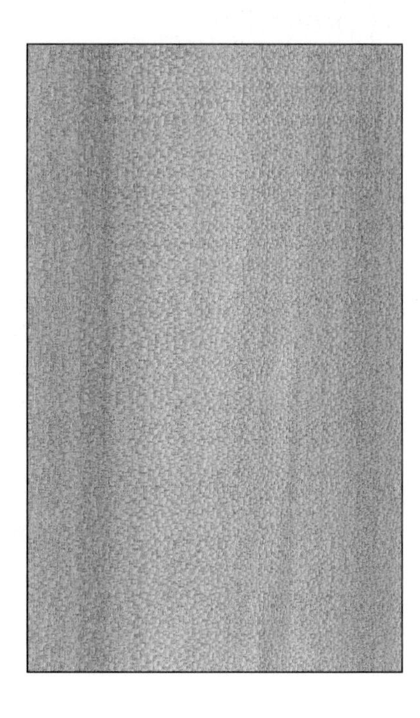

Lithocarpus densiflorus

tanoak

by Jim Flynn

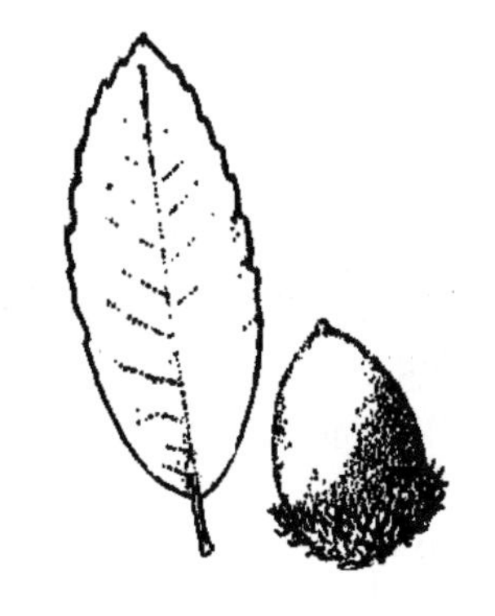

SCIENTIFIC NAME

Lithocarpus densiflorus. Derivation: The genus name is derived from the Greek meaning stone and fruit, referring to the hard acorns. The specific epithet means densely flowered. One of the synonyms for this taxon is *Quercus densiflora* indicating that it was once identified as an oak.

FAMILY

Fagaceae, the beech family.

OTHER NAMES

tanbark-oak and tan-oak. George B. Sudworth's *1897 Checklist of U.S. Trees* lists chestnut oak, California chestnut oak, peach oak and live oak. Before the move out of the *Quercus* genus, Sudworth inserted this footnote in his 1908 edition of *Forest Trees of the Pacific Slope*, "This tree is and always will be an oak to the lumbermen and to the *practical* forester." (Italics mine.)

DISTRIBUTION

Lithocarpus is a large genus of approximately 300 species of evergreen trees and shrubs native to Asia. A single species, *L. densiflorus*, is mostly found along the Pacific coastal ranges from southwest Oregon to southern California and in the Sierra Nevadas to central California.

THE TREE

In appearance, *Lithocarpus* falls between an oak and a chestnut. With flower clusters similar to chestnut and acorns as fruits, tanoak grows to 80 feet or more in height with a main stem diameter of 1 to 2 feet or more. Old virgin forest trees reached ages of 250 years or more. The leaves are covered with thick, pale orange felt when young that persists for three to four seasons. The tiny, pale yellow male flowers, densely packed in narrow catkins, are born in the spring and often again in the autumn. The female flowers and acorns, when mature in their second year, come in stout, stiff spikes below the male catkins. Tannin used in the leather industry is obtained from the thick, scaly bark. Native American communities in California's north coastal range used the acorns as soup, cooked mush or bread. Large trees are now rare. In dense stands, the crown is narrow with upright branches and long, clear trunks; in the open, trees have widespread crowns and branches extending horizontally from short and squat trunks.

THE TIMBER

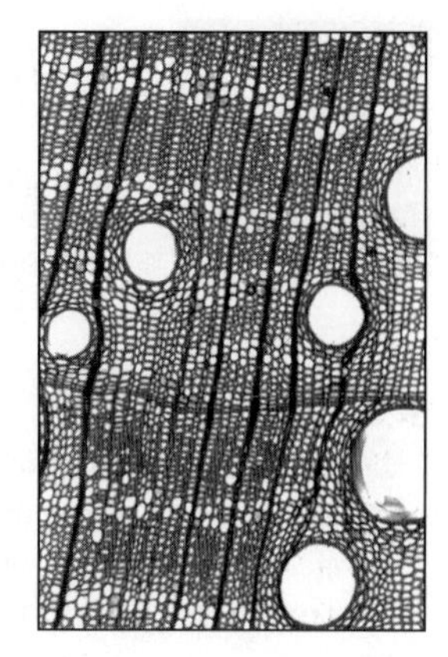

The sapwood is light reddish-brown tinged with red that darkens with age becoming difficult to distinguish from the dark, reddish-brown heartwood. Average reported specific gravity is 0.58 (ovendry weight/green volume), equivalent to an air-dried weight of 45 pcf. There is no distinctive taste or odor. Timber sawn on the radial plane does not have the strong grain pattern as does conventional oak timber but can show interesting yellow and brown streaks. Quartersawn timber exhibits ray flecks similar to those found on similar cuts of oak. The pores are barely visible on the tangential surface, and the wide, narrow rays are less prominent in tanoak.

SEASONING

Average reported shrinkage values (green to ovendry) are 4.9% radial, 11.7% tangential and 17.3% volumetric. It requires fast extraction from the forest and slow, careful drying.

DURABILITY

The heartwood is prone to insect and fungi attack but because of its high tannin content is considered moderately durable.

WORKABILITY

Unfortunately, there is very little technical data published. Straight, long boles will produce workable timber for large items such as furniture and paneling, while timber cut from the wide-crowned trees with many branches and short trunks will be more difficult to machine due to grain deviations.

Tanoak should be similar to the conventional oaks in all cutting and fabricating operations. The wood glues, sands and finishes without undue problems.

USES

Tanoak uses parallel those of oak, including furniture, structural timbers, veneer, cabinets, and craftwork.

SUPPLIES

Because species of hardwood are much sought after in the areas where tanoak thrives, little is available for the commercial market. As with its closely related species, giant chinkapin (*Chrysolepis chrysophylla* WDS 081), IWCS sample suppliers have quantities for collectors and specialty craftwork.

Lophira alata

ekki

by Jon Arno

SCIENTIFIC NAME

Lophira alata. Derivation: The genus name is Greek meaning crest. The specific epithet refers to the winged seed which is similar to the maples. A synonym is *L. procera*.

FAMILY

Ochnaceae, the ochna family.

OTHER NAMES

aba, azobe, bakundu, bongossi, endivi, escore, African oak, red ironwood.

DISTRIBUTION

Ekki is native to west central Africa from the Ivory Coast to the Congo (formerly Zaire) and eastward into the Congo basin.

THE TREE

Unlike most members of its botanical family, this species can attain very respectable dimensions, heights of up to 160 feet with diameters in excess of 5 feet. Since the tree does not produce a pronounced buttress, clear logs of 100 feet in length are obtainable. Ekki is a water-loving tree, growing in swamps and along riverbanks. Its lush foliage and flowers are attractive, but they are invariably hidden from view high in the forest canopy when the species is mature.

THE TIMBER

One of ekki's common names, African oak, may refer to the wood's somewhat coarse texture and its rugged strength; however, this timber is anything but oak-like in appearance. The heartwood is usually very dark in color, ranging from chocolate-brown to deep red occasionally with purple highlights. The grain is often interlocked, giving the radial surface a ribbon-grained effect, and white deposits in the vessels add a speckled appearance to the tangential surface. The pale pink sapwood is generally narrow and clearly demarcated from the heartwood. Ekki has very low surface luster and no pronounced odor or taste. The wood is very strong with excellent bending strength and high resistance to abrasion. With an average reported specific gravity of 0.90 (ovendry weight/green volume), equivalent to an air-dried weight of 73 pcf, ekki richly deserves its common name red ironwood.

SEASONING

Ekki dries slowly and is difficult to season. Its average radial shrinkage (green to ovendry) of 8.4% and tangential shrinkage of 11.0% are not so divergent as to stimulate severe warping, but its overall volumetric shrinkage of 17.0% is very high, and end checking is a serious problem.

DURABILITY

Ekki is very durable, has good weathering properties and is resistant to acids. It is only moderately resistant to termites.

WORKABILITY

Ekki is an absolutely punishing wood to work with hand tools. Its coarse texture does not lend itself to turning and tools dull quickly. Although it has a very high density, its coarse texture allows glues to bond effectively, and it can be dressed to a smooth enough surface to show off its warm color and interesting speckled figure. In the final analysis, however, there are many equally attractive woods which are more stable in use, far easier to work and adequately strong enough for any purpose the typical woodworker would use this wood.

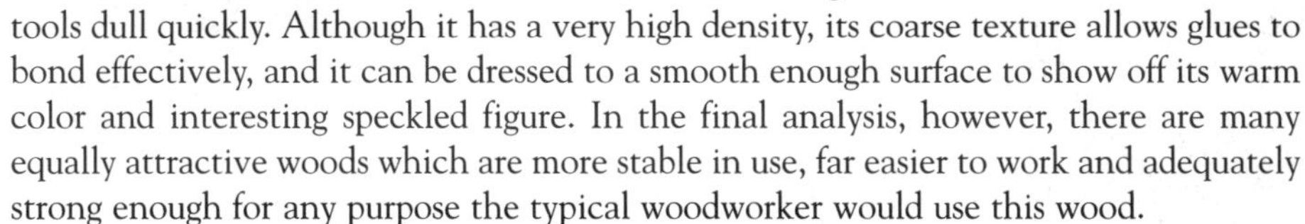

USES

Uses include heavy-duty construction timbers, railroad ties, dock and marine work and flooring.

SUPPLIES

Ekki is still fairly plentiful, but it does not enjoy wide international distribution in lumber form, especially in North America. It is more commonly available in Europe in flooring form and for use in fabricating vats and other containers for liquids.

Lophostemon confertus

brushbox

by Jim Flynn

SCIENTIFIC NAME

Lophostemon confertus. Derivation: The genus name is Latin for crested stamen. The specific epithet is Latin meaning brought together, which refers to the leaves being crowded at the end of the branchlets. A synonym is *Tristania conferta*.

FAMILY

Myrtaceae, the myrtle family.

OTHER NAMES

pelawan, keruntum, selunsur, melabau.

DISTRIBUTION

Brushbox is native to Australia, the Indo-Malayan region and the Philippines and has been introduced in other tropical areas including extensive plantations in Hawaii.

THE TREE

Brushbox is a large evergreen tree which grows to 60 feet in height and 30 inches in diameter with a narrow rounded crown of dense foliage. The grayish-brown bark becomes rough, thick and slightly scaly with long fissures. Its inner bark is light brown, fibrous and tastes slightly bitter. When young, twigs are light green with tiny pressed hairs but as they age turn brown and shed the outer layer leaving raised half-rounded leaf scars. The simple, alternate leaves are crowded at the ends of the branches and vary from ovate to elliptical. The fruit is a round capsule. The tree is often planted along streets but can be overwhelming when mature.

THE TIMBER

The pale brown sapwood is clearly demarcated from the pinkish-brown or grayish-brown heartwood. The texture is medium to fine, and the grain is mildly interlocked and sometimes wavy. The wood usually lacks prominent figure. Occasionally, moon or target ring pattern can be found. Observed in the log's end grain, this rare pattern comes from concentric rings of the lighter color sapwood that are embedded in the darker heartwood. The density will vary depending on where it was grown. Hawaiian plantation stock has a reported specific gravity of 0.67 (ovendry weight/green volume), equivalent to an air-dried weight of 52 pcf, while native Australian species have as high as 0.90, equivalent to an air-dried weight of 73 pcf. The wood has high strength properties with good toughness and wear resistance.

SEASONING

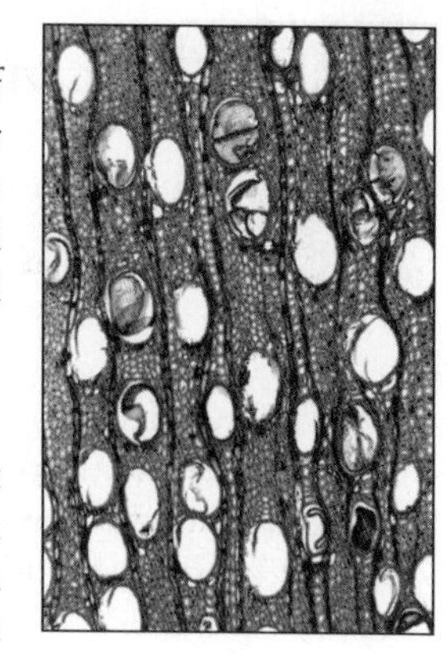

Brushbox does not season well. It will warp and twist badly if it is air dried too rapidly. The best results are obtained by air drying the wood in well made and weighted piles to a 30% moisture content and then kiln drying. Average reported shrinkage values (green to ovendry) are 2.1% to 3.0% radial and 3.5% to 4.0% tangential.

DURABILITY

There are mixed reports on durability. A 1974 U.S. Forest Service report indicated Australian grown wood was much more resistant to decay and termites than wood grown in Hawaii. Current Australian literature estimates that wood in contact with the ground will last between 8 and 15 years. This is probably close to the average for wood grown in both countries but growth and use conditions have a significant effect on this characteristic. It is difficult to pressure treat with preservatives.

WORKABILITY

The wood is relatively free of growth stresses so it saws without splitting or springing away from the saw blade. Because different densities will be encountered, tools must be sharp. The wood machines well, sands and finishes to a smooth surface but is difficult to nail. Gluing presents no problem. It is not a candidate for any application where steam bending is required. It is an excellent wood for turnery.

USES

The best uses are those requiring a hard and durable surface, including flooring, truck beds, pulleys, railway crossties, bridges and construction timber. A considerable quantity of the wood is consumed in making pallets. Because of its weight, it is unsuitable for veneer. An IWCS member in Australia reports that it is a beautiful cabinet timber used in his kitchen.

SUPPLIES

There does not seem to be any worldwide commercial supplies in lumber form. Finished wood such as flooring has been on the market. In growth areas, stocks of unprocessed wood should be plentiful and are in the medium price range.

Lovoa trichilioides

tigerwood

by Max Kline

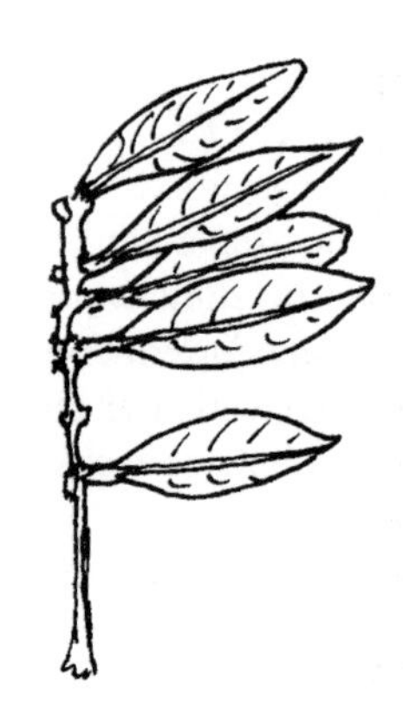

SCIENTIFIC NAME

Lovoa trichilioides. Derivation: The genus name is in recognition of the Lovo River in Africa. The specific epithet is Greek meaning hair-like, referring to the leaves.

FAMILY

Meliaceae, the mahogany family.

OTHER NAMES

African walnut, Lovoa wood, Congo wood, Benin walnut, dibetou, alona wood, Nigerian golden walnut.

DISTRIBUTION

Ivory Coast, Ghana, Gabon, Nigeria, the Congo (formerly Zaire), Uganda, Tanzania.

THE TREE

Tigerwood reaches a height of 130 feet with a diameter of 4 feet. The bole is cylindrical with small buttresses at the base.

THE TIMBER

The heartwood is a bronze shade of yellowish-brown, sometimes marked with dark streaks. The narrow sapwood is white or pale brown and clearly demarcated from the heartwood. The grain is interlocked, while the texture is moderately coarse, but even. When freshly cut, the timber has a slight aromatic fragrance, reminiscent of cedar. Tigerwood is not a true walnut and actually resembles African mahogany (*Khaya* spp. WDS 155) timber more closely. The strength properties are almost as good as those of members of the true walnut family and adequate for the purposes for which the timber is ordinarily used. Average reported specific gravity is 0.45 (ovendry weight/green volume), equivalent to an air-dried weight of 34 pcf.

SEASONING

When heart shake is found in tigerwood logs, the wood may have a tendency to split when drying. Otherwise warping is not excessive when air or kiln drying. Average reported shrinkage values (green to 12% moisture content) are 2.0% radial and 5.0% tangential.

DURABILITY

Under normal conditions, seasoned wood has no durability problems.

WORKABILITY

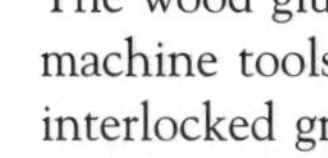

The wood glues easily, and works fairly well with hand and machine tools. If tools are kept sharp, even the wood with interlocked grain will finish cleanly. It takes nails and screws well, stains readily and polishes satisfactorily. If painting is to be done, the wood should first be primed or sealed. To prevent tearing, sharp tools are needed when drilling and turning.

USES

Tigerwood is a decorative timber widely used for furniture, paneling and veneer. It is also used for gunstocks and inlay. When quartersawn, narrow to wide straight stripes provide a handsome figure, closely resembling mahogany.

SUPPLIES

The lumber and veneer are usually in plentiful supply at a moderate price.

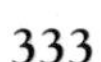

Maclura pomifera

Osage-orange

by Max Kline

SCIENTIFIC NAME

Maclura pomifera. Derivation: The genus name is in honor of William Maclure (1763 to 1840), an American geologist. The specific epithet is indicative of the pomes or apples, referring to the large ball fruits.

FAMILY

Moraceae, the mulberry family.

OTHER NAMES

bowwood, bois-d'arc, bodark, hedge apple, bodock, mockorange, naranjo chino.

DISTRIBUTION

The natural range of Osage-orange is in the southwestern half of Arkansas, southeastern Oklahoma and the eastern half of Texas. It has been extensively planted throughout the prairie region of the Mississippi basin.

THE TREE

Osage-orange is a rapidly growing, unshapely tree with a short trunk. It reaches a height of 50 or 60 feet with a trunk 1 to 1.5 feet in diameter. The twigs have sharp spines. When the yellowish-green ball of fruit, which is about the size of an orange, is crushed, it exudes a bitter, sticky, milky juice. The orange-brown bark of the tree is thin.

THE TIMBER

When freshly cut, the heartwood is golden yellow, sometimes with reddish streaks, but becomes russet-brown upon exposure. The thin sapwood is white and sharply demarcated from the heartwood. The odor and taste are not distinctive. The luster is high. The wood is very hard, heavy, tough and resilient. Average reported specific gravity is 0.76 (ovendry weight/green volume), equivalent to an air-dried weight of 60 pcf. Osage-orange ranks very high in strength properties compared with other North American timber species. It surpasses white oak (*Quercus alba* WDS 230) in strength but not in stiffness.

SEASONING

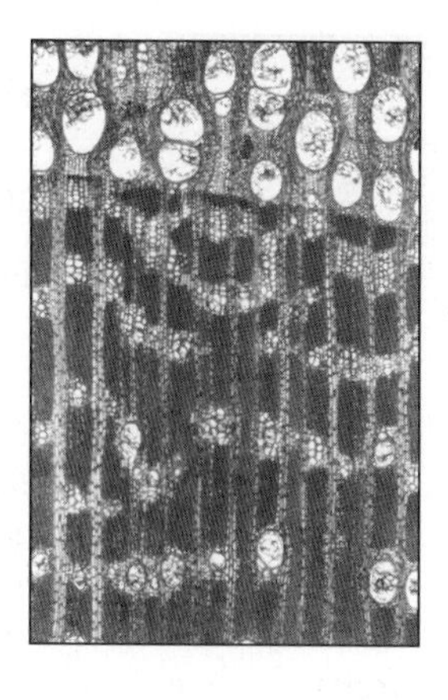

The wood seasons well and average reported volumetric shrinkage (green to ovendry) is 9.2%. After seasoning, it maintains remarkable dimensional stability.

DURABILITY

Osage-orange is considered the most durable of all North American timbers, having a long service life when used as stakes or posts. It is comparatively inert to changes in atmospheric humidity.

WORKABILITY

Because of the hardness of the wood, it is difficult to work and tools require frequent sharpening. It is difficult to nail, but holds screws well, and it is easily glued. When finishing, oils should be avoided, as they will accelerate the color change process.

USES

In the past, Osage-orange was widely used for wheel rims and hubs of farm wagons. Native Americans used it for bows due to its great strength. Because of its durability when in contact with the soil, it makes superior fence posts and railroad ties. It is extensively used to obtain important dyestuffs. The woodworker will be tempted to use it for turnery and novelties. [Ed. note: It is used in musical instruments, especially for fretboards, as a substitute for ebony].

SUPPLIES

Osage-orange is seldom cut into logs or veneer but is available in small quantities to wood collectors at a moderate price.

Magnolia acuminata

cucumbertree

by Jon Arno

SCIENTIFIC NAME

Magnolia acuminata. Derivation: The genus name is in honor of Pierre Magnol (1638 to 1715), a professor of botany and medicine and director of the botanical garden at Montpellier, France. The specific epithet is Latin for acuminate, referring to the pointed leaves.

FAMILY

Magnoliaceae, the magnolia family.

OTHER NAMES

cucumber magnolia, mountain magnolia.

DISTRIBUTION

Cucumbertree is not plentiful anywhere, but its range extends through the Appalachians from western New York to northern Georgia, augmented by a few isolated pockets spread across the southern states as far west as extreme eastern Oklahoma.

THE TREE

Unlike its close relative, yellow-poplar (*Liriodendron tulipifera* WDS 162), which vies with sycamore for being the largest hardwood species in the eastern forests, cucumbertree seldom exceeds about 90 feet in height and about 4 feet in diameter. As its Latin species name (*acuminata*) suggests, its leaves are 6 to 10 inches long, broadly lance-shaped and come to a sharp point. The greenish-yellow flowers are much smaller than those of southern magnolia (*M. grandiflora* WDS 169), but the aggregate, cone-like fruit is similar. Cucumbertree prefers moist, fertile soils and grows in mixed hardwood stands where it usually represents one of the least plentiful members of the community.

THE TIMBER

The wood of cucumbertree is diffuse porous and very finely textured. The heartwood, sometimes streaked with brown, is generally greenish-gray and sharply demarcated from the creamy, almost white sapwood. Close examination of the end grain with a hand lens will reveal fine white lines (terminal parenchyma) separating the annual rings. This feature helps to distinguish the wood of the so-called yellow poplars from that of trembling aspen (*Populus tremuloides* WDS 219) and eastern cottonwood (*P. deltoids* WDS 218), which belong to the true poplar genus, *Populus*. Even though the wood of cucumbertree is slightly stronger than that of yellow-poplar in virtually all respects, both must be ranked as moderately weak. Average reported specific gravity is 0.44 (ovendry weight/green volume), equivalent to an air-dried weight of 33 pcf.

SEASONING

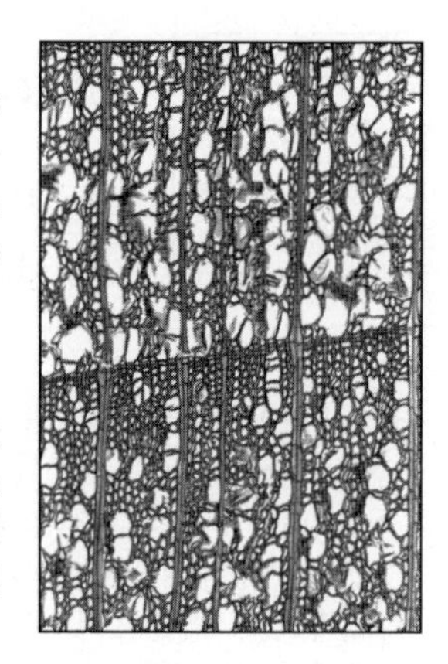

With an average volumetric shrinkage of 13.6% (green to ovendry), cucumbertree is slightly less stable than yellow-poplar (12.7%), but both dry quickly and without much difficulty. Average reported shrinkage values (green to ovendry) are 5.2% radial and 8.8% tangential.

DURABILITY

Durability is poor. Cucumbertree is not appropriate for applications where it is exposed to high moisture levels. The pores do not contain adequate tyloses to prevent rapid absorption of moisture nor is the wood particularly well endowed with natural decay inhibitors.

WORKABILITY

With an average specific gravity of 0.44, cucumbertree is about 10% heavier and harder than yellow-poplar (specific gravity of 0.40), however, both are nice cabinetwoods. Cucumbertree's slightly greater strength and finer texture offers some structural advantages and make it perhaps a slightly better choice for turning, but its greater volumetric shrinkage represents a disadvantage in larger projects such as tables and case goods. Both woods cut, shape and finish very well and are excellent choices for interior projects which will be painted. Their fine, diffuse-porous texture makes it possible to achieve a glass smooth surface using even light-bodied finishes.

USES

Uses include framing for upholstered furniture, plywood corestock, veneered furniture, toys, woodenware, baskets, interior paneling and trim, turnings, carvings and boxes and crates.

SUPPLIES

Because the wood of cucumbertree is marketed along with that of yellow-poplar, it is difficult to be certain which is which unless you harvest it yourself. Cucumbertree is far less common than yellow-poplar.

Magnolia grandiflora

southern magnolia

by Max Kline

SCIENTIFIC NAME

Magnolia grandiflora. Derivation: The genus name is in honor of Pierre Magnol (1638 to 1715), a professor of botany and medicine and director of the botanical garden at Montpellier, France. The specific epithet refers to the large flower.

FAMILY

Magnoliaceae, the magnolia family.

OTHER NAMES

bat tree, big laurel, bull-bay, great laurel magnolia, large flowered evergreen magnolia.

DISTRIBUTION

Extends in a broad band 100 miles wide or less along the Atlantic and Gulf of Mexico coasts from southeastern North Carolina to Mississippi, southeastern Louisiana, and in central Louisiana and eastern Texas. The tree is also found in about half of Florida.

THE TREE

This tree is well shaped and straight, growing 60 to 90 feet in height with a trunk diameter of 2 to 4 feet. The thick and leathery evergreen leaves are dark and glossy green above and silvery underneath. The large, showy flowers have a pleasing fragrance and are very attractive. Charles Sargent once said the magnolia is "the most splendid ornamental tree in the American forests". Southern magnolia is the state tree of Mississippi.

THE TIMBER

The heartwood of southern magnolia is light to dark brown usually tinged with yellow or green and occasionally contains colorful greenish-black or purplish-black streaks. The narrow sapwood is yellowish-white. It has a uniform texture and is generally straight grained. The wood is stiff, hard, moderately low in shrinkage and high in shock resistance. Average reported specific gravity is 0.46 (ovendry weight/green volume), equivalent to an air-dried weight of 35 pcf.

SEASONING

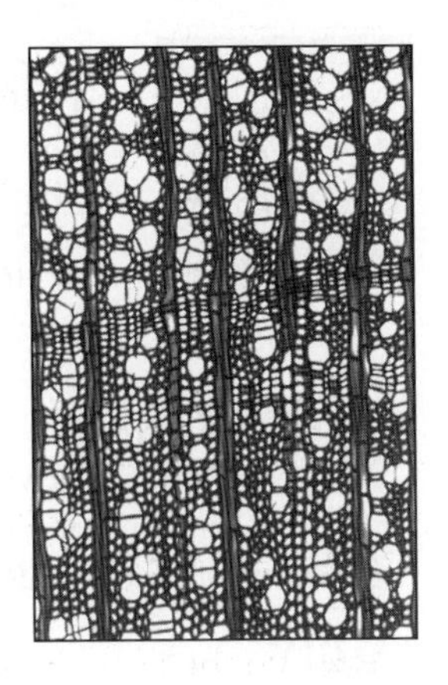

Southern magnolia seasons easily without warp and remains dimensionally stable in use. Average reported shrinkage values (green to ovendry) are 5.4% radial, 6.6% tangential and 12.3% volumetric.

DURABILITY

Southern magnolia is not durable when subjected to conditions favorable to decay.

WORKABILITY

Southern magnolia is easily worked with tools. It takes and holds a good finish or may be finished smoothly in its natural color. Nail holding and gluing properties are satisfactory.

USES

A special use for southern magnolia is in venetian blind slats. It is well suited for this use because of its fine, uniform texture, hardness and ability to remain flat without warping. About two-thirds of the southern magnolia used in manufactured wood products goes into furniture. Other uses are for cabinetry, carving, turnery, woodenware, interior finish, siding, boxes and crates, pallets, handles and flooring. A good quantity is cut into veneer.

SUPPLIES

Southern magnolia is available at a moderate cost in its growth region, but is not widely distributed to other parts of the United States or the world.

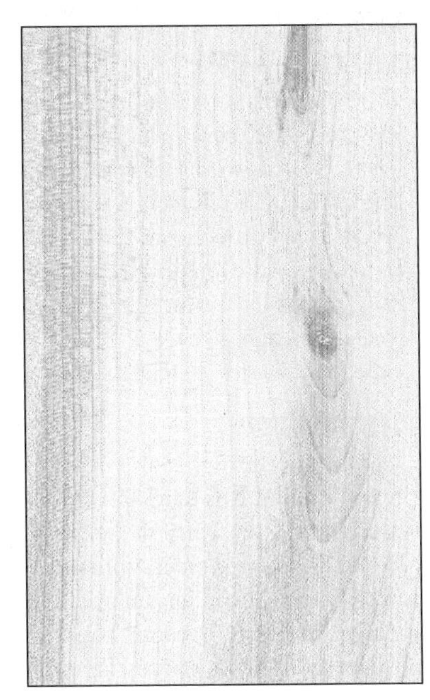

Malus sylvestris

apple

by Jim Flynn

SCIENTIFIC NAME

Malus sylvestris. Derivation: The genus name is the classical Latin name for apple. The specific epithet is Latin for "of the woods".

FAMILY

Rosaceae, the rose family.

OTHER NAMES

There are many species of apple and thousands of varieties. M. *sylvestris* was chosen for this Wood Data Sheet because it is the common apple which the Romans introduced into Britain and is the most prevalent throughout the world.

DISTRIBUTION

Apple trees can be found worldwide in temperate zones and at the higher elevations in tropical areas. To complete the necessary dormant cycle, an apple tree requires a temperature of 45° F or below for 1,000 to 1,200 hours.

THE TREE

The apple tree is an upright or spreading species with dark grayish-brown, irregularly fissured, scaly bark. The twigs are gray to reddish-brown. The simple leaves are ovate or oval, broad cuneate or rounded at the base, crenate-serrate and generally hairy on the underside. The flowers are white and usually suffused with pink. The fruit is a pome. Most trees growing in orchards are severely pruned to allow sunlight on the inner portions and to maintain a low height to facilitate harvesting the fruit. Trees growing in the wild have a more erect profile. Heights vary according to location and cultivation but can reach 50 feet. The U.S. National Register of Big Trees lists a tree in Virginia which measures 70 feet in height with a diameter of 141 inches.

THE TIMBER

Apple wood is light tan with streaks and wide bands of a sharply demarcated brown of various hues. It is hard and close grained, and the pores and rays are imperceptible. The transition from early to latewood is gradual. Average reported specific gravity is 0.61 (ovendry weight/green volume), equivalent to an air-dried weight of 47 pcf. Most wood technology books ignore apple wood completely or state it has little or no commercial value, probably because large and straight saw logs cannot be gathered in sufficient quantity to make harvesting profitable. In the apple-growing Piedmont Plateau region of Virginia, however, some saw mills and kilns specialize in apple wood. Supplies are more than adequate in spite of their relatively short lengths of 8 to 10 feet.

SEASONING

The wood is difficult but not impossible to season. It must be teased along at all stages to prevent checking and warping. Average reported shrinkage values (green to ovendry) are 5.6% radial, 10.1% tangential and 17.6% volumetric. Once seasoned, it is dimensionally stable.

DURABILITY

Apple is not known to be a very durable wood when used in applications subject to moisture.

WORKABILITY

Depending upon the cut, the wood may be difficult to machine because the grain may not lie on the same plane. Sharp tools are needed. While uneven grain patterns add to the beauty, it is a disadvantage in milling. Nevertheless, with proper precautions, the wood is easy to work with machines and hand tools. It glues and finishes well.

USES

The wood from apple trees is used mostly for small objects such as tool handles, woodenware and individual pieces of furniture. Because of its variegated color, it makes beautiful turned bowls. If sections of the trunk can be found with borings from the yellow-bellied sapsucker woodpecker, they can be turned with a very unusual and artistic effect. As a personal note, I made an Appalachian Mountain dulcimer from an apple log by seasoning and resawing it in 0.125-inch slabs. With heat, it bent very well, and I was able to bookmatch pieces with unusual grain patterns and produced a very fine musical instrument.

SUPPLIES

While the wood may not be available commercially at lumberyards, it is relatively easy to find in woodpiles in the vicinity of apple orchards.

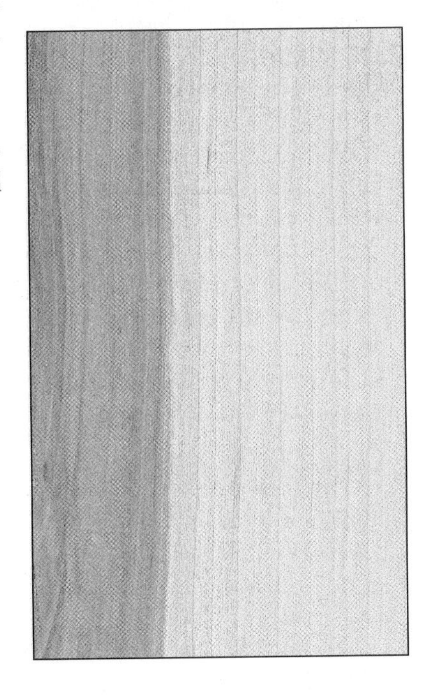

Mammea africana

oboto

by Jon Arno

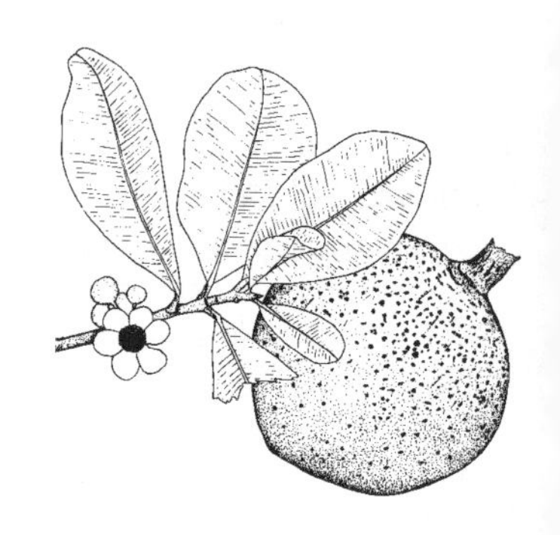

SCIENTIFIC NAME

Mammea africana. Derivation: The genus name is a South American name for the fruit. The specific epithet is Latin meaning the plant is "of Africa".

FAMILY

Guttiferae, the mangosteen family; also Clusiaceae.

OTHER NAMES

bompegya, kaikumba, ologbomodu, aborzok, bokoli.

DISTRIBUTION

Mammea africana is native to the tropical rain forests of west central Africa from Sierra Leone to Angola.

THE TREE

A well formed timber tree, oboto will reach a height of up to 120 feet with a diameter of about 3 feet. Like many rainforest species, the tree produces a swollen, buttressed trunk, but will yield clear logs up to 50 feet.

THE TIMBER

Oboto is a pretty wood, sometimes used as a mahogany substitute. The color of the heartwood is comparable to the dark reddish-brown of true mahogany and is attractively marked or flecked with gum ducts. Oboto, however, lacks the surface luster of mahogany and is considerably coarser in texture. The sapwood is pinkish-brown and clearly demarcated from the heartwood. Radial surfaces may display a ribbon figure as a result of interlocked grain. The wood has no strong odor or taste. In terms of density, oboto has considerable variation in specific gravity, which ranges from 0.53 to 0.70 (ovendry weight/green volume), equivalent to an air-dried weight of 40 to 55 pcf; this, of course, affects its structural properties. On average, it is substantially stronger than American mahogany (*Swietenia macrophylla* WDS 254), especially in terms of bending strength.

SEASONING

Oboto is difficult to season. It has very high shrinkage and will collapse or honeycomb if not dried slowly. Average reported shrinkage values (green to ovendry) are 6.5% radial, 10.0% tangential and 14.1% volumetric. The references do not indicate whether this problem relates only to abnormal wood (stock cut from the swollen trunk base) or is characteristic of the species in general. But, given its high average shrinkage, the latter is probably the case.

DURABILITY

Oboto is durable with respect to decay, but is susceptible to termite attack.

WORKABILITY

While certainly not in a class with the mahoganies, either African (*Khaya* spp. WDS 155) or American (*Swietenia*), oboto is workable, cuts cleanly and polishes to a nice finish, but contains minerals which dull tools. Also, the high gum content tends to clog saw blades, shapers, drills, router bits and sandpaper. While the true mahoganies are renown for their stability, oboto is reported to react to changes in humidity so it is not considered dimensionally stable.

USES

Uses include furniture, millwork and general carpentry. Other species of the Guttiferae family are economically important as sources of gums, oils, dye stuffs, drugs and fruits.

SUPPLIES

Supplies of oboto are adequate. The price is moderate. Availability through normal retail channels is somewhat limited relative to the better-known, imported timbers.

Mangifera indica

mango

by Jim Flynn

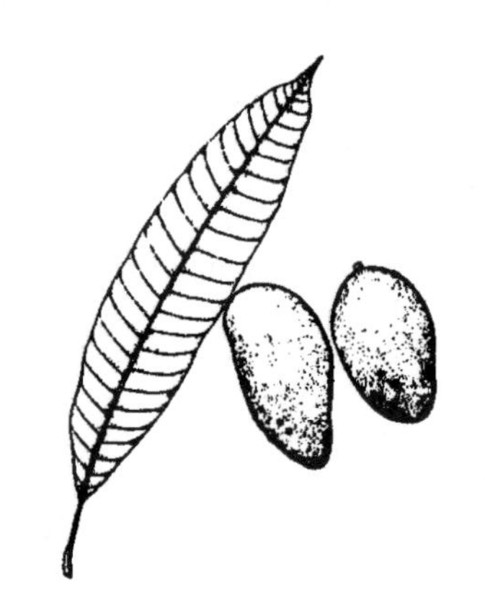

SCIENTIFIC NAME

Mangifera indica. Derivation: The genus name means bearing mangoes, which is from the Portuguese common name. The specific epithet is Latin meaning the plant is "of India".

FAMILY

Anacardiaceae, the cashew family. The genus is in the same family as *Rhus* wherein we find poison ivy, *Rhus radicans*.

OTHER NAMES

Names include manga, mangue, manako, mangot and others resembling the spelling of mango.

DISTRIBUTION

Mango has a wide natural range from India eastward to the China Sea. Due to its tasty fruit, it has been cultivated throughout the tropics. It has naturalized in many areas including southern Florida, the Florida Keys, the West Indies and from Mexico to Peru and Brazil. Hawaii has large stands as do many parts of the world including South Africa.

THE TREE

Widely dispersed, mango is subject to different growing conditions (climate, soil, altitude, moisture, etc.) so only a general description can be given. Many varieties have been designed to improve the quality of the fruit. It is an evergreen, growing to 65 feet in height with a diameter of about 3 feet. In parts of Southeast Asia, the tree can reach heights of 100 feet or more. The large, leathery, dark green, lance-shaped leaves are 6 to 12 inches long and 1.5 to 3 inches wide. The tree produces a large, elliptic fruit with yellow, deliciously edible flesh. The fruit is also used for jams, jellies and beverages. The yellow light-resistant dye that is used in oil paintings is made from the urine of cows that have been fed mango leaves.

THE TIMBER

Mango has an interesting color pattern that seems disorganized. It is generally brown to blondish tan. Interesting patches of darker shades caused by fungal attack occur when seasoning too slowly. Long and short wavy blackish flecks show where pores are exposed on the radial and tangential surfaces. They are quite attractive and create a marbleized appearance on the wood. Frequently, pinkish hues flow in long stripes adding to the mystery of color. The wood had neither taste nor odor. There is significant variance in its specific gravity with an average reported value range (ovendry weight/green volume) of 0.45 to 0.58, equivalent to an air-dried weight of 35 to 45 pcf. The texture is fine to medium, quite similar in appearance and feel to bigleaf mahogany (*Swietenia macrophylla* WDS 254).

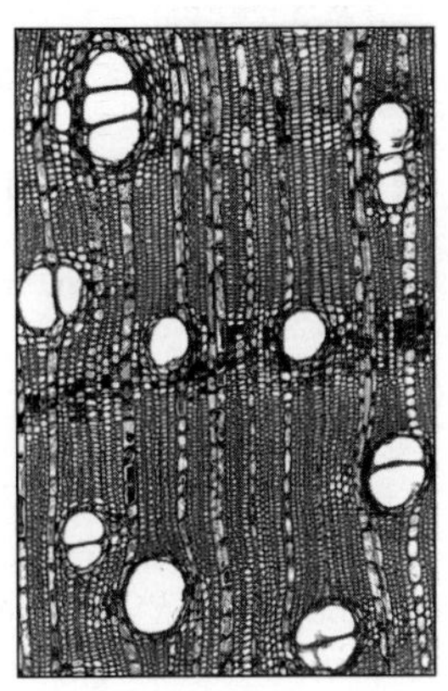

SEASONING

Mango is not difficult to season if it is quickly moved into kilns where it will season with only minor degrade. It will easily succumb to fungal stain if air dried. Many users, however, prefer this stained wood as it adds additional distinctive color. Average reported shrinkage values (green to ovendry) are 3.0% radial, 4.9% tangential and 7.3% volumetric.

DURABILITY

The wood is not durable without treatment. It is subject to fungal and insect damage especially dry-wood termites. Pressure treatment is very effective in combating these problems.

WORKABILITY

Logs are easily converted into dimension timbers. After the wood has dried, wavy grain is apt to tear and chip in the planer and shaper. Drilling, nailing and gluing operations pose no problems. Careful sanding is advised as the wood scratches easily. Mango is stable in use.

USES

Uses include structural timbers, light flooring, furniture, paneling and craft products. It is also used for gunstocks and claimed to be more stable than the conventionally used black walnut (*Juglans nigra* WDS 150). The wood is used for meat chopping blocks, and in the South Pacific it is used for canoes.

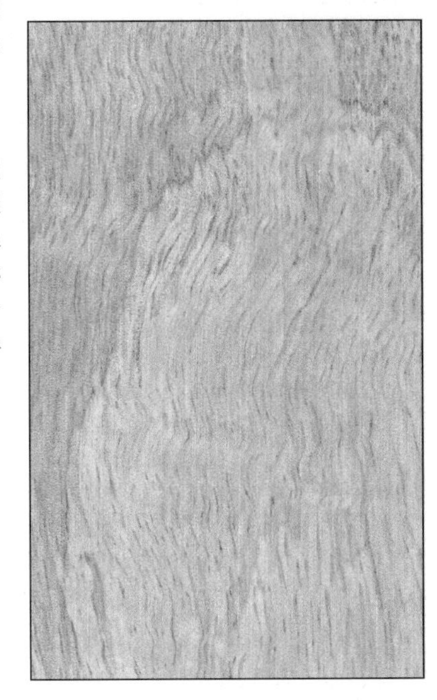

SUPPLIES

Mango has potential value on the commercial market. Perusal of many stock lists of wood importers and suppliers fail to show it as an item of trade. This is not surprising as the species is raised primarily as a source of fruit. There is evidence that the wood is used to a great extent in the area in which the trees grow.

Manilkara bidentata

bulletwood

by Max Kline

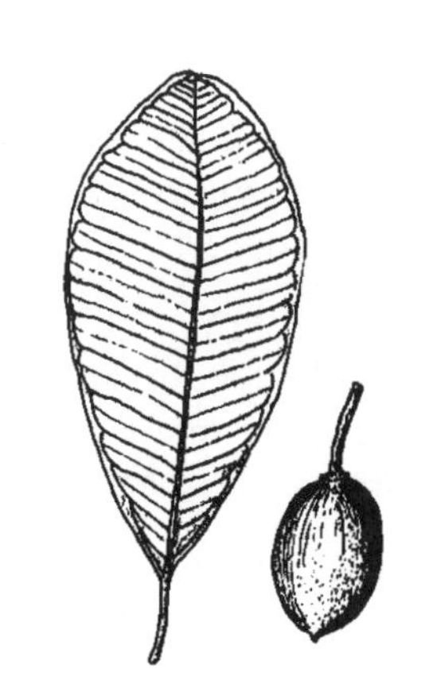

SCIENTIFIC NAME

Manilkara bidentata. Derivation: The genus name is from an Indian vernacular *manyl-kara* peculiar to the Malabar coast of western India. It was used to name a species of this genus and then was applied to all species as a generic name. The specific epithet is Latin for double-toothed, referring to the leaves.

FAMILY

Sapotaceae, the sapodilla family.

OTHER NAMES

red lancewood, balata, wild dilly, sapodilla, bully tree, horseflesh, beefwood.

DISTRIBUTION

Throughout the West Indies, Central America and the northern portion of South America.

THE TREE

Bulletwood is a large tree growing 100 feet in height with a straight trunk up to 4 feet in diameter. The alternate leaves, often darkened by a sooty mold, have petioles from 0.75 to 1.25 inches long. The blades of these leaves are up to 10 inches long and 4.5 inches wide but can be much smaller. There are three to ten flowers together on stalks about 0.5 inch long. The smooth berry has a sweet or gummy pulp and is edible. At one time, bulletwood was the most important tree in Puerto Rico because of its yield of 2 to 5 pints of latex, equivalent to 1 to 3 pounds per tree of dry balata (a rubber-like gum).

THE TIMBER

Bulletwood is a dull plum red to dark reddish-brown and somewhat resembles raw beef in color. The sapwood is whitish or pale brown and distinct but not demarcated from the heartwood. The luster is low. Odor and taste are not pronounced. It is generally extremely hard and heavy with an average reported specific gravity of 0.85 (ovendry weight/green volume), equivalent to an air-dried weight of 69 pcf. The texture is fine, and the grain is straight. Bulletwood is equal or superior to greenheart (*Chlorocardium rodiei* WDS 079) in shock resistance, hardness, shear and bending strength.

SEASONING

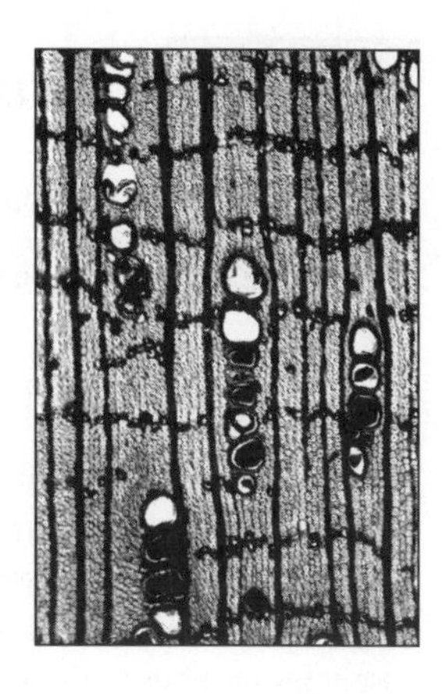

Bulletwood is difficult to season. Air drying must be done slowly, and there is a tendency to check, warp and caseharden. Average reported shrinkage values (green to ovendry) are 6.3% radial, 9.4% tangential and 16.9% volumetric.

DURABILITY

Bulletwood timber is very resistant to termite attack but it is not resistant to marine borer attack. It is extremely impermeable to moisture, surpassing teak (*Tectona grandis* WDS 261) in this respect. It weathers rather poorly, however, and experiences considerable surface checking when exposed to the weather without protection.

WORKABILITY

Bulletwood is moderately difficult to work, but finishes smoothly and takes a beautiful polish. Pre-boring is advised before using nails or screws. Gluing requires special care due to the presence of some oil in the wood.

USES

Its strength, wear resistance and durability make bulletwood very suitable for heavy construction in areas such as bridge decking, boat frames, railroad ties, flooring and post or utility poles. Its bending properties make it suitable for use in boat frames or other bent items. Its fine texture and ability to take a high polish make it suitable for special uses such as shuttles, furniture parts, turnery, fishing rods, musical instruments, umbrella handles and bows.

SUPPLIES

To protect the supply of balata coming from the bulletwood tree, its cutting is limited for lumber and only small quantities are exported. It can be found at high prices, however, from the larger importers in the United States.

Manilkara zapota

sapodilla

by Jim Flynn

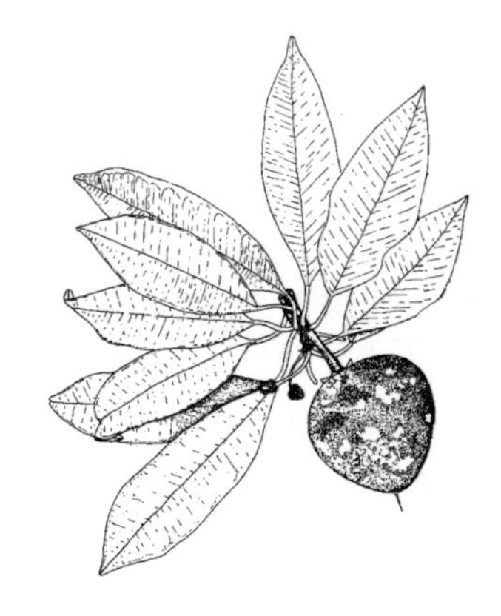

SCIENTIFIC NAME

Manilkara zapota. Derivation: The genus name is from an Indian vernacular *manyl-kara* peculiar to the Malabar coast of western India. It was used to name a species of this genus and then was applied to all species as a generic name. The specific epithet is one of the Spanish names for the fruit. Synonyms are *Sapota zapotilla, Achras zapota, Sapota achras, and Manilkara zapotilla.*

FAMILY

Sapotaceae, the sapodilla family.

OTHER NAMES

chicle, nispero, sapote, zapotillo, dilly.

DISTRIBUTION

Sapodilla is native to the West Indies, Mexico and from there southward to include the northern portions of South America. It is extensively planted in tropical regions of the world. Sapodilla has naturalized in south Florida including the Florida Keys.

THE TREE

In its native range, sapodilla is a large and beautiful evergreen tree growing to 100 feet or more in height with a diameter of as much as 3 feet. The leaves are glossy, leathery and are clustered at the ends of the branchlets. The tree produces an edible fruit known as the sapodilla plum which is very tasty. Until recent years the tree was an important source of chicle, the prime ingredient in chewing gum; a chemical substitute has now taken its place. Today, there is little evidence that *chicleros* still gather the milky latex. In the northern parts of the Yucatan Peninsula in Mexico, the machete-scarred trees now stand untouched and encased in the understory of the forest with their wounds slowly healing. A student of dendrology would have no trouble identifying this species if one of the keys were the machete scars.

THE TIMBER

The wood of sapodilla is noted for its strength and durability. It can be found intact in the lintels and supporting beams in the early Mayan structures. The heartwood is dark red or reddish-brown that is not sharply demarcated from the pinkish sapwood. The wood is hard and heavy with an average specific gravity of 0.84 (ovendry weight/green volume), equivalent to an air-dried weight of 68 pcf. The texture is fine, and the wood has straight grain.

SEASONING

There are no known unusual seasoning problems with sapodilla. Average reported shrinkage values (green to ovendry) are 5.0% radial and 8.0% tangential.

DURABILITY

Sapodilla is highly resistant to decay.

WORKABILITY

Because of its high density, it is not easy to work. The wood has a tendency to splinter. The wood will take a fine finish.

USES

In its growth range, the wood remains a prime source of construction timbers, railway crossties and heavy flooring. It is used in furniture making, tool handles, rulers and shuttles.

SUPPLIES

There is no evidence that sapodilla is exported in any great quantity. Limited amounts can be found in Florida where the tree has naturalized. It is not an expensive timber.

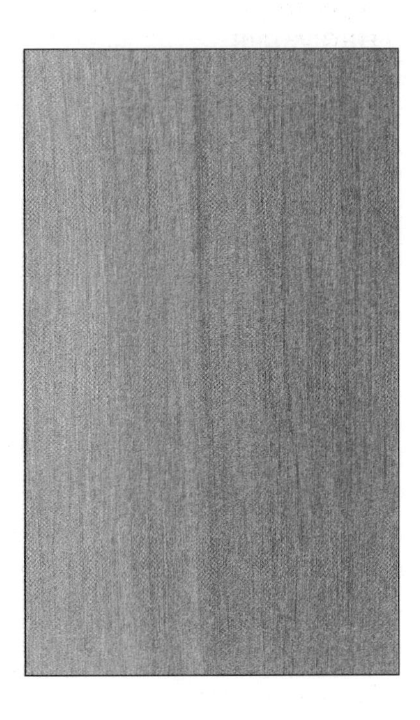

Mansonia altissima

mansonia

by Max Kline

SCIENTIFIC NAME

Mansonia altissima. Derivation: The genus name is in honor of Sir Patrick Manson (1844 to 1922), a Scottish physician. The specific epithet is Latin meaning very tall.

FAMILY

Sterculiaceae, the stericulia or cacao family.

OTHER NAMES

African black walnut, aprono, ofun, bete, koul.

DISTRIBUTION

Scattered in western Africa from the Ivory Coast to Nigeria.

THE TREE

Mansonia reaches a height of 120 feet with a diameter of 2 to 3 feet. The buttressed bole is clear and straight.

THE TIMBER

Resembling black walnut (*Juglans nigra* WDS 150), the heartwood of mansonia varies in color from grayish-brown to chocolate-brown or dark walnut brown, often with a purplish cast. The sapwood is whitish and sharply demarcated from the heartwood. The luster is low to medium. Odor and taste are not distinct. The grain is generally straight, and the texture is rather coarse. Average reported specific gravity is 0.50 to 0.58 (ovendry weight/green volume), equivalent to an air-dried weight of 38 to 45 pcf. About equal to black walnut in strength, mansonia is 30% more resistant to shock and 15% stronger in bending.

SEASONING

Mansonia seasons fairly rapidly, but knots tend to split appreciably. Average reported shrinkage values (green to ovendry) are 4.4% radial, 7.3% tangential and 10.2% volumetric. Very little distortion will be experienced and movement in service is rated as medium.

DURABILITY

The wood is very resistant to decay, but pinhole borers sometimes attack the sapwood.

WORKABILITY

Mansonia works easily with hand and machine tools with little dulling effect on the cutting edges. It has good nailing and gluing properties and takes stain and polish well. For some people, the dust from working dried wood can cause irritation of the skin and mucous membranes.

USES

Uses of mansonia include furniture, cabinets, interior finish millwork, piano cases, gunstocks and decorative veneer. The bark contains a cardiac poison of the digitalis group.

SUPPLIES

Mansonia veneer and solid lumber are available at moderate costs. The irritating effect of the dust limits its use in manufacturing.

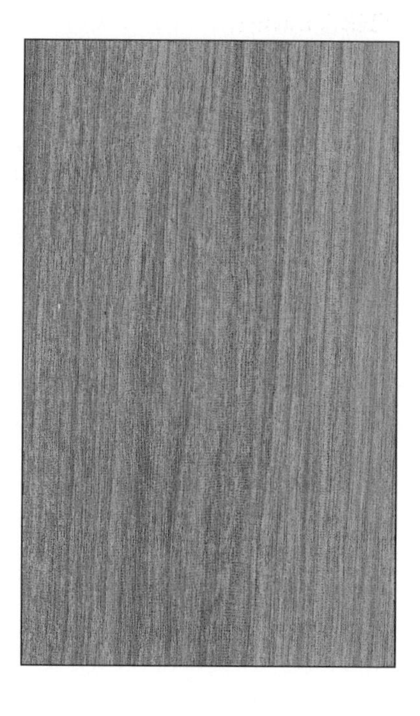

Melaleuca quinquenervia

cajeput-tree

by Jon Arno

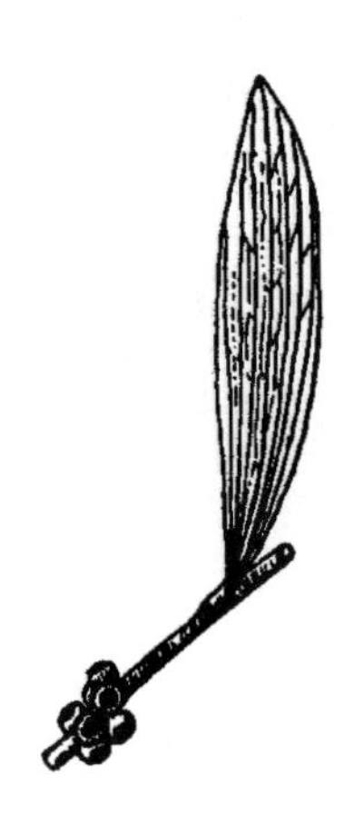

SCIENTIFIC NAME

Melaleuca quinqueneververvia. Derivation: The genus name is from the Greek, black and white, referring to the dark trunk and white branches of one of the species. The specific epithet is Latin for five-nerved, referring to the leaves.

FAMILY

Myrtaceae, the myrtle family.

OTHER NAMES

gelam, paper-bark, malaleuca, punktree, bottlebrush, broad-leaved paperbark, pink tree.

DISTRIBUTION

Cajeput-tree prefers wet, swampy soils and, unlike most tropical species, it will develop dense, pure stands where these conditions prevail. It is native to Malaysia, Indonesia, Papua New Guinea, New Caledonia and parts of eastern Australia. Cajeput-tree has been widely planted in warm climates around the world and is now naturalized in southern California and Florida.

THE TREE

Although capable of reaching heights up to 100 feet and diameters of about 2 feet, cajeput-tree is not an ideal timber producer because the main stem is seldom straight. Once cut, the stumps coppice vigorously and the resulting thickets are of value for both the commercial production of fence posts and for landscaping purposes. The spongy, thick and somewhat flaky bark, pale green leaves and bottle brush-shaped flower spikes provide special character in landscaping applications.

THE TIMBER

With an average reported specific gravity of 0.65 (ovendry weight/green volume), equivalent to an air-dried weight of 51 pcf, the wood is on the heavy side. The texture is extremely fine and lustrous, but the wood lacks figure. Demarcation between the reddish-brown heartwood and lighter, creamy pink sapwood is not distinct. Even though the leaves of cajeput-tree and many other members of the myrtle family are quite aromatic, the dry wood is virtually without odor. The strength is comparable to sugar maple (*Acer saccharum* WDS 010).

SEASONING

Cajeput-tree is very difficult to season. The end grain should be thoroughly sealed with hot wax or glue, and it must be air dried slowly to minimize warping and checking. The average reported volumetric shrinkage is a rather high at 16.2% (green to ovendry) and this, in combination with the fact that it shrinks more than twice as much tangentially (9.5%) as it does radially (4.0%), causes considerable stress as the wood dries.

DURABILITY

Cajeput-tree has excellent weathering characteristics. The heartwood is resistant to attack by termites and marine borers.

WORKABILITY

Due to its fine texture and natural luster, the wood will hold sharp detail and can be polished to a glass smooth finish. These features enhance its value as a reasonably good carving wood, but there are drawbacks. Its rather high density makes it hard to work, and tests reveal it often contains high levels (up to 0.95%) of silica, which will quickly dull blades. Also, it is not a very stable wood and, if it is not properly seasoned, warping and checking can be serious problems.

USES

Cabinetry, carving, plywood veneers and construction timbers account for a portion of its uses. Due to cajeput-tree's excellent weathering characteristics, it is also used as railroad ties and fence posts and in boat building.

SUPPLIES

Cajeput-tree is not as commercially important in international trade as some of its relatives in the *Eucalyptus* genus, but it is a general purpose timber within its native range. Its ability to regenerate by coppicing and the fact that it has been widely planted in warmer regions around the world would suggest it would remain plentiful in the future. It is a low-cost timber.

Melia azedarach

chinaberry

by Max Kline

SCIENTIFIC NAME

Melia azedarach. Derivation: The genus name is a classical Greek name for the ash tree and was transferred by Linnaeus to this genus. The specific epithet is from the Persian name *azad dirakht*, literally noble tree. A synonym is M. *toosendan*.

FAMILY

Meliaceae, the mahogany family.

OTHER NAMES

Persian lilac, beadtree, nimwood, white cedar, chinatree, Chinese umbrella tree, paraiso, pride-of-India.

DISTRIBUTION

A native of China, chinaberry is also found in India, Taiwan and Australia. It was introduced into the United States and grows in the south from North Carolina to Oklahoma and south to Florida and Texas. It can also be found in the West Indies, Mexico, Central America and parts of South America.

THE TREE

An exotic, semi-tropical species, chinaberry is a member of the mahogany family and was introduced into the United States as an ornamental. It grows easily and rapidly in the spring producing profuse, fragrant, lilac-colored flowers. The fruit is a yellowish, cherry-like berry, which forms into clusters and clings to the tree after the leaves have fallen.

THE TIMBER

Chinaberry wood is light brown, pink, light red or light reddish-brown. The luster is high and satiny. Odor and taste are not distinct. The wood is soft and light, with an average reported specific gravity of 0.47 (ovendry weight/green volume), equivalent to an air-dried weight of 36 pcf. The texture is coarse, and the grain is usually straight but attractive on plain and quartersawn surfaces.

SEASONING

Chinaberry air dries easily without serious warping or splitting. Average reported shrinkage values (green to ovendry) are 5.0% radial, 8.5% tangential and 13.5% volumetric.

DURABILITY

The wood is not durable in exterior applications.

WORKABILITY

Chinaberry is easily worked with tools, glues well and takes all kinds of finishes.

USES

At present chinaberry has no commercial value as lumber since it is not an abundant tree. The wood, however, should find a place along with other beautiful North American woods in turnery, woodenware and novelties. In other countries, it has been used for cigar boxes, toys, sporting goods, furniture and interior finish.

SUPPLIES

Since the tree is not widely distributed in the United States, the lumber supply is quite limited and it is rarely sold commercially. The best sources are from wood collectors living in the southeast where the tree is a favorite backyard ornamental because of its pleasant aroma.

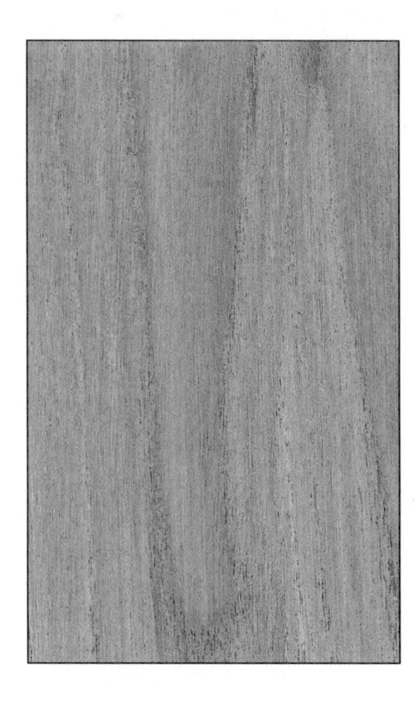

Metopium brownei

chechem

by Jim Flynn

SCIENTIFIC NAME

Metopium brownei. Derivation: The genus name is a Latin word from Greek *metopion*, literally forehead, also reported to be the classical name for an African tree. The specific epithet is in honor of Robert Browne (1773 to 1858), one of the leading English botanists of his time.

FAMILY

Anacardiaceae, the cashew family.

OTHER NAMES

black poison wood, chachin, chechen, coral sumac, poison wood, guao, cedro prieto, bois mulatre.

DISTRIBUTION

Dominican Republic, Cuba, Jamaica, northern Guatemala, Belize and Mexico from the Yucatan to Vera Cruz.

THE TREE

Chechem varies in size from a shrub to a fairly large tree growing to 50 feet in height. Its odd, pinnate leaves have three to seven large, round or obovate leaflets with long petioles. The small, green flowers are borne in large, long-stalked axillary panicles. The fruit is an orange drupe with a resinous pulp. The thin, reddish-brown bark contains a caustic juice, but the wood itself is not poisonous. The tree grows along with sapodilla (*Manilkara zapota* WDS 174) on calcareous soil. In the forests of the Yucatan in Mexico, there is a fair amount of chechem. By being attentive to the black exudate on the bark, the tree can be easily avoided. In 1566 the Spanish Friar Diego De Landa wrote "There is a tree whose sap causes sores when touched, and even its shade is noxious, if one sleeps under it". Today, the shade from this tree is avoided because the dripping sap is known to be as infectious as poison ivy.

THE TIMBER

The heartwood of chechem is a deep brown, almost of the hue of black walnut (*Juglans nigra* WDS 150). The sapwood is yellowish-brown and is sharply demarcated from the heartwood. It is often variegated from red with a greenish tint and a golden luster to plain brown. When slab-cut, many of the pieces show an extremely attractive grain pattern. The wood has no distinctive odor or taste. It is very hard and strong. Average specific gravity is 0.68 (ovendry weight/green volume), equivalent to an air-dried weight of 53 pcf. The texture is fine and uniform.

SEASONING

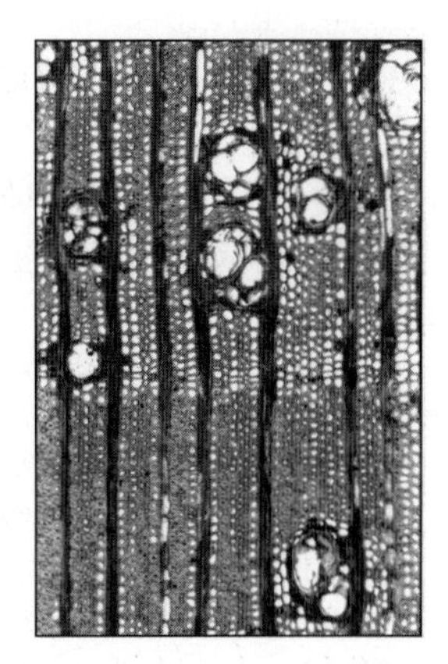

Chechem appears to season well in its local environment but there is not much data available on kiln drying. Once the timber has been cut, the caustic exudates dried, and the wood debarked, it is relatively safe to handle. Shrinkage values are not available.

DURABILITY

There is no reliable data on durability.

WORKABILITY

While the wood is extremely hard, it is workable with sharp tools, takes glue effectively and can be polished to a fine finish. Chechem seems to be similar to hard maple in workability but is a bit more stringy.

USES

Contemporary uses include poles for native structures and furniture making. Timber merchants are exploring other uses, and it appears to have potential for cabinets, turnery and musical instruments.

SUPPLIES

Some chechem is exported through the Forest Production Society of Quintana Roo, Mexico in an attempt to commercialize lesser-known species of wood as substitutes for more exotic and endangered species. This Society represents the banding together of many local farming communities who are working with the Mexican Department of Forestry and its Plan Polito Forestal (Pilot Forestry Program). This program encourages the development of diversified forest industries, which can utilize a variety of local species. It also develops and implements a harvesting program to sustain the forest resource. When wood such as chechem is marketed, consumers should be on the alert for SmartWood Certification, which is a certificate that states the wood has been harvested under approved management conditions and that harvesting has not contributed to tropical forest destruction.

Metrosideros collina

ohia

by Max Kline

SCIENTIFIC NAME

Metrosideros collina. Derivation: The genus name is Greek meaning heart of iron. The specific epithet is Latin meaning pertaining to a hill, which refers to the growth area where the plant is found.

FAMILY

Myrtaceae, the myrtle family.

OTHER NAMES

vuga, anume.

DISTRIBUTION

In the islands of the Pacific, particularly in Hawaii, Fiji and Samoa.

THE TREE

Data for this species is for M. *collina* var. *polymorpha*. The tree, often with stilt roots, may reach a height of 100 feet with a clear bole of 40 to 50 feet and a diameter up to 4 feet. On older trees the bark is loose and shreddy. The generally red flowers form pom-pom like clusters of many individual florets. In Hawaii, ohia is the most abundant of the indigenous trees and may form pure stands. It is one of the first trees to invade fresh lava flows. Eventually, it will be partially replaced by other species, resulting in a mixed forest.

THE TIMBER

The heartwood is reddish to purplish-brown, gradually fading into the pale brown sapwood. The texture is fine to medium. The figure is streaked, and the grain is straight to curly or interlocked. Ohia is lustrous without characteristic taste or odor and is very dense and hard. Average reported specific gravity is 0.70 (ovendry weight/green volume), equivalent to an air-dried weight of 55 pcf.

SEASONING

Ohia timber is prone to warping. Air drying is suggested prior to kiln drying. Average reported shrinkage values (green to ovendry) are 6.9% radial, 12.1% tangential and 19.1% volumetric. The wood is not dimensionally stable.

DURABILITY

The heartwood is not resistant to attack by decay fungi, but has good resistance to termites. The heartwood is resistant to penetration with preservatives.

WORKABILITY

Because of its high density, ohia saws and machines with difficulty. It works well when shaping and boring but rates poorly when planing and turning. It is somewhat brash and brittle.

USES

In the past, the native islanders used ohia for idols, poi boards, poor-quality houses, temple enclosures, canoe gunwales, anvils and weapons. Currently, the wood is used for railroad ties, posts, flooring, pallets, poles and wharf fenders.

SUPPLIES

Ohia is readily available in its growth region, but very little is exported. Wood collectors must secure samples by trading or personal collecting.

Microberlinia brazzavillensis

zebrawood

by Max Kline

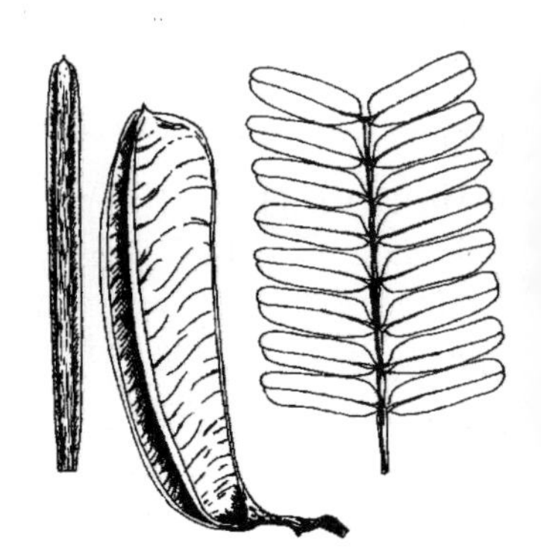

SCIENTIFIC NAME

Microberlinia brazzavillensis. Derivation: The genus name *micro* means small and Berlin, a pupil of Linnaeus. The specific epithet means the plant is "of Brazzaville", a city in the Congo.

FAMILY

Fabaceae or Leguminosae, the legume family; (*Caesalpiniaceae*) the cassia group.

OTHER NAMES

zebrano, okwen, zingana.

DISTRIBUTION

West Africa, mainly in Gabon and Cameroon, sometimes in pure stands along riverbanks.

THE TREE

Reaching 150 feet in height with a straight, cylindrical bole and a diameter of 4 to 5 feet, zebrawood is an impressive tree. Trees growing this large are usually in inaccessible sites requiring native labor to harvest. The bark is 12 inches thick and always trimmed when the tree is felled. The double compound leaves are 6 inches long and 4 inches wide. The small, alternate leaflets are 0.75 inch long and 0.375 inch wide. The fruit is a bivalve-type pod which contains the seeds.

THE TIMBER

Zebrawood is pale golden brown to pinkish-brown usually with pronounced dark brown streaks. It has a high luster, and a distinct characteristic odor is present when the timber is worked. Average reported specific gravity is 0.70 (ovendry weight/green volume), equivalent to an air-dried weight of 55 pcf. It has a somewhat coarse texture. The grain is usually interlocked producing a ribbon figure. This timber is regarded as quite strong but it is not usually used where great strength is required.

SEASONING

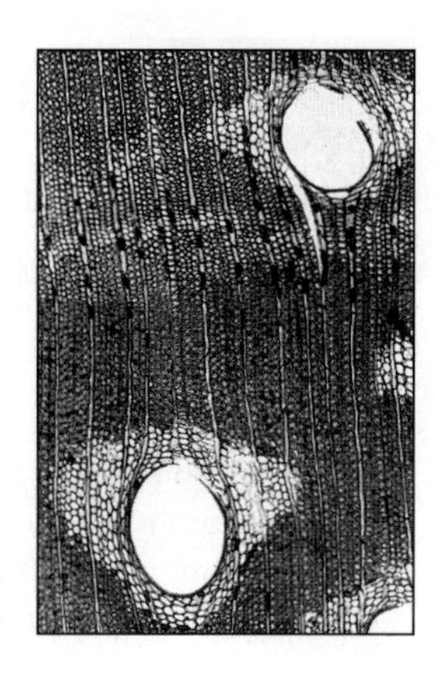

Zebrawood air dries well and once dry changes very little. Checking and cracking are not problems with this species. Average reported shrinkage values (green to ovendry) are 6.8% radial, 11.5% tangential and 16.5% volumetric.

DURABILITY

The heartwood of zebrawood is durable and resistant to termite attack.

WORKABILITY

This wood works well with hand tools. Veneer has the best qualities when quartersawn. It takes a good polish and turns easily.

USES

Because of its striking effect, zebrawood is used extensively in veneer form for crossbanding and inlays and for borders on furniture. It also has limited use in furniture, flooring and general construction.

SUPPLIES

Zebrawood is available for the limited purposes for which it is used. It can be regarded as moderately expensive.

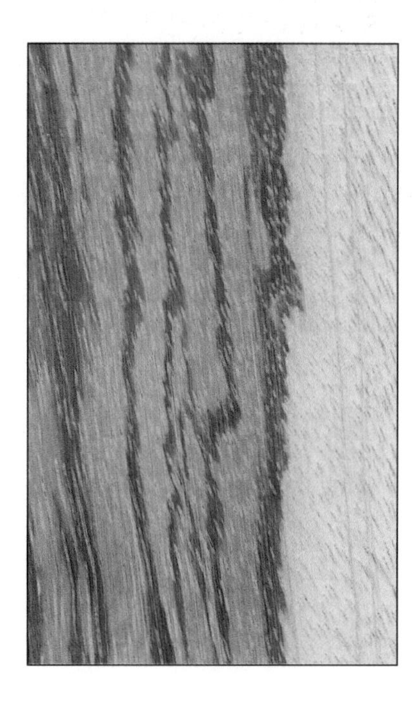

Milicia excelsa

iroko

by Max Kline

SCIENTIFIC NAME

Milicia excelsa. Derivation: The genus name is in honor of Senhor Milicia, an administrator in what is now Mozambique. The specific epithet is Latin for tall or high. A synonym is *Chlorophora excelsa*.

FAMILY

Moraceae, the mulberry family.

OTHER NAMES

mvule, odum, kambala, tule moreira, African teak, African oak, Nigerian teak, oroko, osan.

DISTRIBUTION

Iroko has a wide distribution, extending across the continent of Africa from west to east but principally from the Ivory Coast to Cameroon.

THE TREE

Iroko grows to a height of 160 feet with a diameter of 8 to 9 feet. There are practically no buttresses, and the tree is generally free from branches for the first 70 feet. The bark is pale ash-gray to black. The leaves of the young trees are sandpapery above and pubescent below. The elliptic to ovate leaves are about 12 inches long and 5 inches wide.

THE TIMBER

The heartwood varies in color from light to dark brown, and the clearly demarcated sapwood is a paler shade of brown. Frequently, there are dark streaks and striping. Average reported specific gravity is 0.55 (ovendry weight/green volume), equivalent to an air-dried weight of 42 pcf. Odor and taste are not distinct. The wood is mildly lustrous, and the grain is interlocked. The texture is moderately coarse and even. Probably as a result of injury to the tree, calcium carbonate can be found deposited in cell cavities creating hard, stone-like sections in the wood. Compared to teak (*Tectona grandis* WDS 261), iroko has about the same strength properties, hardness and resistance to applied loads. It is, however, weaker in bending strength and in compression along the grain.

SEASONING

The timber seasons well and rapidly without much degrade. There is only a slight tendency to split and distort. Average reported shrinkage values (green to ovendry) are 2.8% radial, 3.8% tangential and 8.8% volumetric

DURABILITY

Iroko heartwood has unusual durability and preservatives are never required. The sapwood, however, is subject to attack by wood borers.

WORKABILITY

In general, the timber works well. Boards containing the stone-like deposits have an abrasive action on saws and tool edges. Nail and screw holding properties are good. A fair amount of filling is necessary prior to polishing, but a high lustrous finish can be obtained when filled. Gluing can be accomplished easily.

USES

Iroko can be used with confidence where strength is of importance. It is recommended for specialized uses such as countertops and for tables where hard usage is expected. It is used for window frames, sills and doors, plywood, veneer, furniture, cabinets and ship building. Many of its uses are as an alternative to teak since it serves the same purposes at a considerably lower cost.

SUPPLIES

Iroko is not generally available from many sources in the United States but can be obtained from some suppliers at comparatively inexpensive prices.

Millettia laurentii

wenge

by Max Kline

SCIENTIFIC NAME

Millettia laurentii. Derivation: The genus name is in honor of Dr. Millett of Canton, China. The specific epithet is Latin for laurel-like.

FAMILY

Fabaceae or Leguminosae, the legume family; (*Papilionaceae*) the pea or pulse group.

OTHER NAMES

pallissandre, dikela, kiboto.

DISTRIBUTION

Wenge is found in open forests in the southern regions of Tanzania and Mozambique. This species also appears in the Congo region in periodically inundated swampy forests.

THE TREE

Wenge is a medium-sized tree growing 50 to 60 feet in height with a diameter of 30 or 36 inches. It reaches its greatest size in riverine forests where it is nurtured by the moist environment. The yellow or yellowish-gray bark is quite smooth and is a source of kino, a native name for the reddish exudate of condensed tannins. The leaves are alternate, compound, and imparipinnate with a swelling at the base of the rachis. There are between seven and nine pairs of opposite leaflets that are 4 inches long and 2.5 inches wide. The under surface is covered with very fine, white hairs. There is a terminal leaflet. Flowers are lilac-like, and the fruit is a woody, flat pod.

THE TIMBER

The sapwood is whitish, and the dark brown heartwood has fine, close, blackish veining resulting in a handsome appearance, especially on the tangential surface. The luster is low, and odor and taste are not distinct. The texture is coarse, and the grain is straight to slightly wavy. Average reported specific gravity is 0.65 to 0.78 (ovendry weight/green volume), equivalent to an air-dried weight of 51 to 62 pcf. The wood is reported to have good resistance to bending and to shock.

SEASONING

Wenge seasons slowly but without much distortion. Average reported shrinkage values (green to ovendry) are 3.1% radial and 5.8% tangential.

DURABILITY

Wenge is resistant to fungi and termites.

WORKABILITY

This timber is easy to work and veneers well, but polishing is difficult.

USES

Wenge is used for heavy construction, crossties, tool handles, furniture and cabinets. When a handsome and dark appearance is desired for flooring, wenge is especially attractive. The veneer is useful for decorative work in furniture and interior decoration. Woodworkers find wenge interesting for carving and turnery.

SUPPLIES

While there has never been an abundant supply on the North American market, in recent years more is being imported and commercial suppliers are offering it quite frequently. It is a medium-priced timber.

Morus rubra

red mulberry

by Max Kline

SCIENTIFIC NAME

Morus rubra. Derivation: The genus name is the classic Latin name for mulberry. The specific epithet is Latin for red, referring to the fruit.

FAMILY

Moraceae, the mulberry family.

OTHER NAMES

black mulberry, silkworm mulberry.

DISTRIBUTION

The growth range of red mulberry is from a line almost due westward from the northeastern section of Massachusetts to southeastern South Dakota, through all the eastern states, including all of Florida, westward to southwestern Oklahoma and central Texas.

THE TREE

Red mulberry is usually rather small but can reach heights of 70 feet and a diameter of 3 to 4 feet. The trunk often divides near the ground into many spreading branches, forming a broad, spreading crown. The fruit when fully developed is red, but becomes dark purple or nearly black when ripe. It is sweet and juicy and is very attractive to birds. Some people also relish it, as it resembles the blackberry.

THE TIMBER

The heartwood is pale orange, and the wide sapwood is lighter in color. The wood is moderately soft. The grain is coarse, straight, and has no outstanding figure. Its best appearance is on plainsawn lumber. It is tasteless, odorless and ring porous. Average specific gravity is 0.58 (ovendry weight/green volume), equivalent to an air-dried weight of 45 pcf. The timber is rather weak and should not be used where high strength is a requirement.

SEASONING

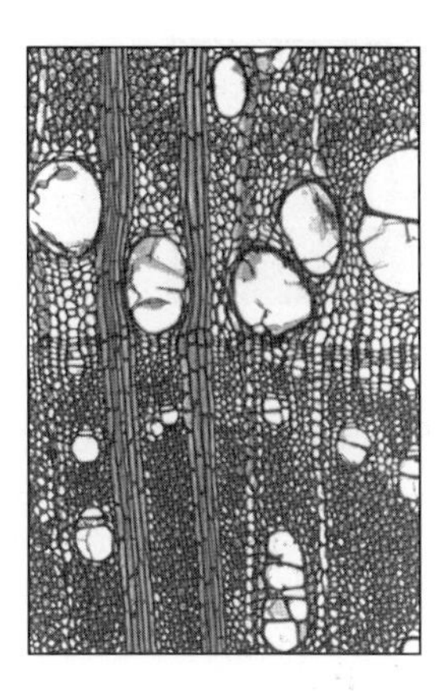

Red mulberry is not difficult to season. While precise shrinkage values are not available, it is reported to hold its place well when dry.

DURABILITY

The wood is decay resistant and quite durable for exterior use.

WORKABILITY

Red mulberry is readily worked with both hand and power tools. It splits easily when nailing but holds screws well. It glues readily.

USES

The wood is not commercially important due to its small size and scattered growth. It finds applications in boat building, fence posts and cooperage due to its durability. It is used to some extent for furniture but has not been widely used for woodenware. Because of its color and texture, however, it should be quite acceptable for use in the home workshop for novelties.

SUPPLIES

The tree has a wide distribution but is not found in thick stands, therefore, it is seldom available from lumber dealers, but can be obtained by wood collectors without great difficulty. [Ed. note: In northern Virginia, red mulberry is considered a pest tree because of the mess caused by dropping fruit and the birds that are attracted to it. Much of the wood is disposed of as firewood and large trees can be had for the asking. It is an outstanding wood for furniture when properly cut and dried.]

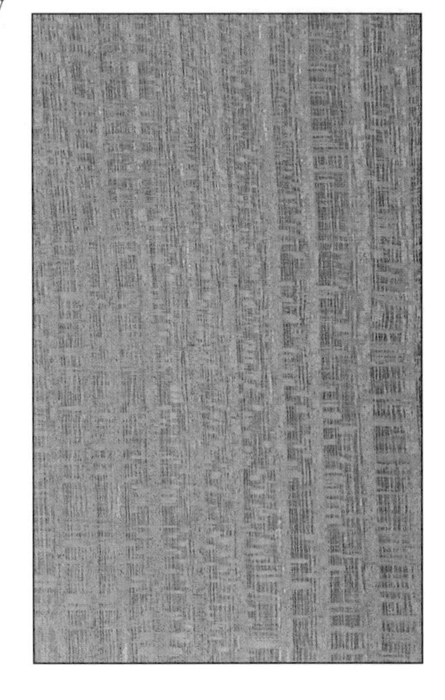

Myroxylon balsamum

balsamo

by Alan B. Curtis

SCIENTIFIC NAME

Myroxylon balsamum including *Myroxylon balsamum* var. *pereirae*. Derivation: The genus name *myron* is Greek meaning a sweet smelling oil and *xylon* meaning wood. The specific epithet is Latin meaning yielding balsam.

FAMILY

Fabaceae or Leguminosae, the legume family; (*Papilionaceae*) the pea or pulse group.

OTHER NAMES

balsam of Peru, incienso, estoraque (Peru), oleo vermelho (trade name in Brazil), Santos mahogany.

DISTRIBUTION

Southern Mexico through Central America (the Pacific side) and in South America from Colombia, Peru, Brazil, south to Argentina.

THE TREE

Balsamo trees have straight trunks and in Central America usually grow 50 to 65 feet in height occasionally reaching 100 feet. In Peru, specimens have been recorded with heights up to 180 feet and trunks clear of limbs up to three-fourths of the height. Diameters average 14 to 36 inches but are rarely larger. The leaves are compound with 5 to 11 leaflets, and the flowers are whitish with many clustered together in long racemes. It grows in humid low lands and at elevations up to 1,800 feet.

THE TIMBER

The sapwood and heartwood are sharply demarcated. The heartwood is reddish-brown becoming deep red to purplish-red upon exposure. The wood varies from uniform to striped. The luster is medium to high, and the texture is fine to medium. The grain is typically interlocked and ripple marks are present. The wood has a pleasant spicy scent when cut. Average reported specific gravity is 0.74 to 0.81 (ovendry weight/green volume), equivalent to an air-dried weight of 59 to 65 pcf.

SEASONING

Average reported shrinkage values (green to ovendry) are 3.8% radial, 6.2% tangential and 10.0% volumetric. My experience with this wood (from a tree grown in Florida) is that it is very stable and easy to dry without loss. Wood from Peru is reported to check when drying.

DURABILITY

The heartwood is very durable and highly resistant to attack by fungi. The sapwood is susceptible to insect attack.

WORKABILITY

Balsamo is moderately difficult to work. There is more than the usual dulling of cutters, although the wood does not contain silica. It turns nicely and finishes very smoothly. The wood has a high natural luster. It does not stain well. When finished it may resemble West Indies mahogany.

USES

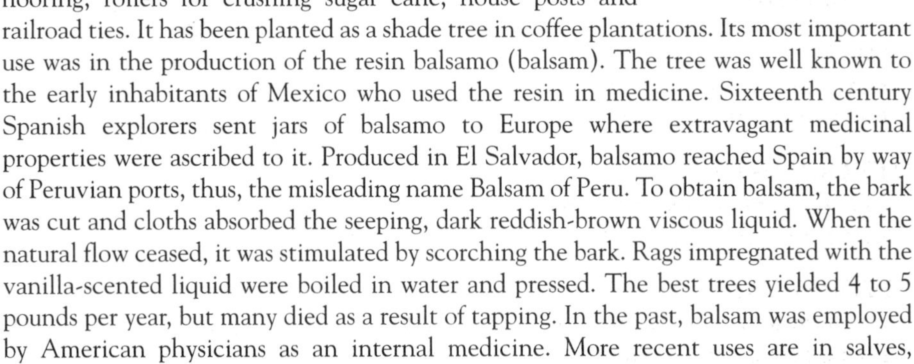

Uses include fine furniture, cabinetry, turnery, interior trim, flooring, rollers for crushing sugar cane, house posts and railroad ties. It has been planted as a shade tree in coffee plantations. Its most important use was in the production of the resin balsamo (balsam). The tree was well known to the early inhabitants of Mexico who used the resin in medicine. Sixteenth century Spanish explorers sent jars of balsamo to Europe where extravagant medicinal properties were ascribed to it. Produced in El Salvador, balsamo reached Spain by way of Peruvian ports, thus, the misleading name Balsam of Peru. To obtain balsam, the bark was cut and cloths absorbed the seeping, dark reddish-brown viscous liquid. When the natural flow ceased, it was stimulated by scorching the bark. Rags impregnated with the vanilla-scented liquid were boiled in water and pressed. The best trees yielded 4 to 5 pounds per year, but many died as a result of tapping. In the past, balsam was employed by American physicians as an internal medicine. More recent uses are in salves, antiseptics and cough syrup flavoring. In Europe it has been used as a fixative in perfume manufacture. The tree has been planted in West Africa, India and Sri Lanka (Ceylon) for resin production.

SUPPLIES

Rarely seen in world markets, supplies have been depleted over much of its natural range. This is regrettable, as the wood is quite attractive. The tree could be grown in plantations, and its wood would command a high price from woodworkers. The tree has been planted and successfully grown in botanical gardens in south Florida and in Hawaii.

Nauclea diderrichii

opepe

by Max Kline

SCIENTIFIC NAME

Nauclea diderrichii. Derivation: The genus name *naus* is Greek meaning ship and *kleio* meaning to close, which refers to the valves of the fruit. The specific epithet is in honor of M. R. Diderrich (1867 to 1925), a Belgian explorer and mining engineer who first collected the plant.

FAMILY

Rubiaceae, the madder family.

OTHER NAMES

kusia, badi, bilinga, akondoc, kilingi, lusia.

DISTRIBUTION

Widely distributed from Sierra Leone to the Congo region and eastward to Uganda. Opepe is sometimes found in pure stands.

THE TREE

Opepe grows to a height of 160 feet with a diameter at breast height of 6 feet. There are no buttresses, and the bole is long and clean up to 80 or 100 feet. The bark is light brown to grayish-yellow with shallow, longitudinal fissures. The opposite, broadly elliptic leaves are 6 inches long and 4 inches wide. The grayish-brown fruit, which contains brown, ovoid seeds, has white flesh and is 1 to 1.5 inches in diameter.

THE TIMBER

A striking feature of opepe is the uniform yellow or orange-brown color of its heartwood. This color together with the ribbon grain produces an attractive appearance. The sapwood is frequently uniformly pink. The taste is slightly bitter. The odor of freshly cut wood is slightly fragrant. The luster is medium to high, and the texture is rather coarse. Average reported specific gravity is 0.63 (ovendry weight/green volume), equivalent to an air-dried weight of 49 pcf. An alkaloid in the wood may be toxic to woodworkers. The strength properties of opepe compare favorably with most other imported species showing about 15% to 20% more strength than teak (*Tectona grandis* WDS 261).

SEASONING

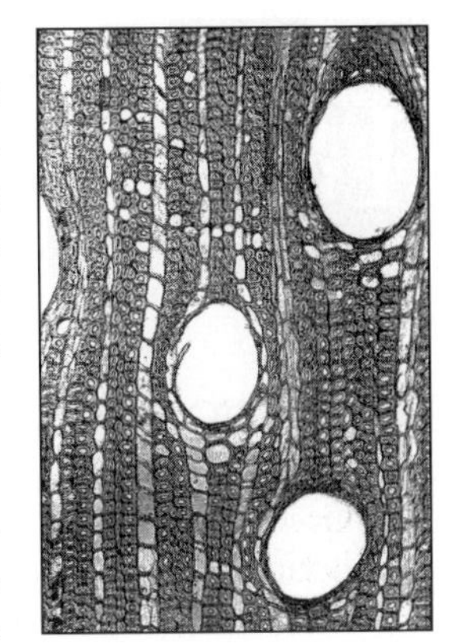

Flatsawn lumber may develop considerable degrade in the form of minute surface cracks. End coating is suggested. Quartersawn lumber dries more rapidly with little checking or warping. Average reported shrinkage values (green to ovendry) are 4.5% radial, 8.4% tangential and 12.6% volumetric. When dry, the stability factors are good and the wood holds shape very well.

DURABILITY

It is reported that opepe timber is resistant to African termites. The heartwood is rated as very durable, while the sapwood is liable to attack by powder post beetles. The heartwood is also somewhat resistant to preservative treatment.

WORKABILITY

Opepe works with moderate ease in most hand and machine operations. In planning, the cutting angle of the knives should be lowered to avoid a tendency for the grain to pick up. A slow feed rate during sawing is recommended. The timber has a tendency to split when nailing but takes screws well. It stains readily and with a moderate amount of filling finishes effectively. Glue adheres well but the wood is unsuited for steam bending.

USES

Since the wood is very durable, it is often used for piling and dock work and for decking on wharves and jetties. Because of its good wearing qualities, it is used for flooring and the appearance is pleasing and decorative. Limited uses are found for furniture and cabinet parts, interior work, carving, turnery, veneer and plywood.

SUPPLIES

Opepe is plentiful and only moderately expensive. The sizes are usually good, and the quality is often excellent.

Nesogordonia papaverifera

danta

by Max Kline

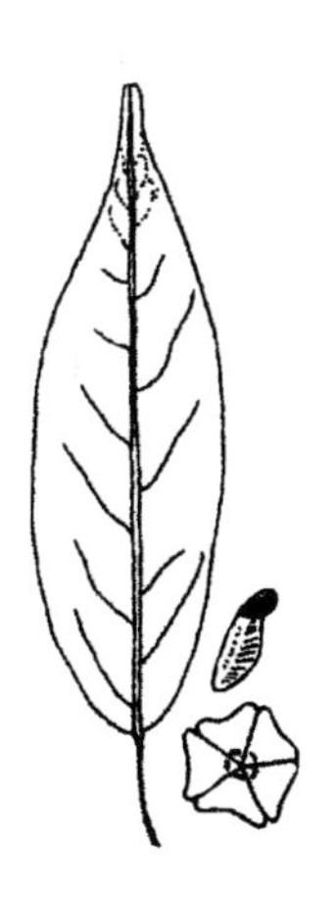

SCIENTIFIC NAME

Nesogordonia papaverifera. Derivation: The genus name is from the Greek *nesos* meaning island and is in honor of James Gordon (d. 1781), a correspondent of Linnaeus and a nurseryman in London. The specific epithet is a Latin name for wild poppy.

FAMILY

Sterculiaceae, the stericulia or cacao family.

OTHER NAMES

kotibe, otutu, ovoue, dante, oro, tsanya, eprou.

DISTRIBUTION

A tree of mixed deciduous forests, danta is found from Sierra Leone to Cameroon and northern Gabon.

THE TREE

Danta may reach a height of 90 to 120 feet. Trunk diameters above the short buttresses are 2.5 to 3.5 feet. The usually straight bole is clear for 40 to 80 feet. The tree has brown bark. The elliptic leaves average 3 inches long and 1 inch wide.

THE TIMBER

This timber is reddish-brown with a lustrous surface similar to dark mahogany. It has interlocked grain, and a fine texture with a greasy feel. Average reported specific gravity is 0.65 (ovendry weight/green volume), equivalent to an air-dried weight of 51 pcf. The presence of small sound pin knots and dark streaks of scar tissue sometimes mar the appearance. The luster is medium. The wood has a musty odor when worked. The taste is not distinct. The interlocked grain produces a fine ribbon figure on quartersawn surfaces. The heartwood is sharply defined from the lighter colored sapwood that is 2 to 3 inches wide. Danta has high bending strength, medium resistance to shock loads and a low stiffness factor. Its good strength properties make is suitable for structural purposes.

SEASONING

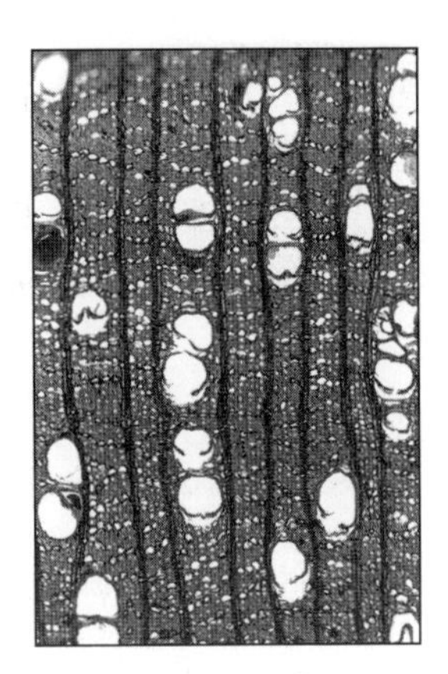

Danta dries slowly but well, with little degrade. Overly rapid drying should be avoided as there is a tendency for the wood to warp and caseharden with some ribbing of the surface and knot splitting during kiln drying. Average reported shrinkage values (green to ovendry) are 5.4% radial, 8.2% tangential and 12.4% volumetric. In service it is relatively dimensionally stable.

DURABILITY

The heartwood is rated as durable and fairly resistant to fungal and termite attack. Powder post beetles, however, can attack the sapwood. The timber is only moderately durable to marine borers. The heartwood is very resistant to preservative treatment.

WORKABILITY

The timber is easy to turn and carve. In machine operations, the interlocked grain affects the workability causing some tearing of grain when planing. To avoid this, a cutting angle of 15° is suggested. The blunting effect on cutters is sometimes appreciable. Danta can be peeled to produce a low-grade veneer. Because there is a tendency for the wood to split, pre-boring is advised when using nails and screw. Glues adhere satisfactorily. Drilling and mortising should be done at slightly lower than normal speeds. The response to all of the usual types of finishing is good, and in particular the timber will stain and polish well.

USES

Danta is used for general construction, flooring, joinery, turnery, boat building, tool handles, gunstocks, plywood and furniture. It is an etching timber for graphic arts. Its high resistance to abrasion makes it particularly satisfactory for heavy pedestrian traffic such as for ballroom and gymnasium floors. When making miniatures, it should be considered as a duplicate for bigleaf mahogany (*Swietenia macrophylla* WDS 254).

SUPPLIES

Danta has been exported in relatively small quantities from West Africa. Supplies are available for small orders such as those amounts needed for miniature or model work. The price is in the moderate range.

Nothofagus dombeyi

coigue

by Jim Flynn

SCIENTIFIC NAME

Nothofagus dombeyi. Derivation: The genus name is Latin meaning not a true beech. The specific epithet is in honor of Joseph Dombey (1742 to 1796), an 18th century French botanist who travelled in Chile and Peru.

FAMILY

Nothofagaceae, the southern beech family.

OTHER NAMES

anis, coihue, coyan, hualo, rauli, roble ruili, lengue, nire, roble. This species may be marketed as Chilean beech.

DISTRIBUTION

The trees in the genus *Nothofagus*, meaning false beech, are known as Southern Hemisphere beeches. They are scattered about primarily in Australia, New Zealand and South America. Coigue is native to Chile as is a similar species, *N. procera*, which bears the common name ratili. Coigue may be found as far south as 38° south latitude and from there northward along the Chilean coast and into the river valleys and to the high cordilleras in the northern Llanquihue Lake country. It grows mainly on poor soil. A somewhat similar distribution of this species can be found in central and southern Argentina. The genus may have had its origin in prehistoric Antarctica.

THE TREE

Frequently a big tree, coigue often grows 130 feet in height with a diameter of 2 to 3 feet. Sometimes, tree diameters up to 6 to 8 feet are found depending upon the growing site. When found in pure stands, the boles may be as high as 60 feet without branches. The species is an evergreen. The seeds, or beech-nuts, are similar to those found on their northern relatives in the genus *Fagus*. Coigue is a fast grower and is often found in pure stands at elevations of 2,500 to 4,000 feet.

THE TIMBER

Coigue wood is similar in texture and feel to the northern beeches, especially *Fagus grandifolia* (American beech WDS 119) and *F. sylvatica* (European beech). One key element, the ray flecks, which are so prominent in the northern beeches, is not noticeable in the *Nothofagus* genus. Coigue heartwood is very attractive pinkish-brown, and the luster is high. The grayish-white sapwood is wide and not clearly demarcated from the heartwood. The growth rings are well defined resulting in pleasing grain patterns on finished lumber. The wood has no distinctive odor. It is not as strong as the northern beeches. Average specific gravity is in the range of 0.45 to 0.53 (ovendry weight/green volume), equivalent to an air-dried weight of 34 to 40 pcf.

SEASONING

Coigue is generally very difficult to season and has a pronounced tendency to distort and collapse. Dry kiln schedules for different thicknesses should be adjusted accordingly. Average reported shrinkage values (green to ovendry) are 4.1% radial and 6.6% tangential.

DURABILITY

The heartwood of coigue is rated as variable with some references placing it in the medium category, that is, having a service life between 8 and 15 years when in contact with the ground. It is moderately resistant to impregnation with preservatives.

WORKABILITY

Coigue is fairly easy to work and not too severe on tooling. It works more easily than hard maple or cherry. I had the pleasure of experimenting with a piece of coigue from Chile and found it to cut easily. I was especially pleased to watch the nice curly shavings coming from my hand plane. It glues well, sands easily and when polished it was difficult to distinguish from cherry. The wood is reported to possess good bending qualities and nails and screws can be used satisfactorily. Preboring nail and screws holes in thin pieces is advised.

USES

The species has a wide range of uses including light structural timbers, flooring, ship and boat building, furniture, boxes and crates and inside trim. Coigue appears to qualify as a general all-purpose hardwood.

SUPPLIES

Coigue appears to be reaching world markets and is available at prices comparable to competitive hardwoods such as maple and birch.

Nyssa sylvatica

blackgum

by Max Kline

SCIENTIFIC NAME

Nyssa sylvatica. Derivation: The genus name means water nymph. The specific epithet is Latin for "of the woods".

FAMILY

Cornaceae, the dogwood family; also Corylaceae.

OTHER NAMES

black tupelo, sour gum, tupelo, stinkwood, tupelo gum, yellow gum tree, pepperidge, wild pear tree.

DISTRIBUTION

Extends through all of the eastern states from Maine to Michigan, Illinois and Missouri, and south to Texas.

THE TREE

Blackgum is usually symmetrical, growing to a height of 60 to 120 feet with a trunk diameter from 2 to 4 feet. Its crown is well rounded, and the tree has a single, long main stem. The bark is quite thick and deeply ridged. It is an attractive ornamental with gorgeous scarlet or purple autumnal foliage. Its plum-like fruit is attractive to birds. Blackgum prefers very moist or swampy soils.

THE TIMBER

The heartwood is pale brownish-gray or yellow to light brown with a very wide, lighter colored sapwood. The wood is generally uniform in texture, has close interlocked grain and is without luster. Average reported specific gravity is 0.46 (ovendry weight/green volume), equivalent to an air-dried weight of 35 pcf.

SEASONING

Due to its above average amount of interlocked grain, the wood will twist and warp very badly when seasoning unless carefully handled. Average reported shrinkage values (green to ovendry) are 5.1% radial, 8.7% tangential and 14.4% volumetric.

DURABILITY

Mature trees are often subject to heart rot. Blackgum has no natural ability to resist decay, but can be successfully treated with preservatives.

WORKABILITY

Blackgum is difficult to split and nail. It must be worked with care, but glues well and can be finished very satisfactorily. Finished surfaces are smooth and bright.

USES

In the past, blackgum was used for ox yokes and chopping bowls because of its toughness. Today it is used for boxes and crates, factory flooring, rollers, crossties, paper pulp, baskets, cigar boxes, caskets, sashes, doors, blocks, gunstocks and furniture. The more select logs are cut into veneer for berry boxes and baskets.

SUPPLIES

Blackgum is readily available, especially in the eastern United States. The price is inexpensive.

Ochroma pyramidale

balsa

by Jon Arno

SCIENTIFIC NAME

Ochroma pyramidale. Derivation: The genus name means ochre-colored. The specific epithet is Latin for pyramid-shaped. A synonym is *O. lagopus*.

FAMILY

Bombacaceae, the bombax family.

OTHER NAMES

corcho, enea, lana, pau de balsa, tami.

DISTRIBUTION

Balsa is widely distributed in tropical America from southern Mexico to southern Brazil and from the West Indies to Peru. Ecuador accounts for a large proportion of the balsa in international commerce.

THE TREE

This species can reach heights of 80 feet with diameters of up to 4 feet. Balsa is an extremely fast grower, and on ideal sites can reach heights of over 60 feet with diameters in excess of 2 feet in only 5 years. The fruit is a capsule that contains downy seeds. Balsa trees are very fragile and even slight injuries will cause the tree to produce inferior, hard fibers.

THE TIMBER

Balsa is the lightest of all commercially important timbers. Because it is used for purposes where low density is a desirable characteristic, balsa is also one of the few species where the sapwood, rather than the heartwood, represents the primary product. Average reported specific gravity ranges from 0.10 to 0.17 (ovendry weight/green volume), equivalent to an air-dried weight of 7 to 12 pcf. The best stock comes from fast-grown trees. Typical balsa wood is straight grained, whitish gray or sometimes pinkish-white in color and extremely soft and spongy. When measured in absolute terms, balsa is exceptionally weak. In relation to its weight, however, it is very strong. On a strength-to-strength basis, balsa is about twice as strong as spruce in resisting compression parallel to the grain. This is remarkable considering that, in the days of wood aircraft, spruce was noted for its superior strength to weight ratio.

SEASONING

Balsa seasons very quickly. Because of it susceptibility to blue stain, checking and warping, it should be kiln dried. Average reported shrinkage values (green to ovendry) are 3.0% radial, 7.6% tangential and 10.8% volumetric.

DURABILITY

Durability is very poor. Balsa is susceptible to termites and borers and has low resistance to blue stain and decay.

WORKABILITY

Almost every woodworker, at one time or another, has worked with balsa. It is perhaps the easiest of all woods to cut, shape and sand. Being diffuse porous and virtually void of any density differential, its texture is exceptionally uniform. A block of balsa wood can be carved as effortlessly, or perhaps even more effortlessly, as soap. But, because balsa crushes so easily, it is important to keep cutting tools sharp.

USES

In addition to model making, balsa has many commercial and industrial uses as insulation for heat, sound and vibration. It is also used for flotation, cushioning and core stock. It has been used for centuries as the preferred wood for making rafts.

SUPPLIES

Because of its extensive natural range, rapid growth and the fact that it is now plantation grown, balsa is plentiful and will probably remain so. On a board footage basis, it is fairly expensive, but it is normally used on projects requiring such small quantities that price is not a particularly significant factor.

Ocotea bullata

stinkwood

by Max Kline

SCIENTIFIC NAME

Ocotea bullata. Derivation: The genus name was coined by Jean Aublet (1723 to 1778) and presumably Latinized from a French Guinea vernacular name. The specific epithet means blistered, which is in reference to the foliage.

FAMILY

Lauraceae, the laurel family.

OTHER NAMES

stinkhout, Cape laurel, umnukane (African).

DISTRIBUTION

South Africa, predominantly in forests of Cape Province.

THE TREE

Stinkwood grows to a height of 60 to 80 feet with a clean, straight bole 3 to 5 feet in diameter.

THE TIMBER

The color of the wood ranges from bright yellow to yellowish-brown or green to chocolate-brown or almost black. The black wood is often mottled with yellow and usually has darker streaks. The wide sapwood is not always clearly defined from the heartwood but instead gradually merges into it. The name stinkwood is from the extremely pungent and unpleasant odor of freshly cut timber, which fades on exposure or drying. It has a high natural sheen or golden luster. The average specific gravity (ovendry weight/green volume) ranges from 0.57 to 0.66, equivalent to an air-dried weight of 44 to 52 pcf. Some stinkwood can be considered moderately heavy but it depends on the color of the wood. The light-colored wood has a lower density, while the very dark wood has the maximum weight. The texture is moderately close and even, and the grain is typically interlocked often producing a ribbon figure. Planed surfaces have a silky appearance. The strength of stinkwood compares favorably with U.S. domestic hardwoods of like specific gravity.

SEASONING

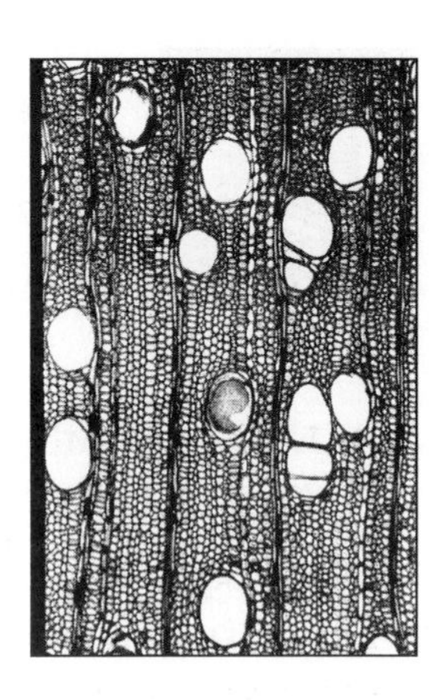

Stinkwood air dries very slowly and does not warp unduly, however, it has a tendency to check or split. It may be kiln dried with satisfactory results provided initial temperatures are kept low. Shrinkage can be in the order of 8% tangentially. The light-colored wood dries more rapidly with less tendency to degrade. After seasoning, dimensional stability is good.

DURABILITY

Stinkwood shows a high degree of resistance to fungal infection, and beetle damage will occur only in the sapwood. For its intended uses, preservative treatments are not commonly used, as the resistance to impregnation would add little to its natural durability.

WORKABILITY

This timber is not easy to work as it severely dulls hand and machine tools. Best surfaces are obtained by the use of a scraper or sandpaper. During sawing, some operators find the sawdust causes irritation of their nasal passages. Occasionally splitting will occur during sawing. In quartersawn wood there is a tendency to chip out or flake and lowering the cutting angle during planing is advisable. Worked edges remain sharp, which means the species is reasonably good for ornamental turnery. Nail and screw holding properties are good, although thin stock requires pre-boring for nails. Glues adhere well. Finished stinkwood will darken on exposure to air and light.

USES

This species is the most highly prized furniture wood in South Africa and commands a higher price than any wood for this purpose. It is also used for cabinetry, panels, carving and turnery.

SUPPLIES

No stinkwood is exported from Africa. In fact, a friend there reports that a ban on the cutting of this species has been imposed whereby felling of the trees may not be undertaken for the next 200 years. Wood collectors with a sample should consider themselves fortunate.

Ocotea porosa

imbuya

by Max Kline

SCIENTIFIC NAME

Ocotea porosa. Derivation: The genus name was coined by Jean Aublet (1723 to 1778) and presumably Latinized from a French Guinea vernacular name. The specific epithet means porous, its application unclear. A synonym is *Phoebe porosa.*

FAMILY

Lauraceae, the laurel family.

OTHER NAMES

embuia, imbuia, Brazilian walnut, canella imbuia.

DISTRIBUTION

Southern Brazil.

THE TREE

Imbuya grows at altitudes of 2,500 to 4,000 feet and reaches a maximum height of 130 feet with a trunk diameter of about 6 feet. It is an evergreen tree but sheds most of its old leaves in August or September. The fruits mature in January and after falling to the ground provide food for swine.

THE TIMBER

Frequently figured and variegated in color, the heartwood is yellowish-brown to chocolate-brown. When green the wood has a spicy resinous scent, which fades as the wood dries. It is rather fine textured and moderately hard. The straight to wavy grain has a ribbon figure. Average reported specific gravity is 0.53 (ovendry weight/green volume), equivalent to an air-dried weight of 40 pcf.

SEASONING

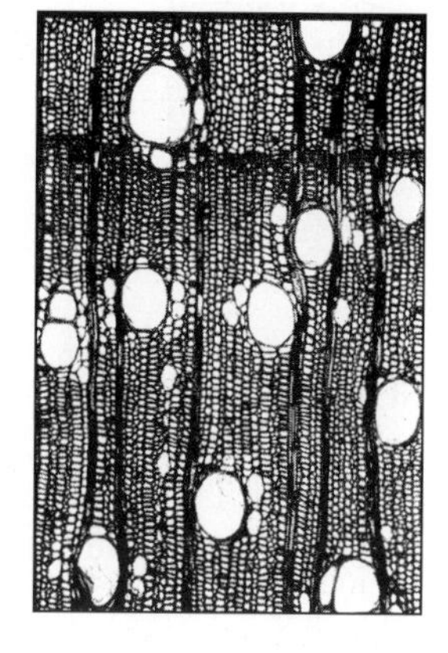

Although fairly easy to season, imbuya should be carefully stacked to avoid warping. When dry, the wood is dimensionally stable. Average reported shrinkage values (green to ovendry) are 2.7% radial, 6.0% tangential and 9.0% volumetric.

DURABILITY

Imbuya is resistant to decay and insect attack.

WORKABILITY

Imbuya is easy to work and takes a high polish. The wood can be selected to match any kind of walnut although it will probably be harder than wood in the *Juglans* species. It turns well but care should be taken when planing to avoid chipping. When sawn, the fine dust is particularly irritating to some people and may cause sneezing.

USES

In southern Brazil, particularly in Sao Paulo, imbuya is one of the most important woods for high-grade furniture, floorings and joinery. Figured material is sliced into decorative veneer and is a substitute for black walnut (*Juglans nigra* WDS 150).

SUPPLIES

Usually the local demand in Brazil just about equals the supply so not much is exported. Occasionally there are quantities available in the United States and England, especially in veneer form which sells for a medium price.

Dimension timber is in the higher price range.

Olea europaea

olive

by Jon Arno

SCIENTIFIC NAME

Olea europaea. Derivation: The genus name is the classical Latin name for the olive. The specific epithet means the plant is "of Europe".

FAMILY

Oleaceae, the olive family.

OTHER NAMES

olivo, olivier, olbaum.

DISTRIBUTION

This species is native to the Mediterranean region including southern Europe, the Middle East and North Africa. Olive requires a period of cool winter temperatures to induce flowering, but it also demands the bright sun and low humidity characteristic of a Mediterranean-type climate. It has been widely planted where these conditions exist such as in southern California.

THE TREE

Olive is a short, gnarly branched tree that seldom exceeds 25 feet in height. It is extremely long lived and 1,500 years or more is not uncommon. The trunk can reach diameters of 2 to 3 feet. The leaves are a soft, grayish-green color, and the bark is dark brown and deeply furrowed on mature trees. The fruit, of course, is the familiar olive, which requires some processing before it is actually edible.

THE TIMBER

This species produces virtually no long straight lumber because of its size and growth tendencies. Its close relative, east African olive (*O. hochstetteri*), however, can reach heights of 80 to 100 feet and provides some timber for international commerce. Olive wood is fine textured, close grained and relatively hard with an average specific gravity of about 0.70 (ovendry weight/green volume), equivalent to an air-dried weight of 55 pcf. The heartwood is light yellowish-tan exhibiting occasional dark brown streaks. The usually very narrow sapwood is creamy yellow and clearly demarcated from the heartwood. The annual rings are noticeable, but the tangential figure is subdued and somewhat less important to the wood's attractive appearance than is the contrast between the dark brown streaks and yellowish heartwood. This is a strong wood that has exceptionally good resistance to abrasion.

SEASONING

Olive dries slowly and has a tendency to check. Reaction wood is common in stock cut from branches, and this material is especially prone to warping. Average reported volumetric shrinkage is 20% (green to ovendry).

DURABILITY

Durability is in the moderate range. Olive is susceptible to termites, but somewhat resistant to fungi.

WORKABILITY

Olive wood is hard but otherwise very pleasant to work. It is difficult to saw across the grain, but its fine, even texture is predictable and yields smoothly to a sharp cutting edge. The wood polishes very well and has a natural waxy feel. Carvers and turners should be certain the wood is thoroughly seasoned to avoid checks or warping in the finished piece.

USES

Olive has been used since ancient times for small woodenware objects, spoons, bowls, boxes, carvings and turnings. Both plain stock and burl are used for inlays in furniture and small decorative items. Olive makes excellent, but very expensive, flooring.

SUPPLIES

While olive is relatively plentiful in areas where the trees are cultivated for oil and olives, it is not readily available through the timber trade. Olive burl veneer, lathe billets and carving flitches are occasionally marketed, but they are expensive. This is a species that is perhaps best acquired by foraging or through trading.

Olneya tesota

desert-ironwood

by Max Kline

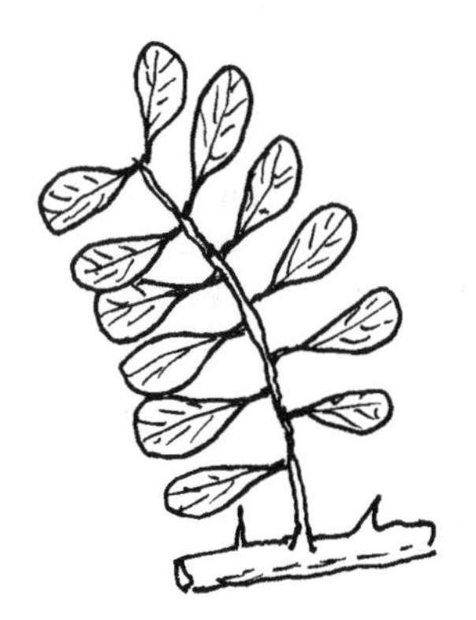

SCIENTIFIC NAME

Olneya tesota. Derivation: The genus name is in honor of Stephen Thayer Olney (1812 to 1878), a businessman and botanist of Rhode Island. The specific epithet is a Native American name.

FAMILY

Fabaceae or Leguminosae, the legume family; (*Papilionaceae*) the pea or pulse group.

OTHER NAMES

tesota, ironwood, Sonora ironwood, palo de hierro, Arizona-ironwood.

DISTRIBUTION

In the low depressions of the Colorado River, south of the Mojave Mountains, south western Arizona, westward to southern California and southward into lower California and Mexico. The largest trees are found in Sonora, Mexico.

THE TREE

Desert-ironwood is a typical desert tree, which reaches a height of 18 to 25 feet with a short, thick trunk that is up to 18 inches in diameter. The branches have hard, sharp spines. The thin, scaly, flaky bark is deep reddish-brown or grayish-green. The tree bears a small purplish flower that grows in clusters and resembles pea blossoms.

THE TIMBER

Desert-ironwood is the second heaviest wood found in North America with an average specific gravity of 0.88 (ovendry weight/green volume), equivalent to an air-dried weight of 72 pcf. The heartwood is deep chocolate-brown, almost black, mottled with yellowish-red. The lemon-colored sapwood is narrow. Desert-ironwood has a distinctive odor, and the dust is unpleasant in the nostrils. It takes a fine polish, and the luster is high. The wood is brittle but in hardness it cannot be excelled. Some species present a very beautiful colored figure.

SEASONING

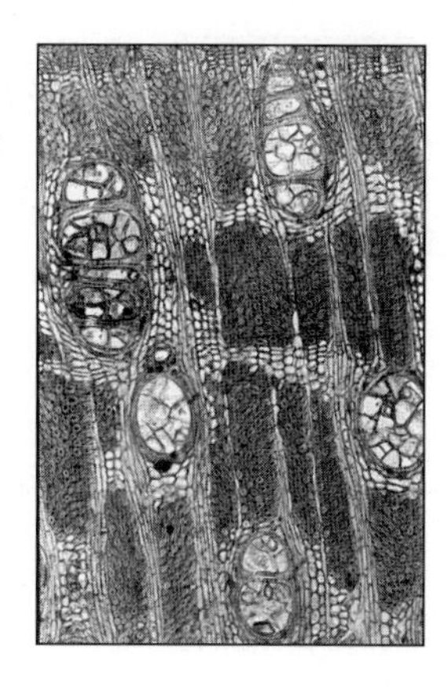

Desert-ironwood is exceedingly difficult to season without the development of numerous cracks and checks, some of which cannot be seen on exterior surfaces. Shrinkage values are not available.

DURABILITY

The wood is very susceptible to worm infestation.

WORKABILITY

It is almost impossible to work desert-ironwood with hand tools due to its extreme hardness. With power saws or lathes, it can be slowly machined or cut if the tools are sharp. It finishes well.

USES

Due to the small supply of desert-ironwood available, its uses are very limited. Native Americans made arrowheads from it. Today it is used for novelties, canes and tool handles. It has been used extensively for firewood because it burns slowly and makes good coals.

SUPPLIES

Desert-ironwood is not a commercial wood species and availability is thus dependent on the wood collector. As the wood is generally very defective and worm infested, yielding little clear wood except in the older stumps, only small supplies are found among collectors. Sound pieces are never available in large dimensions, limiting its usefulness for craftwork to only small objects.

Ostrya virginiana

eastern hophornbeam

by Jon Arno

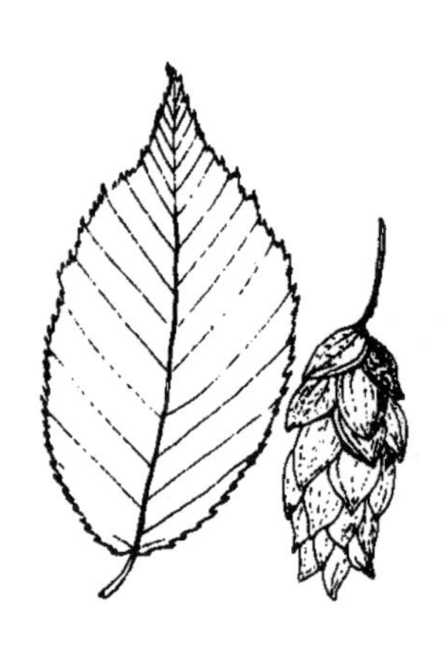

SCIENTIFIC NAME

Ostrya virginiana. Derivation: The genus name *ostrua* has been Latinized from the Greek, a tree with very hard wood and very likely the related European hornbeam. The specific epithet is Latin and means the plant is "of Virginia".

FAMILY

Betulaceae, the birch family.

OTHER NAMES

ironwood, hophornbeam, hornbeam.

DISTRIBUTION

Eastern hophornbeam prefers dry upland soil and is therefore more common in hilly country, but it has a very expansive range extending from the Great Plains to the Atlantic Ocean and from central Florida to Ontario. It is also found in isolated pockets down into Mexico.

THE TREE

At a glance, the leaves and general appearance of eastern hophornbeam are similar to American elm (*Ulmus americana* WDS 274). Eastern hophornbeam, however, is a much smaller tree, seldom exceeding 40 to 50 feet in height and 1 to 1.5 feet in diameter. The bark is reddish gray-brown and somewhat shaggy or stringy. The seeds form in clusters, each one packaged in its own little parchment-like bag, which makes them look very much like the hops used in brewing.

THE TIMBER

Eastern hophornbeam is a very hard and heavy wood. With an average reported specific gravity of 0.63 (ovendry weight/green volume), equivalent to an air-dried weight of 49 pcf, it is one of the densest woods native to temperate North America. The heartwood is light yellowish-tan, sometimes tinged with red. The sapwood is usually wide and whitish in color. The wood is diffuse porous with very fine pores dispersed throughout the growth ring and often grouped in short radial chains. The rays are very plentiful, but fine and indistinct to the naked eye. The annual rings form wavy bands, but they are not conspicuous enough to give the wood a particularly showy figure. This species is very strong and has exceptional resistance to abrasion.

SEASONING

Eastern hophornbeam is moderately difficult to season. The wood is not prone to severe checking, but its extreme density makes it very slow to dry. Average reported shrinkage values (green to ovendry) are 8.2% radial, 9.6% tangential and 18.6% volumetric.

DURABILITY

This timber is best used in dry locations to avoid fungi and insect attack. Effects of treatment with preservatives are not known.

WORKABILITY

Eastern hophornbeam's extreme density gives it characteristics comparable to stone. The fine but very plentiful rays create a fabric-like texture, which makes it almost equally difficult to cut or plane in any direction. Expect copious clouds of bluish-gray smoke when using power tools on this species even when it is slowly fed into wide kerf, carbide-tipped blades. Using nails or screws is impossible without first drilling perfectly matched pilot holes. Eastern hophornbeam might best be described as a poor man's lignumvitae (*Guaiacum officinale* WDS 132).

USES

This species has its aficionados among carvers and turners, but its primary modern use is as firewood (at which it excels.) Historically, it was much used for splitting wedges, mallet heads, wagon axles and tool handles.

SUPPLIES

Eastern hophornbeam is plentiful within its native growth range, but it is not a commonly available commercial timber.

Oxandra lanceolata

lancewood

by Jon Arno

SCIENTIFIC NAME

Oxandra lanceolata. Derivation: The genus name is Latin meaning sharp flower. The specific epithet is Latin for lanceolate, referring to the shape of the leaves.

FAMILY

Annonaceae, the annonia or custard-apple family.

OTHER NAMES

yaya, haya prieta, bois de lance.

DISTRIBUTION

Lancewood is native to the islands of the Caribbean, most commonly Cuba, Hispaniola, Puerto Rico and Jamaica, with a few separate but similar species found in the Amazon basin.

THE TREE

Lancewood is a slender forest tree with leathery leaves. It seldom exceeds 50 feet in height and when mature has a diameter of 18 inches.

THE TIMBER

With an average reported specific gravity of 0.81 (ovendry weight/green volume), equivalent to an air-dried weight of 65 pcf, lancewood is a very hard, fine textured and extremely elastic wood. Although the heartwood in mature trees tends to turn dark grayish-brown, it is the pale yellowish sapwood which is most commonly encountered, since it is of the most commercial importance. Lancewood is moderately lustrous, usually straight grained and without odor or taste when dry. This wood is very strong and has good resistance to abrasion.

SEASONING

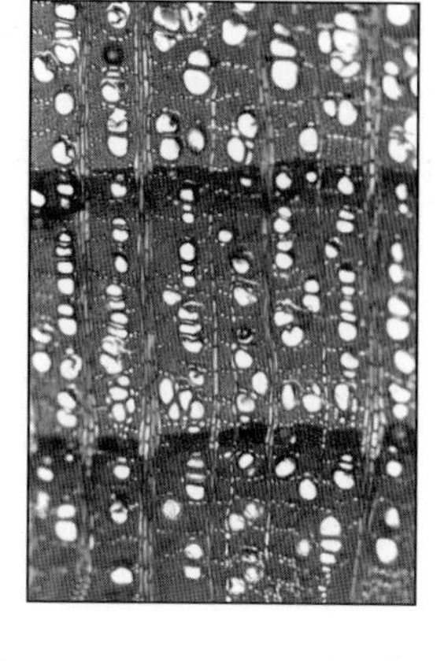

This species is moderately difficult to dry because it has a rather high average volumetric shrinkage of 15.4% (green to ovendry) and will check badly if the end grain is not thoroughly sealed. Its radial shrinkage of 6.2% and tangential shrinkage of 9.6%, while high, are less divergent than in most woods and warping is not normally a problem.

DURABILITY

The wood is considered nondurable.

WORKABILITY

Lancewood polishes very well and is an excellent wood for turning. It is very hard to work with hand tools and, although finer in texture, it cuts and shapes somewhat like hickory. Its use as a cabinetwood, however, is drastically limited, since it is seldom available in stock that is much wider than 5 or 6 inches.

USES

Uses for lancewood include turnery, billiard cues, tool handles and textile machinery. Comparable to flowering dogwood (*Cornus florida* WDS 088) it is used for shuttles. Like hickory, lancewood was once important to the auto industry where it was used for spokes and body trim. It was also once a popular wood for fishing rods and archery bows, but it has been replaced in most of these applications by modern synthetics. Because of its similar appearance and uses, lancewood is sometimes confused with degame (*Calycophyllum candidissimum* WDS 056), which is also native to the Caribbean region, but these two woods are not closely related in that degame belongs to the madder family, Rubiaceae.

SUPPLIES

Considering the decline in the commercial use of this wood, supplies are adequate. But it is no longer easy to find through retail hardwood outlets in the United States, since much of the supply comes from Cuba. When available, the price is high.

Oxydendrum arboreum

sourwood

by Jim Flynn

SCIENTIFIC NAME

Oxydendrum arboreum. Derivation: The genus name is Greek for sour tree, referring to the acid taste of the leaves. The specific epithet means tree-like. Synonyms are *Andromeda arborea* and *Lyonia arborea.*

FAMILY

Ericaceae, the heath family.

OTHER NAMES

sorrel-tree, lilly-of-the-valley-tree, sour gum, arrowwood.

DISTRIBUTION

The native range of sourwood includes a wide swath of the eastern United States from southern Pennsylvania west to Indiana and then south to the western parts of Florida and eastern Louisiana. The tree was introduced in England in 1752 and can be found in many parts of Europe.

THE TREE

This beautiful tree is the only species in the genus *Oxydendrum.* It grows to 60 feet or more in height with a narrow, oblong crown. The main stem is usually straight attaining a diameter of 20 inches or more. The thick bark, which is deeply furrowed with broad scaly ridges, is easy to spot in the wintertime and is a helpful identification aid. The simple, alternate, deciduous leaves are 5 to 7 inches long and 1 to 3 inches wide. They have a striking orange or scarlet color in the autumn and are sour tasting. The creamy white flowers grow in long spike-like clusters. Sourwood honey is a much sought after ambrosia in the southern Appalachian Mountains. This tree never forms pure stands and is found scattered among oaks, pines, hickories and sweetgum. Below these impressive forest stalwarts grow the azalea, mountain laurel, cinnamon clethra, blueberry and elder. Many Appalachian Mountain songs refer to this tree including "Sourwood Mountain".

THE TIMBER

Looking at a piece of maple and sourwood, placed side by side, it is hard to tell them apart at a quick glance. The most noticeable feature is the soft and attractive pinkish-brown color of sourwood. The sapwood is very wide (up to 80 or more growth rings) with a yellowish-brown to light pinkish color. The heartwood is brown with tinges of red and pink becoming dull with exposure. Average reported specific gravity is 0.50, equivalent to an air-dried weight of 38 pcf. The wood is diffuse porous. The luster is high, and the grain lines on radial and tangential surfaces are subdued.

SEASONING

Sourwood has a marked tendency to warp, and the average reported shrinkage values (green to ovendry) are 6.3% radial, 8.9% tangential and 15.2% volumetric.

DURABILITY

Durability is rated as low, and sourwood is subject to insect, bacterial and fungal attack. The timber should not be used where exposed to the weather unless adequately protected by preservatives.

WORKABILITY

There are no unusual problems in working sourwood with either hand or machine tools. The usual prerequisites for sharp tools demanded by hardwoods are advised. Due to the diffuse-porous wood, sharp and crisp edges on machined wood are assured. It glues satisfactorily and sands and finishes very well.

USES

Because of the relative scarcity of the wood in the commercial marketplace, it finds many unpublicized uses where the wood is abundant. It makes excellent handles for agricultural tools, bearings for machinery and, at one time, was used for sled runners, a testimony of the wearability of the wood. Because the wood grows in mixed hardwood stands, it is suspected that much of it is cut and combined with other species and sent to pallet and crate manufacturing facilities. Firewood and pulp are other uses.

SUPPLIES

Because of the fragrant and beautiful late spring flowers, the trees are becoming a favorite ornamental. Again, this points out the prospect for getting useful and excellent supplies of wood in urban areas as a result of pruning and storm damage. In the native growth areas, supplies can always be found if one searches.

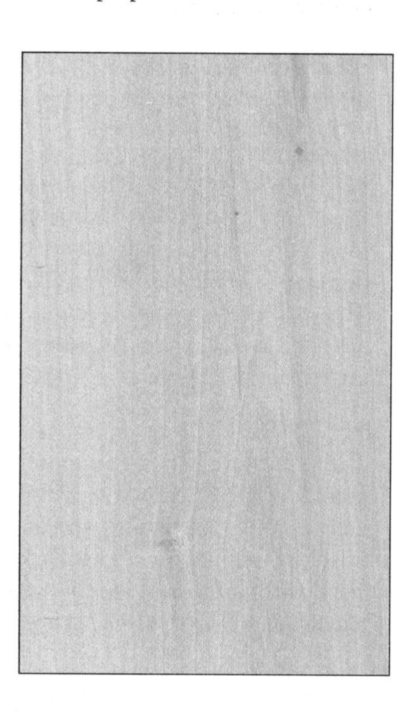

Parapiptadenia rigida

angico

by Jim Flynn

SCIENTIFIC NAME

Parapiptadenia rigida. Derivation: The genus name is Greek meaning "along side falling glands", referring to the anther. The specific epithet is Latin for rigid or stiff. A synonym is *Piptadenia rigida*.

FAMILY

Fabaceae or Leguminosae, the legume family; (*Mimosaceae*) the mimosa group.

OTHER NAMES

Angico appears to be the only trustworthy common name for *Parapipitadenia rigida*. There are over 80 species in this genus and many of the common names have crossed over to several of the species. There seems to be insupportable evidence that curupay is another common name, but this name is used in Argentina for *Anadenanthera macrocarpa* which was formerly *Piptadeniia macrocarpa*.

DISTRIBUTION

South America, mostly in Brazil. Definitive range boundaries for this particular species of *Parapiptadenia* are not readily available.

THE TREE

Angico is a large tree that can reach a height of up to 100 feet with a trunk diameter of 4 feet. Often, its bole is clear of branches up to 50 feet making it an ideal timber to buck into usable saw logs. It is not heavily buttressed. The leaves are bipinnate, and its flowers are small and white and often greenish in color. The fruit is a thin flat legume. The bark is an important source of tanning material. These tall and well formed trees are often used as street plantings both for shade and decoration. The tree grows best in mixed hardwood forests at low altitudes in moderately deep soil.

THE TIMBER

The yellow or pinkish-gray sapwood is not well defined. The reddish-brown heartwood bears some resemblance to old-growth mahogany (*Swietenia*) but is much denser. The grain is often interlocked and displays dark uniform stripes. The texture is medium to fine, and the luster medium. No odor or taste is detectable. Angico is a hard and heavy wood with an average specific gravity of 0.95 (ovendry weight/green volume), equivalent to an air-dried weight of 78 pcf.

SEASONING

Angico seasons easily when cautious, slow-drying techniques are used. Average reported shrinkage values (green to oven-dry) are 3.8% to 4.0% radial, 8.5% to 9.5% tangential and 14.0% volumetric.

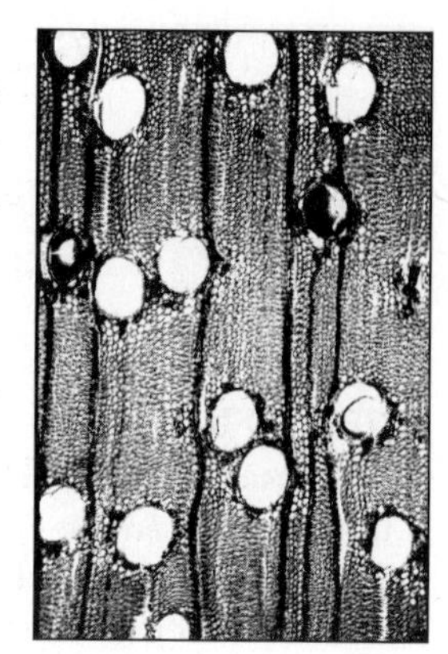

DURABILITY

The wood has a good record of durability. Heartwood posts in contact with the ground have an approximate life span of 15 to 20 years. It is resistant to fungal and insect attack and weathers very well.

WORKABILITY

Angico works well with both hand and machine tools. Because of its interlocked grain, it may, on occasion, be difficult to plane. It is best finished by using a sealer and a final sanding which produces an excellent high polished surface. When using a hand-held cabinet scraper on the quartersawn surface some interesting curls are formed due to the differing densities of the grain. All in all, even though a bit heavy, angico is a nice wood to work.

USES

Angico timber has many uses. It is a well suited for structural timbers, heavy-duty flooring and mine timbers. It is also used for boat building, furniture and turnery.

SUPPLIES

Angico is available on the commercial market and is in the medium price range. Caution should be exercised, however, when purchasing under a common name only. There is a wide variety in the woods of the other species of *Parapiptadenia*.

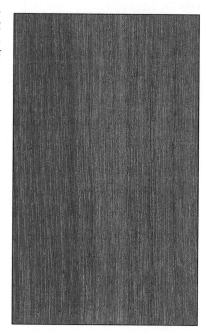

Paulownia tomentosa

royal paulownia

by Max Kline

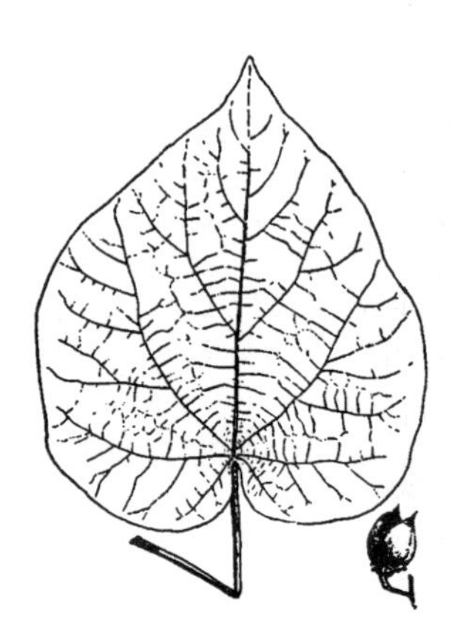

SCIENTIFIC NAME

Paulownia tomentosa. Derivation: The genus name is in honor of Anna Paulowna (1795 to 1865), the daughter of Czar Paul I of Russia. The specific epithet is Latin for tomentose or densely soft-hair.

FAMILY

Scrophulariaceae, the figwort family.

OTHER NAMES

princesstree, blue catalpa, empress tree.

DISTRIBUTION

Royal paulownia is a native of China and Japan but has naturalized since its introduction in the United States. It now covers a range from southern New York, southward to Florida and Texas and extends northward into Arkansas and Oklahoma.

THE TREE

Royal paulownia can reach 60 to 80 feet in height with a diameter of 2 to 4 feet. The trunk is usually short and divides into branches to form a wide-spreading, flat-topped crown. The leaves are very large and heart-shaped. The tree is one of the few that produces fragrant, blue and violet colored flowers that last only a few days in the spring. The empty, dark seedpods remain on the bare branches throughout the winter. The rather thick bark is dark grayish-brown with a network of fissures.

THE TIMBER

The wood of royal paulownia is soft, lightweight, ring porous and straight grained. The heartwood is purplish-brown or light gray with brown streaks, and the narrow sapwood is nearly white. The wood exhibits a satiny luster. Average reported specific gravity is 0.35 (ovendry weight/green volume), equivalent to an air-dried weight of 27 pcf.

SEASONING

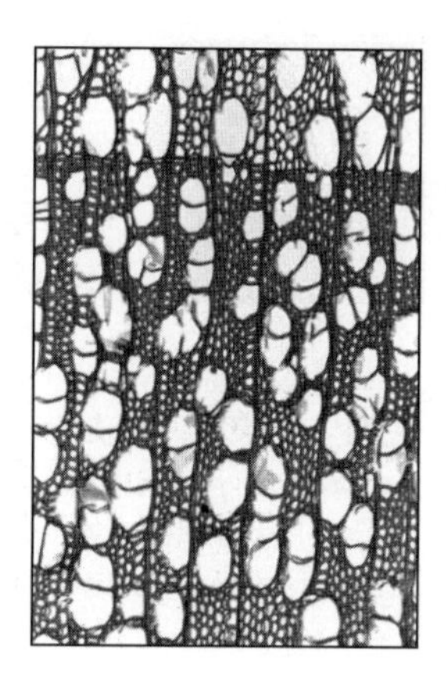

Royal paulownia timber is easily air dried without serious degradation or checking. Average reported shrinkage values (green to ovendry) are 2.2% radial and 4.0% tangential.

DURABILITY

The wood is considered nondurable.

WORKABILITY

Royal paulownia is easily worked with tools but has a dulling effect on saws due to the presence of silica. It has a tendency to tear unless tools are kept very sharp. A satin luster is obtained by polishing, and the lumber holds nails and screws well without splitting. All normal finishing materials can be applied.

USES

Royal paulownia is of no commercial use in the United States but in recent years some quantities have been exported to the Orient, where it is highly valued. It can be used successfully for many purposes where a light soft wood is satisfactory, especially for home workshop projects.

SUPPLIES

Quantities are limited in the United States but local residents occasionally cut some trees and woodworkers can obtain the wood. The recent exportation to the Orient has caused the price to approach that of black walnut (*Juglans nigra* WDS 150) on the commercial market. [Ed. note: Several ventures to grow this tree for commercial enterprises in some of the mid-Atlantic states have been tried and failed. While the tree is fast-growing, there does not seem to be enough of a market to make these investments profitable.]

Peltogyne paniculata

purpleheart

by Max Kline

SCIENTIFIC NAME

Peltogyne paniculata. Derivation: The genus name is Greek meaning a shielded organ. The specific epithet refers to the clustered flowers with each flower borne on a separate stalk.

FAMILY

Fabaceae or Leguminosae, the legume family; (*Caesalpiniaceae*) the cassia group.

OTHER NAMES

violetwood, pau roxo, pau ferro, coracy, amaranth.

DISTRIBUTION

Ranges from Sao Paulo in Brazil to Trinidad and Panama.

THE TREE

In the Amazon rain forests, the tree reaches a height of 100 to 120 feet. The clear bole is straight, cylindrical and 50 feet in length. The tree is covered with smooth grayish-black bark.

THE TIMBER

Purpleheart sapwood is pinkish-cinnamon with light brown streaks and is from 2 to 4 inches wide in mature trees. The heartwood is a dull brown color when freshly cut but oxidizes to a violet purple color when exposed to light. When exposed to sun and rain, the purple color will become black. The luster is medium, and the straight-grained wood is fine and even in texture. The odor and taste are not distinct. It is hard and heavy, with an average reported specific gravity ranging from 0.67 to 0.91, equivalent to an air-dried weight of 52 to 74 pcf. The mechanical properties of purpleheart are intermediate between those of greenheart (*Chlorocardium rodiei* WDS 079) and oak. One outstanding property of the wood is its ability to withstand sudden shock.

SEASONING

This timber seasons well and fairly rapidly with little degrade. In thick planks, there is some difficulty in extracting the moisture from the center. Average reported shrinkage values (green to ovendry) are 3.2% radial, 6.1% tangential and 9.9% volumetric.

DURABILITY

This timber is highly resistant to decay.

WORKABILITY

Purpleheart must be worked slowly through machines and all cutter tools must be of high-speed steel to produce fine cabinetwork. Some tearing occurs when planing when the grain is interlocked. The wood has a tendency to split when nailed. Most finishes can be used satisfactorily, but to preserve the rich natural color wax is often the only coating applied. The wood is dimensionally stable is use.

USES

Because of its good mechanical properties and durability, purpleheart is used for heavy outdoor construction such as bridges and dock work. It is reported to have good acid resistance and can be used in chemical plants for vats and filter press plates. As a flooring material, it has excellent abrasion resistance. An important use in Brazil is for making spokes for cartwheels. When exported, its biggest use is for billiard cue butts. Other uses in the United States include decorative veneer, inlay, marquetry, tool handles and general cabinetry. A dye produced from the wood is used for textile fabrics.

SUPPLIES

Purpleheart timber and veneer is available in small quantities. It has not found wide demand in the United States, and its price may be considered as moderately costly.

Pericopsis elata

afrormosia

by Max Kline

SCIENTIFIC NAME

Pericopsis elata. Derivation: The genus name is from the Greek *perikope*, meaning a cutting. The specific epithet is Latin for tall. A synonym is *Afrormosia elata*.

FAMILY

Fabaceae or Leguminosae, the legume family; (*Papilionaceae*) the pea or pulse group.

OTHER NAMES

kokrodua, yellow satinwood, African satinwood, Benin satinwood, redbark, devil's tree, African-teak, asamela.

DISTRIBUTION

In the semi-deciduous forests of the Ivory Coast, Ghana, Cameroon, and the Congo (formerly Zaire).

THE TREE

Afrormosia reaches a height of 160 feet with a diameter of 4 to 5 feet. The bole is usually straight and unbuttressed, and large trees may be free of branches for 100 feet.

THE TIMBER

Afrormosia is a high-quality timber with an attractive appearance resembling a fine-grained teak (*Tectona grandis* WDS 261). When freshly cut, the heartwood is yellowish-brown but it loses the yellow upon exposure. The sapwood is narrow and slightly lighter in color than the heartwood. The grain is straight or slightly interlocked. The texture is fine, and the luster is medium. Odor and taste are not distinct. Average reported specific gravity is 0.57 (ovendry weight/green volume), equivalent to an air-dried weight of 44 pcf. Under damp conditions, the wood is liable to stain when in contact with iron. It is appreciably stronger and harder than teak, but it is only moderate in bending properties. It is tough and shock resistant.

SEASONING

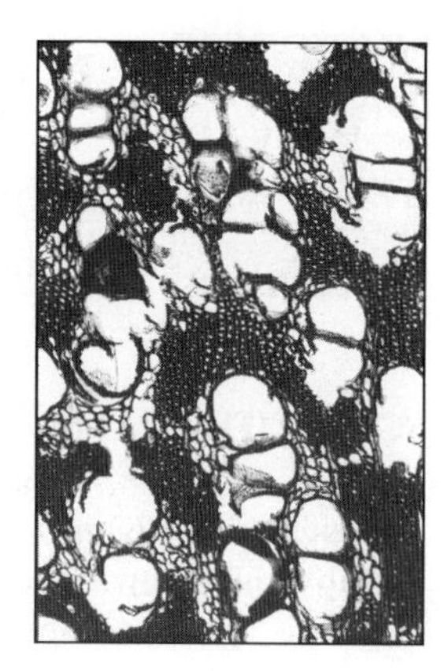

Afrormosia timber seasons fairly well but at a slow rate with little degrade. Average reported shrinkage values (green to ovendry) are 3.0% radial, 6.4% tangential and 10.7% volumetric.

DURABILITY

Afrormosia is resistant to fungi, insects and marine borers and is classified among the top ranking timbers in durability properties.

WORKABILITY

The wood works fairly well with machine and hand tools. The blunting effect on cutting edges is much less pronounced than with teak. When the grain is interlocked, machined surfaces are liable to tear unless a lower cutting angle is used to provide a clean finish. When drilling, the work must be well supported at the tool exit to prevent break away. Good results are obtained with stains and polishes. When nailing there is a marked tendency to split. It can be glued satisfactorily.

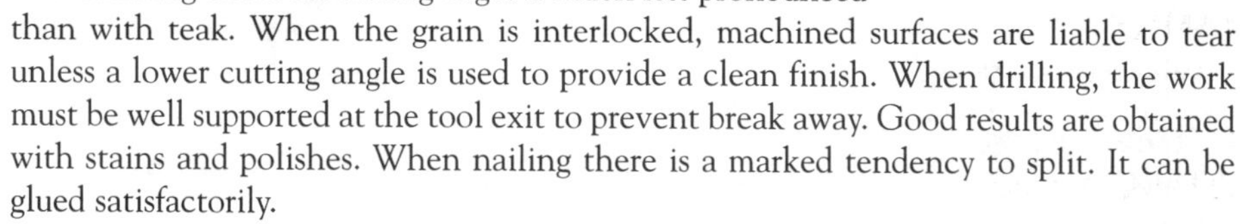

USES

Afrormosia has been used as a teak substitute in many industries, especially where durability and stability are required. Principal uses are in shipbuilding, interior trim, furniture, decorative veneer, flooring and high-class joinery. It should not be used where moisture levels are high due to its discoloration when in contact with iron.

SUPPLIES

Appreciable quantities are available in the form of lumber, flitches and logs at lower prices than demanded for teak or many other comparable hardwoods. [Ed. note. This species has been placed on the endangered species list thus restricting international trade.]

Persea americana

avocado

by Jim Flynn

SCIENTIFIC NAME

Persea americana. Derivation: The genus name is the ancient Greek name for an unidentified Egyptian tree with fruit growing directly from the stem. The specific epithet means the plant is "of America".

FAMILY

Lauraceae, the laurel family.

OTHER NAMES

There are probably as many different common names for this tree as there are countries in which it grows. The tree is called persea, alligator-pear, aquacate (Spanish), apricot (Virgin Islands), alageta (Guam), bata (Palau) and palta (South America).

DISTRIBUTION

Avocado is native to Mexico, Guatemala and Honduras. Widely cultivated in tropical and subtropical regions of the world, avocado has naturalized in many countries. Throughout these regions, including Hawaii, Florida and California, there are many extensive orchards of avocado trees. Archeological records from Central America suggest the tree has been cultivated for at least 8,000 years.

THE TREE

Avocado is not an exceptionally large tree. It reaches heights from 30 to 60 feet with diameters of about 30 inches. It is apt to be found in cultivated orchards where it is pruned to allow ready harvesting of its valuable fruit. The shiny yellowish-green fruit is 4 to 5 inches long, 3 to 4 inches in diameter and pear-shaped although sometimes it is more rounded. It has a delightfully edible flesh surrounding its large egg-shaped seed. At one time avocado was known as midshipman's butter because the pulp yields an oil that can be used as a substitute for olive oil. Avocado is a deciduous tree with a brownish-gray bark that is slightly rough and fissured. The inner bark is orange-brown, slightly spicy and gritty to the taste. The twigs are green, angular and finely hairy, and become brown over time. The alternate simple leaves are clustered near the end of the twigs on yellowish-green petioles, and the blades are slightly aromatic when crushed. Flowers appear near the end of twigs when trees are leafless or nearly so.

THE TIMBER

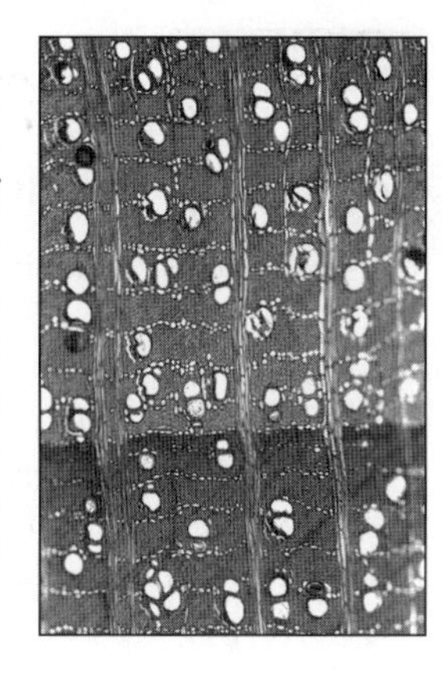

Because of its diminutive stature, avocado is not among the big-name timber producers. The sapwood is wide and cream-colored, and the heartwood is a light pinkish-tan. Because of the many varieties of the species and its worldwide distribution, it is possible to encounter much variety in the wood's density, color and grain pattern. A standard-sized sample of a California-grown species has an average reported specific gravity of 0.53 (ovendry weight/green volume), equivalent to an air-dried weight of 40 pcf. Generally, the wood is straight grained with a subdued pattern. There is no discernible taste or odor. Luster and texture are medium.

SEASONING

Avocado should be dried carefully to lessen its tendency to warp. Average reported shrinkage values (green to ovendry) are 4.8% radial, 9.5% tangential and 13.5% volumetric. Once dry, it appears quite stable.

DURABILITY

Avocado is not considered durable and is subject to dry-wood termites. No data is available on effective preservation methods.

WORKABILITY

The wood is easy to work and produces crisp lumber that maintains its edges after machining. No problems should be expected when sanding or finishing, and it glues effectively. It is quite similar to various species of alder (*Alnus* spp.).

USES

There are no recorded glamorous uses for this wood. It appears suitable for turnery, cabinetry and use in fine furniture.

SUPPLIES

Considering that avocado is a profitable crop and the trees bear fruit for at least 50 years, there is little profit in marketing the wood commercially. Because the wood is common worldwide, however, collectors are able to locate samples at wood auctions and IWCS meetings.

Persea borbonia

redbay

by Jim Flynn

SCIENTIFIC NAME

Persea borbonia. Derivation: The genus name is the ancient Greek name for an unidentified Egyptian tree with fruit growing directly from the stem. The specific epithet is an old generic name for persea.

FAMILY

Lauraceae, the laurel family.

OTHER NAMES

shorebay, swamp redbay, sweetbay, smooth redbay. In the early part of the 18th century, it was known as Carolina laurel tree, *Laurus carolinensis*. Taxonomically, this species has been classified as consisting of two varieties, *P. borbonia* var. *borbonia* (typical) redbay and *P. borbonia* var. *humilis*, silkbay or two distinct species, *P. borbonia* and *P. palustris*. The latter classification appears to be gaining favor.

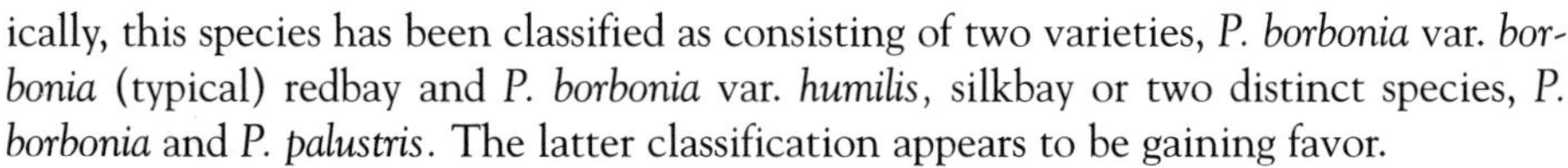

DISTRIBUTION

With spotty appearances in Maryland and Virginia, its range begins in southern Delaware and continues in a wide swath along the Atlantic coastal plain through Florida, with outliers in the Bahamas, and then to southeastern Mississippi with occasional appearances in Louisiana and Texas.

THE TREE

Redbay will grow as a shrub or to tree height. Larger specimens are found along stream banks, in swamps and in moist rich soils, where it thrives with longleaf pine, red maple (*Acer rubrum* WDS 009), blackgum (*Nyssa sylvatica* WDS 188) and oaks. The bole is often crooked and bears large limbs 12 feet or so above the ground. Although rare, maximum heights are about 60 feet. On prime specimens, trunk diameters are 15 inches or more. The oblong to oblong-lanceolate evergreen leaves are 3 to 4 inches long and red when young but turn green as they mature. The leaves are often used for flavoring soups. Growing in red cups in groups of two or three, the oval berries are dark blue. The deeply (0.5 inch or more) furrowed bark has a distinctive reddish cast which aids in identification. It is often planted as an ornamental.

THE TIMBER

The wood is very strong although somewhat brittle. It is close grained and an attractive rose color with darker pink or brown streaks. The sapwood is light tan and quite wide. The billet used to prepare this Wood Data Sheet was grown in southern Virginia and had a diameter of 8.5 inches with 1.5-inch-wide sapwood. The luster is fair but dependent upon the cut. A sweet-smelling odor will permeate the shop when the wood is cut and may become overpowering. Average reported specific gravity is 0.60 (ovendry weight/green volume), equivalent to an air-dried weight of 46 pcf, but it will vary according to growth conditions.

SEASONING

Redbay seasons well if an end sealer is applied immediately after it is cut. Splitting the wood through its length helps avoid checking. Slow drying is recommended. Average reported shrinkage values (green to ovendry) are 4.8% radial, 9.5% tangential and 13.5% volumetric.

DURABILITY

Little data can be found. A statement in a publication dated 1846 implied the wood was durable when used in shipbuilding.

WORKABILITY

There are no problems encountered in sawing and machining. It is dense enough to leave clean and sharp edges on finished work. Sharp tools are necessary, especially when planing, as some grain may be twisted thus resulting in chipping. It sands and glues well and takes a fine polish especially if finished naturally.

USES

No longer available in quantities sufficient to warrant commercial recognition, the use of redbay appears to be localized. It is suitable for interior cabinets, furniture and craftwork. Furniture restorers and antique collectors please note redbay was used for high-quality furniture before mahogany came into general use in the United States.

SUPPLIES

One has to travel along the Atlantic and Gulf seaboards to find this wood. IWCS member Ike Behar reports seeing it on trucks along with other species bound for paper-pulping mills. Having some familiarity with pallet making operations in the Carolinas, I suspect it is also mixed with other hardwoods used in box and pallet manufacturing.

Phyllostachys edulis

moso bamboo

by Jim Flynn

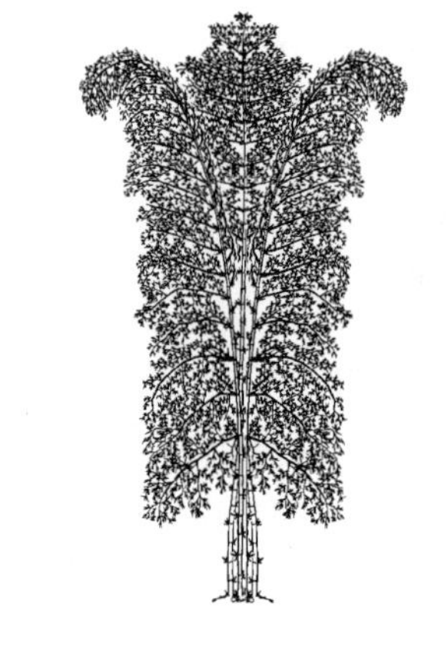

SCIENTIFIC NAME

Phyllostachys edulis. Derivation: The genus name *phyllon* is Greek meaning leaf and *stachys* meaning spiked. The specific epithet means edible, referring to new growth shoots.

FAMILY

Gramineae, the grass family; also Poaceae.

OTHER NAMES

mousou-chiku, tortoise-shell bamboo, bamboo, kikko-chiku, mao zhu.

DISTRIBUTION

Native to temperate Asia and east China, moso bamboo is cultivated in Japan and elsewhere.

THE TREE

Bamboo is botanically classified as a grass but has fascinating wood-like characteristics. Many of these bamboos, estimated to be over 1,000 different species, are referred to as tree grasses and woody bamboos. Moso bamboo is one of the great woody bamboos and thrives in the Hunan Province in south central China. Cultivated in its natural environment and plantations, harvest growth cycles are 3 to 3.5 years, after the culms (stems) have grown to 40 feet or more with diameters averaging 3 inches. These culms are hollow except at nodes where branches develop and display the familiar jointed or ringed appearance. While bamboo is a flowering plant, it may take from 15 to 50 years to blossom and produce seed. Often whole populations of a similar species blossom at the same time and die. Propagation is assured by the underground growths called rhizomes that spread rapidly with long roots. This underground network may grow 25 feet or more in a year and be difficult to control.

THE TIMBER

A culm of moso bamboo 35 feet in height, 3 to 4 inches in diameter and hollow only begins to look like conventional timber when it is cut, slivered, dried, laminated and glued into stringers. It is very dense. A standard 3- by 6- by 0.5-inch sample had a specific gravity of 0.59 (ovendry weight/green volume), equivalent to an air-dried weight of 46 pcf. Bamboo contains silica in its cell walls. It is hard, strong, elastic and easy to split. Moso bamboo is an attractive straw color with delicate linear lines of brighter and darker hues. On wood that has been carbonized by controlled heating, the bamboo takes on a mellow brown oak-like color and the vascular strands are emphasized. Where each growth appears, an attractive and unique figure is developed, equivalent to a knot.

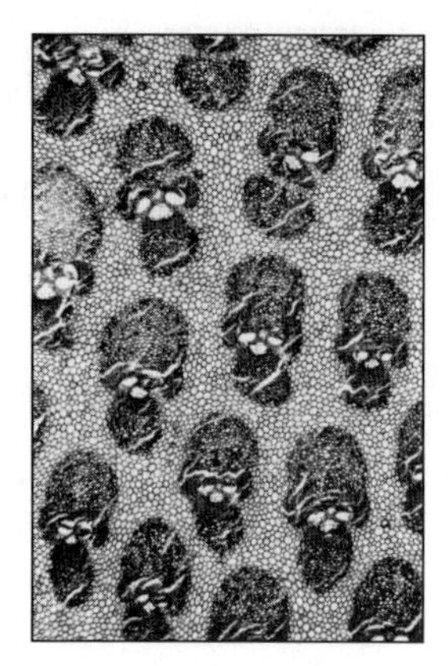

SEASONING

In its natural shape, moso bamboo loses some of its girth and is apt to develop linear splitting, but this does not degrade the stock. When converted to flooring, furniture parts and the like, carefully controlled drying is accomplished with only minor degrade.

DURABILITY

Moso bamboo is not considered suitable for use in contact with the ground or in locations with abnormal moisture amounts. Insect and fungi infections can be controlled by chemical means. Processed moso bamboo in laminated, timber-like structures is treated to prevent insect and mildew damage and to conform to fire prevention standards.

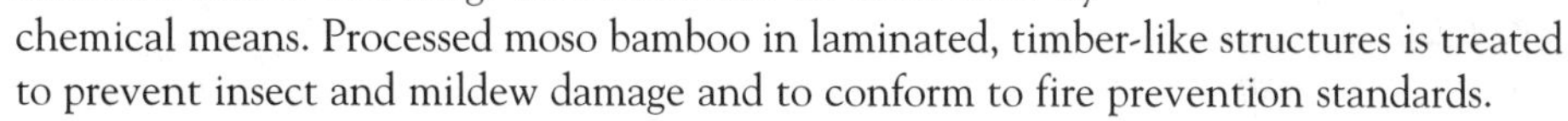

WORKABILITY

When working with some laminated pieces of moso bamboo as milled flooring, pieces measuring 24 by 3.625 by 0.625 inch were fed into a Delta 12-inch surface planer to reduce thickness to 0.5 inch. The cuts were clean and smooth. Trimming the edges to remove the tongue and grooves with a bandsaw and then jointing them with a Delta 6-inch jointer squared the pieces nicely with clean sharp edges. Nailing tests with standard flooring nails proved remarkably easy with no splitting. The chips and sawdust had a sandy feel unlike regular wood. Machine sanding especially on the end grain requires care lest friction burns result. Gluing and finishing obviously have been perfected.

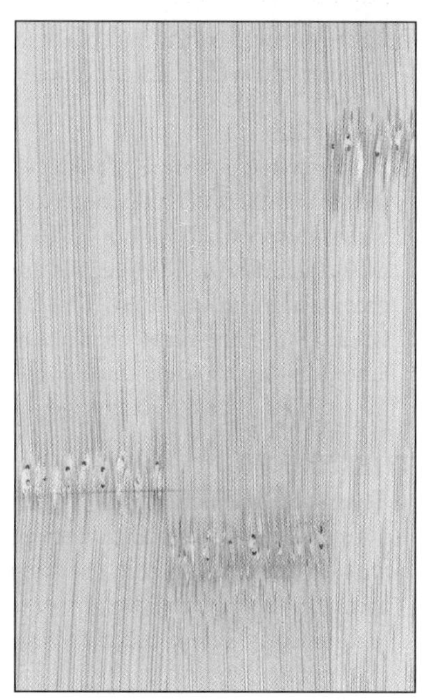

USES

Uses include flooring, shelters, furniture and fishing poles as well as the notorious sharpened and tempered spikes used along footpaths in some combat zones.

SUPPLIES

Commercial supplies of moso bamboo in the form of unprocessed culms are readily available worldwide. Laminated bamboo in many forms, especially milled and finished flooring, can be found in most hardwood flooring emporiums at a high cost.

Picea mariana

black spruce

by Chuck Holder

SCIENTIFIC NAME

Picea mariana. Derivation: The genus name *picea* is from the Latin pix (pitch), referring to a pine, later applied to spruces as the genus name. The specific epithet *mariana* is the Latin form of "from Maryland" but in a broad sense refers to northeastern North America. This species is not native in Maryland.

FAMILY

Pinaceae, the pine family, a temperate Northern Hemisphere family of trees and shrubs with some nine genera and 210 species.

OTHER NAMES

bog spruce, eastern spruce, swamp spruce, epinette noire.

DISTRIBUTION

Black spruce is widespread in Canada and is found growing in every province and territory as well as the northeastern states in the United States. It defines the northern limit of tree growth, and black spruce trees near this limit are often found to be up to 90 years old with a bole diameter of 12 inches.

THE TREE

Black spruce is a slow-growing, small to medium-sized conifer that normally grows 50 feet in height with a diameter of 12 inches. Maximum growth is 80 feet in height with a diameter of 20 inches. The tree has a straight tapering bole and an irregularly cylindrical crown. The leaves are 0.25 to 0.5 inch long, linear, four sided, dull bluish-green and blunt pointed. The 0.75 inch wide by 1.5 inches long, ovoid cone is purple when young but turns brown when mature. They persist for up to 30 years in clusters. The 0.25- to 0.5-inch-thick bark is grayish-brown to reddish-brown with an olive-green inner layer. Black spruce is the official tree of Newfoundland and Labrador, Canada.

THE TIMBER

The wood of black spruce is straight grained and fairly fine textured. It is nearly white with little or no contrast between the heartwood and the sapwood. It is lightweight with an average reported specific gravity of 0.38 (ovendry weight/green volume), equivalent to an air-dried weight of 28 pcf. Stronger than white spruce, it is classified as medium in strength but above average in stiffness.

SEASONING

Black spruce seasons easily and shrinkage when drying is moderate. Average reported shrinkage values (green to ovendry) are 4.1% radial, 6.8% tangential and 11.3% volumetric.

DURABILITY

Black spruce is not resistant to situations favoring decay and is very resistant to impregnation with preservatives.

WORKABILITY

Black spruce machines and holds nails well. It is easy to glue and can be painted satisfactorily.

USES

Black spruce is used for lumber, pulpwood and construction plywood. It is also used for food containers since it is almost odorless and tasteless.

SUPPLIES

In ample supply, black spruce is marketed with other eastern spruces such as white spruce and red spruce (*Picea rubens* WDS 205) because the wood of the three species is very similar. It is in the low-cost price range.

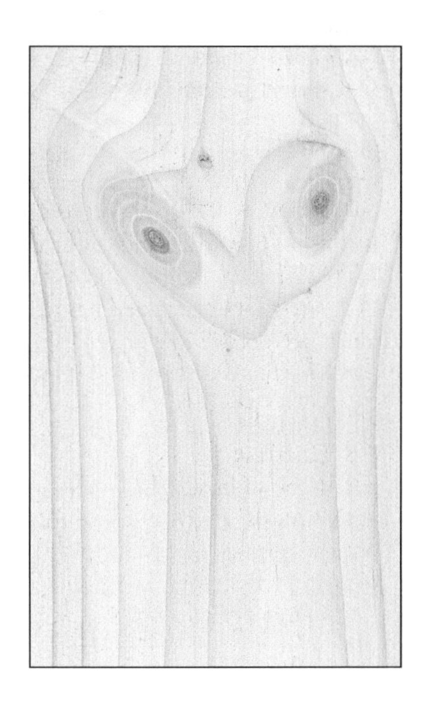

Picea rubens

red spruce

by Chuck Holder

SCIENTIFIC NAME

Picea rubens. Derivation: The genus name *picea* is from the Latin pix (pitch), referring to a pine, later applied to spruces as the genus name. The specific epithet *rubens* is Latin for red, from the Celtic rub.

FAMILY

Pinaceae, the pine family, a temperate Northern Hemisphere family of trees and shrubs with some nine genera and 210 species.

OTHER NAMES

eastern spruce, yellow spruce, he-balsam, epinette rouge.

DISTRIBUTION

Red spruce grows throughout Canada's Maritime Provinces and stretches south as far as North Carolina in the United States.

THE TREE

Red spruce is a medium-sized conifer normally growing to 80 feet in height with a diameter of 24 inches. Maximum growth is 162 feet in height and 57 inches in diameter. Trees grown in the open develop a broadly conical crown extending nearly to the ground. In the forest, the crown is somewhat pagoda-shaped and restricted to the upper portion of the tree on its long cylindrical bole. The shiny, linear leaves are 0.625 inch long, four-sided and yellowish-green. The chestnut brown, ovoid-oblong cones are 1.25 to 2 inches long and fall off the tree during the first winter or following spring. The bark is 0.5 inch thick and has irregular grayish to reddish-brown scales while the inner layers are dull yellow or reddish-brown. The tree reaches maturity in about 200 years and lives for 400 years. Red spruce is the official tree of Nova Scotia, Canada.

THE TIMBER

Like most spruces, red spruce is strong for its weight. Average reported specific gravity is 0.37 (ovendry weight/green volume), equivalent to an air-dried weight of 27 pcf. It is nearly white or cream in color with little or no contrast between the sapwood and the heartwood. It is moderately long fibered, odorless and slightly resinous. The wood is straight grained, fine textured and above average in stiffness.

SEASONING

Red spruce seasons easily with moderate shrinkage. Average reported shrinkage values (green to ovendry) are 3.8% radial, 7.8% tangential and 11.8% volumetric.

DURABILITY

Red spruce is not resistant in situations favoring decay and is very resistant to impregnation with preservatives.

WORKABILITY

Red spruce is easy to work with hand or machine tools. It sands well, takes a good finish and holds paint and nails well.

USES

An important commercial species, red spruce is highly valued for pulp, lumber, plywood, containers and sounding boards for musical instruments.

SUPPLIES

In ample supply at reasonable prices, red spruce is marketed with other eastern spruces such as white spruce and black spruce (*Picea mariana* WDS 204) because the wood of these three species is very similar.

Picea sitchensis

Sitka spruce

by Jim Flynn

SCIENTIFIC NAME

Picea sitchensis. Derivation: The genus name *picea* is from the Latin pix (pitch), referring to a pine, later applied to spruces as the genus name. The specific epithet refers to Sitka Island (now Baranof Island) in southeastern Alaska where the species was discovered and named by Europeans in 1892.

FAMILY

Pinaceae, the pine family, a temperate Northern Hemisphere family of trees and shrubs with some nine genera and 210 species.

OTHER NAMES

coast spruce, tideland spruce, yellow spruce, Menzies spruce.

DISTRIBUTION

This North American species hugs the northern Pacific coast, seldom growing inland more than 50 miles. It thrives best in moist maritime climates where there is sufficient rain throughout the year, relatively mild winters and cool summers. The tree grows from sea level to elevations from 3,000 to 5,000 feet. It ranges from southern Alaska (Kodiak Island and Cook Inlet) through southeastern Alaska, western British Columbia, the southwestern corner of Yukon, and to western Washington, western Oregon and northwestern California.

THE TREE

Sitka spruce is a vigorous fast-growing tree. It is the largest and most imposing of all the spruces. Mature trees grow 125 to 175 feet in height and generally 3 to 6 feet in circumference. It is not unusual for trees to be 750 to 800 years old. In dense stands, the tree produces a long clear trunk which is often buttressed at the base. The horizontal branches have many drooping branchlets which hold the cylindrical, light brown cones that are 2 to 3.5 inches long. The somewhat diamond-shaped needles are 0.625 to 1 inch long, sharp-pointed, flattened and thin with two whitish bands on each side. One of the largest trees on record is near Seaside, Oregon, and measures 16.7 feet in diameter, 216 feet in height and has a crown of 93 feet. Blowdown causes the most serious damage, but attacks by insects, disease organisms and animals also take their toll. Sitka spruce is the state tree of Alaska.

THE TIMBER

The wood has one of the highest strength-to-weight ratios. It has a fine and uniform texture and generally a very straight grain. The sapwood is creamy white to light yellow and blends gradually into the heartwood, which is pinkish-yellow to brown. Average reported specific gravity is 0.37 (ovendry weight/green volume), equivalent to an air-dried weight of 28 pcf. The annual rings are distinct which is important when quarter-sawing musical instrument soundboards.

SEASONING

Sitka spruce is extremely easy to season and kiln dry. Average reported shrinkage values (green to ovendry) are 4.3% radial, 7.5% tangential and 11.5% volumetric.

DURABILITY

When long service life is required, it must be treated. This is accomplished by use of a water-diffusion process since the wood is resistant to pressure impregnation of preservatives. It has low resistance to decay.

WORKABILITY

This wood is one of the easiest to cut, glue and finish. This is especially true for old-growth, knot-free pieces.

USES

Because of its long strong fibers and the ease with which it can be processed, the wood makes excellent pulp for the manufacture of high-grade printing and bond paper. Its most demanding application was for aircraft construction during WWII, which continues on a smaller scale today. Because of the wood's natural resonance, it serves extremely well as musical instrument soundboards. It is also used for ladders, oars, masts and planking for boats and turbine blades for wind-energy conversion systems (windmills).

SUPPLIES

Sometimes difficult to isolate commercially, it is often marketed with white spruce, true firs and other softwood species. An exception to this is what is used for musical instruments. This highly specialized and selective process has a market of its own. Entrepreneurs often seek individual trees that show promise of fine and straight grain and go to great lengths to harvest them.

Pinus aristata
bristlecone pine

Pinus longaeva
intermountain bristlecone pine

by Jim Flynn

SCIENTIFIC NAME

Formerly, *Pinus aristata* comprised two varieties of bristlecone pine, *P. aristata* var. *aristata* and *P. aristata* var. *longaeva*. Taxonomists have now determined that each is a distinct species. Because of the similarity of the wood, this Wood Data Sheet will apply to both. The genus name is the classical Latin name for pine. The specific epithet *aristata* means awned, referring to the long slender prickles on the cones. The specific epithet *longaeva* means long-lived.

FAMILY

Pinaceae, the pine family, a temperate Northern Hemisphere family of trees and shrubs with some nine genera and 210 species.

OTHER NAMES

Pinus aristata is also known as Colorado bristlecone pine, Rocky Mountain bristlecone pine and hickory pine. Names for *P. longaeva* are ancient pine, Great Basin bristlecone pine and western bristlecone pine.

DISTRIBUTION

There are no extensive stands. The trees exist in many spotty locations throughout the southwestern United States. The Agriculture Research Service's database places *P. aristata* in north and central Arizona, Colorado and northern New Mexico. *P. longaeva* grows in southern California, Nevada and Utah.

THE TREE

Bristlecone pine is one of the longest living trees. Trees in the White Mountains of eastern California have been measured by core samples to be 4,500 years old. Some of the deadwood on the ground is reputed to be over 10,000 years old. The cold and rocky environs demand that the tree fight for life and it becomes twisted, tough and gnarled, rarely growing to any great height. Trees have grown to heights of 40 feet or more from seeds blown to lower, more friendly elevations. The reddish-brown bark of aged trees is thin and furrowed. The twigs are light orange becoming black as the tree matures and bear long tufts of foliage at the tips. The stout and usually curved dark green leaves are shiny underneath, 1 to 1.5 inches long and grow in bundles of five. The cones are 3 to 3.5 inches long with short stalks and thick scales that are dark chocolate-brown at the end. The umbo has a long, fragile, bristle-like prickle. *Pinus aristata* is the state tree of Nevada.

THE TIMBER

This wood is pale orange-brown, rather brittle and unevenly grained. Because of its contorted growth pattern, straight even-grained stems of reasonable length are hard to find. Small slabs exhibit interesting and varied grain patterns. The wood weighs 35 pcf, equivalent to a specific gravity of 0.46 (ovendry weight/green volume).

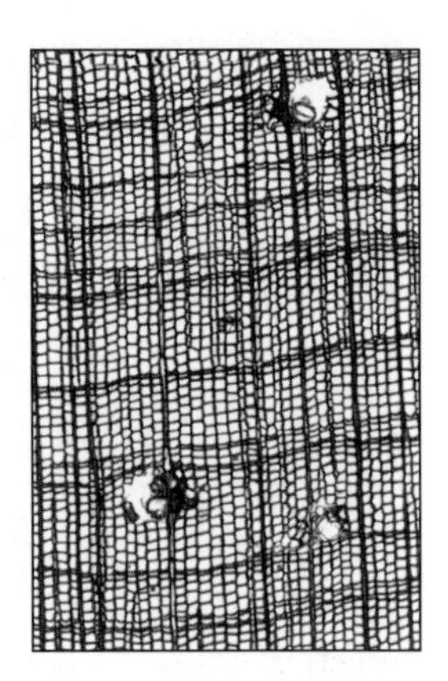

SEASONING

Seasoning data and shrinkage values are not available.

DURABILITY

The wood from trees at high elevations is extremely durable because of the low temperatures and lack of moisture which inhibits bacteria growth.

WORKABILITY

The wood is easy to cut and work with hand tools. Because the grain may not lie on the same plane over any distance, it is difficult to surface without tearing. Sharp tools are needed. It glues and finishes well.

USES

Older records state the wood was used for firewood and mine timbering. The tree is cultivated in "lowlands" as an ornamental. The wood has potential for crafts and turnery. A fascinating use of the tree and wood is in dendrochronology. Knowledge of the dates for major events, e.g., volcanic eruptions, can be correlated to the growth rings and other events paced off and dated. This highly specialized field has resulted in a Tree-Ring Research Laboratory at the University of Arizona. The Water Resources Division of the U.S. Geological Survey maintains a Tree-Ring Laboratory at the National Center in Reston, Virginia. Bristlecone pine plays an important role in this research.

SUPPLIES

Many bristlecone pines are in state and national parks and forests and protected from cutting. Others are in such remote and inaccessible places that for all practical purposes they are protected. There are no known commercial suppliers except possibly in local areas.

Pinus caribaea

Caribbean pine

by Jim Flynn

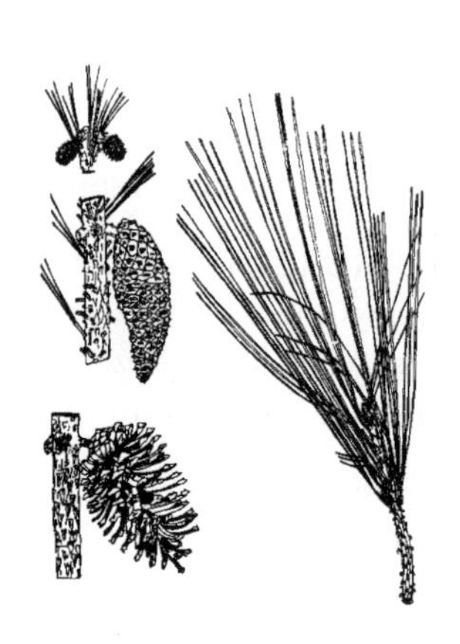

SCIENTIFIC NAME

Pinus caribaea. Derivation: The genus name is the classic Latin name for pine. The specific epithet means the plant is "of the Caribbean".

FAMILY

Pinaceae, the pine family, a temperate Northern Hemisphere family of trees and shrubs with some nine genera and 210 species.

OTHER NAMES

In Latin America, Caribbean pine is generally known as pino. In Mexico, Guatemala, Honduras and Nicaragua, the tree is called ocote. Prior to 1940 the U.S. Forest Service considered *P. caribaea* as one of the slash pines that included the common names Cuban pine, yellow slash pine, swamp pine, pitch pine as well as slash pine. Since then the so-called *P. caribaea* growing in the southeastern portion of the United States has been reclassified as *P. elliotta.* This latter species retains the common name slash pine as well as several others.

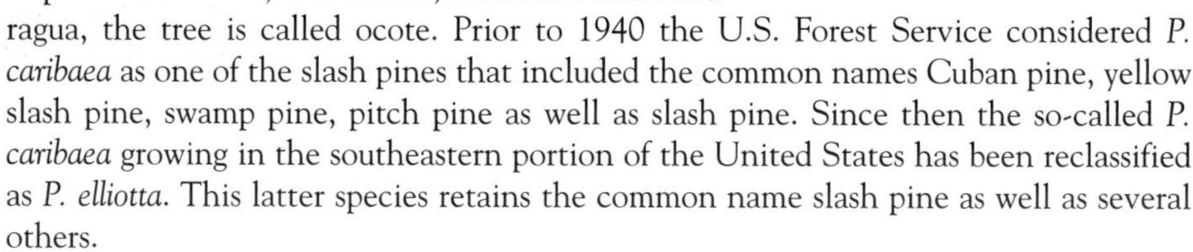

DISTRIBUTION

This species, along with a variety *P. caribaea* var. *hondurensis*, is native to the lower geographical levels of Cuba, Belize, Honduras, Nicaragua, Guatemala, the Bahamas and in the state of Quintana Roo in Mexico. The species is widely cultivated worldwide.

THE TREE

Because of its ability to thrive in warm climates as a plantation crop, Caribbean pine is now well known both north and south of the equator. The trees grow to 100 feet or more in height with a trunk diameter of 30 to 40 inches. The clear, straight boles are up to 70 feet in length producing ideal poles and sawlogs. The dark green, stout, lustrous needles are 8 to 12 inches long and grow two and three to a cluster.

THE TIMBER

The light yellow, 1- to 2-inch-wide sapwood is clearly distinct from the golden-brown to reddish-brown heartwood. The wood has a strong resinous odor and a greasy feel. The texture is coarse, and the luster is medium. The grain is usually straight, and growth rings are prominent. There is wide variance, attributable to growing condition, in the air-dried weight which may range from 25 to 53 pcf. Concomitantly, the specific gravity may range from 0.34 to 0.68 (ovendry weight/green volume). Care must be exercised when selecting timber for use in structural applications where strength is important.

Caribbean pine is heavier than slash pine (*P. elliottii*), but the mechanical properties of these two species are rather similar.

SEASONING

The timber air dries slowly and thick stock has a tendency for end splitting. Low-density plantation wood dries rapidly with minimum checking and only slight warping. Average reported shrinkage values (green to ovendry) are 6.3% radial, 7.8% tangential and 12.9% volumetric.

DURABILITY

Caribbean pine sapwood is prone to blue stain. The heartwood is generally rated moderately durable. Resistance to insect attack varies with the resin content as does the resistance of the heartwood to penetration by preservatives. The sapwood easily accepts preservatives in either open-tank or pressure-vacuum systems.

WORKABILITY

Caribbean pine is easy to work. The biggest problem is the necessity to stop machining operations to remove the accumulated resin from the equipment. Sandpaper will also clog. It accepts glue satisfactorily and can be nailed and fastened with screws. Use of metal cabinet scrapers is recommended to achieve a glass smooth surface before finishing.

USES

This wood has many applications throughout the world. It is a good light and heavy construction timber. Innumerable uses in carpentry, flooring, boat building and similar work abounds. Illustrative of the commercial value of this species are its uses for utility poles, railroad crossties, plywood and pulp and paper products.

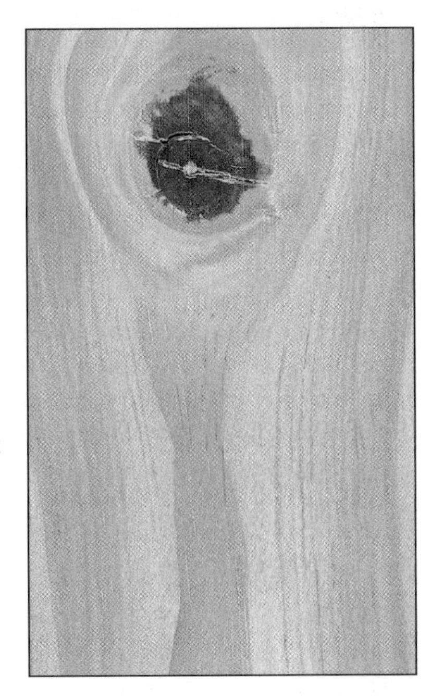

SUPPLIES

Because of the abundance of southern yellow pine, Caribbean pine is not widely used in the United States. Plantation-grown stocks, however, as well as those in the native zone are obtainable in many places throughout the world. Generally this is an economical timber, but prices will vary depending on location.

Pinus contorta

lodgepole pine

by Chuck Holder

SCIENTIFIC NAME

Pinus contorta. Derivation: The genus name *pinus* is from the Greek name *pinos* for the pine trees. The specific epithet *contorta* is Latin for contorted or twisted. *P. contorta* var. *latifolia*, mainly referred to here is Latin for with broad thin leaves.

FAMILY

Pinaceae, the pine family, a temperate Northern Hemisphere family of trees and shrubs with some nine genera and 210 species.

OTHER NAMES

The common name lodgepole pine is derived from the use of the tree by native people in constructing their lodges. Other names include Rocky Mountain lodgepole pine, black pine and pin tordu.

DISTRIBUTION

Widely distributed throughout western North America, several varieties are recognized. The range of shore pine (var. *contorta*) is along the Pacific coast. Sierra lodgepole pine (var. *murrayana*) is found in the southern Cascades, Sierra Nevadas and mountains of Baja California. Lodgepole pine (var. *latifolia*) is found in the Rocky Mountains and intermountain regions. A fourth subspecies, *bolanderi*, is a shrub confined to Mendocino County, California. Lodgepole pine is the most common and abundant tree in the Rocky Mountains and foothills regions of Alberta. It forms dense even stands after fire, and integrates with jack pine where the species overlap.

THE TREE

Lodgepole pine is a conifer growing to 100 feet in height with a diameter of 24 inches. Maximum growth is 135 feet in height and 44 inches in diameter. Its typically straight boles of little taper are long, clean and slender. The evergreen needles are twisted and stiff in bundles of two. They are 1 to 3 inches long, dark green to yellowish-green and grow in dense clusters toward the end of the branches. The cones are 0.75 to 2 inches long curving backward toward the base of the branches. They remain closed for many years, with a sharp spine at the tip of each scale. The thin, yellowish-brown bark is somewhat scaly.

THE TIMBER

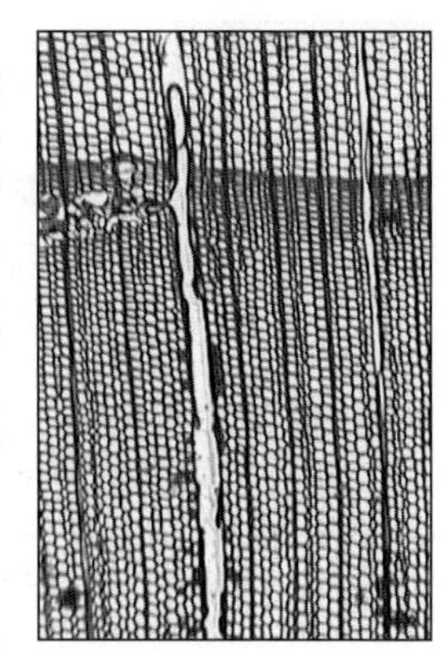

The wood is white to yellowish-brown. It is moderately light with an average reported specific gravity of 0.38 (ovendry weight/green volume), equivalent to an air-dried weight of 28 pcf. The wood is soft to medium hard with more or less uneven grain. As the species name implies, spiral grain is present among and within standing trees and is a cause of cross grain in products cut from this species. Flatsawn wood often shows a dimpled pattern. It is considered to be of medium strength.

SEASONING

Lodgepole pine seasons readily. Average reported shrinkage values (green to ovendry) are 4.3% radial, 6.7% tangential and 11.1% volumetric.

DURABILITY

The wood has low durability and should be treated with preservatives if used in conditions favorable to decay.

WORKABILITY

The wood is easy to work with hand or machine tools. It sands well, takes a good finish and holds paint well.

USES

Lodgepole pine makes excellent heavy construction timber, and this is its principal use. After pressure treatment, it is used for railway ties, poles and mine timbers. It is also used for boxes and crates, poles and pulpwood.

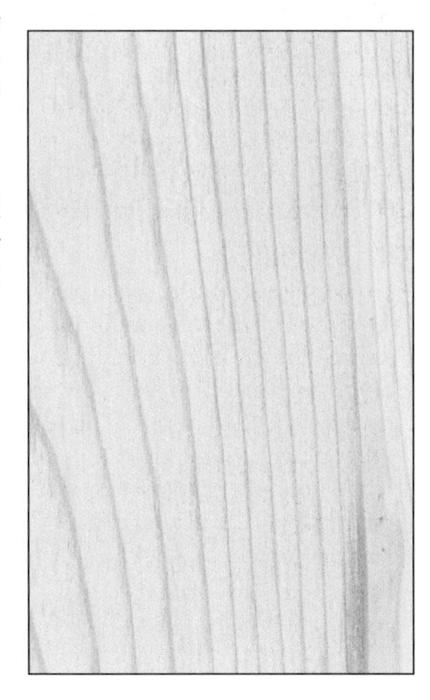

SUPPLIES

Lodgepole pine is available commercially and marketed with white spruce and alpine fir for general construction in the spruce/pine/fir (SPF) group.

Pinus lambertiana

sugar pine

by Jon Arno

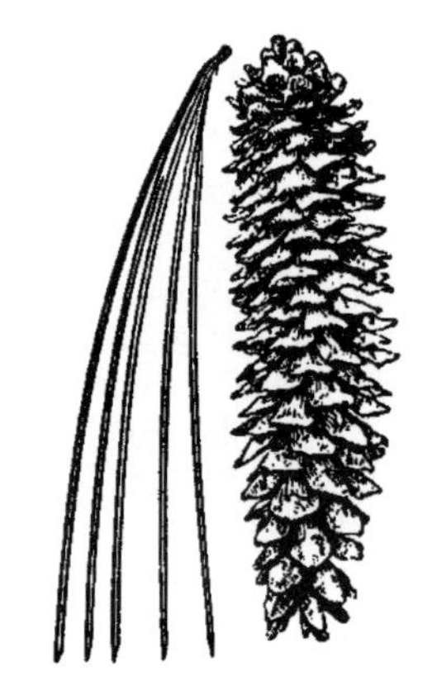

SCIENTIFIC NAME

Pinus lambertiana. Derivation: The genus name is the classic Latin name for pine. The specific epithet is in honor of Aylmer Bourke Lambert of England (1761 to 1842), an author of a classic illustrated work on the genus *Pinus*.

FAMILY

Pinaceae, the pine family, a temperate Northern Hemisphere family of trees and shrubs with some nine genera and 210 species.

OTHER NAMES

California sugar pine.

DISTRIBUTION

Sugar pine is native to California and Oregon where it occurs primarily on the more moist western slopes of the Cascade, Sierra Nevada and Coast Range Mountains. In the southern extremes of its range, it is found only at high elevations of 8,000 to 10,000 feet.

THE TREE

Sugar pine is the largest in the *Pinus* genus, attaining diameters in excess of 10 feet and heights well over 200 feet. The needles form in groups of five and are about 3 inches long. The cones are cylindrical, supple and up to 2 feet long. With its reddish-brown plated bark, tall clear trunk and bluish-green crown, this is a majestic tree when mature.

THE TIMBER

Given the size of the tree, sugar pine produces a great deal of large, often perfectly clear lumber. The heartwood is light brown, sometimes with a reddish hue, while the sapwood is creamy white. The wood is straight grained, very uniform in texture and often displays a faint, but very attractive, tangential figure liberally flecked with large, dark brown resin canals. The resin canals are far larger and more conspicuous than in other pines, making them a useful characteristic for identifying the species. Sugar pine scores poorly in terms of strength. Lacking dense bands of latewood and with an average reported specific gravity of only 0.34 (ovendry weight/green volume), equivalent to an air-dried weight of 25 pcf, sugar pine is a soft and relatively weak timber.

SEASONING

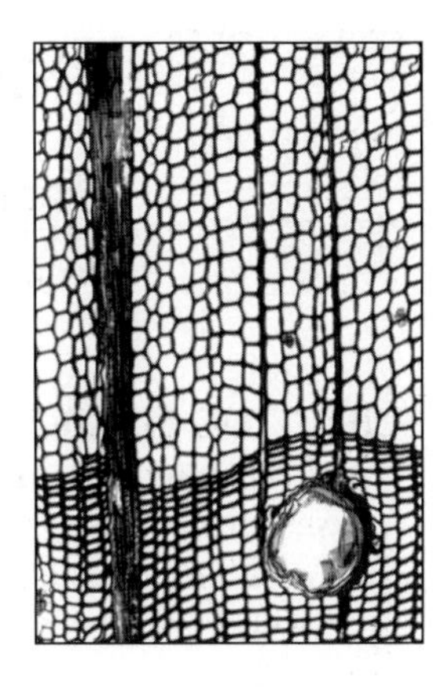

Sugar pine seasons exceptionally well with little warping or checking. Average reported shrinkage values (green to ovendry) are 2.9% radial, 5.6% tangential and 7.9% volumetric.

DURABILITY

Durability is average for a white pine, but poor compared to some of the other Pacific coast softwoods such as western redcedar (*Thuja plicata* WDS 268) or redwood (*Sequoia sempervirens* WDS 247). It should not be selected for applications where this feature is important.

WORKABILITY

Sugar pine is an extremely pleasant wood to work. Given its uniform texture and low density, it cuts and shapes easily and predictably. It glues well, sands well and is very stable in use. Its resinous sweet scent fills the shop with a very pleasant aroma. Some caution, however, is advisable in selecting finishes. Because of its high resin content, turpentine-based sealers may soften and bubble around knots and on the end grain.

USES

Top grades of sugar pine are used for millwork, pattern making and cabinetry. Some is used for general construction and lower grades find use in pallets and crating.

SUPPLIES

The easily accessible stands of sugar pine were heavily exploited in the late 19th and early 20th centuries. Modern logging techniques now make it possible to get to the less accessible stands and consequently supplies are adequate. Expect to pay a slight premium over other white pines, especially for wide clear stock.

Pinus pungens

Table Mountain pine

by Jim Flynn

SCIENTIFIC NAME

Pinus pungens. Derivation: The genus name is the classic Latin name for pine. The specific epithet refers to the sharp point, from the peculiar, stout, hooked spines on the cone.

FAMILY

Pinaceae, the pine family, a temperate Northern Hemisphere family of trees and shrubs with some nine genera and 210 species.

OTHER NAMES

hickory pine, mountain pine, prickly pine. Commercially, Table Mountain pine is considered a minor species of the great southern pines, which include longleaf pine (*Pinus palustris*), slash pine (*P. elliottii*), shortleaf pine (*P. echinata*) and loblolly pine (*P. taeda*).

DISTRIBUTION

Table Mountain pine can be found in the Appalachian Mountain region from Pennsylvania southwest to eastern West Virginia, Virginia, east Tennessee, the western Carolinas and northern Georgia. It is also found in New Jersey, Maryland and Delaware. Toward the southern end of its range, it can be found in pure stands.

THE TREE

This rugged pine is similar in outline and appearance to Scotch pine (*Pinus sylvestris*). Commonly, Table Mountain pine is 30 to 40 feet in height and occasionally reaches 60 feet. Its thick brown bark is fissured into large, loose, scaly plates. The stout, short, sharp-pointed, yellowish-green needles are broad, flat, rigid, 2 to 2.5 inches long and grow in clusters of twos (sometimes threes) crowded in confused masses on the twigs. The young sheaths are about 0.5 inch long. The ovoid cones are about 3 to 4 inches long with very thick scales armed with a strong reflexed prickle about 0.25 inch long. The cones are persistent and remain on the tree for 10 to 20 years.

THE TIMBER

As typical of the southern pines, Table Mountain pine is a soft, yellowish-white colored wood with light brown heartwood. It has a pronounced grain, and the transition from earlywood to latewood is semi-abrupt. Frequent knots are encountered. Specific data on strength is not available. As it most approximates the less-dense species of the southern pines (shortleaf and loblolly), the wood can be considered a moderately heavy pine, but weaker than the denser southern pines. Average reported specific gravity is 0.49 (ovendry weight/green volume), equivalent to an air-dried weight of 37 pcf.

SEASONING

Table Mountain pine seasons easily, and average reported shrinkage values (green to ovendry) are 3.4% radial, 6.8% tangential and 10.9% volumetric.

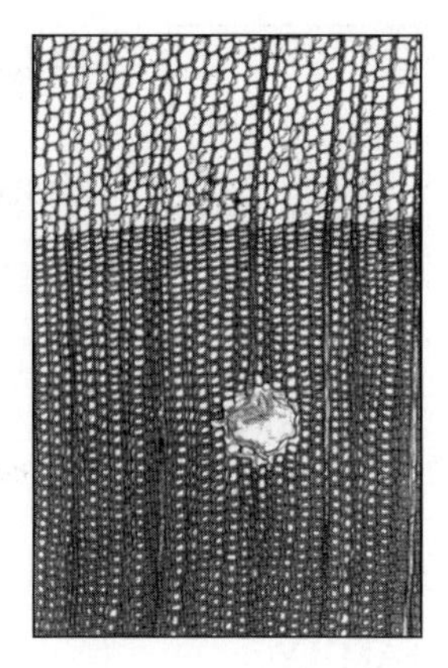

DURABILITY

Natural durability of Table Mountain pine is low. The sapwood is easily impregnated with wood preservatives.

WORKABILITY

This typically straight-grained, medium-textured wood is relatively difficult to work with hand tools. It is rated high in nail holding capacity, but does not accept glue readily.

USES

While older references state that the wood is not of much use except to make a poor grade of charcoal, this is no longer true. Table Mountain pine is marketed commercially with the less-dense southern pines as construction lumber where strength is not important. It is widely used in the manufacture of wood pulp for paper. It also makes a good addition to a wood collectors sample inventory. Because of its Appalachian Mountain heritage, the author is using some to build an Appalachian Mountain dulcimer.

SUPPLIES

Because Table Mountain pine is not marketed as such, and is combined with the major species of southern pines, only by individual search can it be acquired.

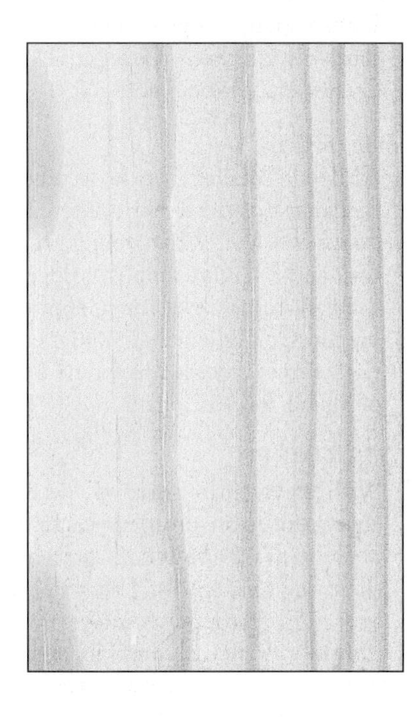

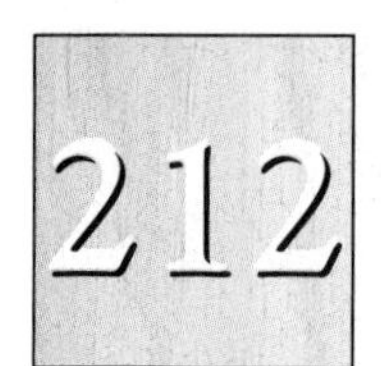

Pinus strobus

eastern white pine

by Jon Arno

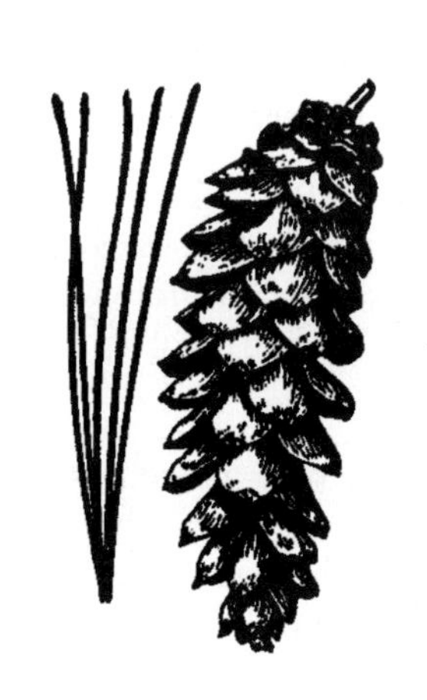

SCIENTIFIC NAME

Pinus strobus. Derivation: The genus name is the classic Latin name for pine. The specific epithet is Latin for pine cone.

FAMILY

Pinaceae, the pine family, a temperate Northern Hemisphere family of trees and shrubs with some nine genera and 210 species.

OTHER NAMES

northern white pine, Weymouth pine, cork pine, pumpkin pine, pin blanc.

DISTRIBUTION

Ranges from the Great Plains to the Atlantic seaboard and from east central Canada to the mountains of Guatemala. Although in mixed stands in the northern portion of its range, hardwoods on better growing sites in milder climates displace it. Below the Great Lakes region, it is generally isolated to higher elevations along the Appalachians as far south as northern Georgia. Distribution through the mountains of Mexico and Guatemala is sporadic.

THE TREE

Although smaller than sugar pine (*Pinus lambertiana* WDS 210), eastern white pine is the largest of the eastern pines, achieving heights well over 100 feet and diameters in excess of 4 feet. When young, the tree is rather anemic-looking with smooth, pale gray bark and a skimpy supply of fine, light green needles five to the sheath. As it matures, the bark turns dark brown, almost black, and is deeply furrowed. With the main stem supporting a billowy, emerald green crown high above its neighbors, it is an impressive timber tree. Eastern white pine is the official tree of Ontario, Canada and the state tree of Maine and Michigan.

THE TIMBER

With an average reported specific gravity of 0.34 (ovendry weight/green volume), equivalent to an air-dried weight of 25 pcf, eastern white pine is the softest and lightest of the pines. Named cork pine by lumberjacks, the logs floated high in the water on their way to sawmills. The wood can be distinguished from the hard or "yellow" pines by its lighter, creamy, almost white color and far less prominent annual rings. Also, the transition from earlywood to latewood is gradual with the latewood much narrower and less dense. It can be distinguished from sugar pine which has a sweeter scent, much larger resin canals and is usually slightly darker or more buff-brown in color. Eastern white pine heartwood occasionally displays tannish-brown or soft orange streaks.

SEASONING

The wood will air dry easily and quickly. The greatest risk of degrade is blue staining, and it is important to stack the boards in stickered piles to allow airflow through the pile as soon as the logs are cut. The wood is rather weak and has low abrasion resistance. Average reported shrinkage values (green to ovendry) are 2.1% radial, 6.1% tangential and 8.2% volumetric.

DURABILITY

Durability is poor. It is not a good choice for exterior applications.

WORKABILITY

The wood is soft, extremely uniform in texture and very easy to work. It planes well, glues easily, accepts fasteners without difficulty or pilot holes and is quite stable in use. Unlike the yellow pines, its figure remains soft and mellow even under dark stain. When staining, there is less tendency for the grain to swell unevenly and consequently fewer coats of varnish are required to achieve a smooth surface. It does not turn particularly well and is too weak for durable spindles. It will hold fairly sharp detail in carving.

USES

Uses include millwork, sashes, panel doors, interior trim and paneling, cabinetry, furniture, match sticks, pattern making and some plywood veneer. Lower grades are used for general construction, roof boards, sheathing and crating. It was used for masts in wood war ships, and in winter a tea was brewed from the needles to prevent scurvy.

SUPPLIES

Supplies are limited but recovering. It is still very difficult to find clearer grades. Because this species tends to branch out in whirls at intervals of about 18 to 24 inches, even these lower grades yield plenty of short, clear pieces which are ideal for small projects.

Piscidia piscipula

Florida fishpoison-tree

by Jim Flynn

SCIENTIFIC NAME

Piscidia piscipula. Derivation: Both the genus name and specific epithet are Latin for fish and kill which refer to the use of the bark and foliage in stupefying fish. A synonym is *P. erythrina*.

FAMILY

Fabaceae or Leguminosae, the legume family; (*Papilionaceae*) the pea or pulse group.

OTHER NAMES

fishfuddletree, Jamaica dogwood, fish-poison tree, habin (Mexico). There are many other names throughout its range.

DISTRIBUTION

This tree flourishes in the coastal areas of south Florida including the Florida Keys. It extends to the Bahamas, Cuba and Haiti. Also in southern Mexico, including the Yucatan Peninsula, Belize and Caribbean offshore islands of Honduras.

THE TREE

Florida fishpoison-tree is a stout erect tree reaching a medium height of 40 to 50 feet with a fairly large trunk that is 2 to 3 feet in diameter. The gray, 0.125-inch-thick bark is blotched with an olive color and small square scales. The crown is irregular with twisted branches growing at uneven angles. The evergreen leaves are odd-pinnate with 5 to 11 dark green leaflets. The fruit is a light-brown pod that is 4 inches long with four papery scalloped wings. The pod contains several reddish-brown flat seeds. The tree is easy to propagate. Freshly cut saplings and branches that are stuck in the ground for posts will readily grow. In the West Indies, the Carib Indians used the bark of the roots and young branches as well as powdered leaves to sprinkle on water where fish habituate. The fish would then be stunned, allowing them to be easily gathered, but not killed or contaminated to a point where they would not be edible.

THE TIMBER

Florida fishpoison-tree wood is very hard, heavy and close grained. The heartwood is a clear yellowish-brown color, sometimes streaked with darker brown, and the wide sapwood is lighter in color. The wood has a tendency to darken with age. When seasoned, it has no distinctive taste or odor. The grain is often wavy. The specific gravity is in the range of 0.80 to 0.90 (ovendry weight/green volume), equivalent to an air-dried weight of 64 to 73 pcf.

SEASONING

Because the wood is not high on the list of commercial timbers, there is an absence of technical data on this characteristic and shrinkage values are not available. Because it is a very dense wood, however, normal precautions for proper seasoning should be taken including not drying too fast and maintaining the proper humidification of saw logs.

DURABILITY

Florida fishpoison-tree is reported to be very durable in contact with the ground.

WORKABILITY

The wood is not easy to work without sharp tools. It finishes well and takes stains and wax easily. It is well worth the effort to use this beautiful wood.

USES

Because of its durability, it has been used for boatbuilding, railroad crossties and other heavy construction applications. Unfortunately, it also makes very good firewood and charcoal. Woodworkers use it for turnery, small novelties and furniture.

SUPPLIES

Florida fishpoison-tree is not available commercially. Because Florida has many restrictions related to cutting trees, care must be taken in selecting locations where the wood can be legally cut with the appropriate environmental considerations. Nevertheless, the wood is often available at IWCS auctions.

Platanus occidentalis

sycamore

by Max Kline

SCIENTIFIC NAME

Platanus occidentalis. Derivation: The genus name is the classical Latin and Greek name of *Platanus orientalis*, oriental planetree, from the Greek word for broad, referring to the leaves. The specific epithet is Latin and means the plant is "of the Western Hemisphere".

FAMILY

Platanaceae, the sycamore family.

OTHER NAMES

American plane tree, buttonball, buttonwood, ghost tree, water beech.

DISTRIBUTION

Sycamore grows throughout most of the eastern half of the United States from southern Maine to southeastern Nebraska, south into Texas and along the Gulf of Mexico to northern Florida.

THE TREE

Sycamore averages 60 to 120 feet in height with a diameter of 2 to 5 feet. Its conspicuous white bark, mottled with varying shades of green and brown, cause it to be called the ghost tree of the forest. On the older trees, the bark is 2 inches or more thick and due to the many fissures has a scaly or shaggy appearance. As the fruit matures, a dense round seedball is formed which hangs from a slender stem, giving rise to the common name buttonball tree.

THE TIMBER

The heartwood is various shades of reddish-brown, and the sapwood is a somewhat lighter shade. When quartersawn, the mottled texture with conspicuous rays is quite attractive. Though firm, tough and strong, the wood is not very heavy with an average reported specific gravity of 0.46 (ovendry weight/green volume), equivalent to an air-dried weight of 35 pcf. The texture is medium, and the grain is usually irregular. Sycamore wood does not impart taste, odor or stain to a substance with which it comes in contact.

SEASONING

Sycamore has a marked tendency to warp unless properly dried. Kiln drying can be done satisfactorily provided a suitable drying schedule is used. Average reported shrinkage values (green to ovendry) are 5.0% radial, 8.4% tangential and 14.1% volumetric.

DURABILITY

Sycamore is a perishable wood and is easily consumed by decay when in contact with the ground.

WORKABILITY

This wood must be worked with care due to the interwoven fibers. It may be turned with ease on the lathe and finishes smoothly when sharp tools are used. Nail and screw holding properties are good, but thin stock may split unless pre-bored. Glue adheres well, and drilling and mortising create no problems. Response to stain of all types is good, but care is needed in finishing to achieve good results.

USES

Because of its toughness, sycamore is the favorite wood for butcher blocks. The lumber is used for boxes and crates, trunk slats, brush backs, woodenware, slack cooperage and vehicle manufacture. The veneers are used for berry boxes and baskets for fruit and vegetables. Quartersawn lumber often is used in furniture manufacture because of its attractive appearance.

SUPPLIES

The lumber is readily available and may be considered inexpensive.

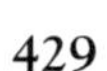

215 *Plathymenia reticulata*

vinhatico

by Max Kline

SCIENTIFIC NAME

Plathymenia reticulata. Derivation: The genus name means broad-shaped. The specific epithet is Latin for netted, referring to the leaf veins. This is the only species in the genus.

FAMILY

Fabaceae or Leguminosae, the legume family; (*Mimosaceae*) the mimosa group.

OTHER NAMES

amarello, Brazilian mahogany, yellowwood, candela.

DISTRIBUTION

In eastern Brazil from the lower Amazon to Sao Paulo.

THE TREE

At the northern limits of its range, vinhatico is an unarmed tree of small to medium size, but in the south it is large and can reach heights of 125 feet with diameters of 3 feet. It is free of branches for 60 to 70 feet. The grayish bark tends to be shaggy. The leaves are bipinnate with numerous small leaflets. The large, thin flat legumes have several seeds with a membranous wing. Vinhatico is scattered throughout the forest with an average of only about 3 per acre.

THE TIMBER

The narrow, yellowish-white sapwood is well defined from the heartwood that is an uncommon shade of orange-brown, often with a figuring of darker streaks. Upon exposure to air, the color changes to a deep reddish-brown. The luster is high and satiny. Odor and taste are not distinct. The wood is light and soft to moderately hard with an average reported specific gravity of 0.38 (ovendry weight/green volume), equivalent to an air-dried weight of 28 pcf. The texture is medium, and the grain is straight to decidedly wavy. The strength factor is good, although it is unlikely vinhatico will be used for applications requiring high strength.

SEASONING

Vinhatico must be thoroughly seasoned before it is put into use or it may move quite freely. It air dries easily without degrade, and kiln drying presents no difficulty. Shrinkage values are not available.

DURABILITY

The timber is classified as moderately resistant to both fungal and insect attack and does not need a preservative treatment.

WORKABILITY

Vinhatico is easy to work, finishing smoothly with golden luster. To achieve satisfactory results, thin-edged sharp tools are essential to cut the fibrous timber, which shows a tendency to tear in sawing. Pre-boring for nails is not necessary and glue will adhere satisfactorily. It is not a recommended species for turning. If properly filled, the surface will take a good polish. Drilling and mortising cause no problems.

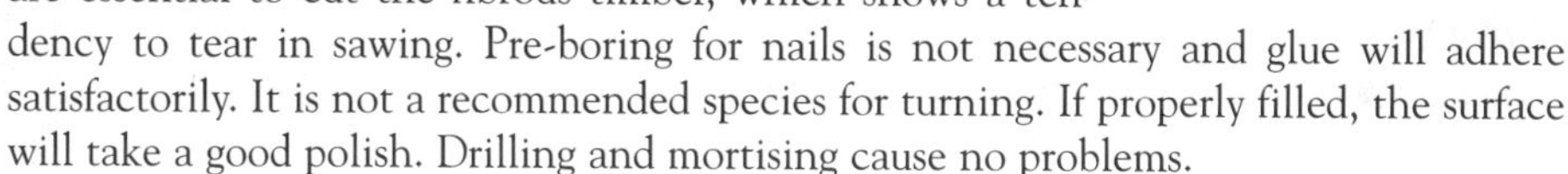

USES

In Brazil, the lumber is in great demand, and the natives use it for dugout canoes. Other uses vary from cabinetry to furniture, interior trim, doors, parquet flooring, veneer and plywood. It also finds some use in boat building and construction.

SUPPLIES

Vinhatico appears on commercial inventories in the United States and elsewhere.

Podocarpus totara

totara

by Max Kline

SCIENTIFIC NAME

Podocarpus totara. Derivation: The genus name is derived from the Greek *podo* meaning foot and *karpos* meaning fruit, which refers to the fleshy fruit stalks. The specific epithet is a local name.

FAMILY

Podocarpaceae, the podocarp family.

OTHER NAMES

Totara has no other names.

DISTRIBUTION

Totara is found in the forests of the north and south islands of New Zealand.

THE TREE

Although its average height is 70 feet with diameters of 2 to 5 feet, totara occasionally reaches a height of 130 feet. Totara is a softwood tree that is found in dry valleys but occasionally may grow on mountainsides up to elevations of 2,500 feet.

THE TIMBER

The wood has a uniform reddish-brown color with growth rings that are not clearly defined. Average reported specific gravity is 0.43 (ovendry weight/green volume), equivalent to an air-dried weight of 32 pcf. It has unusually fine, uniform texture and straight grain. Totara timber is comparatively weak in static bending properties. It is also reputed to have low shock resistance in bending, but its compressive strength is relatively good. For this reason, totara is more suited for use as columns and posts rather than for beams and joists.

SEASONING

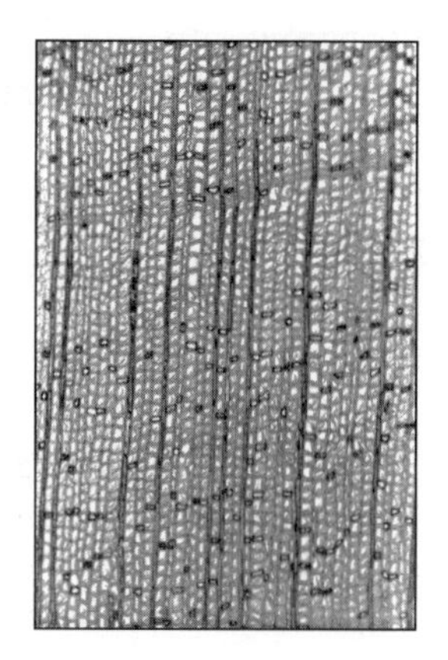

Totara seasons well and quickly. Average reported shrinkage values (green to ovendry) are 2.8% radial and 5.1% tangential. After proper seasoning, it is dimensionally stable in use.

DURABILITY

The heartwood is quite resistant to decay but can be attacked by the common furniture beetle.

WORKABILITY

Totara works readily and easily in all hand and machine operations. Because of its fine texture, totara is easy to cut into veneer. It turns well and holds nails and screws without difficulty. Finishing procedures require special treatment because of the presence of resin.

USES

In New Zealand large quantities are used for the planking of boats, barges, pontoons and other sea work. It is the only softwood generally concluded to be resistant to attack from marine borers. It is also suitable for flooring, exterior work and use in contact with the ground. The burls are unusually attractive for small turnery projects by the woodworker.

SUPPLIES

Totara is not commercially available in the United States. Wood collectors are able to secure small amounts from members in New Zealand.

Populus balsamifera

balsam poplar

by Jon Arno

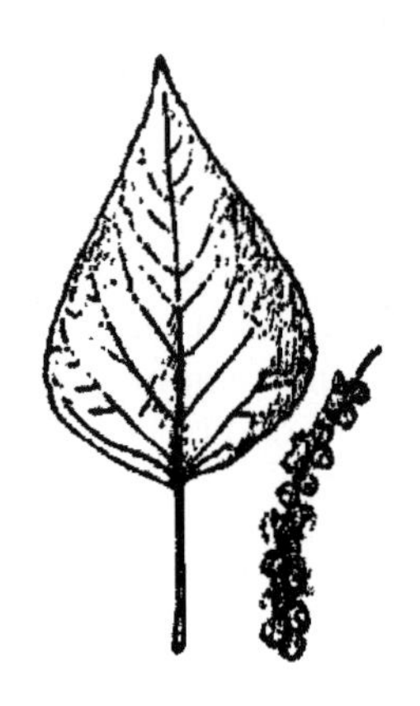

SCIENTIFIC NAME

Populus balsamifera. Derivation: The genus name is the classical Latin name. The specific epithet means balsam-bearing, in reference to the odor of balsam. A synonym is *P. trichocarpa.*

FAMILY

Salicaceae, the willow family. This family consists of two genera and about 350 species of which some 40 species are poplars, including 15 species native to North America.

OTHER NAMES

balm, tacamahac, balm-of-Gilead, black poplar.

DISTRIBUTION

Balsam poplar is a cold-tolerant tree, which is most plentiful in Canada. Its range extends from Newfoundland to Alaska, dipping down into the contiguous United States along the Appalachians and Rockies and through the St. Lawrence basin from Minnesota to Maine. It is found mostly on low, moist sites or where competition with conifers for sunlight has been otherwise diminished (on logged or burned-over land).

THE TREE

The light gray, leathery bark becomes darker and develops fissures as the tree matures. In terms of shape and size, balsam poplar resembles the cottonwoods and aspens to which it is closely related. The leaves are dark green above and paler gray below and are longer and more pointed than those of most other poplars. The rust-red buds become fragrant and sticky when they begin to swell in the spring, attracting wildlife as the bitter cold of winter subsides.

THE TIMBER

Balsam poplar lumber is seldom segregated from other poplars, cottonwoods and aspens in the trade and its wood differs only subtly from these related species. Demarcation between the almost chalk white sapwood and ash gray heartwood is more pronounced than in eastern cottonwood (*Populus deltoids* WDS 218). The wood is coarser textured, less lustrous, slightly less dense and more figured than aspen. While usually described as a diffuse-porous wood, close examination of the end grain of balsam poplar with a hand lens reveals a clear gradation in pore size from the larger earlywood pores to the smaller latewood (summerwood) pores. Average reported specific gravity is 0.31 (ovendry weight/green volume), equivalent to an air-dried weight of 23 pcf.

SEASONING

Balsam poplar dries quickly but is more susceptible to warp than aspen. Average reported shrinkage values (green to ovendry) are 3.0% radial, 7.1% tangential and 10.5% volumetric. It is a weak wood and has a low resistance to abrasion. It should not be used where these features are critical such as for beams or flooring.

DURABILITY

Durability is poor.

WORKABILITY

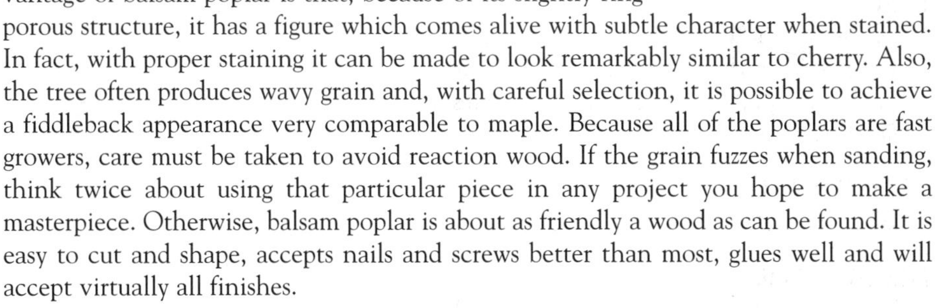

Balsam poplar is a very pleasant wood to work. The great advantage of balsam poplar is that, because of its slightly ring-porous structure, it has a figure which comes alive with subtle character when stained. In fact, with proper staining it can be made to look remarkably similar to cherry. Also, the tree often produces wavy grain and, with careful selection, it is possible to achieve a fiddleback appearance very comparable to maple. Because all of the poplars are fast growers, care must be taken to avoid reaction wood. If the grain fuzzes when sanding, think twice about using that particular piece in any project you hope to make a masterpiece. Otherwise, balsam poplar is about as friendly a wood as can be found. It is easy to cut and shape, accepts nails and screws better than most, glues well and will accept virtually all finishes.

USES

Balsam poplar pulp is used for paper production, and the wood is used for crating, core stock for plywood, veneer, ice cream sticks and fruit baskets. Some lumber is produced, generally for low-grade purposes such as sheathing and tongue and groove paneling.

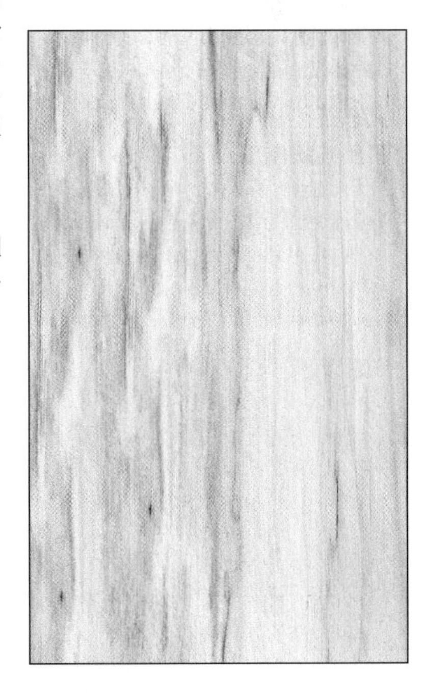

SUPPLIES

Supplies are plentiful, but it is seldom marketed specifically as balsam poplar. The price is inexpensive.

Populus deltoides

eastern cottonwood

by Max Kline

SCIENTIFIC NAME

Populus deltoides. Derivation: The genus name is the classical Latin name. The specific epithet is Latin meaning deltoid or triangular, referring to the shape of the leaves.

FAMILY

Salicaceae, the willow family. This family consists of two genera and about 350 species of which some 40 species are poplars, including 15 species native to North America.

OTHER NAMES

Carolina poplar, southern cottonwood.

DISTRIBUTION

From Quebec to southern Manitoba and North Dakota, south to Florida and Texas.

THE TREE

Eastern cottonwood grows 100 feet in height with a trunk diameter of 3 to 4 feet. It grows rapidly along watercourses. The broad leaves on its profuse branches are leathery in appearance and a bright green color. This tree is the state tree of Kansas and Nebraska.

THE TIMBER

The sapwood is grayish-white merging gradually into the brownish heartwood. The grain is fairly straight. The wood is uniform in texture and has a dull luster. Average reported specific gravity is 0.37 (ovendry weight/green volume), equivalent to an air-dried weight of 27 pcf. Green wood has some odor, which disappears when thoroughly seasoned. It has no taste. Eastern cottonwood is tough and strong when its light weight is considered.

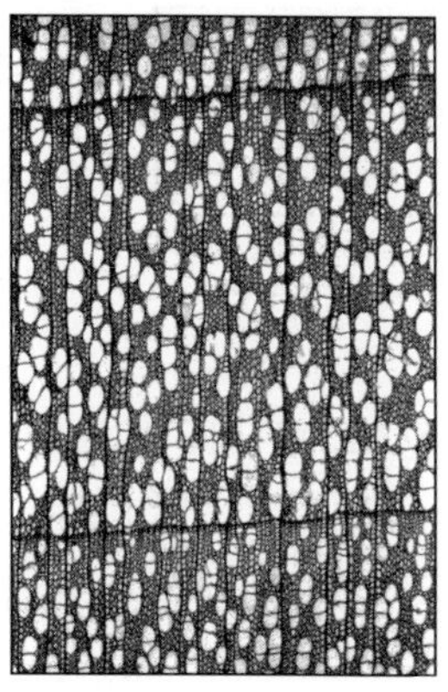

SEASONING

Eastern cottonwood timber is difficult to season as it can warp severely during drying unless carefully dried using an appropriate drying schedule. Average reported shrinkage values (green to ovendry) are 3.9% radial, 9.2% tangential and 13.9% volumetric.

DURABILITY

Eastern cottonwood is not very durable. The wood can be treated with preservatives and used for poles and posts.

WORKABILITY

Eastern cottonwood is easily worked with tools and is a favorite wood for boxes and crates because it takes stencil ink so well. It nails without splitting, but nail and screw holding properties are poor. It is easily glued and takes paint well.

USES

The principal uses are for boxes and crates, packing cases, paper pulp and excelsior. It is also used for the inexpensive parts of furniture, poultry and apiary supplies (bee hive and honey sections), kitchen cabinets, food pails and butter tubs. In its growth region, it is used for posts, poles and fuel. The tree is frequently planted on the Great Plains for shelter and ornamental purposes, as it is remarkable for rapid growth during its first 40 years.

SUPPLIES

Ample quantities are available at a low cost.

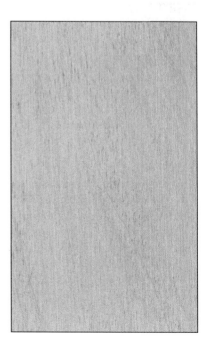

Populus tremuloides

trembling aspen

by Chuck Holder

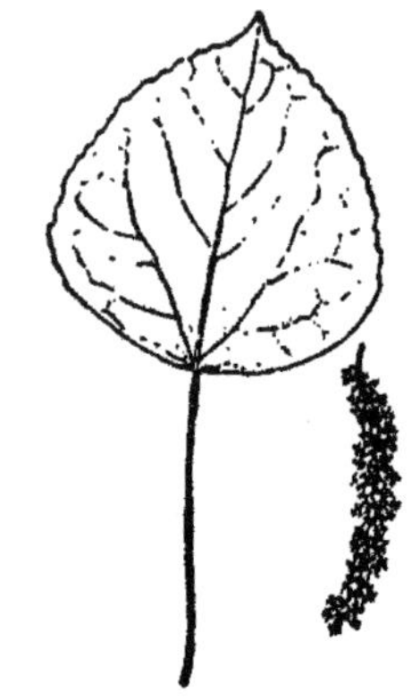

SCIENTIFIC NAME

Populus tremuloides. Derivation: The genus name *populus* is the classical Latin name of the poplars possibly originating from ancient times when poplar was called *arbor populi* meaning the tree of the people since it was used to decorate public places in Rome. The specific epithet *tremuloides* is a combination of the Latin *tremulus* meaning trembling and Greek *oides* meaning resembling.

FAMILY

Salicaceae, the willow family. This family consists of two genera and about 350 species of which some 40 species are poplars, including 15 species native to North America.

OTHER NAMES

quaking aspen (preferred common name in the United States), golden aspen, mountain aspen, quiverleaf aspen, popple (used in the wood trade), trembling poplar, peuplier faux-tremble, alamo blanco.

DISTRIBUTION

Trembling aspen is the most widely distributed tree species in North America. It occurs throughout Canada and the northern half of the United States, often growing in pure stands. Large stands of trembling aspen, consisting of thousands of trees, have been found that are all clones derived from a single aspen seedling. These are thought to have originated soon after the Pleistocene ice sheet melted, placing them among the largest and oldest organisms in the world.

THE TREE

Trembling aspen is a medium-sized deciduous (hardwood) tree that usually grows to 60 feet in height with a diameter of 24 inches. Maximum growth is 110 feet with a diameter of 39 inches. The oval to nearly round leaves are 1 to 2 inches wide with fine irregular teeth. The upper surface is deep green but underneath is paler. The leaf stalks are usually longer than the blade and are flattened and slender causing the leaves to tremble in the slightest breeze. The smooth bark is pale green to chalky white on younger trees, becoming darker and furrowed with diamond-shaped horizontal marks as the tree matures.

THE TIMBER

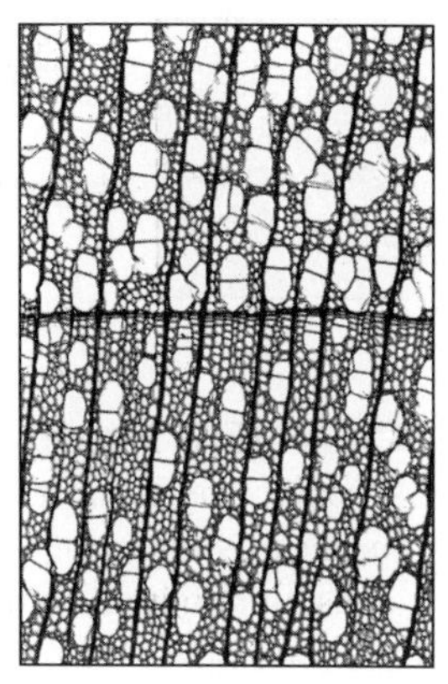

Indistinguishable from largetooth aspen, trembling aspen wood is white to creamy or grayish-white with short fibers and relatively low strength. It is straight grained with a fine and even texture and faint growth ring figure. Trembling aspen weathers to a light gray with a silvery sheen. The wood is light weight with an average reported specific gravity of 0.35 (ovendry weight/green volume), equivalent to an air-dried weight of 26 pcf. It is comparable to white spruce in strength and has a high resistance to wear for such a low-density wood.

SEASONING

Trembling aspen seasons satisfactorily with moderate shrinkage. Average reported shrinkage values (green to ovendry) are 3.5% radial, 6.7% tangential and 11.5% volumetric.

DURABILITY

Under conditions favorable to decay, trembling aspen is not durable.

WORKABILITY

Trembling aspen is easy to work with hand and machine tools. Care is required in machining to produce quality surfaces. There is a tendency for the wood to sand "woolly". Nail holding properties are good with little danger of splitting. It is moderately easy to glue.

USES

Trembling aspen is largely used for pulp and panel products such as waferboard. It is increasingly being used in the manufacture of oriented strandboard and other forms of particleboard. The lumber is used for furniture, interior trim, pallets and boxes and crates.

SUPPLIES

Trembling aspen is available commercially at a low cost and is marketed with largetooth aspen as aspen poplar or popple.

Prioria copaifera

cativo

by Jim Flynn

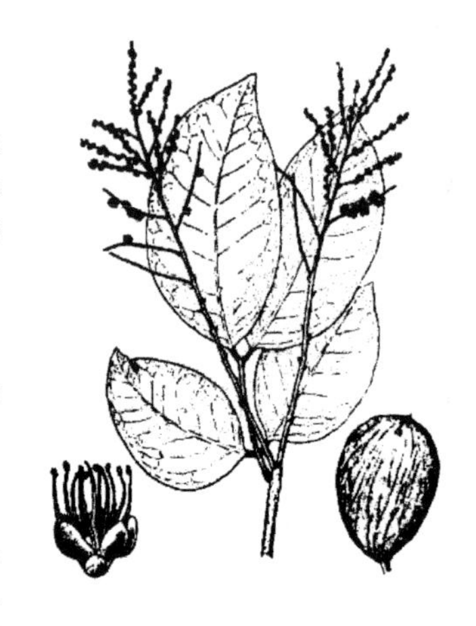

SCIENTIFIC NAME

Prioria copaifera. Derivation: The genus name is Latin meaning early or prior. The specific epithet *copaiba* is the Tupi Indian name for the tree and *fera* means yielding, thus *copaifera* means a tree yielding resin.

FAMILY

Fabaceae or Leguminosae, the legume family; (*Caesalpiniaceae*) the cassia group.

OTHER NAMES

cautivio, trementino, amansa-mujer, copachu, muramo, curacai, floresa, Spanish walnut, tabasara, taito, canime, camibar. The village of Cativá, near Colon in Panama, derives its name from the tree.

DISTRIBUTION

Native to Jamaica, Nicaragua, Costa Rica, Panama and Colombia.

THE TREE

This is the only species in the genus *Prioria*. It is an unarmed tree with pinnate leafs, usually with two pairs of pellucid-dotted leathery leaflets. Small, cream-colored, scented flowers are borne on panicled spikes. The large, flat, woody, one-seeded pods hang in small clusters from the ends of the upper branches. This species is one of those rare trees that grow in Central America in almost pure stands. The main habitat lies on the upward side of coastal mangrove swamps and in river valleys. Heights of 100 feet or more with diameters of 24 to 36 inches have been reported with clean boles two-thirds of the way before the crown. There are little or no buttresses.

THE TIMBER

The wood is generally quite uniform in color and varies from gray to pinkish or distinctly reddish. The heartwood generally consists of a small core of black wood, which is much harder than the rest of the wood. This may be attributable to some traumatic event. A zone of irregularly pigmented black or brown lines may surround the central core, and sometimes only pigmented lines are present. The sapwood on mature timber may be as wide as 10 inches or more and varies from a pale pink to buff or a reddish color. The grain is straight, and the texture is uniform and comparable to that of the fast-grown, lightweight bigleaf mahogany (*Swietenia macrophylla* WDS 254). The luster is low, and there is neither distinguishable taste nor odor. Average reported specific gravity is 0.40 (ovendry weight/green volume), equivalent to an air-dried weight of 30 pcf.

SEASONING

Cativo logs must be processed or protected with preservatives quickly after they are felled in order to avoid loss from insect and fungal attack. The darker material, due to its density, requires careful seasoning to prevent collapse. Because of the gum content, high temperatures must be used during the final stages of kiln drying. Average reported shrinkage values (green to ovendry) are 2.4% radial, 5.3% tangential and 8.9% volumetric.

DURABILITY

Without preservatives, cativo is highly susceptible to termites, pinhole borers and marine borers. When treated, it weathers moderately well and has very good dimensional stability.

WORKABILITY

The timber often contains tension wood, and positioning the logs on the saw deck prior to cutting must be carefully planned. This is particularly troublesome when rip-sawing as it pinches the saw blade and causes tool overheating. The wood machines well but sharp tools are a must to avoid a tendency toward "woolly" surfaces. It severely blunts the edges of cutting tools. The wood has a fair to good steam bending rating, and it peels well in veneer cutting operations. It produces a highly lustrous finish and glues well.

USES

Cativo has a wide range of uses including furniture, veneer, pulpwood, boxes and crates, interior trim matches and pattern making.

SUPPLIES

This timber is generally available on the commercial market worldwide.

Prosopis glandulosa

honey mesquite

by Max Kline

SCIENTIFIC NAME

Prosopis glandulosa. Derivation: The genus name is an ancient Greek plant name used by Dioscorides apparently for burdock. The specific epithet is Latin for glandular, referring to the petioles with glands at the base.

FAMILY

Fabaceae or Leguminosae, the legume family; (*Mimosaceae*) the mimosa group.

OTHER NAMES

mesquite, common mesquite, western honey mesquite, velvet mesquite, algarobo, honey pod.

DISTRIBUTION

Mesquite is found in mesas, canyons and the desert plains of the more arid regions of southern California, Nevada, Arizona, New Mexico, Texas, Kansas and Oklahoma, south into Mexico, Central America, the West Indies, Venezuela and Colombia.

THE TREE

Honey mesquite can grow as either a small tree with a short trunk that is 20 to 50 feet in height or as a large shrub. It is a very slow growing, poorly shaped, scrawny-looking tree with an exceptionally deep taproot. The branches have small spines or thorns, and the fragrant, yellowish-green flowers appear in clusters that are 2 to 4 inches long.

THE TIMBER

The color of honey mesquite is light to dark cocoa or chocolate brown with a golden luster. The very narrow sapwood is yellowish-cream or lemon yellow. The odor of fresh cut wood is fragrant, suggesting violets. Taste is not distinct. The grain is fine, wavy and close, quite frequently interlocked or cross-grained. It has a beautiful luster and is hard and heavy. Average reported specific gravity is 0.82 (ovendry weight/green volume), equivalent to an air-dried weight of 66 pcf. The texture is medium to coarse. The strength properties of mesquite are quite high, but it is somewhat brittle. It does, however, have high elasticity.

SEASONING

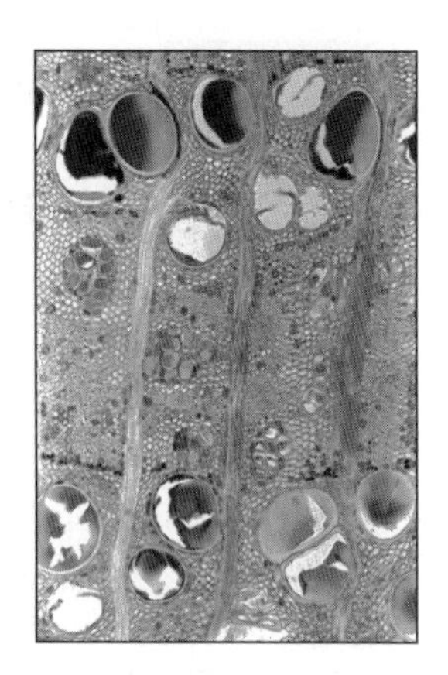

Honey mesquite timber can be air dried without serious degrade. Average reported shrinkage values (green to ovendry) are 1.6% radial, 3.2% tangential and 4.8% volumetric.

DURABILITY

The timber is very resistant to decay.

WORKABILITY

When cross grain is present, honey mesquite is somewhat difficult to work, but otherwise no problems should be expected. It is very stable, sands and polishes well and finishes to a high polish with excellent results. It turns exceptionally well.

USES

Honey mesquite provides one of the best woods for fuel in the desert and for fence posts. It is also valuable for railroad crossties, vehicle construction and for paving blocks. Its full potential has not been realized by the woodworker as beautiful novelties can be made with superior color and grain. A gum exudes from the tree trunk and is used for making mucilage. The bark is used in tanning, and the roots supply fiber for cordage and coarse fabric. The seed pods are a source of forage for livestock and are sometimes used by Native Americans as a food staple when ground into meal. They can also be fermented to make a kind of beer.

SUPPLIES

Honey mesquite is not a commercial timber, but wood is available either by personal cutting, trading or buying from wood collectors who may have access to it because of their proximity to its natural growth range. [Ed. note: Recently, honey mesquite has become available in limited quantities and has been advertised in some of the well-known wood and craft magazines in the medium price range for similar craft woods.]

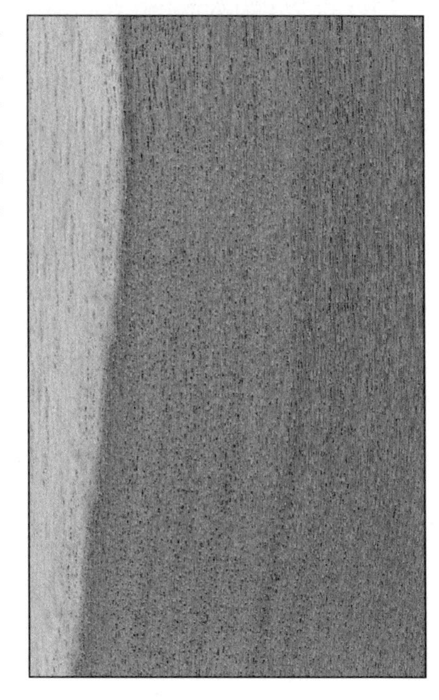

Prunus padus

European bird cherry

by Chuck Holder

SCIENTIFIC NAME

Prunus padus. Derivation: The genus name *Prunus* is the classical Latin name for the plum tree. The specific epithet *padus* is the Greek name for a wild cherry tree. *P. padus* var. *commutata* is also referred to here. *Commutata* is Latin for changed or changing.

FAMILY

Rosaceae, the rose family. *P. padus* is in the same subgenera (*Padus* or wild cherry) as chokecherry (*P. virginiana*) and black cherry (*P. serotina* WDS 223) which provides highly valued craft and furniture wood.

OTHER NAMES

bird cherry, Haag tree. The common name Mayday tree is used in western Canada due to its early flowering, but should not be confused with the name Mayday tree that refers to the English hawthorn (*Crataegus laevigata*). Also the common name bird cherry is not to be confused with pin cherry (*P. pennsylvanica*) and bitter cherry (*P. emarginata*) both of which are also referred to as bird cherry and are members of the subgenus *Mahaleb* or bird cherry.

DISTRIBUTION

Native to Europe (including Britain) and Asia. The variety *commutata* is of Manchurian origin. It is used as a city and ornamental garden tree in many places in North America.

THE TREE

European bird cherry is a small deciduous tree, which in 20 years can grow to 30 feet (max. 50 feet) in height with a crown that is 15 to 25 feet wide and a short bole up to 20 inches in diameter. It has a tendency to produce constantly forking branches, and it suckers profusely from the roots. The elliptic, dark green leaves are 4 inches long and 2.5 inches wide and turn red or yellow in the autumn. They taper to a point and are finely toothed. The dark gray bark is smooth and has with a disagreeable odor. The flowers have five white petals that are 0.375 inch wide on erect or nodding terminal racemes that are up to 6 inches long. The fruit is a rounded, 0.3125 inch, glossy black berry. The tree is vulnerable to attack by the forest tent caterpillar as well as aphids.

THE TIMBER

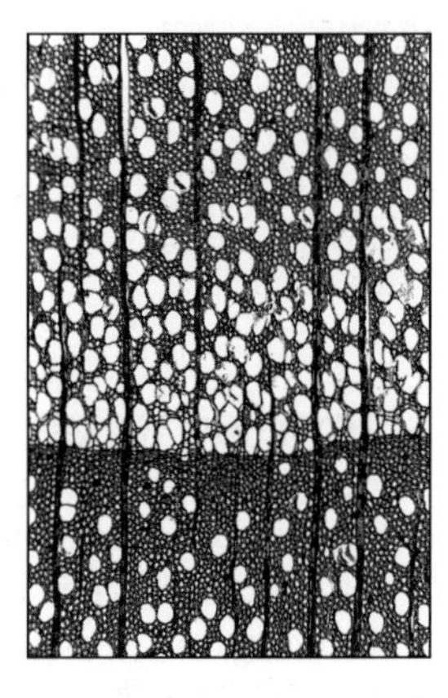

Very little is published on the characteristics of the wood. An interesting reference to *P. padus* was found in the 1902 edition of Gamble's *Manual of Indian Timbers* to the effect that the wood often displays a handsome grain, it deserves to be better known, and that it was scarcely ever used. As a close relative of black cherry, there are analogies one can draw from published data on that wood. I had a 22-year-old Mayday tree (*P. padus* var. *commutata*) in my yard which succumbed to some form of disease. I now have some small boards from its short bole (4 feet 11 inches in diameter) that have been air dried for 4 years. The heartwood is light brown, but not as pinkish-brown as black cherry. The 0.25-inch-wide growth rings provide some very distinctive color and grain patterns in the wood. A standard 3- by 6- by 0.5-inch sample had a specific gravity of 0.52 (ovendry weight/green volume), equivalent to an air-dried weight of 40 pcf.

SEASONING

European bird cherry, when air dried, seasons well. The ends should be well waxed, but even so, it tends to warp appreciably. Shrinkage values are not available.

DURABILITY

Confirmed in one published source, the wood appears durable and stable in use.

WORKABILITY

As with black cherry, European bird cherry works well, saws cleanly and planes and turns well. It takes glue well and will take a very fine finish.

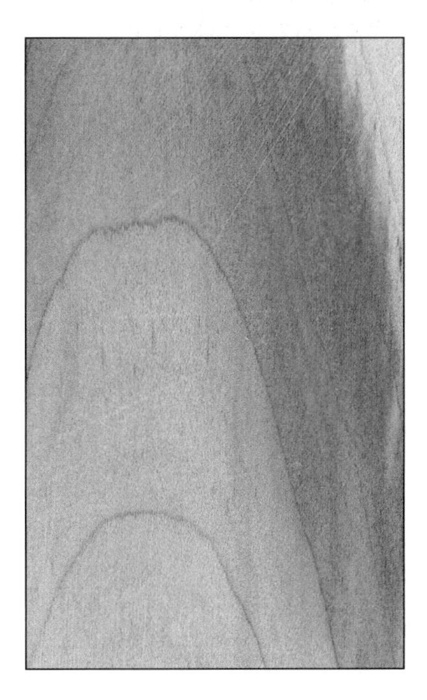

USES

European bird cherry is used locally for small furniture and craftwork. Some particularly beautiful turned work can be obtained from this wood.

SUPPLIES

It is not available commercially. Befriend a local arborist and ask them to keep an eye out for unwanted specimens.

Prunus serotina

black cherry

by Max Kline

SCIENTIFIC NAME

Prunus serotina. Derivation: The genus name is the classical Latin name for the plum tree. The specific epithet means late, referring to the relatively late-maturing fruit.

FAMILY

Rosaceae, the rose family.

OTHER NAMES

cabinet cherry, rum cherry, wild cherry, wild black cherry.

DISTRIBUTION

Black cherry grows from Nova Scotia to Maine and Minnesota, south to eastern Texas and east to central Florida. It is also found in western and southern Mexico and Guatemala.

THE TREE

Black cherry is a handsomely shaped tree, growing to a height of 60 to 80 feet with a diameter of 2 to 3 feet. It prefers deep, rich soil with uniform moisture. The pea-sized dark red cherries are edible and have a slightly bitter taste. They are sometimes used for jellies and a beverage called rum cherry.

THE TIMBER

The narrow sapwood is whitish to pale reddish-brown but is not always clearly defined from the heartwood. The heartwood is variable in color from light yellowish or pinkish-brown to dark reddish-brown but is normally uniform in a given specimen. It has a rich luster usually with a straight grain but frequently pieces are found with dark wavy streaks of striking beauty. The crotches and burls are highly prized for figured veneers. The figure on quartersawn surfaces is beautiful. Average reported specific gravity is 0.47 (ovendry weight/green volume), equivalent to an air-dried weight of 36 pcf. This species has strength properties that are in many respects excellent, in fact superior to those of some woods classed as structural timbers.

SEASONING

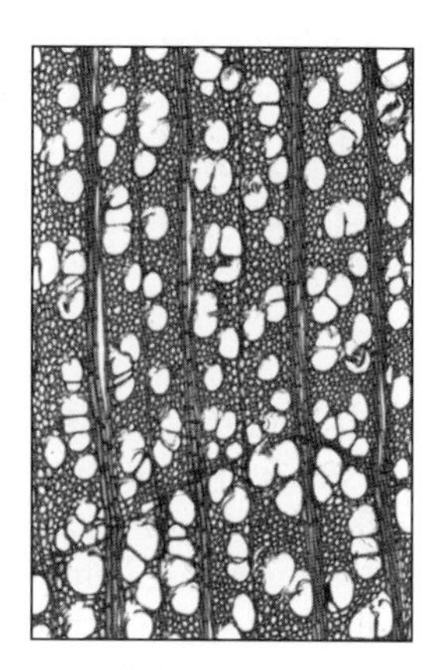

Black cherry seasons mildly and well, normally drying at a better than average rate. It can be air or kiln dried with equally good results. It has a tendency to warp, but this can be corrected by weighting the seasoned stock. Average reported shrinkage values (green to ovendry) are 3.7% radial, 7.1% tangential and 11.5% volumetric.

DURABILITY

Beetle damage is found in black cherry, and the wood is not resistant to fungal attack. For the purposes for which black cherry is used, durability is unimportant.

WORKABILITY

In general, this species works well, saws cleanly, planes excellently and is a useful timber for turnery purposes. Screw holding properties are good and glue adhere well. It takes all kinds of finishes very well, turning darker and richer with age. It often bears a strong resemblance to genuine mahogany and for this reason is frequently called New England mahogany.

USES

Black cherry was used extensively in the 18th century for manufacturing furniture, cabinets and interior trim. It was also used as backings for electro types and zinc etchings, woodenware, veneer, musical and scientific instruments, novelties, tool handles and furniture. Black cherry has regained its popularity as a cabinetry wood and is being used despite its increasing cost.

SUPPLIES

Although supplies are not as plentiful as in the past due to extensive cutting, black cherry is available at ever increasing prices on the North American market. The largest production comes from the New York, Pennsylvania and West Virginia mountain areas.

Pseudotsuga menziesii

Douglas-fir

by Max Kline

SCIENTIFIC NAME

Pseudotsuga menziesii. Derivation: The genus name is from the Greek *pseudo* meaning false and the Japanese *tsuga* meaning hemlock, referring to the relationship to *Tsuga*. The specific epithet is in honor of Archibald Menzies (1754 to 1842), a Scottish physician and naturalist, who discovered it in 1793 at Nootka Sound on Vancouver Island, British Columbia.

FAMILY

Pinaceae, the pine family, a temperate Northern Hemisphere family of trees and shrubs with some nine genera and 210 species.

OTHER NAMES

Douglas spruce, Douglas yew, Oregon pine, yellow Douglas-fir, Puget Sound pine, red fir, red spruce.

DISTRIBUTION

The range extends from British Columbia and western Alberta southward through Washington, Oregon, northern California, Idaho, western Montana through Wyoming, Nevada, Utah, Colorado, Arizona, New Mexico and into Mexico and western Texas. It attains the best growth in Washington, Oregon and British Columbia in the region between the coast and the Cascade Mountains.

THE TREE

Douglas-fir reaches a height of over 200 feet with a clear bole for over a third of its height. It has a diameter of 3 to 12 feet. The largest trees may be 400 to 1,000 years old. The reddish-brown bark may be 10 to 12 inches thick on large mature trees. Douglas-fir is the state tree of Oregon.

THE TIMBER

The heartwood is yellowish to very light reddish-tan with an orange hue. The narrow, almost pure white sapwood seldom exceeds 3 inches in width. Average reported specific gravity ranges from 0.45 to 0.46 (ovendry weight/green volume), equivalent to an air-dried weight of 34 to 35 pcf. The grain is normally straight, although there is a tendency toward curly or wavy characteristics. The grain is attractive when manufactured into rotary-cut veneer because of the large growth rings in rapidly grown Douglas-fir. Compared with other North American woods, Douglas-fir has the highest strength to weight ratio.

SEASONING

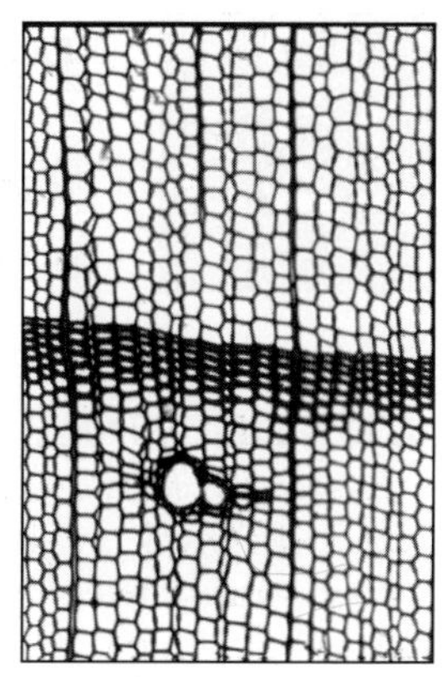

Douglas-fir seasons rapidly with only slight degrade. Average reported shrinkage values (green to ovendry) are 4.8% radial, 7.5% tangential and 11.8% volumetric. High initial temperatures in drying can be tolerated.

DURABILITY

Douglas-fir heartwood is moderately resistant to decay. The wood is moderately difficult to treat with preservatives.

WORKABILITY

The wood is brittle and splits easily. It works well with machine tools but is rather difficult to work with hand tools unless they are in excellent condition. Glues adhere satisfactorily. It holds latex paints relatively well but not oil-based paints.

USES

Douglas-fir is used mostly for building and construction purposes in the form of lumber, marine fendering, piles and plywood. Considerable quantities are used for railroad crossties, cooperage stock, mine timbers, poles and fencing. Douglas-fir is also used in the manufacture of products such as sashes, doors, laminated beams, general millwork, pallets and boxes and crates. Small amounts are used for flooring, furniture, ship and boat construction and tanks.

SUPPLIES

Since the tree is fast growing over a wide area of North America, supplies are very adequate. The cost of the timber falls into the low to medium range. Douglas-fir is remarkably free of knots, strong and light, which makes it one of the world's best known softwood timbers.

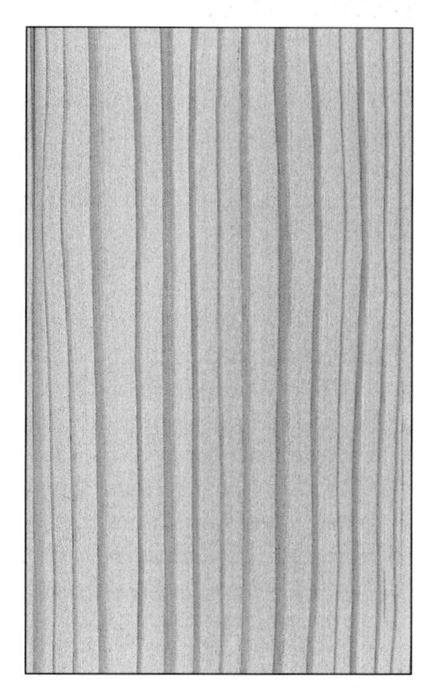

Pterocarpus dalbergioides

Andaman padauk

by Max Kline

SCIENTIFIC NAME

Pterocarpus dalbergioides. Derivation: The genus name *ptero* is Greek meaning wing and *carpus* meaning fruit, which refers to the fleshy fruit stalks. The specific epithet refers to the genus *Dalbergia*.

FAMILY

Fabaceae or Leguminosae; the legume family; (*Papilionaceae*) the pea or pulse group.

OTHER NAMES

padauk, Andaman redwood, vermilion wood, red narra, yellow narra, maidon.

DISTRIBUTION

Andaman padauk occurs in the deciduous and semi-deciduous forests of the Andaman Islands where it is the principal species.

THE TREE

Andaman padauk is a large tree attaining a height of 80 to 120 feet with a diameter of 3 to 5 feet. The bole is straight and cylindrical and can be 40 feet or more to the first branch. It should not go unnoticed that this species of padauk was the only timber that was considered valuable in the Andaman forest when extraction first started. A small party with an elephant would disappear into the forests for a several days and return dragging along a fine handhewn padauk square. The supply of these trees has been diminishing ever since.

THE TIMBER

The narrow sapwood is whitish or yellowish-gray and is clearly defined from the heartwood. Although most commonly a crimson shade with darker streaks, the heartwood is extremely variable ranging from yellowish-brown to dark red. With time the color tends to tone to reddish-brown. The odor of freshly cut wood is faintly aromatic, and the taste is indistinct. Planed surfaces of the wood are mildly silky to highly lustrous. Average reported specific gravity is 0.63 (ovendry weight/green volume), equivalent to an air-dried weight of 49 pcf. The texture is moderately coarse but uniform. The grain is usually irregular or wavy producing a ribbon figure. Andaman padauk is very strong wood, equal in most respects to teak (*Tectona grandis* WDS 261).

SEASONING

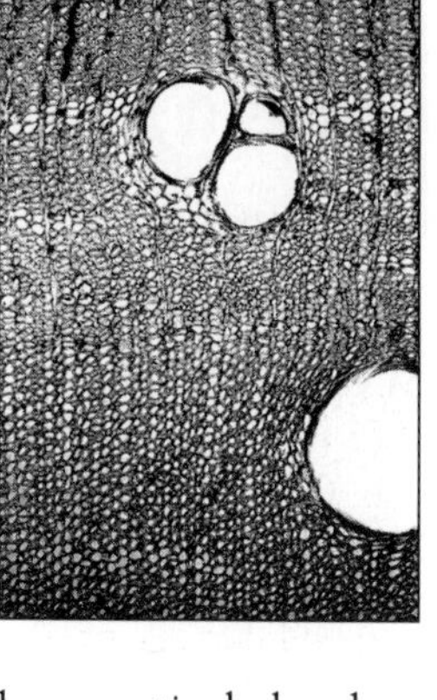

Andaman padauk can be either air or kiln dried with little tendency to warp or split. Average reported shrinkage values (green to ovendry) are 3.3% radial, 4.4% tangential and 6.4% volumetric. To prevent the development of fine surface cracks, the trees are girdled and left to die before felling.

DURABILITY

Pinhole borer damage can be found in the sapwood, but the resistance to insect attack is fairly high. The timber is also resistant to fungal attack.

WORKABILITY

Andaman padauk is not the easiest wood to work due to the alternate layers of hard and soft grain, which can cause problems particularly when planing. It turns well and the edges remain sharp. Glue adheres satisfactorily. A fair amount of filling is necessary prior to finishing because of the large pores present in the timber. Nail and screw holding properties are good, but pre-boring for nails is advisable.

USES

In the past, the Pullman Company made much use of Andaman padauk for trim in dining, smoking and sleeping cars. Due to its good wearing properties, it was used for counters and flooring. Today its cost has restricted it to a furniture and cabinetwood where it finds use in turnery, carving, veneer and interior fittings.

SUPPLIES

Supplies of Andaman padauk are limited and the price may be considered costly.

Pterocarpus indicus

narra

by Jim Flynn

SCIENTIFIC NAME

Pterocarpus indicus. Derivation: The genus name *ptero* is Greek meaning wing and *carpus* meaning fruit, which refers to the fleshy fruit stalks. The specific epithet means the plant is "of India".

FAMILY

Fabaceae or Leguminosae, the legume family; (*Papilionaceae*) the pea or pulse group.

OTHER NAMES

There are at least 70 species of *Pterocarpus* growing in pan-tropical regions. Their common names are

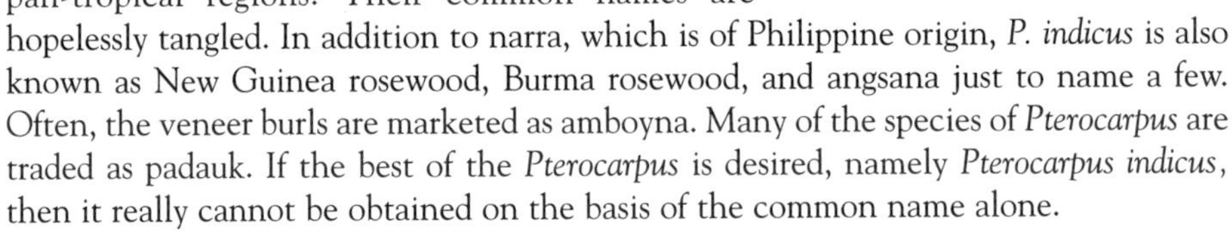

hopelessly tangled. In addition to narra, which is of Philippine origin, *P. indicus* is also known as New Guinea rosewood, Burma rosewood, and angsana just to name a few. Often, the veneer burls are marketed as amboyna. Many of the species of *Pterocarpus* are traded as padauk. If the best of the *Pterocarpus* is desired, namely *Pterocarpus indicus*, then it really cannot be obtained on the basis of the common name alone.

DISTRIBUTION

This tree is native to Malaysia, Indonesia and the Philippines. It is, however, widespread throughout Southeast Asia and into Australia.

THE TREE

Narra is a great, handsome, spreading tree growing to heights of 80 feet or more with diameters sometimes as large as 6 feet. It is one of the finest of the tropical shade trees. It has a broad crown of long droopy branches that nearly touch the ground similar to the weeping willow. The pinnate leaves are dark glossy green on top and the underside is light dull green. The tree is essentially an evergreen but loses its leaves during climatic stress and quickly forms new ones. The seedpods are circular and winged. The bark exudes a bright red sap when slit or wounded. Narra is the national tree of the Philippines.

THE TIMBER

Narra is a hard and strong wood with a full spectrum of colors ranging from pale yellow to deep blood red, all of which can be found in the same species. At one time, the wood from trees native to the Philippines was sorted according to color. The lighter woods were marketed as narra amarilla and the darker shades as narra encarnada. The grain is mostly straight but sometimes crossed, wavy or interlocked thus producing interesting patterns. A faint cedar-like aroma can be detected when working with the wood. There is a wide variety in the density of the wood depending upon growing

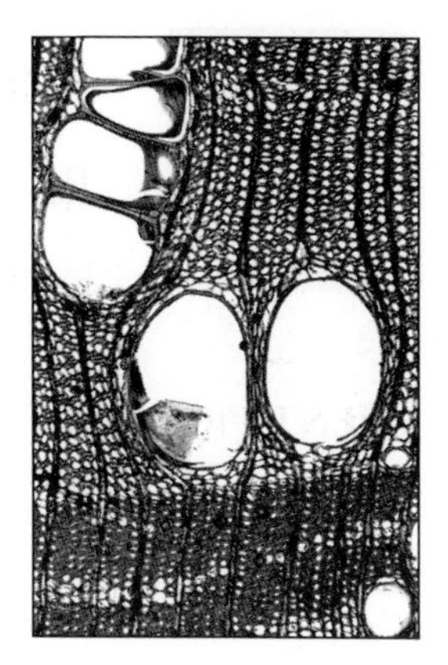

conditions, but average reported specific gravity is 0.52 (ovendry weight/green volume), equivalent to an air-dried weight of 40 pcf. Chips of wood can impart a blue and yellow fluorescence to water.

SEASONING

Narra seasons very well with minimal checking and warping. Average reported shrinkage values (green to ovendry) are 2.8% radial and 4.0% tangential.

DURABILITY

The timber is extremely durable. The average service life of wood in contact with the ground is about 25 years. It is, however, subject to attack by powder post beetles.

WORKABILITY

Narra wood has been and remains a favorite of many native woodworkers in its growth area, especially in the Philippines. The ease with which it can be worked with unsophisticated tools has long been demonstrated. It machines well with both hand and machine tools but may exhibit some difficulty in planing if interlocked grain is present. It nails and glues well and takes an excellent finish and high polish.

USES

A wide range of uses for narra has been recorded including flooring, boat building, furniture, musical instruments, veneer, carvings and turnery.

SUPPLIES

Care must be exercised in acquiring narra because this common name is associated with so many other woods of the genus *Pterocarpus*. Many wood suppliers stock this wood under a variety of common names and some of them, to their credit, indicate the scientific name of *Pterocarpus indicus*. Supplies are not especially plentiful and are on the high end of the price range.

Pterocarpus macrocarpus

Burma padauk

by Max Kline

SCIENTIFIC NAME

Pterocarpus macrocarpus. Derivation: The genus name *ptero* is Greek meaning wing and *carpus* meaning fruit, which refers to the fleshy fruit stalks. The specific epithet is Latin for big fruit.

FAMILY

Fabaceae or Leguminosae, the legume family; (*Papilionaceae*) the pea or pulse group.

OTHER NAMES

mai pradoo, pradoo.

DISTRIBUTION

Burma padauk occurs in the deciduous and mixed forests in the hills of Myanmar (Burma), often associated with stands of teak (*Tectona grandis* WDS 261) and Indian laurel (*Terminalia elliptica* WDS 262).

THE TREE

Burma padauk grows to a height of 60 to 80 feet with a diameter of 2 to 3 feet. Usually, the tree has a clean bole 20 to 40 feet in length. It is found sparsely distributed with an average of only 20 to 30 trees per 100 acres of forestland. This rather handsome tree has a spreading round crown, long drooping branches and small pinnate leaves. It bears a profusion of yellowish, sweet-scented flowers in March and April, followed by small, circular winged pods.

THE TIMBER

Burma padauk is yellowish-red to brick red with occasional darker lines which on exposure tone to attractive golden brown. The interlocked grain produces a narrow ribbon-grained figure. It is moderately coarse in texture, and the luster is medium to high. The odor of freshly cut wood is fairly aromatic. Average reported specific gravity is 0.75 (ovendry weight/green volume), equivalent to an air-dried weight of 60 pcf. Burma padauk is one of the strongest and hardest of Southeast Asian timbers. In this respect, it surpasses by a considerable margin the strength properties of either Andaman padauk (*Pterocarpus dalbergioides* WDS 225) or teak.

SEASONING

Except for a tendency to develop fine surface cracks, Burma padauk timber dries slowly without much degrade. Average reported shrinkage values (green to ovendry) are 3.4% radial, 5.8% tangential and 8.4% volumetric. It is well known that it is dimensionally stable in service.

DURABILITY

The wood is unusually resistant to termites. Wood set in the ground for 15 years showed no signs of fungal decay when removed.

WORKABILITY

Seasoned timber is quite refractory and difficult to cut into lumber. It cuts well on the veneer slicer, however, and produces a smooth and glossy surface. The wood glues well and sands easily without fuzzing. It takes a beautiful finish and high natural polish. It is very resistant to denting, a feature which is favorable for furniture construction. The wood holds screws firmly but pre-boring is necessary.

USES

Many of the uses of Burma padauk are similar to those of teak but it is preferred where greater strength is required. Uses include cartwheels, boat frames, oil presses, tool handles, parquet flooring, furniture and cabinetry. In Myanmar, it is the wood most used for harps and has also proved satisfactory for drumsticks, billiard tables and cues.

SUPPLIES

Burma padauk veneer and lumber are available from importers but are costly.

Pterocarpus soyauxii

African padauk

by Max Kline

SCIENTIFIC NAME

Pterocarpus soyauxii. Derivation: The genus name *ptero* is Greek meaning wing and *carpus* meaning fruit, which refers to the fleshy fruit stalks. The specific epithet is in honor of Hermann Soyaux (b. 1852), the leader of a botanical expedition in Africa (1873 to 1876).

FAMILY

Fabaceae or Leguminosae, the legume family; (*Papilionaceae*) the pea or pulse group.

OTHER NAMES:

comwood, barwood, corail, yomo, vermilion, bois rouge, African coralwood, muenge.

DISTRIBUTION

This species occurs in Gabon, Congo (formerly Zaire), Cameroon, Nigeria and the Ivory Coast.

THE TREE

With a straight and well shaped trunk, African padauk reaches a height of 70 to 100 feet with a breast height diameter of 50 inches. It yields a bright red dye, which is used by African natives to make a cosmetic that they smear on their bodies for religious festivals.

THE TIMBER

The heartwood is bright orange-red to blood red, and the sapwood is grayish-white. The odor of freshly cut wood is faintly aromatic, and the luster is medium to high. Average reported specific gravity ranges from 0.55 to 0.67 (ovendry weight/green volume), equivalent to an air-dried weight of 42 to 52 pcf. The grain is irregular to wavy. It has a moderately coarse texture with large pores. This wood possesses good strength, is flexible, shock resistant and is resistant to compression and dents.

SEASONING

African padauk seasons fairly slowly with minimal degrade. Average reported shrinkage values (green to ovendry) are 3.3% radial, 5.2% tangential and 7.6% volumetric. It is dimensionally stable is use.

DURABILITY

African padauk timber is classified as very durable.

WORKABILITY

The wood saws and planes easily to a very smooth surface. It glues well and takes an excellent finish.

USES

Before the use of aniline dyes, the dye from African padauk was in demand by the dyeing industry. Currently the timber is used by the veneer and lumber trades. It is also used to a limited extent for fancy turnery and for high-quality tool handles. It is an excellent flooring timber, suitable for heavy traffic in public buildings. The bulk of the exported material is manufactured into veneer, which is used in furniture, cabinets, inlay, novelties and decorative panels. The red color effects that are found in the timber are well adapted for making panels of Oriental design.

SUPPLIES

African padauk veneer and lumber are available but the price is high.

Pyrus communis

pearwood

by Max Kline

SCIENTIFIC NAME

Pyrus communis. Derivation: The genus name is the classical Latin name for pear. The specific epithet is Latin for common.

FAMILY

Rosaceae, the rose family.

OTHER NAMES

pear tree, common pear.

DISTRIBUTION

This tree has wide distribution throughout Europe, parts of Asia and parts of the United States.

THE TREE

There are many cultivars to this species as well as most of the other species of pear. This presents a variety of shapes and forms from shrub-like to full-grown trees. *Pryus communis* can reach heights of 50 feet or more with trunk diameters of 1 to 2 feet. The dark green, leathery leaves have serrated margins and are 3 inches long with long stalks. The fragrant, five-petalled flowers are known for their beauty. The tree survives best in a cool climate.

THE TIMBER

The wood is pinkish-brown. The sapwood and heartwood are usually poorly defined, although in old trees the heartwood is often darker and distinct. The grain is straight, and the texture is very close and uniform. Average specific gravity is 0.58 (ovendry weight/green volume), equivalent to an air-dried weight of 45 pcf. It has no distinctive smell. Planed surfaces have a mild, silky sheen. It is similar to oak in strength properties, but tougher and more difficult to split.

SEASONING

When air drying, pearwood timber dries slowly and shows a marked tendency to distort unless heavy weighting of the timber stocks is used. It kiln dries satisfactorily if mild initial temperatures are employed. Shrinkage values are not available. After drying, the wood is dimensionally stable.

DURABILITY

Pearwood is not sufficiently durable for outdoor use without a preservative treatment.

WORKABILITY

Pearwood glues very easily and stains and finishes well. Because of the uniform quality of its fibers, pearwood can be worked with tools in any direction and sands without producing a fuzzy surface. It turns well and is excellent for carving. Nail and screw holding properties are good. It peels well for veneer.

USES

A unique use for pearwood veneer is for marquetry. Because it can be uniformly dried, it is sometimes stained black and used as a substitute for ebony. It is used for mechanical instruments, rulers, wood engravings, printing blocks, tool handles and for fancy turnery. In the past, in Europe, it was made into fine furniture and cabinets. [Ed. note: Pearwood is one of the finest tonewoods for stringed musical instruments, especially esteemed in Europe.]

SUPPLIES

The most superior pearwood comes from Germany and France. Supplies are very limited as only the oldest trees unfit for bearing fruit are cut for lumber and not much of this is exported from Europe. Usually the only imported wood available in the United States is in the form of veneer, and it is expensive. Domestic supplies are also quite scarce, and the quality and appearance of this wood is not as attractive as that from the European market. [Ed. note: Recent ads indicate that pearwood from Europe is now being sold in the United States at a reasonable price.]

Quercus alba

white oak

by Max Kline

SCIENTIFIC NAME

Quercus alba. Derivation: The genus name is the classic Latin name for oaks, derived from the Celtic fine and tree. The specific epithet is Latin for white.

FAMILY

Fagaceae, the beech family.

OTHER NAMES

eastern white oak, forked-leaf white oak, stave oak, ridge white oak.

DISTRIBUTION

The range extends from southern Quebec, southern Ontario, Minnesota and Nebraska, south to Florida and Texas.

THE TREE

White oak is one of the most valuable and useful of the North American forest trees. It is usually 70 to 80 feet in height with a trunk diameter of 2 to 3 feet. The tree is sturdy and rugged, and few hardwoods are more handsome. It has a high straight stem with a broad round crown of wide-spreading branches. The tall trunk produces a large quantity of the finest lumber. In the autumn, the leaves turn to bright yellow or red and later pale brown. They usually cling to the twigs throughout the winter. Squirrels and birds relish the acorns. White oak is the state tree of Connecticut, Illinois and Maryland.

THE TIMBER

The heartwood is light tannish-brown, and the narrow sapwood is nearly white. Average reported specific gravity is 0.60 (ovendry weight/green volume), equivalent to an air-dried weight of 46 pcf. It is very strong and hard with a moderately fine grain. Quartersawn oak reveals a large number of rays. The grain is usually straight. White oak ranks fairly high in all strength properties and is classified as very good for steam bending. Its wear resistance is outstanding.

SEASONING

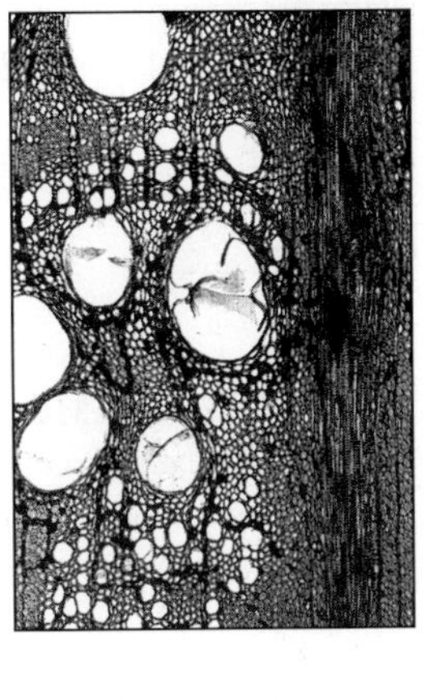

White oak timber seasons fairly slowly and tends to check and split but movement in service is only medium. Average reported shrinkage values (green to ovendry) are 5.6% radial, 10.5% tangential and 16.3% volumetric.

DURABILITY

The heartwood is renowned for its durability and is commonly used outdoors without preservative treatment. Fire and gypsy moths are the greatest enemies to the growing tree.

WORKABILITY

White oak timber may be worked fairly easily, taking a smooth finish. It can be glued satisfactorily.

USES

Uses range from fine cabinetry, interior trim, general millwork, flooring and veneer for paneling to heavy construction work such as bridges, ships, railroad cars and motor vehicle parts. Because of its impermeability, the timber is used extensively for liquor barrels and other containers. The tannic acid in the wood causes unsightly discoloration from corrosion when iron, steel or lead materials are in contact with the wood under damp conditions. Therefore, use of non-ferrous fastenings is recommended for assembly purposes.

SUPPLIES

White oak veneers are plentiful. Lumber is available at a moderate cost when compared to other hardwoods.

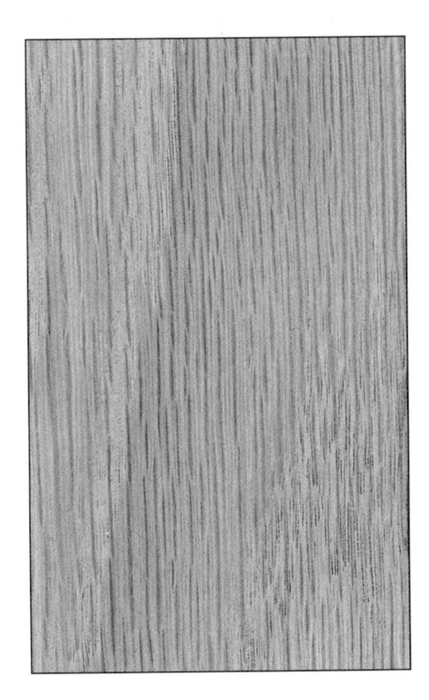

Quercus arkansana

Arkansas oak

by Jim Flynn

SCIENTIFIC NAME

Quercus arkansana. Derivation: The genus name is the classic Latin name for oaks, derived from the Celtic fine and tree. The specific epithet means the plant is "of Arkansas" in recognition of where it was discovered.

FAMILY

Fagaceae, the beech family.

OTHER NAMES

Arkansas water oak, water oak.

DISTRIBUTION

The range maps in the *Atlas of U.S. Trees, Vol. 4* show the most accurate, but spotty, distribution of this species. The largest grouping is in southwestern Arkansas, it skips Mississippi completely, a small colony is in Tammany Parish, Louisiana, three isolated outcroppings are in the Florida panhandle and a few pockets of trees grow in Georgia and Alabama. Arkansas oak was first found by B. F. Bush near the village of McNab in Hempstead County, Arkansas. Dr. Charles C. Mohr in Alabama apparently discovered the species in 1880, but he supposed it to be a hybrid of *Quercus marilandica* (blackjack oak) and *Q. nigra* (water oak). In 1924, Dr. William Trelease published a monograph on American oaks and treated *Quercus arkansana* as a hybrid apparently referencing Dr. Mohr's work. In 1925, Ernest Palmer took issue with the alleged hybridization in the *Journal of the Arnold Arboretum*. Palmer proved Arkansas oak was a distinct species. The official list of trees of Alabama does not show *Q. arkansana* but we know it to be there.

THE TREE

Arkansas oak is a member of the red-black oak group. With a narrow rounded crown, Arkansas oak reaches 80 feet in height with a circumference up to 11 feet. The thick black bark is deeply furrowed. The small leaves are 2 to 3 inches long, broadest above the middle and slightly three-lobed. The acorns have short (less than 0.5 inch) stalks and are very broad near the base, rounded at the tip and generally 0.33 inch long and 0.5 inch thick. Palmer, in the above cited reference, states that there is no species of American oak that is truer to type or more easily recognized in the field than the Arkansas oak, and it is a relic nearing extinction because of the peculiar ecological conditions under which it has survived.

THE TIMBER

Arkansas oak has the usual characteristics of all red oaks. The sapwood is white and averages about 1.5 inches in width. The heartwood is brown, often tinged with red. Milled lumber cannot be separated from other species of red oak. Average specific gravity is about 0.55 (ovendry weight/green volume), equivalent to an air-dried weight of 42 pcf. Second-growth timber is generally harder than old-growth stock.

SEASONING

Red oaks typically suffer from considerable shrinkage when drying. Normal drying procedures undertaken with care ensure quality wood. While precise shrinkage values are not available, they are estimated to be within the range of those for white oak (*Quercus alba* WDS 230).

DURABILITY

The heartwood is not resistant to decay when moist so it is a poor choice for use in damp weather areas unless treated with preservatives.

WORKABILITY

Since ancient times practically all oaks have been the choice for combining strength and beauty. While Arkansas oak is a hard species, it is easily worked with sharp tools. Because it is a red oak, the latewood pores are larger than those in the white oak group. This makes the grain pattern on flat surfaces more distinct. Use of a paste filler is generally desired as the first step in finishing. It has good bending qualities.

USES

Major uses include furniture, flooring, fuel and caskets. With preservative treatment, uses include railroad ties, mine timbers and fence posts.

SUPPLIES

As is the case in practically all of the American hardwood forests, there is no attempt to separate oaks other than "white from red". Consequently, there is no tally as to the specific species on the market. Arkansas oak may be obtained in the field.

Quercus robur

English oak

by Max Kline

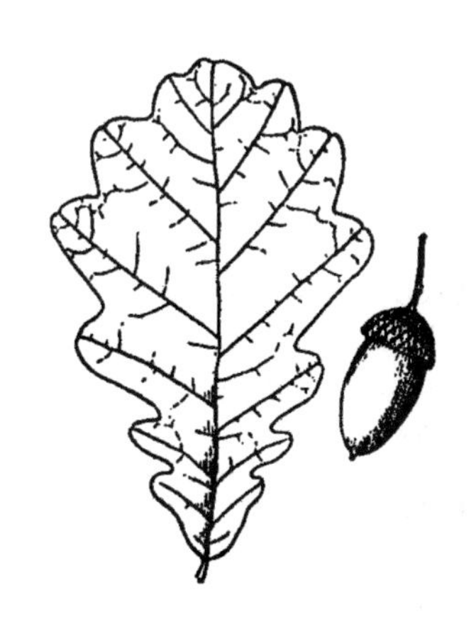

SCIENTIFIC NAME

Quercus robur. Derivation: The genus name is the classic Latin name for oaks, derived from the Celtic fine and tree. The specific epithet is an ancient Latin name referring to the strength of the wood.

FAMILY

Fagaceae, the beech family.

OTHER NAMES

Since English oak comes from two species, namely *Q. robur* and *Q. petraea*, and it is difficult to distinguish between them, they will be treated as one timber in this Wood Data Sheet. Other common names are European oak, polladro, pedunculate oak, sessile oak, durmast oak, French oak, bog oak and pollard oak.

DISTRIBUTION

English oak is found throughout western and central Europe. Major sources of supply are France, Poland, Yugoslavia and the Baltic countries. It is the most common forest tree in Britain, especially in Wales.

THE TREE

English oak reaches a height of 60 to 100 feet with a diameter of 4 to 6 feet although it can be considerably larger. It grows in pure stands and is often the dominant species in mixed forests. When growing in the open it branches quite low, but under forest conditions it forms a straight clear bole up to 50 feet in length.

THE TIMBER

The sapwood is 1 to 2 inches wide and is light colored and distinct from the yellowish-brown heartwood. The grain is usually straight except in the vicinity of knots. Plainsawn English oak has a well marked growth ring figure like other ring-porous hardwoods. Trees that have been attacked by a fungus produce a variety of wood known as European brown oak. When this timber is dried, the fungus dies and there is no danger of further decay. Average specific gravity is 0.58 (ovendry weight/green volume), equivalent to an air-dried weight of 45 pcf.

SEASONING

English oak is not an easy wood to season. Drying is a slow process even in the kiln and care must be taken to avoid excessive degrade. While precise shrinkage values are not available, they are estimated to be within the range of those for northern red oak (*Quercus rubra* WDS 233).

DURABILITY

Beetles of various types can damage the wood. When used for such purposes as fencing, the wood should be treated with a preservative.

WORKABILITY

English oak is not particularly difficult to machine. It holds nails and screws firmly, but pre-boring is advisable. It stains well and gives excellent results when painting, varnishing, waxing and polishing. It can be glued satisfactorily.

USES

English oak, especially the brown variety, maintains a prominent place among the decorative woods for paneling and joinery. Owing to strength, durability and beauty, it has been the traditional timber for construction in most European countries. Large quantities are used for shipbuilding, flooring, caskets, casks for beer and wine, carriages and general construction work.

SUPPLIES

The brown variety of English oak is sought for special uses and effects and is available only in very small quantities at a high price in the United States. The more common English oak is not imported into the United States because its uses are the same as those of our domestic oaks.

Quercus rubra

northern red oak

by Jon Arno

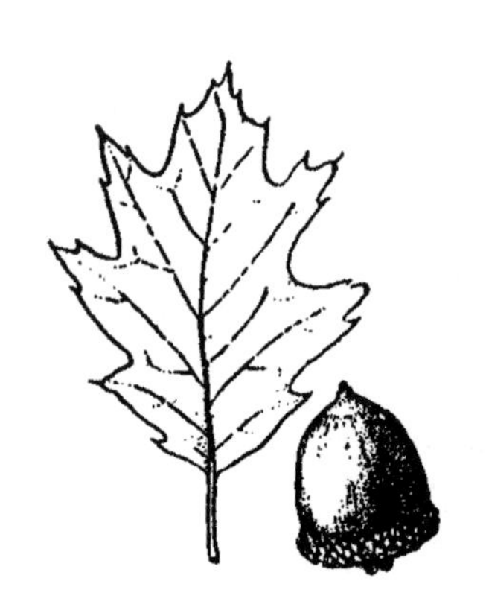

SCIENTIFIC NAME

Quercus rubra. Derivation: The genus name is the classic Latin name for oaks, derived from the Celtic fine and tree. The specific epithet is Latin for red.

FAMILY

Fagaceae, the beech family.

OTHER NAMES

eastern red oak, gray oak, mountain red oak, chene rouge.

DISTRIBUTION

Northern red oak has an expansive range extending from southern Quebec to central Alabama and from extreme eastern Nebraska to the Atlantic Ocean.

THE TREE

Typically, northern red oak is smaller than white oak (*Q. alba* WDS 230) reaching heights of 60 to 70 feet and diameters of up to about 3 feet. The leaves have pointed lobes and, on mature trees, the almost black bark is deeply furrowed. Northern red oak is the official tree of Prince Edward Island, Canada and the state tree of New Jersey.

THE TIMBER

The heartwood is light reddish-tan. The sapwood, usually about 2 inches wide, is almost white in color. Like most oaks, the rays are prominent, but they are generally shorter, narrower and darker in color than those of white oak. As a result, quartersawn northern red oak is far less showy. With an average reported specific gravity of 0.56 (ovendry weight/green volume), equivalent to an air-dried weight of 43 pcf, it is also somewhat softer than white oak (specific gravity of 0.60). Because northern red oak is extremely ring porous, timber from slow-growing trees will be less dense and more porous than that produced by fast-growing trees. Northern red oak is slightly more elastic than white oak, but it is weaker with respect to shear parallel to the grain and hardness. Both are strong timbers and appropriate for applications where strength is important.

SEASONING

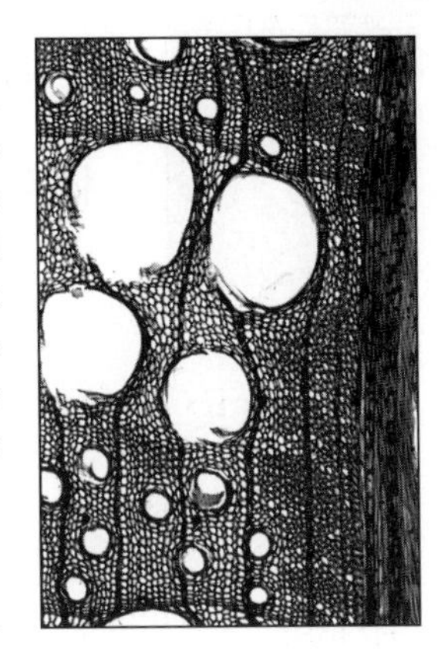

Northern red oak experiences above average shrinkage while drying, so the pile should be well weighted to prevent the upper layers from warping. Average reported shrinkage values (green to ovendry) are 4.0% radial, 8.6% tangential and 13.7% volumetric. Otherwise, it is not difficult to air dry. It dries substantially faster than white oak because, unlike white oak, its vessels (pores) are relatively free of tyloses. The plentiful rays tend to inhibit excessive checking, but it is always wise to coat the end grain.

DURABILITY

Unless treated with preservatives, northern red oak is nondurable and subject to fungi and insect attack.

WORKABILITY

This species is hard, but it machines reasonably well for a ring-porous wood. Because of the prominent rays, care should be taken to segregate quartersawn and flatsawn stock using one or the other consistently in a given project. Otherwise, once stained, edge-glued seams will become glaringly obvious. Like other oaks, its high tannin content can be irritating to the skin. These same features, however, while potentially negative, offer great flexibility with respect to the types of finishes which can be achieved. Due to its exceptionally open grain, use of a light-colored filler followed by a darker stain produces the once popular limed look. The high tannin content allows it to be ammonia fumed to an almost black "Jacobean" finish. Also, careful selection of quartersawn stock yields a ray-dominated look that is truly unique.

USES

The wood of northern red oak is indistinguishable from that of southern red oak, and these two species (combined with other species) are marketed as red oak and used for mine timbers, cabinetry, flooring, millwork, plywood, railroad ties and many other purposes.

SUPPLIES

Supplies of northern red oak are plentiful. It is one of the most, if not the most, commonly available U.S. domestic hardwood. Because it is also one of the fastest growing species of oak, prospects are good that it will remain plentiful and relatively inexpensive for years to come.

Quercus virginiana

live oak

by Jim Flynn

SCIENTIFIC NAME

Quercus virginiana. Derivation: The genus name is the classic Latin name for oaks, derived from the Celtic fine and tree. The specific epithet means the plant is "of Virginia".

FAMILY

Fagaceae, the beech family.

OTHER NAMES

Virginia live oak, encino (Spanish). There are several varieties of *Q. virginiana*, the principle ones are *Q. virginiana* var. *virginiana* (live oak), *Q. virginiana* var. *fusiformis* (Texas live oak) and *Q. virginiana* var. *geminata* (sand live oak).

DISTRIBUTION

The range of live oak starts on the coastal plain in the eastern United States near the Atlantic coast from southeast Virginia south to southern Georgia and southern Florida, including the Florida Keys and from there west to southern and central Texas. It is also found in southwest Oklahoma and in the mountains of northeast Mexico.

THE TREE

Live oak is a fast-growing, large, evergreen tree with massive and often contorted spreading branches. The crown is rounded and extremely wide. The main stem is generally short and stout with its lower branches within 6 feet of the ground. The tree grows 40 to 50 feet in height with a circumference of 3 to 4 feet. It is resistant to saltwater spray. Specimens have been found that are nearly 300 years old. Earlier estimates that the trees lived to be over 900 years old have now been proven inaccurate. In many places, the tree is host to epiphytic plants including Spanish moss. Live oak is the state tree of Georgia.

THE TIMBER

Live oak is one of the heaviest domestic U.S. trees. Average reported specific gravity is 0.80 (ovendry weight/green volume), equivalent to an air-dried weight of 64 pcf. The wood is yellowish-brown with an uneven, contorted grain often exhibiting small, non-continuous checking. Extremely interesting grain patterns are displayed when slab sawn.

SEASONING

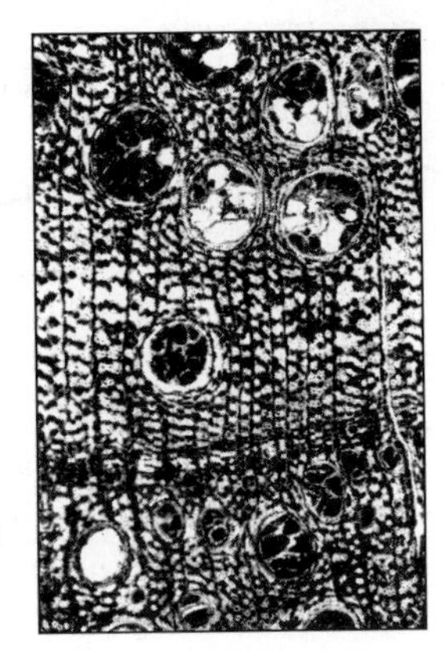

The wood must be protected from drying too rapidly when first cut. Protection from the hot sun by coating the end grain and stickering is essential as well. Stocks of live oak have been stored by immersing in water and have been found suitable for use after 200 years. Average reported shrinkage values (green to ovendry) are 6.6% radial, 9.5% tangential and 14.7% volumetric.

DURABILITY

This wood is extremely durable and has a high resistance to decay in marine environments.

WORKABILITY

Live oak is extremely hard to work. Early ship builders from New England who harvested the wood would bring the molds (patterns) of ship timbers with them so they could shape the wood as the tree was cut. With the tools available at the time, it was found that these huge timbers were almost impossible to cut in the northern shipyards when the wood was dry.

USES

Live oak was the preferred timber used for framing ships in Colonial days. The U.S. Navy had live oak forests set aside as a reserve. Today, the tree is most valued as an ornamental and secondary uses include fence posts, structural timbers, firewood and some ship framing. The acorns are edible, and it is reported that Native Americans obtained sweet oil from them.

SUPPLIES

Live oak is not known to be on the commercial market to any extent. Limited supplies, however, can be obtained at very reasonable prices in the areas where the trees grow in abundance.

Rhizophora mangle

mangrove

by Jon Arno

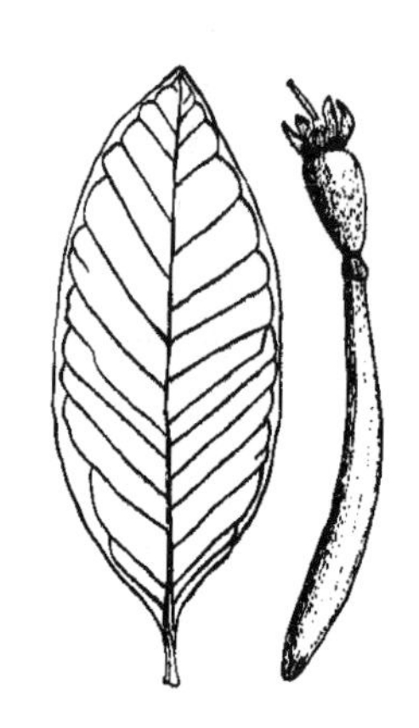

SCIENTIFIC NAME

Rhizophora mangle. Derivation: The genus name is from the Greek meaning rootbearing, referring to the prominent, arching prop roots. The specific epithet is the common Spanish name for mangrove.

FAMILY

Rhizophoraceae, the mangrove family.

OTHER NAMES

mangle colorado, candelon, purga, red mangrove.

DISTRIBUTION

This species is one of the most widespread and is found on brackish coastal mud flats in both the Americas and Asia where the climate is frost free. It is especially common along the coast and in the river estuaries of extreme southern Florida and through Latin America to southern Brazil. It is also found on the Pacific coast from Baja California to Peru.

THE TREE

Mangrove stands on a base of aerial roots and has sinuous branches with gray bark supporting glossy evergreen leaves and pale yellow flowers. The seeds are vipiparous (germinate while still attached to the parent tree) and fall into the tide as small, spike-shaped saplings which drift off to colonize new territory. In its typical habitat, however, this species tends to form impenetrable thickets clogged with debris and alive with insects and snakes. Although seldom allowed to grow to full maturity on ideal sites, it is capable of reaching heights of 100 feet with diameters of over 30 inches. Clear logs 30 to 40 feet are occasionally harvested, but are the exception as the tree usually sets out numerous low branches to compete for sunlight.

THE TIMBER

At a glance, the wood looks somewhat like black cherry (*Prunus serotina* WDS 223). Light red when first cut, the heartwood turns reddish-rust brown as it dries. The grain is normally straight and tends to produce a striped figure, sometimes with purple highlights in the darker bands. Except in extremely mature specimens, the sapwood may account for half or more of the log's volume and is generally yellowish-gray or light pink near the bark, transitioning to darker shades of pink as it blends into the heartwood. It has low surface luster and virtually no taste or odor. While both the color and fine texture resemble black cherry, its average specific gravity of 0.89 (ovendry weight/green volume), equivalent to an air-dried weight of 72 pcf, makes it almost twice

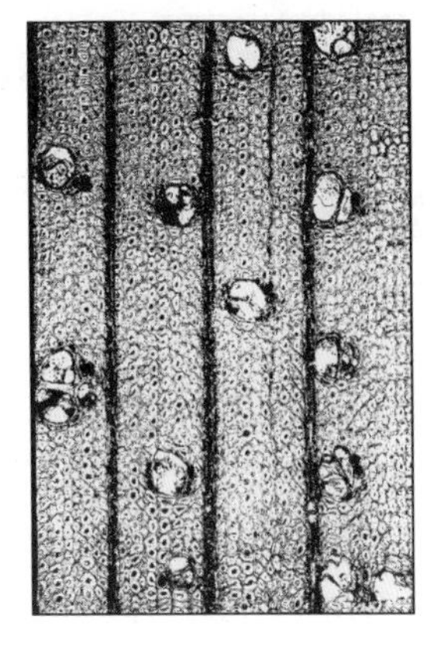

as heavy. With a maximum crushing strength of 10,750 psi at 12% moisture content, mangrove is a very rugged timber and well suited for heavy construction.

SEASONING

This species is difficult to season, tends to end check and is prone to severe warping. At 14.3%, its average volumetric shrinkage (green to ovendry) is quite high. Also, its radial shrinkage of 5.0% and tangential shrinkage of 10.7% contribute to considerable drying stress.

DURABILITY

Mangrove is durable with respect to decay caused by fungi, but susceptible to marine borers and termites.

WORKABILITY

It is difficult to work with hand tools, but its fine texture makes it a reasonably good choice for turning. It will hold sharp edges and can be brought to a smooth surface with scrapers or abrasives. Although it has an attractive natural color, it lacks the volumetric stability to be a first rate wood for cabinetry.

USES

Timber applications include heavy construction, railroad ties, knees and ribs in boat building, turnery and some cabinetry. Tannin constitutes between 20% and 30% of the bark's ovendry weight so it is a valuable commodity in leather processing. Mangrove is an excellent firewood and large quantities are converted into charcoal.

SUPPLIES

This species readily replenishes itself and is very plentiful throughout most of its range. In recent years, however, it has been hard hit by real estate developers in southern Florida who fill in tidal flats to create valuable shore front property. Laws have now been passed to protect these areas, as they provide a vital breeding ground for birds and marine life.

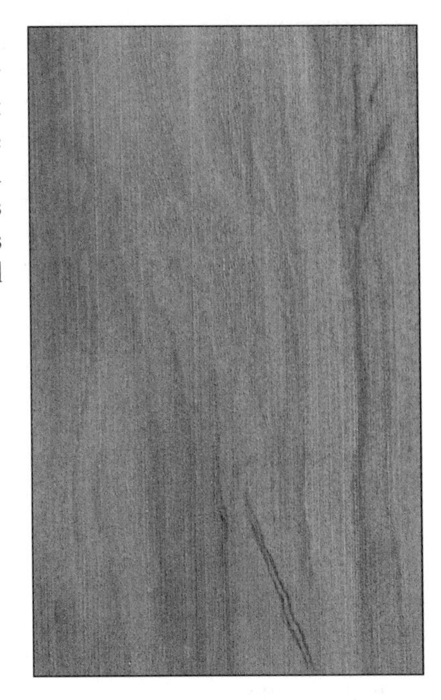

Rhus typhina

staghorn sumac

by Max Kline

SCIENTIFIC NAME

Rhus typhina. Derivation: The genus name is the classical Greek and Latin name for the type of species, Sicilian sumac, *Rhus coriaria*. The specific epithet *typha* means cattail, referring to the hairy twigs.

FAMILY

Anacardiaceae, the cashew family.

OTHER NAMES

hairy sumac, sumac, velvet sumac, vinegar tree, American sumac.

DISTRIBUTION

From Nova Scotia westward to North Dakota, eastern Iowa and in practically all states in the eastern half of the United States from Maine to eastern Kentucky and Tennessee, northwestern Georgia and northern Alabama.

THE TREE

Staghorn sumac, the largest of the native sumacs, usually grows as a large shrub and is often found in thickets. Occasionally it reaches a tree height of 35 to 40 feet with a trunk up to 12 inches at the base. It has fern-like foliage, and in May and June the pale yellow or yellowish-green flowers appear. In August they have turned to crimson bunches of velvety fruit that often remain through the winter. The tree is beautiful in the autumn when the leaves turn red, purple and yellow. After the leaves fall, the velvety branches give the appearance of the antlers of a stag in the velvet, from which the name staghorn sumac is derived. Tannin may be obtained from the bark and leaves.

THE TIMBER

The wood is light weight, soft and brittle with very prominent ring pores. Average reported specific gravity is 0.45 (ovendry weight/green volume), equivalent to an air-dried weight of 34 pcf. The heartwood is greenish-orange or orange-green, sometimes golden yellow streaked with tints of brown and green. The narrow sapwood is almost white. The heartwood is one of the best for showing fluorescence under black light. Staghorn sumac has a high satiny luster and is quite attractive.

SEASONING

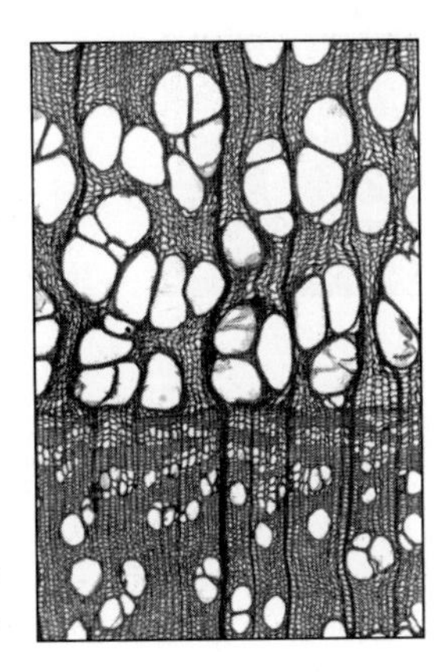

Staghorn sumac air dries easily. It loses moisture readily without developing cracks or defects. When freshly cut, a thick sticky pitch exudes from the area between the bark and the wood but upon seasoning the tackiness disappears. Shrinkage values are not available.

DURABILITY

Staghorn sumac is not a durable timber.

WORKABILITY

The wood is easily worked with tools, but because of its softness tools must be kept very sharp. It has a tendency to fray when turned which requires considerable sanding to a finished surface before removal from the lathe.

USES

Staghorn sumac has little commercial use and is mainly used by hobbyists for carving and turnery. It makes a beautiful contrast to other woods when used in such articles as lamps, candle holders and bowls.

SUPPLIES

Staghorn sumac cannot be purchased commercially but must be collected by the wood collector when the opportunity presents itself. It is well worth the effort to use this species because of its unusual color and grain appearance.

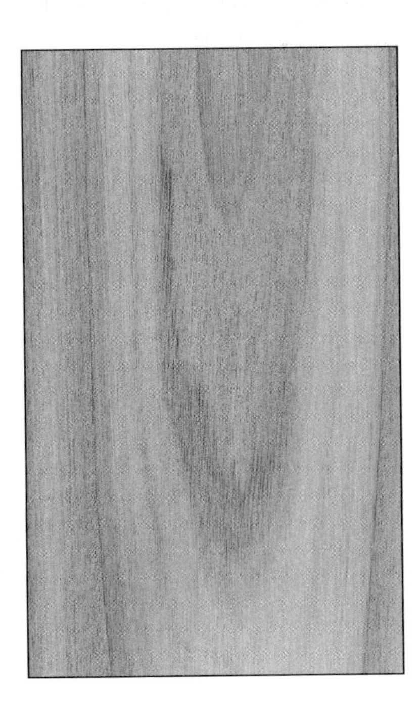

Robinia pseudoacacia

black locust

by Max Kline

SCIENTIFIC NAME

Robinia pseudoacacia. Derivation: The genus name is in honor of Jean Robin (1550 to 1629) and his son Vespasian Robin (1579 to 1662), herbalists to the Kings of France and the ones who first cultivated black locust in Europe. The specific epithet is an old generic name meaning false acacia.

FAMILY

Fabaceae or Leguminosae, the legume family; (*Papilionaceae*) the pea or pulse group.

OTHER NAMES

yellow locust, false acacia, acacia, white locust.

DISTRIBUTION

The natural range is the Appalachian Mountains from Pennsylvania to Alabama and parts of Arkansas and Missouri, but black locust has been planted in most states and southern Canada.

THE TREE

Black locust is a small to medium-sized tree. It ranges from 40 to 75 feet in height with a trunk 12 to 36 inches in diameter. Its spreading root system makes it especially desirable for preventing erosion. It grows rapidly and matures at an early age.

THE TIMBER

The heartwood of black locust varies from greenish-yellow to dark or golden brown, sometimes with a tinge of green, turning to a russet shade upon exposure to the air. It is odorless, and taste is not distinct. The luster is fairly high. Average reported specific gravity is 0.66 (ovendry weight/green volume), equivalent to an air-dried weight of 52 pcf. The grain is fairly straight and prominent, and the texture is uneven. Black locust timber is stronger and stiffer than white oak (*Quercus alba* WDS 230). It has bending properties comparable to those of ash and beech.

SEASONING

Black locust seasons slowly and tends to warp. Average reported shrinkage values (green to ovendry) are 4.6% radial, 7.2% tangential and 10.2% volumetric. After seasoning, black locust holds its place well with little shrinkage or swelling.

DURABILITY

The sapwood of black locust is perishable but the heartwood is resistant to decay. The trees are frequently attacked by a wood-boring insect, which may kill them.

WORKABILITY

Black locust is difficult to work with hand tools but is easily machined. It finishes very smoothly and will take a high polish. It has a moderate dulling effect on cutting edges and is not easy to nail. It can be glued satisfactorily.

USES

Black locust was the favored wood for insulator pins used on power pole cross arms. It was ideal in the manufacture of wagon wheel hubs in the days of the prairie schooners. Additional uses include tool handles, ship construction, boxes and crates and to some extent woodenware and novelties. Large quantities are consumed as fence posts because of its durability.

SUPPLIES

Stocks are plentiful in the eastern United States and elsewhere but prices vary depending on location.

Roystonea regia

Cuban royal palm

by Jim Flynn

SCIENTIFIC NAME

Roystonea regia. Derivation: The genus name is in honor of General Roy Stone (1836 to 1905), a U.S. Army Engineer who rendered service to Puerto Rico during the Spanish-American War. The specific epithet means royal. Recent taxonomic changes in this genus, as reported by the USDA in the GRIN database, have resulted in relegating *R. elata* and *R. floridana,* the Florida royal palms, as synonyms of *R. regia.*

FAMILY

Palmae, the palm family; also Arecaceae.

OTHER NAMES

The uniting of several taxa into one taxon brings with it the common names associated with the synonyms, including Florida royal palm, royal palm, palmier royal, konigspalme, palmeir real, chaguaramo and palma real de Cuba.

DISTRIBUTION

Roystonea consists of a small group of about 16 species. These are endemic to tropical regions of the Americas from northern South America through the Caribbean. Cuban royal palm is native to Cuba and the southern portions of Florida and is widely planted elsewhere as an ornamental.

THE TREE

Humboldt referred to the palms as "the princes of the vegetable kingdom". Cuban royal palm is certainly one of them. This elegant and stately palm is the national tree of Cuba. It can be found in semi-deciduous forests and is often more abundant in secondary vegetation. Its silhouette forms part of the Cuban national emblem. Growing 100 feet in height with an irregularly thick trunk and an occasional bulge, it looks like a concrete column. The deep green, pinnately compound leaves stand out like a crown on the tall stem. They are 10 feet long and each leaflet is approximately 3 feet long. This palm is of great utility in Cuba as the fruit yields valuable oil for making soap and is a principal food in the diet of hogs. The leaves are valuable for sheathing roofs and sides of rural shelters as well as wrappers for food and tobacco. It is widely planted along streets and forms impressive statuary adding to the tropical ambiance.

THE TIMBER

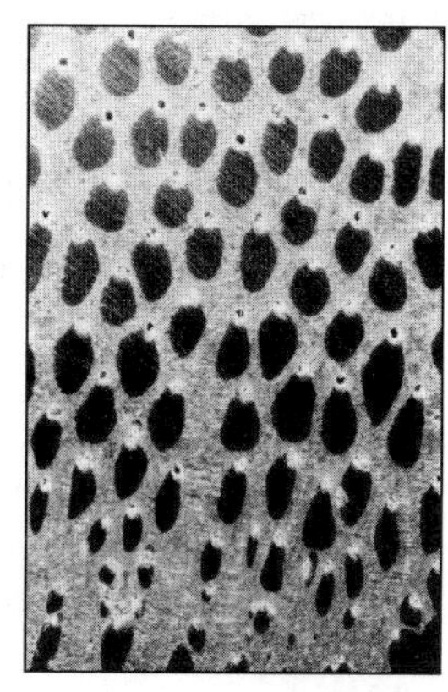

Most species of palms develop a similar wood structure. The outer portion of the main stem has tough fibro-vascular bundles imbedded in a mass of parenchyma tissue. These stringy bundles contain strong and hard individual strands that are very dense at the outer circumference of the trunk and diminish in density toward the center, which is frequently soft and mushy. It has a specific gravity of 0.94 (ovendry weight/green volume), equivalent to an air-dried weight of 77 pcf. I confirmed my calculation by "sinking" my sample in a container of water.

SEASONING

Shrinkage values are not available. It really is of little importance as it is used in applications not demanding seasoned wood. One exception is when the wood is used as flooring.

DURABILITY

Cuban royal palm is reported to be durable.

WORKABILITY

This is a tough wood on machine tools and saw blades. Nevertheless, it can be cut, hacked, split and machined but tools will require extensive sharpening when finished. The wood takes an excellent polish and very attractive finish. Epoxy is a good glue to use. Pre-boring is advised when using nails and screws.

USES

This timber is not considered a commercial product but is very important to those engaged in subsistence farming. The wood is used for pilings, posts and house frames. Split sections of the trunk are used for walking sticks, umbrella handles and fishing rods as well as very durable and attractive flooring. An office in the Centro de Investigacion Forestal in Cuba is paneled with it to illustrate its potential. Thin slices of wood from the cross-section are often used by marquetarians.

SUPPLIES

While the supplies are limited, it can often be found when storm damaged trees are removed in its growth area. Several other species of palms are now becoming available in craft supply outlets.

Sabal palmetto

cabbage palmetto

by Jim Flynn

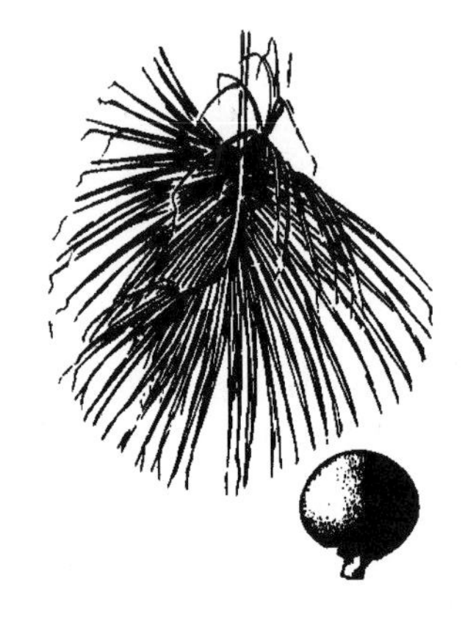

SCIENTIFIC NAME

Sabal palmetto. Derivation: The genus name is probably an old Native American name. The specific epithet *palmito* is Spanish meaning a small palm.

FAMILY

Palmae, the palm family; also Arecaceae.

OTHER NAMES

Carolina palmetto, common palmetto, palmetto, cabbage-palm, swamp cabbage.

DISTRIBUTION

This species of palm is native to the southeastern part of the United States and the Bahamas. Its range extends northward from the Florida Keys through its epicenter in south-central Florida to Cape Fear, North Carolina. From North Carolina south to the Florida line, it hugs the coastline usually within 20 miles of the ocean. In Florida it extends through Gainesville and across the peninsula to the Gulf Coast.

THE TREE

In spite of the diminutive Spanish name, cabbage palmetto often reaches heights of 80 feet with trunk diameters of 18 inches. The brown to grayish trunk is nearly straight and covered with the remains of leaf bases (boots). The fan-shaped leaves are 5 to 6 feet long and have a prominent midrib. The leaflets are unarmed. The name cabbage palmetto is derived from its edible terminal which tastes somewhat like cabbage. Removal of the bud will kill the tree. This was an important tree to the Seminole Indians who thatched their huts with the leaves and made meal for bread from the fruit. Currently, young cabbage palmetto fronds are collected and shipped worldwide each spring for use on Palm Sunday. Cabbage palmetto is the state tree of Florida and South Carolina.

THE TIMBER

This wood is pale brown and occasionally nearly black. It has numerous hard fibrovascular bundles that appear on the transverse plane as tiny 1-cm (0.394 inch) dots and long streaks up to 50 cm (19.69 inches) on both the tangential and radial planes making it difficult to distinguish between the two. This is because this species does not exhibit growth rings. The outer 2 inches of the stem is denser than the interior. Specific gravity on a standard size sample was 0.39 (ovendry weight/green volume), equivalent to an air-dried weight of 29 pcf. Old records indicate that tree trunks were ideal for building fortifications because the wood did not produce lethal splinters when struck by cannonballs.

SEASONING

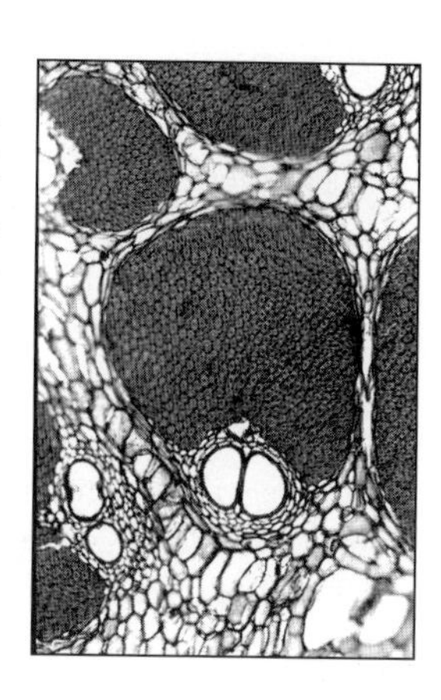

There is very little information recorded on this characteristic and shrinkage values are not available. It would appear that the density of the vascular bundles would have a significant effect on shrinkage properties, but this is conjecture.

DURABILITY

The only evidence available on this characteristic is that the round timbers are effective pilings for wharves due to their resistance to marine borers. Former uses for transporting water by hollowing out the cores and for building posts seem to indicate that the wood has some resistance to decay.

WORKABILITY

The timber can be readily cut with a chainsaw with a sharp chain, but the wood quickly dulls cutting tools. Nevertheless it can be slabbed, squared, planed and worked into lumber. It sands and finishes well. The less dense wood can be glued with commonly available adhesives but the denser stock is best glued with epoxy. No data is available to the author on turning and other uses.

USES

Because of the unusual grain pattern, especially that of the end grain, the wood is useful for small craft items, especially tabletops. Flooring and canes and walking sticks are other uses.

SUPPLIES

There is an ample supply of cabbage palmetto in the growth area because of the many storm damaged trees as well as those removed for construction of roads and homes. IWCS regional meetings in Florida generally provide members with more than they can handle.

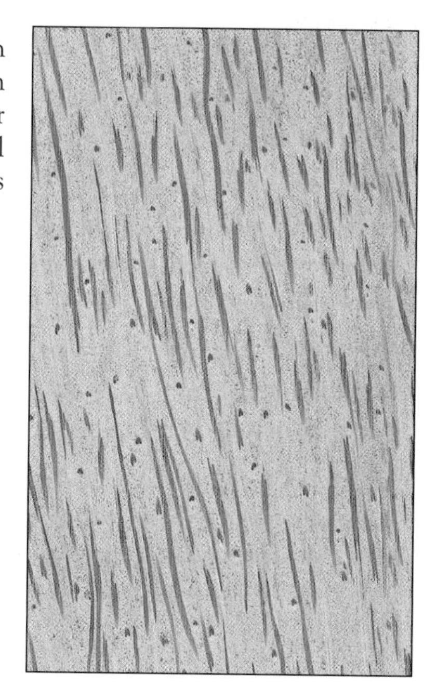

Salix nigra

black willow

by Jim Flynn

SCIENTIFIC NAME

Salix nigra. Derivation: The genus name is derived from two Celtic words, *sal* and *lis*, which means near water. The specific epithet *nigra* refers to the black bark on the mature tree.

FAMILY

Salicaceae, the willow family. This family consists of two genera and about 350 species.

OTHER NAMES

In the southern parts of the United States, it is generally referred to as swamp willow. Other names include Gooding willow, western black willow, Dudley willow and sauz (Spanish).

DISTRIBUTION

Black willow is a tree of the eastern United States. The natural range also includes a small portion of Canada northwest of Vermont and in a small coastal area of New Brunswick. There are several scattered and isolated stands throughout northern Mexico. It is absent from most of Florida. The largest commercial stands can be found in the Mississippi River Valley and its tributaries. Black willow thrives best in wet areas that are not permanently flooded.

THE TREE

There is great variation in the size of the tree depending upon where it grows. In the Mississippi Valley, the tree's maximum height can be as much as 140 feet with a diameter of 48 inches. Black willow reaches maturity at about 55 years. The oldest known black willow without decay was 70 years old. The lanceolate leaves are 3 to 6 inches long and 0.375 to 0.75 inch wide. On older trees, the bark is dark brown to nearly black with flaky ridges. The roots must have continuous access to water during the growing season. It does not tolerate shade.

THE TIMBER

Black willow wood is one of the lightest of the U.S. domestic hardwoods with an average reported specific gravity of 0.36 (ovendry weight/green volume), equivalent to an air-dried weight of 27 pcf. It is extremely weak as a structural wood. The color varies considerably; the heartwood is pale reddish-brown to grayish-brown and the sapwood is whitish tan to light tan. Its grain is interlocked, and the wood is uniform in texture.

SEASONING

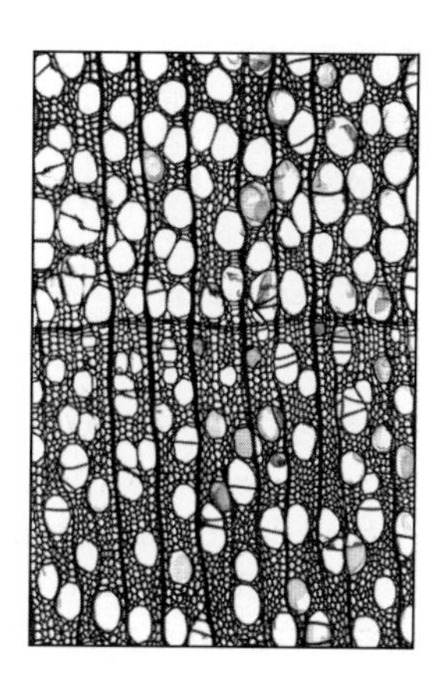

Care must be taken to prevent warping during seasoning. Average reported shrinkage values (green to ovendry) are 3.3% radial, 8.7% tangential and 13.9% volumetric. Once properly seasoned, however, black willow has good dimensional stability.

DURABILITY

Black willow is not rated as very durable. But because of its pliancy, availability and low cost, it has been used extensively in reinforcing levees on the banks of the Mississippi, Ohio and Missouri Rivers.

WORKABILITY

Because of its interlocked grain, black willow is very difficult wood to machine. It glues well and takes a good finish.

USES

The major uses of black willow are in the millwork and household furniture industries. Packing cases and boxes, pulping, artificial limbs, picture frames, polo balls and venetian blinds are some of the uses for this wood. Because of its low tendency to check, it is considered a desirable wood for carving.

SUPPLIES

Black willow is readily available on the commercial market.

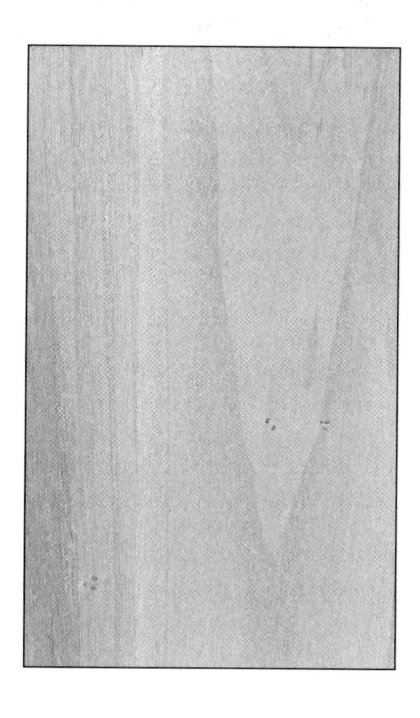

Samanea saman

monkey-pod

by Jim Flynn

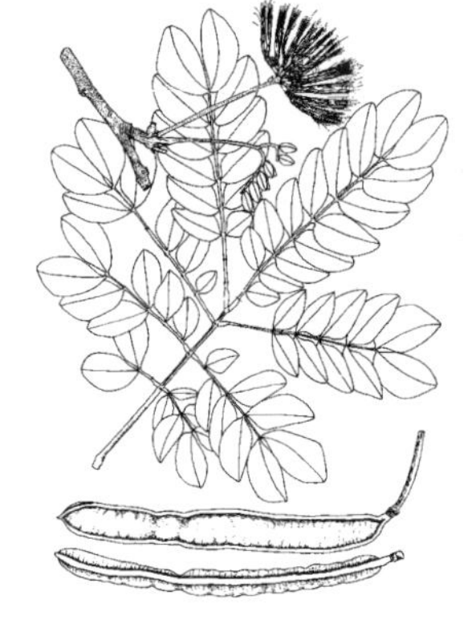

SCIENTIFIC NAME

Samanea saman. Derivation: The genus name is from the Spanish *saman*, derived from a French Caribbean vernacular, *zamang*, rain tree. The specific epithet is of the same origin. The synonyms *Pithecellobium saman* and *Albizia saman* appear in many older references.

FAMILY

Fabaceae or Leguminosae, the legume family; (*Mimosaceae*) the mimosa group.

OTHER NAMES

The most widely used common name is raintree. *Silvics of North America*, USDA Handbook 654, explains: "The leaflets close up at night or when under heavy cloud cover, allowing rain to pass easily through the crown". Saman is used throughout Latin America and mimosa is used in the Philippines. Additional names are dormilon (Puerto Rico), algarrobo (Cuba, Mexico, Guatemala), cenicero (El Salvador, Costa Rica), samaguare (Colombia), lara, carabali (Venezuela) and huacamayo-chico (Peru).

DISTRIBUTION

Native to Mexico (Yucatan Peninsula) and Guatemala south to Peru, Bolivia and Brazil, it has naturalized throughout the West Indies, Mexico and in other tropical regions. Monkey-pod may have been introduced into Puerto Rico and Guam as early as the 16th century. The tree was reportedly introduced into Hawaii in 1847 by a businessman with two seeds, both of which germinated and are possibly the progenitors of all trees there.

THE TREE

When grown in the open, it is a beautiful shade tree with a large trunk and very broad arched crown of dense foliage. When forest grown, the main stem is somewhat narrower and free of branches for 25 feet or more. In prime growth areas, the tree will grow to 100 feet or more in height with trunk diameters of 3 to 4 feet. The bipinnate leaves have many diamond-shaped paired leaflets that fold together at night and on cloudy days. The flower heads are a mass of thread-like stamens, pink in the outer half and white in the inner half. Cattle and horses eat the flattened blackish pods that are filled with rich sugar pulp. The seeds are used to string leis. The thick, gray, rough bark is furrowed into long plates or corky ridges.

THE TIMBER

The heartwood is dark walnut to dark chocolate-brown and turns light to golden-brown with darker streaks when seasoned. The narrow sapwood is yellowish and clearly differentiated from the heartwood. The texture is medium to coarse, and the luster is medium. The wood is without distinctive taste or odor. Density varies widely according to climate and soil. Average specific gravity is 0.50 (ovendry weight/green volume), equivalent to an air-dried weight of 38 pcf.

SEASONING

Relatively difficult to air dry, it dries without checking. It develops a moderate to severe warp but is stable once seasoned. Average reported shrinkage values (green to ovendry) are 2.5% radial and 3.0% tangential.

DURABILITY

The wood is rated durable to very durable in resistance to attack by white- and brown-rot fungus and is resistant to attack by dry-wood termites. Durability will vary based on where the wood was harvested.

WORKABILITY

Machining characteristics vary according to the grain structure of sawlogs. Timber from short and stubby stems may have many knots and twisted grain whereas that from trees grown in dense forests would have the opposite. Because of its high density and low shrinkage, it is an ideal wood for turning bowls. It can be turned green and left to season with very little degrade from drying. It glues well and will take an outstanding finish.

USES

Monkey-pod is an ideal craftwood. It is widely used for turned bowls for the Hawaiian tourist trade. Because of the inadequate supply in Hawaii, these bowls are produced in the Far East. Furniture, cabinets, boat trim, miscellaneous trim and decorative items are other prime uses.

SUPPLIES

A standard item in world commerce, the quality of the wood will vary depending upon where it was grown.

Santalum album

sandalwood

by Jon Arno

SCIENTIFIC NAME

Santalum album. Derivation: The genus name is derived from Persian for sandalwood. The specific epithet means white.

FAMILY

Santalaceae, the sandalwood family.

OTHER NAMES

East Indian sandalwood, white saunders, yellow saunders, yellow sandalwood.

DISTRIBUTION

Santalum album is native to southern India, although similar species are wide spread from Indochina and Australia to Hawaii.

THE TREE

Sandalwood is parasitic. Although it is capable of producing its own food via photosynthesis, sandalwood attaches itself to the roots of other trees as a source of minerals and water. The tree normally grows to a height of 35 to 40 feet with diameters up to about 8 inches. An old reference reports diameters of up to 14 inches. While the species may have the genetic capability to reach a fairly large size, it is so prized for its wood and oil that it is seldom allowed to attain maximum size.

THE TIMBER

The yellowish-tan heartwood generally makes up about one-third of the log's volume and darkens to reddish-brown or brick orange as it seasons. The sapwood is creamy white and clearly demarcated from the heartwood. The wood's texture is very fine, and the pores and rays are not visible to the naked eye. Sandalwood is very hard and heavy, with an average reported specific gravity of 0.75 (ovendry weight/green volume), equivalent to an air-dried weight of 60 pcf. The grain is sometimes wavy (fiddleback), but the wood's most prominent features are its oily, almost sticky feel and strong, spicy odor. It is strong by most standards, but not normally used in anything large enough to challenge its structural properties.

SEASONING

Sandalwood is relatively easy to season. Like most high-density woods, it dries slowly but with little degrade. Shrinkage values are not available.

DURABILITY

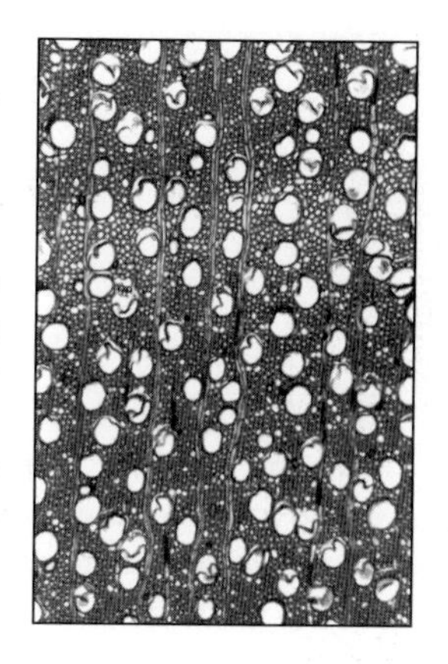

Sandalwood is extremely durable, probably owing to its plentiful volatiles, fine texture and density.

WORKABILITY

For those who enjoy the firm resistance presented to a cutting edge by wood with a high density, sandalwood works very well. It holds sharp detail and is an excellent choice for carving and turning. Due to its high oil content (5% to 7%), the wood will quickly polish to a satin smooth and slightly lustrous surface. Wood with a pronounced and fragrant scent brings an added pleasure to the shop and sandalwood is no exception. Since the oil is used in both cosmetics and medicine, and the references do not stress any severe toxic effects from exposure to the dust, sandalwood does not appear to present the kind of health hazard typical of so many of the highly scented woods. The odor of sandalwood is nonetheless potent enough to become overpowering when the wood is put under a belt sander. Also, the references do not indicate any special cautions with respect to choice of glues or finishing materials. Owing to its high oil content, however, it might be advisable to test your choices on a few scraps before proceeding; sandalwood is not something you want to waste.

USES

Even though there are now synthetic substitutes, the oil distilled from heartwood chips is still in great demand worldwide for perfume, cosmetics, soap and medicine. Sandalwood incense is also an important part of religious ceremonies in India, China and elsewhere in the Orient. The wood is used for carvings, turnery, combs, walking sticks and fine furniture of the small and precious sort such as jewelry boxes, display cases and chests. Some species produce a sweet edible fruit that is a brownish-black drupe about the size of a small plum.

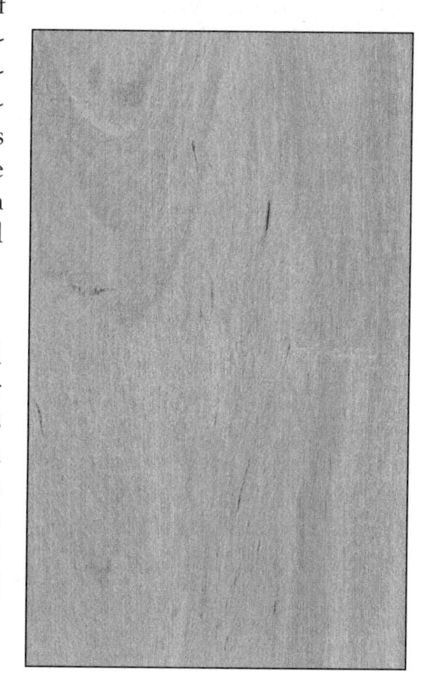

SUPPLIES

While various species of sandalwood are plentiful throughout an extensive range, the non-timber demands placed on them are intense. The tree is being nurtured in what might be described as a form of plantation silviculture, but its parasitic nature makes this a more tricky proposition than with other conventional forestry species. The wood is scarce in international commerce and very expensive.

Sapindus saponaria

western soapberry

by Jim Flynn

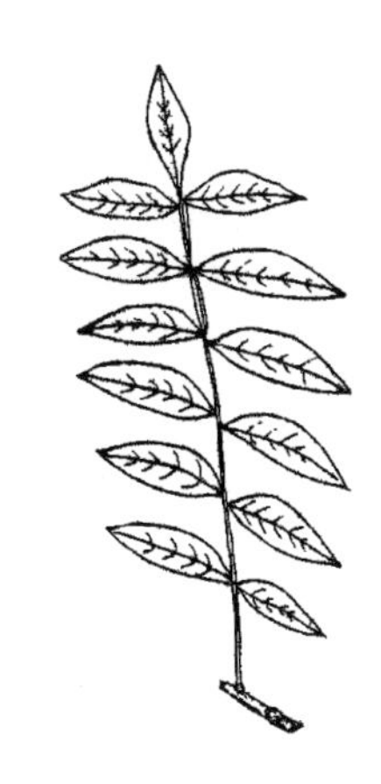

SCIENTIFIC NAME

Sapindus saponaria var. *drummondii*. Derivation: The genus name is from the Latin word *sapo* meaning soap and *indicus* meaning Indian. The specific epithet means "of soap". The varietal eponym is in honor of Thomas Drummond (1780 to 1835), a Scottish botanical explorer in North America. Drummond was one of Sir Joseph Hooker's (England) plant collectors and was one of the earliest botanists to visit Texas. In 1833 his first letter to Hooker was written from "Town of Velasco, Mouth of the Rio Brazos" (on the Gulf coast due south of present Houston). Two years later Drummond died in Cuba probably from afflictions acquired in Texas where he suffered poor health. *S. drummondii* and *S. indicus* are synonyms.

FAMILY

Sapinaceae, the soapberry family.

OTHER NAMES

wild chinatree, cherioni, jaboncillo, amole de bolita, palo blanco, Indian soap-plant.

DISTRIBUTION

Western soapberry can be found sometimes in widely scattered small pockets and, on occasion, in more concentrated stands. The range includes southeastern Missouri, west to Kansas and southern Colorado and from there south to central and southern Arizona and east to Trans-Pecos and southern Texas and Louisiana. Northern Mexico also contains populations of this species, but the greatest concentrations are in Texas.

THE TREE

Western soapberry is a small tree normally reaching 40 feet in height with a trunk diameter of 16 inches. The U.S. National Register of Big Trees recognizes a specimen growing in Corpus Christi, Texas as the champion. It is 62 feet in height with a trunk 126 inches in circumference. The grayish-brown bark is smooth when the tree is young but wrinkles and becomes scaly as the tree matures. The trunk, on large specimens, is apt to be buttressed. The compound leaves have between 4 and 9 pairs of lance-shaped leaflets, which can be alternate or opposite. The poisonous fruit, a round yellow berry 0.5 inch in diameter, contains one or more dark brown seeds. These berries contain a substance called saponin. When the berries are macerated and mixed with water, it produces suds which is an effective substitute for soap. Early references indicate this soap was used for washing hair and fine fabrics such as silk as well as for other more mundane applications. The seeds, when scattered on the surface of water, are used for stunning fish.

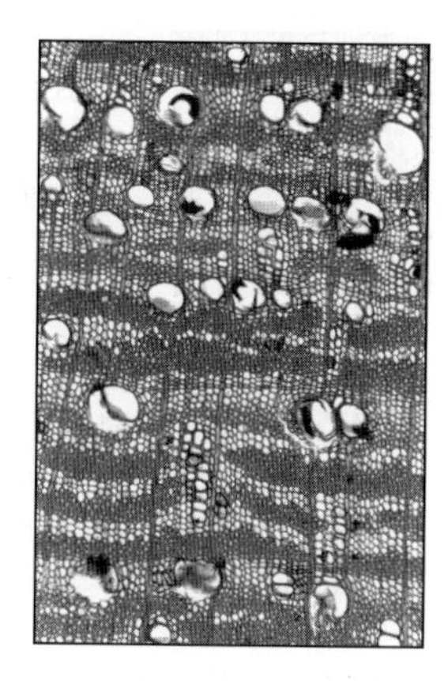

THE TIMBER

The predominant feature of this wood is the soft yellow color that is often streaked with brown. The heartwood is not noticeably distinct from the sapwood. The luster is low. There is little discernible taste or odor. The wood is ring porous so the grain shows in graceful patterns, especially on quartersawn surfaces. Average specific gravity is 0.88 (oven-dry weight/green volume), equivalent to an air-dried weight of 72 pcf.

SEASONING

Because western soapberry is not commercially significant, there is little data available on this characteristic and shrinkage values are not available. It most likely requires careful and slow drying to prevent checking and warping.

DURABILITY

Western soapberry is not considered to be a durable wood.

WORKABILITY

Its overall workability is poor. Due to the hardness and often uneven grain, woodworkers should be cautious of using machine tools to cut and mill the timber. It glues satisfactorily, sands well and takes a nice finish. Because the green wood is pliant, it can be riven along the grain with the annual rings and divided into strips for basket making.

USES

One of the traditional uses for western soapberry was for baskets for gathering cotton. It was also used for fabricating frames for packsaddles in Texas. Besides these uses, few other specific uses are recorded. Nevertheless, it appears to be suitable for manufacturing small craft items and items of furniture as well as for turnery.

SUPPLIES

Within the native growth range, supplies should be attainable. IWCS members often have wood available for exchange.

Sassafras albidum

sassafras

by Max Kline

SCIENTIFIC NAME

Sassafras albidum. Derivation: The genus name was apparently a Native American name used by the Spanish and French in Florida in the middle of the 16th century. The specific epithet means white.

FAMILY

Lauraceae, the laurel family.

OTHER NAMES

cinnamon-wood, saxifrax-tree, red sassafras.

DISTRIBUTION

The natural range is from the coast of southern Maine to southern Ontario, Michigan, Iowa and Kansas, south to Florida and Texas.

THE TREE

Sassafras trees vary in size from little more than a crooked shrub to a slender tree less than 40 feet in height. At its best, it reaches a height of 75 to 90 feet with a trunk diameter of 2 to 5 feet. Many trees survive as long as 700 to 1,000 years. Sassafras has the distinction of having three different shaped leaves on a single branch. In the springtime, it has beautiful small yellow or greenish flowers, and in autumn its foliage turns shades of red, orange and yellow.

THE TIMBER

In appearance sassafras wood resembles white ash (*Fraxinus americana* WDS 121) or American chestnut (*Castanea dentata* WDS 065). The heartwood is pale brown, deepening to dull orange-brown upon exposure. The narrow, yellowish-white sapwood is not very sharply demarcated from the heartwood. The wood is soft, brittle, straight, coarse grained and has a spicy taste and a slightly characteristic aromatic scent. Average reported specific gravity is 0.42 (ovendry weight/green volume), equivalent to an air-dried weight of 31 pcf. The texture is coarse, and the luster is medium. It is one of the few low-density hardwoods with an interesting grain pattern.

SEASONING

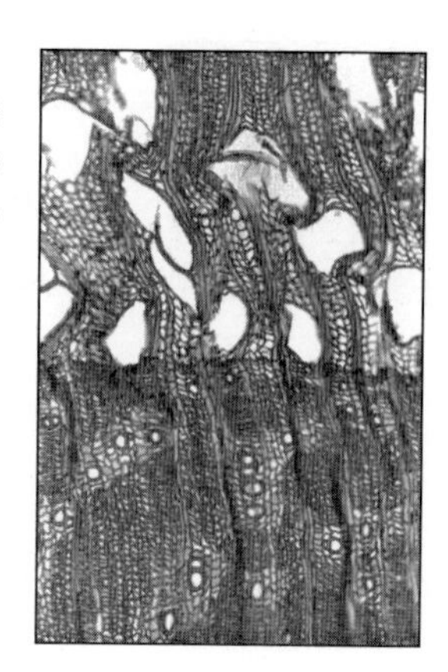

Sassafras is very easy to season, but develops some checks during drying. Average reported shrinkage values (green to ovendry) are 4.0% radial, 6.2% tangential and 10.3% volumetric.

DURABILITY

The timber is very durable when exposed to dampness, making it suitable for fences and house sills.

WORKABILITY

Sassafras wood may be easily worked and takes a good finish. It must be nailed with care, screws hold well and it glues with ease. After seasoning, it is dimensionally stable in use.

USES

In lumber form, the uses of sassafras are limited to inexpensive furniture, boxes, slack cooperage, posts, fence rails and kindling. The production of sassafras oil is the largest indirect product of the bark from the tree root, which is the source of the commercial oil used to scent or flavor various products such as soap and medicines. Small roots are a common article of commerce and are used for making sassafras tea, a drink reputed to be a spring tonic. [Ed. note: This is now suspected of being a carcinogen.] The orange color obtained from the bark was used by early settlers to dye homespun articles.

SUPPLIES

Not much sassafras lumber is available, as the majority of trees are small in size. When cut, sassafras is often mixed at the mill with other species such as ash. The price is moderate.

Schefflera morototoni

yagrumo macho

by Jim Flynn

SCIENTIFIC NAME

Schefflera morototoni. Derivation: The genus name is in honor of Jakob Christian Scheffler (1718 to 1790), a German botanist. The specific epithet is an unspecified vernacular *morototon* meaning matchwood. A synonym is *Didymopanax morototoni*.

FAMILY

Araliaceae, the ginseng family.

OTHER NAMES

There are probably 50 or more common names for this tree, and they are sometimes applied to other species in this genus. In commerce, the wood is known as morototon. Other names include mangabe, gargoran, pava, pavo, yagrume, pana cimarrona, cordovan and yarumo.

DISTRIBUTION

Yagrumo macho is a well-known pioneer species throughout the tropical Americas. The range is extensive from 17° north to 25° south latitude covering wet and moist forests of the West Indies and continental tropical America from Mexico through Colombia and Venezuela to Brazil and Argentina. The species was introduced into Jamaica and has been planted in Florida. The tree is quite common in Puerto Rico and often grows with trumpet-tree (*Cecropia peltata* WDS 069).

THE TREE

In ideal conditions, this common evergreen forest tree grows to 100 feet or more in height. Trunk diameters of 30 inches are not uncommon. The compound leaves are composed of 7 to 10 long-stalked, oblong or ovate, entire, accuminate leaflets. On older large trees, the leaflets are pale beneath and covered with dense cottony hairs. On young plants, the leaflets are green on both sides and instead of cottony hairs have rough bristly hairs on the upper surface. The tree bears a fleshy fruit (berry) nearly year around. The gray or light brown trunk is smooth with many faint horizontal rings and large leaf scars. When the movement known as "Let Nature Take its Course" began in the Caribbean Basin to arrest the indiscriminate cutting of tropical hardwoods and encourage natural regeneration, yagrumo macho was one of 18 species determined to be a good candidate for the program because of its fast growth.

THE TIMBER

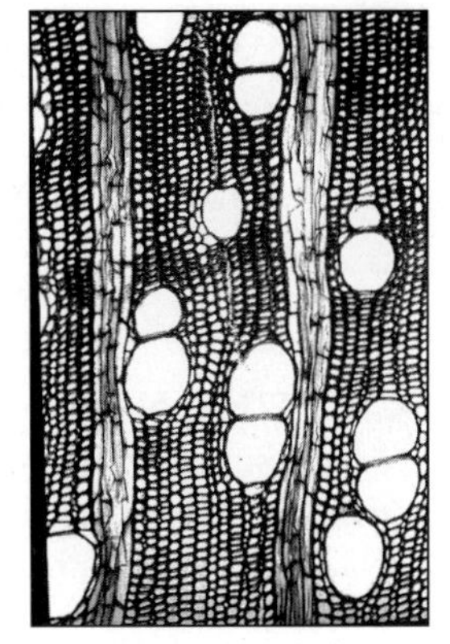

The wood is grayish or brown. The luster is medium and density varies from rather light and soft to heavy and hard. The specific gravity is approximately 0.40 for wood of average density, equivalent to an air-dried weight of 30 pcf. The texture is medium to fine, and the grain is usually straight. The mechanical and physical properties of the wood are somewhat better than yellow-poplar (*Liriodendron tulipifera* WDS 162). Yagrumo macho is a suitable substitute for pine and spruce in general carpentry.

SEASONING

As is the case with many tropical woods, yagrumo macho air dries rapidly but suffers much degrade in the process. This is one of the reasons why export markets do not flourish, and there is so much reliance upon North American softwoods which come precisely dimensioned in wholesale lots for general building purposes. Average reported shrinkage values (green to ovendry) are 6.4% radial, 9.8% tangential and 5.6% volumetric.

DURABILITY

Yagrumo macho is very susceptible to attack by dry-wood termites, other insects and decay. Preservation with conventional chemicals can be achieved but the results vary widely.

WORKABILITY

Yagrumo macho can be planed, shaped, mortised, sanded, glued and finished reasonably well. Turning, boring and spindle shaping is reported to be poor. It accepts nails and has an excellent resistance to splitting when using screws.

USES

The wood is especially suitable for the manufacture of boxes and crates. In areas where there is an abundance of large-sized trees, it is used for general carpentry and interior construction. Wood harvested that is lighter weight can be used as a substitute for heavier grades of balsa (*Ochroma pyramidale* WDS 189).

SUPPLIES

Much of the wood is used locally so it is difficult to obtain. It does not seem to be a common item in foreign trade. The best opportunity for obtaining yagrumo macho is by visiting its growth area.

Schinopsis quebracho-colorado

quebracho

by Max Kline

SCIENTIFIC NAME

Schinopsis quebracho-colorado. Derivation: The genus name is Latin meaning having the appearance of the plants in the *Schinus* genus. The specific epithet is in the Spanish vernacular meaning a very hard and red tropical tree. A synonym is *S. lorentzii*.

FAMILY

Anacardiaceae, the cashew family.

OTHER NAMES

brauna, barauna, quebracho macho, quebracho moro.

DISTRIBUTION

Northern Argentina, western Paraguay and a small portion of Bolivia.

THE TREE

Of scrubby growth with a bent and twisted trunk, quebracho reaches 30 to 50 feet in height with diameters of 1 to 3 feet. Trees are scattered, averaging about 5 per acre. They are felled with heavy axes, and the bark and sapwood are removed to reduce the weight and prevent attack by beetles.

THE TIMBER

The heartwood of quebracho is light red, deepening to brick red sometimes with black streaks. The heartwood is clearly demarcated from the yellowish sapwood. The odor is not distinctive but the taste is astringent. Quebracho is extremely hard, heavy and strong. Average reported specific gravity is 1.00 (ovendry weight/green volume), equivalent to an air-dried weight of 83 pcf. The luster is low to medium, and the texture is fine and uniform. The grain is irregular or wavy.

SEASONING

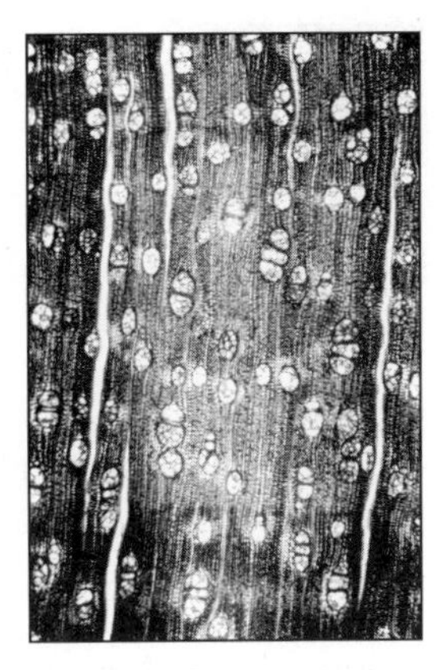

Quebracho lumber is extremely difficult to season as checking and warping are severe. Shrinkage values are not available.

DURABILITY

Quebracho is highly durable after drying. Unless the bark is immediately removed after felling, it is extremely susceptible to attack by beetles which deposit their eggs in the bark and can completely destroy a log in 6 months.

WORKABILITY

Quebracho is difficult to machine because it becomes flinty when dry but it splits readily. It takes a high natural polish. Some people can contract a form of dermatitis when handling the sawdust or even by touching the leaves and branches of the tree.

USES

In the past, the greatest use of quebracho was the extraction of the 20% to 30% tannin contained in the heartwood. The logs were exported to the United States and Europe where the extracting was done. The development of synthetic tanning agents has largely eliminated this market. In South America, the wood is extensively used for fuel and for fence posts, telegraph poles, bridge timbers and railway ties. It is also used for wood block paving, heavy construction and cart axles. Quebracho is seldom cut into thin boards.

SUPPLIES

It is difficult to obtain quebracho at the present time because so little is exported. Wood collectors do sometimes manage to acquire small pieces.

Sequoia sempervirens

redwood

by Max Kline

SCIENTIFIC NAME

Sequoia sempervirens. Derivation: The genus name is in honor of Sequoyah (also spelled Sequoia) or George Guess (1770? to 1843), a Native American inventor of the Cherokee alphabet. The name was unexplained by its author, an Austrian linguist and botanist. The specific epithet is Latin for evergreen.

FAMILY

Taxodiaceae, the redwood family.

OTHER NAMES

coast redwood, California redwood, sempervirens, Humboldt redwood.

DISTRIBUTION

Redwood occurs in a narrow broken belt, 10 to 35 miles wide, extending southward from southern Oregon for nearly 500 miles along the Pacific slope of the coast range.

THE TREE

The redwood is unusually tall with a heavily buttressed trunk that is clear of limbs for 100 feet or more. It has an open, well-rounded crown with heavy, short, drooping branches. The cones are scarcely 1 inch long. The dense fibrous bark is about 12 inches thick and is dark red. The average tree diameter is between 5 and 10 feet. Redwood is the state tree of California.

THE TIMBER

The heartwood varies from light cherry-red to dark reddish-brown or mahogany, while the narrow sapwood is almost white. The quality of the wood varies greatly. Some of it is very soft, fine grained and uniform in texture but some is coarse grained. Average reported specific gravity is 0.36 (ovendry weight/green volume), equivalent to an air-dried weight of 27 pcf. The wood has no distinctive odor or taste. Burls are common, reaching diameters of 6 feet, and have rich color and a great diversity of grain patterns. Redwood exhibits its greatest strength in end-wise compression making it well suited for columns.

SEASONING

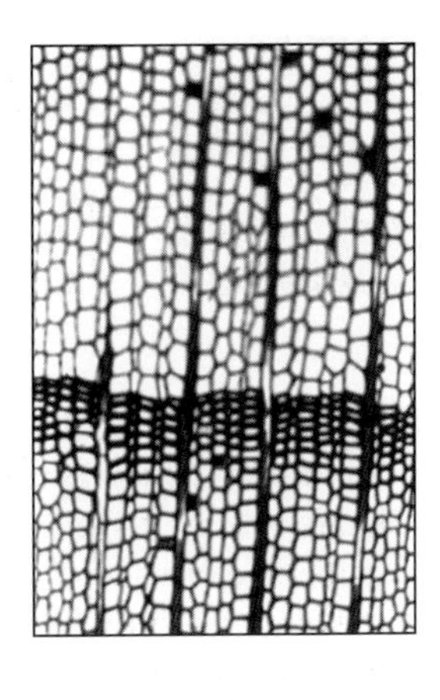

Freshly cut lumber contains a large amount of water, but it dries without excessive shrinkage and damage from checking is small. Average reported shrinkage values (green to ovendry) are 2.4% radial, 4.6% tangential and 6.9% volumetric. Once seasoned, redwood is comparatively inert to changes in humidity.

DURABILITY

None of the ordinary wood-rotting fungi grow in redwood timber, and the tree is free from fungus diseases. Few insects cause harm and it is highly resistant to attacks by termites. The heartwood displays outstanding durability when in contact with humid or moist conditions.

WORKABILITY

Redwood works moderately well with tools, glues well and nails easily, but nail and screw holding properties are poor. It takes and holds paint exceptionally well.

USES

A large part of redwood lumber is used in the form of planks, dimension boards, posts and joists for construction of homes, flooring, decks, fences, greenhouses, porches, tanks, silos, shingles, caskets and outdoor furniture. The beautiful burls are well adapted for use by the home woodworker for providing elegant and outstanding giftware. The bark is used in making novelties and as a constituent of roofing paper and other insulating purposes.

SUPPLIES

Environmental restrictions have seriously reduced the supply of this timber and prices are relatively high.

Sequoiadendron giganteum

giant sequoia

by Alan B. Curtis

SCIENTIFIC NAME

Sequoiadendron giganteum. Derivation: The genus name is in honor of Sequoyah (also spelled Sequoia) or George Guess (1770? to 1843), a Native American inventor of the Cherokee alphabet and the Greek word *dendron* meaning tree. The specific epithet means giant.

FAMILY

Taxodiaceae, the redwood family.

OTHER NAMES

bigtree, Sierra redwood, Wellingtonia (British).

DISTRIBUTION

Native to the western slope of the Sierra Nevadas of central California, it is not widespread in the forests, but is found in groves at elevations of 4,000 to 8,000 feet. Fossil evidence indicates that is was once widespread in North America, Europe and Asia.

THE TREE

Giant sequoias are the world's largest trees in volume and weight and among the largest in trunk diameter. The massive boles are from 10 to 36 feet in diameter and 150 to 300 feet in height. The cinnamon-red fibrous bark may be as much as 2 to 3 feet thick and protects the trees against fire during their 2,000 to 3,000 years of life. The long-pointed crown is characteristic when young, but older trees often have broken and damaged crowns as a result of lightning strikes and wind. This conifer has fine evergreen foliage and produces a small cone that contains seeds. When young the tree is a fast grower in both height and diameter.

THE TIMBER

The heartwood is purplish reddish-brown and often has a sheen. The sapwood is whitish and occupies a wide band of 100 to 200 annual rings. The heartwood is light weight with an average specific gravity of 0.29 (ovendry weight/green volume), equivalent to an air-dried weight of 21 pcf. The wood is soft, weak, brittle, non-resinous and without odor or taste. It is straight grained. Old trees have a very fine grain with 30 to 40 or more rings per inch.

SEASONING

Giant sequoia is easy to air dry, and there is seldom any twisting or warping. While precise shrinkage values are not available, they are estimated to be within the range of those for redwood (*Sequoia sempervirens* WDS 247). Dried wood is dimensionally stable.

DURABILITY

Researchers have found a few trees they believe fell more than 1,000 years ago. Although the sapwood has rotted away, the large cylinders of heartwood are sound and firm. The wood is very durable in contact with the soil due to its high tannin content. Sawn lumber weathers well.

WORKABILITY

It is very easy to work with hand or power tools, but is too soft to be of interest to most turners. It is easy to glue.

USES

In the past, this wood was hand split for fence posts, shakes, shingles and vineyard stakes and was sawn for siding and general construction lumber. It is not used for structural purposes. Today some logs and pieces cut a century ago and left in the forest are being salvaged and cut into thin boards for use as decorative paneling. It is an attractive ornamental and is widely planted in the western United States and Europe.

SUPPLIES

Giant sequoia is no longer commercially exploited and nearly all of the remaining large trees are located in national parks or preserves. Salvage of old, downed logs occurs occasionally on private lands and is the source of wood offered for sale. Giant sequoia is not to be confused with redwood, which only grows along the northwest coast of California and into southwest Oregon. Redwood is available on the commercial market and is used for decking, siding and decorative paneling. The wood of giant sequoia and redwood closely resemble one another.

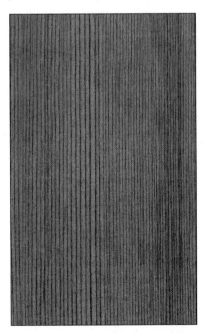

HISTORY

Giant sequoia was discovered by a California gold miner in 1852 while he was hunting a grizzly bear. It was introduced into England the following year. In the late 19th century many trees were felled. These felled trees were often too large to handle or so badly damaged in felling that they had little value. Today, the advent of portable sawmills and chainsaw mills enables their recovery.

Shorea spp.

lauan

by Jon Arno

SCIENTIFIC NAME

Shorea spp. Derivation. This genus name is in honor of John Shore, a Governor of Bengal.

FAMILY

Dipterocarpaceae, the dipterocarp family.

OTHER NAMES

The genus *Shorea* is sometimes identified under the genera *Parashorea* and *Pentacme*. Other names include Philippine mahogany, meranti, seraya and almon.

DISTRIBUTION

Shorea is wide spread in tropical Asia with various species ranging from India to the Philippines and south to Papua New Guinea and eastern Indonesia.

THE TREE

Most species of *Shorea* are sparsely branched trees growing up to 200 feet in height with straight trunks 3 to 6 feet in diameter above the buttress. The flowers form in racemes. The fruit is a winged nut. The leathery leaves are evergreen.

THE TIMBER

The lauans are, by and large, all straight-grained coarse-textured woods, but they are extremely variable with respect to color and density. In fact the lumber trade separates them into groups based on appearance characteristics rather than species. They are marketed as dark red meranti, light red meranti, white lauan, red lauan, etc. Depending upon the group, colors may vary from ash gray to dark reddish-brown. Average reported specific gravity ranges from 0.34 to 0.68 (ovendry weight/green volume), equivalent to an air-dried weight of 25 to 53 pcf. In other words, some of the white lauans are softer than eastern white pine (*Pinus strobus* WDS 212), while the harder, dark red merantis are denser and heavier than white oak (*Quercus alba* WDS 230). Wood that is red exhibits its greatest strength in end-wise compression making it well suited for columns. The strength of this wood varies with density. The so-called Balau group, the densest of the *Shorea*, is the strongest and is comparable to the North American white oaks.

SEASONING

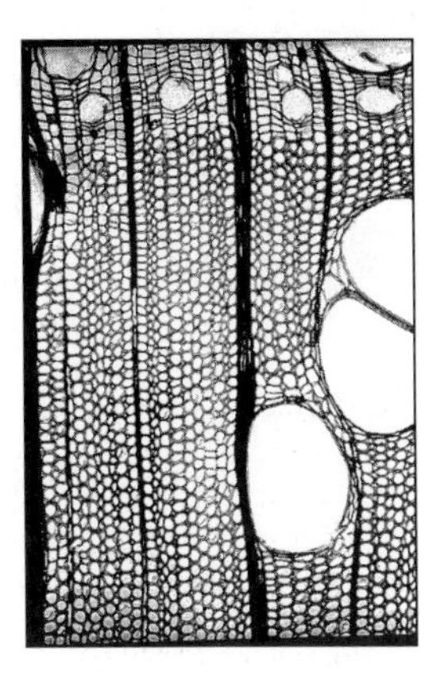

Again, due to the great diversity among the lauans, their seasoning properties differ widely. Generally, the denser groups are more difficult and slower to season. The lauans experience relatively high shrinkage when drying and both warping and checking can be problems. Depending on the species, average reported shrinkage values (green to ovendry) range from 3.0% to 4.6% radial, 6.9% to 11.4% tangential and 7.7% to 14.3% volumetric.

DURABILITY

The lighter lauans are not durable. They blue stain easily and are susceptible to borers. The darker, more resinous groups are durable enough to be used in marine applications such as ship decking and framing, but are not as durable as teak (*Tectona grandis* WDS 261).

WORKABILITY

Due to its coarse texture, lauan is stringy and frays or splinters when crosscut. It is difficult to finish to a smooth surface without filling, and its figure is somewhat bland in comparison to American mahogany. Some species are extremely resinous (the source of dammar, a resin used in varnish) and will gum up tools. Also, some species have a high silica content and will quickly dull blades. The lauans are not particularly stable when exposed to changes in humidity. Lauan is not a true mahogany and should not be confused with the American or African species in the genera *Swietenia* and *Khaya*.

USES

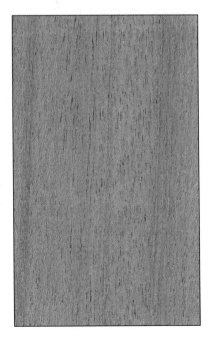

Lauan is a very versatile timber and supports a vast lumber and wood processing industry. It is used for millwork, plywood, doors, cabinetry and many architectural purposes.

SUPPLIES

Supplies of lauan are plentiful. It is a major timber in international commerce, especially with respect to veneer. In recent years, it has ranked number one among the hardwoods used in the manufacture of hardwood plywood. Low cost and almost universal availability in every shape and size make lauan one of the world's most popular woods.

Sorbus americana

American mountain-ash

by Jim Flynn

SCIENTIFIC NAME

Sorbus americana. Derivation: The genus name is the classical Latin name of *Sorbus domestica,* service mountain-ash of Europe. The specific epithet means the plant is "of America".

FAMILY

Rosaceae, the rose family.

OTHER NAMES

mountain-ash, mountain sumac, roundwood, dog-berry, wine tree, American rowantree, rowan berry, service tree, life of man.

DISTRIBUTION

From Newfoundland south to South Carolina and extending west to the Mississippi River.

THE TREE

There are 75 species throughout northern and higher elevations of North America, Europe and Asia. Three species are native to North America. The U.S. National Register of Big Trees lists the champion in West Virginia State Park, West Virginia. It is 62 feet in height and has a trunk circumference of 80 inches. Ordinarily, it is small, often bushy and shrub-like, growing under other taller trees. With branches ascending to form a rounded top, the main stem is short, usually less than 30 to 40 feet in height, with a diameter about 10 to 12 inches. The alternate compound leaves are 6 to 10 inches long with 11 to 17 paired leaflets on a stout stem. The lance-shaped leaflets are wedge-shaped at the base, sharp-pointed at the tip and finely toothed. The flowers are small (about 0.75 inch), white and borne in flat-topped clusters. The bright red, glossy berries often remain on the tree during the winter and, while not considered palatable, can be made into jelly. The tree is often used as an ornamental, although the European mountain-ash (*Sorbus aucuparia*) is preferred for this use.

THE TIMBER

The references state that this wood has no commercial possibilities probably because of limited supplies. It has a pleasing light-brown heartwood with paler brown sapwood. A piece I obtained from the Blue Ridge Mountains of Virginia is beautifully striped with various shades of tan, yellow and brown. The wood does not have a reputation for being strong, and it is hard to locate the origin of this alleged fault. Average specific gravity is 0.54 (ovendry weight/green volume), equivalent to an air-dried weight of 41 pcf. The wood has neither taste nor odor. It is diffuse porous and has medium to high luster.

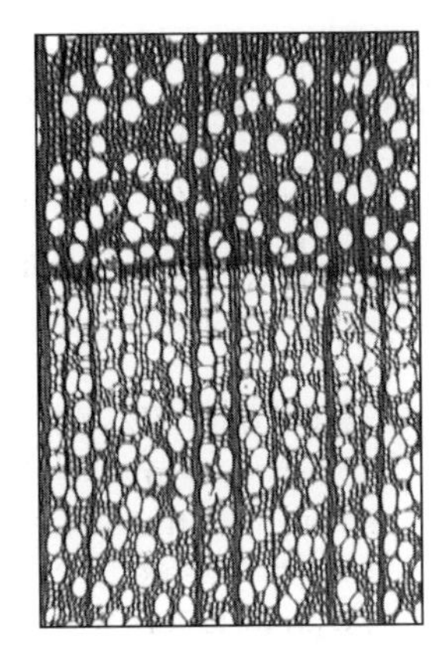

Because it is not a large tree and has a short life, one cannot expect to obtain large sawlogs.

SEASONING

Typically, small trees must be handled carefully by sealing the end grain with a wax emulsion as soon as it is cut. It should not be subjected to drastic changes in humidity until milled. Air drying in well-weighted, stickered piles is recommended. Data on kiln drying and shrinkage values are not available.

DURABILITY

This wood is probably not durable in applications exposed to the weather.

WORKABILITY

I found the wood extremely easy to work. It cuts well on the band saw and ran through a surface planer without trouble. Sanding produced a bright and lustrous surface with sharp edges. It glues as well as any other commonly used wood. While I have not applied stain or finish to the wood, it appears, by touch, to be adaptable to almost any type of finish.

USES

Uses include small pieces of furniture, crafts and turnery. Because it is not available in large quantities, the article should be matched to the wood rather than the reverse.

SUPPLIES

Two cautionary notes are required. American mountain-ash is not to be considered among the more popular species of ash on the commercial market, which are in the genus *Fraxinus* and are very different woods. Further, an Australian wood on the market called mountain ash is *Eucalyptus regnans* (WDS 116), a very different wood. Supplies of American mountain-ash are obtained by individuals searching for the wood or finding an ornamental tree that is being removed. If searching for ornamentals, care should be taken to correctly identify this species as distinct from European mountain-ash.

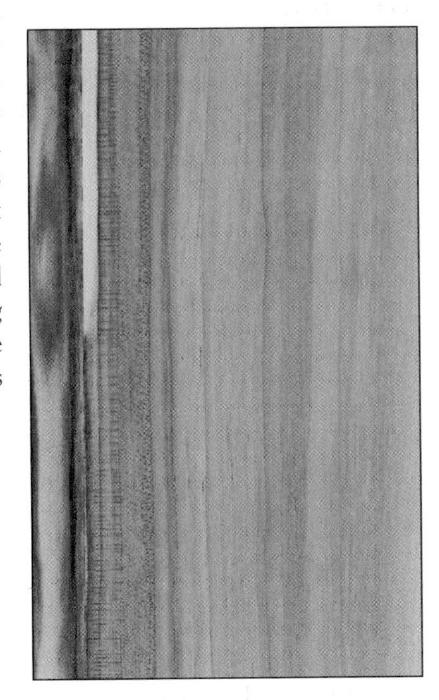

Spondias mombin

jobo

by Max Kline

SCIENTIFIC NAME

Spondias mombin. Derivation: The genus name is Greek for an old species of plum. The origin of the specific epithet is obscure and is probably a local name. A synonym is *S. lutea*.

FAMILY

Anacardiaceae, the cashew family.

OTHER NAMES

hog plum, Jamaica plum, Yucatan plum, Spanish plum, hobo, mombin, acaja.

DISTRIBUTION

Throughout most of the West Indies and from southern Mexico to Peru and Brazil.

THE TREE

Jobo grows to 130 feet in height with diameters to 48 inches. The boles are coarsely furrowed and can be 60 to 80 feet in length. Jobo is widely planted throughout the tropics for its edible fruit.

THE TIMBER

Jobo is generally uniformly cream to buff in color but sometimes contains darker streaks. The luster is medium. Odor and taste are not distinct. It is a light and soft wood. Average reported specific gravity is 0.40 (ovendry weight/green volume), equivalent to an air-dried weight of 30 pcf. The grain is straight to slightly wavy, and the texture is medium. This timber is tough and strong when its low weight per cubic foot is considered.

SEASONING

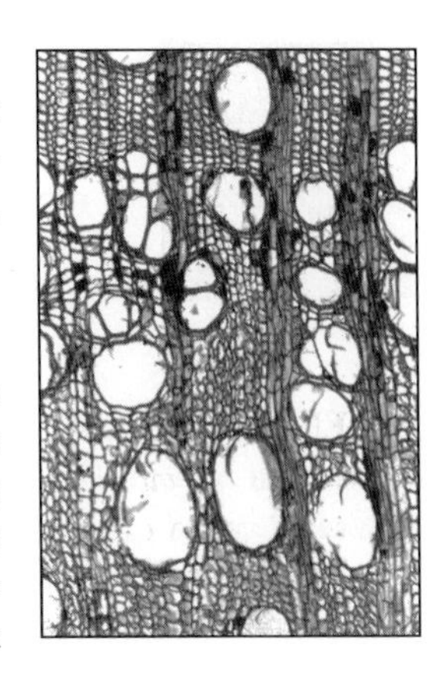

Jobo air dries rapidly but warps readily and exhibits slight checking. Average reported shrinkage values (green to oven-dry) are 2.7% radial, 4.7% tangential and 7.5% volumetric. The wood is subject to blue stain.

DURABILITY

Jobo timber is so susceptible to decay and insect attack when air dried that it is considered nearly worthless for use as lumber. When the logs are promptly processed after cutting and kiln dried to minimize deterioration, a useful product is obtained. Logs are particularly prone to blue stain. Preservatives are readily absorbed using pressure-vacuum or open tank systems.

WORKABILITY

Jobo is easily worked and generally finishes smoothly. Occasionally fuzzy grain may develop in some operations. Jobo has good nail holding properties.

USES

Jobo is used for boxes and crates, general carpentry, millwork, utility plywood, paper pulp and match sticks. The tree is planted as live fencing.

SUPPLIES

Because the timber has little value, not much jobo is exported from the tropics. During WWII, Yale University was engaged in a Navy project to explore the use of tropical woods to manufacture wood boats. At the end of the project, much of the excess wood was donated to IWCS and distributed to its members. This included quantities of jobo. Most of the wood harvested is used locally and would be in the low-cost category.

Sterculia apetala

chica

by Jim Flynn

SCIENTIFIC NAME

Sterculia apetala. Derivation: The genus name *stercus* was derived from Roman mythology meaning manure, which was applied to plants and was thought to cause the odor of the leaves and fruits of some species of *Stericulia*. The specific epithet is Latin for without petals. A synonym is *S. carthagensis*.

FAMILY

Sterculiaceae, the stericulia or cacao family.

OTHER NAMES

Sterculia apelata is known exclusively in Panama as panama. Panama is an Indian name from which the city of Panama derived its name. According to historians, the indigenous word panama means land where butterflies and fish abound or, simply, abundant. Other names include anacaguita (Puerto Rico), bellota (Mexico), sunsun (Venezuela), camajura (Colombia), castano (chestnut) in most of Central America, camaruca (Cuba) and huayra caspi (Peru).

DISTRIBUTION

Plants in this genus can be found in tropical climates throughout the world. The distribution of this species is from southern Mexico and Central America, including the West Indies, extending to Peru and Brazil.

THE TREE

Chica is widely planted as a shade tree and for its edible seeds and attractive flowers which are used in honey production. This giant tree can reach 130 feet in height with a trunk diameter of 80 inches or more. The thick trunk develops prominent narrow buttresses. The tree has a broad dense crown. The lobed leaves have long stalks and in some respects look like large maple leaves. The large flowers are borne in auxiliary panicles and do not have petals. The fruit consists of a cluster of five pod-like carpels containing large, brown chestnut-like seeds. The interior of these pods is covered with stiff bristles that penetrate the flesh and cause intense irritation. The seeds are useful for a variety of applications including a non-drying oil that is used in the food industry and in the manufacture of soap. When ground and mixed with water, they are used as a beverage or to flavor chocolates. They are also used to fatten pigs. The bark is used as a remedy for malaria, chest ailments and colds. Chica is the national tree of Panama.

THE TIMBER

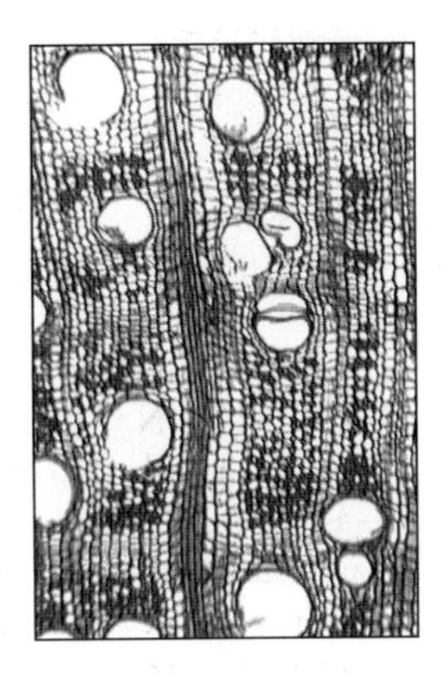

The heartwood varies between light brown, reddish-brown and yellowish-brown. It is not sharply demarcated from the yellowish sapwood. The luster and texture are medium. The wood has no distinctive taste or odor. The rays are very prominent, sometimes 20 cells wide and very high. Average reported specific gravity is 0.33, equivalent to an air-dried weight of 24 pcf.

SEASONING

Chica is a good candidate for air drying if done slowly. The stacks must be well-stickered or the slabs stood upright out of the direct rays of the sun. Average reported shrinkage values are 3.7% radial, 8.3% tangential and 11.8% volumetric.

DURABILITY

The timber has very little resistance to attack by decay fungi or insects and is also subject to blue stain. There are conflicting reports as to the ability of the wood to absorb preservatives.

WORKABILITY

As is characteristic of a light, low-density wood, few problems are encountered in working the timber with hand tools or machinery. Because a sample was not available to the author, tests for gluing and finishing could not be performed. A piece of S. *pururiens* was on hand, however, and exhibited satisfactory results.

USES

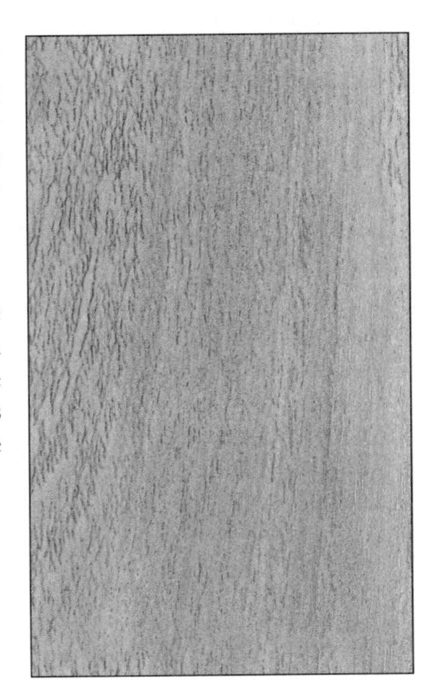

Throughout its growth range, chica has many uses including boxes and crates, interior paneling and plywood. In Peru, the wood is also used for furniture, musical instruments and, because of its lightness, cores for the interior of heels for women's shoes.

SUPPLIES

Supplies of chica do not appear on lists from the major hardwood timber dealers. Those dealing in Central and South American timbers may have stocks available at times but the best source is probably by personal contact with friends in the growth area.

Styphnolobium affine

Texas sophora

by Jim Flynn

SCIENTIFIC NAME

Styphnolobium affine. Derivation: The genus name implies that the leaf lobes are similar to those of the genus *Styphelia*. The specific epithet is Latin for akin to or neighboring, but it is unclear what this relationship implies. A synonym is *Sophora affinis*.

FAMILY

Fabaceae or Leguminosae, the legume family; (*Papilionaceae*) the pea or pulse group.

OTHER NAMES

coralbean, pink sophora, Eve's-necklace, bear-berry, pink locust, beaded locust.

DISTRIBUTION

Species of *Styphnolobium* can be found in areas with milder temperature in North America and Asia. Texas sophora is native to the extreme northwestern part of Louisiana, southeastern Oklahoma and extends into central Texas, including Dallas, Kerrville, Austin and San Antonio. Often, it appears in small groves on hillsides and along streams. It is not considered plentiful. An attractive ornamental, it may be found beyond its natural range in urban venues.

THE TREE

Texas sophora is not exceptionally large and may be a round-topped shrub or a small tree. Its average height is about 25 feet with a main stem diameter about 8 to 10 inches. The U.S. National Register of Big Trees lists the champion in Leakey, Texas. This specimen is 33 feet in height and has a main stem 60 inches in circumference. The branches are slightly zigzag and bright green when young but turn orange-brown as the tree matures. The tree is unarmed. The compound, feather-like leaves consist of 13 to 15 elliptical-oval, entire, dark green and lustrous leaflets. The flowers, appearing from April to June, are rosy-white in beautiful drooping clusters. A prominent feature is the seed pods. They are pinched in at each seed, giving it the appearance of a string of beads. From this the common name Eve's-necklace is derived.

THE TIMBER

The wood is ring porous and resembles honey locust (*Gleditsia triacanthos* WDS 126). The color may vary from light red to brownish-yellow. The wide sapwood is light yellow. The rays are very fine and better seen with a 5X magnifying glass. There is no odor or taste. The wood is extremely hard. A standard 3- by 6- by 0.5-inch sample has a specific gravity of 0.80 (ovendry weight/green volume), equivalent to an air-dried weight of 64 pcf. Being doubtful, I placed the sample in a container of water and it just bordered on sinking.

SEASONING

Shrinkage values are not available. An 8-inch piece of sophora that was sent to me by an IWCS member in Texas showed absolutely no sign of warp, indicating the drying process worked very well.

DURABILITY

Because the wood is not commercially important, there is little interest in researching this characteristic.

WORKABILITY

This heavy tough wood is difficult to work. If flat lumber is desired, it must be carefully run through the sawmill so the grain flow is as straight as possible, thus minimizing problems in down-stream milling. There is apt to be tear out in planing when the angle of the grain changes. Nevertheless, the wood sands well, glues satisfactorily and accepts a fine finish.

USES

Record and Hess, in *Timbers of the New World*, disposed of this by stating "Of no special utility and without commercial possibilities". I am not sure about this. Most luthiers, when encountering a new piece of wood, grasp it at a nodal point between their thumb and index finger, tap the wood with their knuckle and hold the wood near an ear. They are looking for a tap tone. Texas sophora has a splendid tone, and I would not hesitate to use it on an experimental musical instrument. Tone bars on a xylophone may also be a good application. In addition, the wood should be useful for small furniture items, craft work and perhaps turnery.

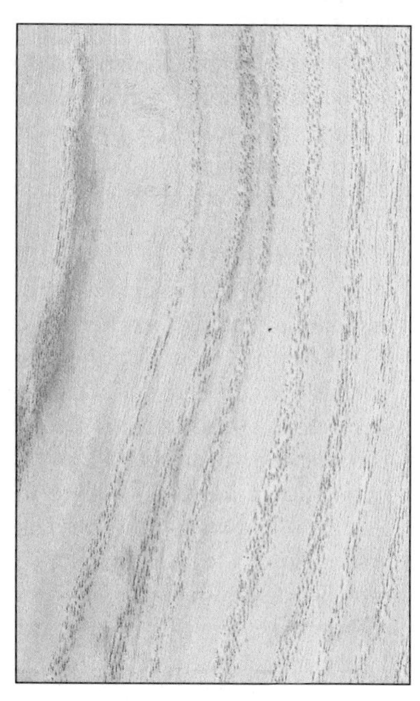

SUPPLIES

While the natural range of the species is restricted, outcroppings exist in other areas. The best source for small quantities is through members living in the growth areas.

Swietenia macrophylla

bigleaf mahogany

by Jon Arno

SCIENTIFIC NAME

Swietenia macrophylla. Derivation: The genus name is in honor of Gerard von Swieten (1700 to 1772), a German physician. The specific epithet is Latin meaning large leaf.

FAMILY

Meliaceae, the mahogany family.

OTHER NAMES

genuine mahogany, American mahogany, true mahogany, caoba (Spanish), acajou (French), Honduras mahogany.

DISTRIBUTION

Although commonly called Honduras mahogany, because it was first shipped to England from that colony as early as the late 17th century, this species is native from southern Mexico to Brazil.

THE TREE

Growing to heights in excess of 150 feet with diameters of over 6 feet, bigleaf mahogany is one of the more impressive species in the rain forests of Central and South America. Despite its great size, however, it tends to be sparsely distributed. Forest tracts containing populations of only 3 to 4 mature trees per acre constitute rich and commercially valuable stands of this species.

THE TIMBER

Since bigleaf mahogany is such a popular wood, it is often used as a comparative standard in describing other woods. In reality, though, this timber is extremely variable. Its color may range from light grayish-tan to "mahogany" red. Also, with a range in average reported specific gravity of 0.39 to 0.56 (ovendry weight/green volume), equivalent to an air-dried weight of 29 to 43 pcf, it can be as soft as yellow-poplar (*Liriodendron tulipifera* WDS 162) or harder than most red oaks. Because of this variability, a lot of look-alike woods have been successfully marketed as mahogany. Although not an absolutely reliable clue, true mahogany can usually be identified by its storied rays. On the flatsawn surfaces, short but dark ray flecks end to form wavy horizontal bands across the board. Owing to its extreme variability in terms of density, its strength properties are also quite variable. At the low extreme, bigleaf mahogany is barely adequate for use in some furniture applications. It is not particularly elastic, but it has a good strength-to-weight ratio.

SEASONING

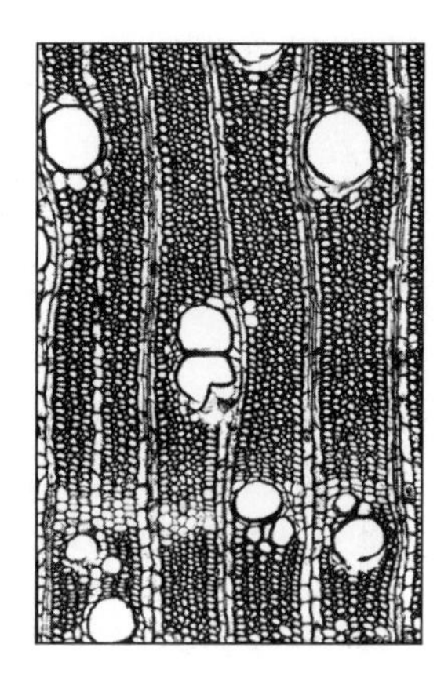

With a volumetric shrinkage of only 7.8% (green to ovendry), this species is one of the most stable of all commercially important cabinetry woods. Additional average reported shrinkage values (green to ovendry) are 3.0% radial and 4.1% tangential. Because it is not highly susceptible to blue staining, it is easy to air dry with little risk of degrade.

DURABILITY

The durability of bigleaf mahogany is excellent. It has superb weathering properties and is used extensively in boat building.

WORKABILITY

Bigleaf mahogany ranks among the finest cabinetry woods in the world. Its working characteristics are outstanding in nearly all woodworking processes, including cutting, shaping, turning and sanding. Due to its moderately coarse texture, filling may be necessary in order to achieve a glass smooth finish, but bigleaf mahogany accommodates virtually all finishing methods.

USES

Uses include fine furniture, interior trim, paneling, cabinetry, turning, carving, model making, veneer and boat building.

SUPPLIES

At the moment, this species is readily available and moderately priced. Unfortunately, there is no way it will remain so given current levels of demand. Initial efforts to grow this species on a plantation basis have not been entirely successful in that, when densely planted, it is susceptible to insect attack. Efforts to cultivate it outside its native range (and away from its natural predators) may prove successful. At the very least, though, I suspect we can anticipate a gap in its availability during the 21st century that will be at least decades in duration. The facts suggest this species should be taken out of service as a high volume, general construction timber and conserved for only the most precious applications.

Syringa vulgaris

lilac

by Max Kline

SCIENTIFIC NAME

Syringa vulgaris. Derivation: The genus name is from the Greek *syrinx* meaning pipe because pipe stems were made from the hollow stems of the plant. The specific epithet is Latin meaning common.

FAMILY

Oleaceae, the olive family.

OTHER NAMES

There are hundreds of varieties and cultivars of this species all bearing their individual common names. It would not be possible to include them here.

DISTRIBUTION

Lilac is cultivated in practically every state in the northern half of the United States from the Atlantic to the Pacific coast and extending well south into the more temperate states. *Syringa* is a genus of about 30 species from Asia and southeast Europe. Because of their showy fragrant flowers, this plant can be found in many places throughout the world.

THE TREE

A number of lilac species are cultivated in the United States. Usually these are large shrubs, but lilac occasionally reaches a tree size of 15 to 25 feet in height with a short trunk from 3 to 6 inches in diameter. Originally from Europe, lilac is a beautiful flowering shrub and is a favorite for its hardy flowers which thrive well in the northern U.S. climate. *Syringa vulgaris* is known as the common lilac but there are many variations planted as ornamentals with large fragrant white, pink or rose flowers.

THE TIMBER

Lilac wood is close grained and of moderate density. It varies in color with the location of growth. The wood is a pleasing yellowish-cream color with purplish streaks. The sapwood is very narrow, and the growth rings are conspicuous. It has a pleasant aromatic odor when worked. A standard 3- by 6- by 0.5-inch sample had a specific gravity of 0.73 (ovendry weight/green volume), equivalent to an air-dried weight of 58 pcf.

SEASONING

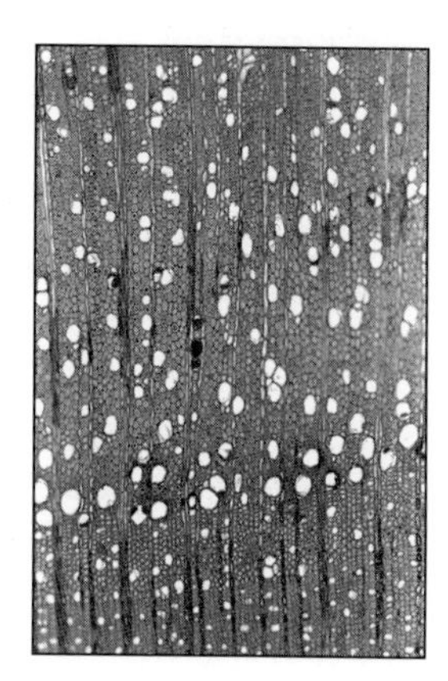

Lilac wood is difficult to season without checking. The most satisfactory method is to cut the stem lengthwise down the center, remove the outer bark, wax the ends and air dry slowly. Shrinkage values are not available.

DURABILITY

Many older lilacs are attacked by worms, which are destructive to the wood.

WORKABILITY

This wood is easily worked. It takes a fine polish and turns and carves well. Natural finishes are recommended to enhance the beautiful color and texture of the wood.

USES

Lilac has no commercial use but should find favor among woodworkers for making very unusual novelties such as jewelry, candle holders or pepper shakers. It can be used to make attractive jewelry.

SUPPLIES

There is no lilac available from lumber suppliers but the collector should always be on the alert for a tree that can be obtained for the asking.

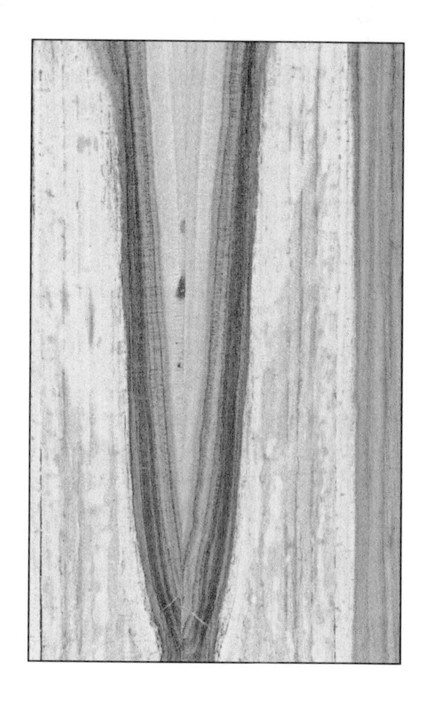

Tabebuia donnell-smithii

primavera

by Max Kline

SCIENTIFIC NAME

Tabebuia donnell-smithii. Derivation: The genus name *tabebuia* is a South American Indian name. The specific epithet is in honor of Donnell-Smith (1829 to 1928), a botanist. A synonym is *Cybistax donnell-smithii*.

FAMILY

Bignoniaceae, the bignonia or trumpet creeper family.

OTHER NAMES

durango, palo blanco, copal, Cortez, white mahogany, roble, aragan, penda.

DISTRIBUTION

Southwestern Mexico, the Pacific coast of Guatemala and El Salvador and north central Honduras.

THE TREE

Primavera is a large-sized tree often reaching 100 feet in height with a long smooth trunk that is 4 feet in diameter. The bark is pale gray. The large leaves are opposite, deciduous, digitally compound, long petioled and usually entire. The large, showy, yellow flowers, which appear before the new leaves, are most beautiful. The pods are pendant, long and cylindrical and contain numerous broadly winged seeds.

THE TIMBER

The timber is yellowish-white to light yellowish-brown and often striped. The luster is fairly high, and the texture is medium to rather coarse. It is odorless and tasteless. Average reported specific gravity is 0.40 (ovendry weight/green volume), equivalent to an air-dried weight of 30 pcf. The grain is straight to finely and attractively wavy. The general appearance of primavera resembles that of Ceylon satinwood (*Chloroxylon swietenia* WDS 080) and while sometimes called white mahogany, it is entirely unrelated botanically to the mahogany family. The strength properties of primavera are more than adequate for the purposes for which it is used.

SEASONING

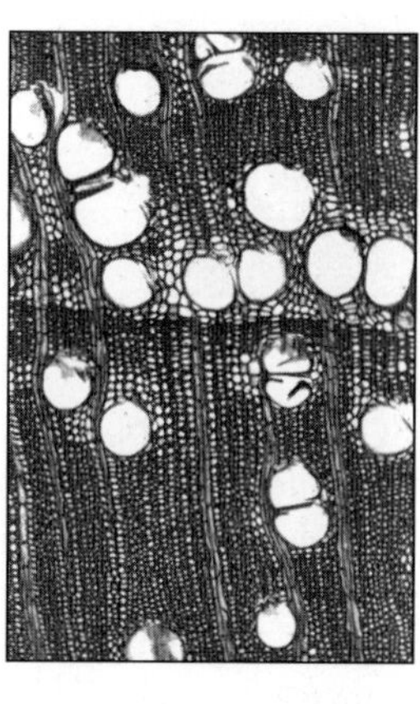

Primavera kiln dries easily and rapidly with little degrade due to warping or checking. Average reported shrinkage values (green to ovendry) are 3.1% radial, 5.1% tangential and 9.1% volumetric.

DURABILITY

The wood may be classified as sufficiently durable for interior work but is susceptible to fungal attack if used externally.

WORKABILITY

When interlocked grain is present, care is necessary in the surfacing of primavera, and it is not highly recommended for turnery purposes. In most other operations, the lumber will

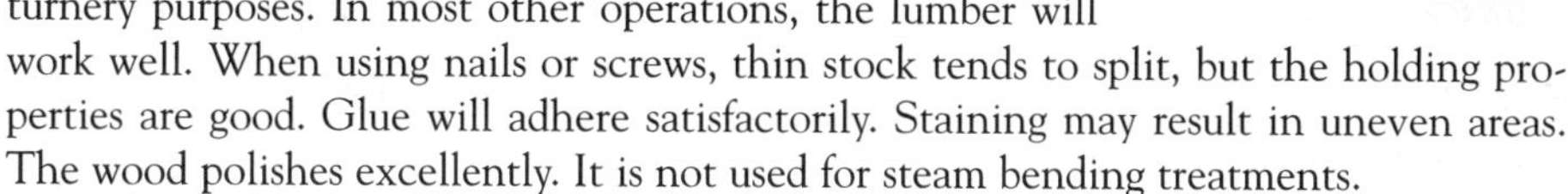

work well. When using nails or screws, thin stock tends to split, but the holding properties are good. Glue will adhere satisfactorily. Staining may result in uneven areas. The wood polishes excellently. It is not used for steam bending treatments.

USES

Due to its dimensional stability, working ease and pleasing appearance, primavera is used for solid furniture, paneling, interior trim patterns, veneer, cabinetry and millwork.

SUPPLIES

Veneer is scarce but the lumber is available at a high price in the United States. Timber is seldom exported to Europe.

Tabebuia spp.

ipe

by Jim Flynn

SCIENTIFIC NAME

Tabebuia spp. (Lapacho group). The genus name *tabebuia* is a South American Indian name. There is a large and varied population of trees classified in the genus *Tabebuia* and, over the years, there have been many changes in their botanical nomenclature. They are now classified into four distinct groups: white cedar, roble, lapacho and miscellaneous. In this Wood Data Sheet, we are concerned with the species contained within the Lapacho group. This may include, but is not limited to *T. heptaphylla*, *T. chrysanthak*, *T. serratifolia* and *T. guayacan*.

FAMILY

Bignoniaceae, the bignonia or trumpet creeper family.

OTHER NAMES

The most common of the common names is ipe. Other names are bethabara, lapacho, madera negra, amapa, Cortez, guayacan, guayacan plovillo, flor amarillo, Surinam greenheart (not to be confused with Demerara greenheart), tachuario, lapacho negro, pui and poui.

DISTRIBUTION

Species of *Tabebuia* in the Lapacho group grow throughout tropical South and Central America and to some extent in the islands of the Lesser Antilles. The tree does not seem to be sensitive to any particular growing site and is found on ridge tops, riverbanks and in marshy areas.

THE TREE

Usually these trees grow straight to heights of 140 to 150 feet with well formed trunks measuring 2 to 3 feet in diameter. Some boles are clear to the 60-foot level and their bases may or may not be buttressed. The leaves are typically digitate and deciduous. The flowers are mostly yellow but sometimes pink, red or violet depending upon the species observed. The fruit is a long woody capsule with winged seeds.

THE TIMBER

The relatively wide sapwood is yellowish-white and sharply differentiated from the heartwood. When freshly cut, the heartwood is grayish and darkens upon exposure to a grayish-green or brownish-olive. Because there is no control over the specific species of *Tabebuia* reaching the marketplace, there may be a wide variety of physical characteristics of the wood. Generally, the timber has a fine but often uneven and interlocked grain. There is neither distinct odor nor taste to the wood. The vessels of the heartwood contain a yellowish-green deposit called lapachol, which has the appearance of sulphur. In the presence of alkaline, this substance turns a deep red. Average reported specific gravity is 0.85 to 0.97 (ovendry weight/green volume), equivalent to an air-dried weight of 69 to 80 pcf.

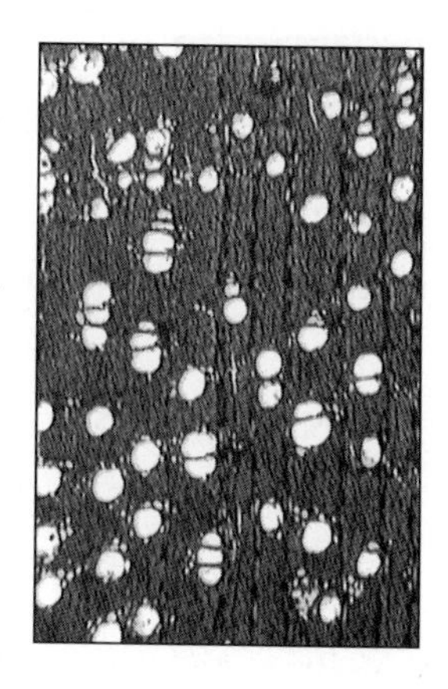

SEASONING

The timber seasons fairly easily and dries with very little warping and checking. After seasoning and manufacture, it is dimensionally stable. Average reported shrinkage values (green to ovendry) are 6.6% radial, 8.0% tangential and 13.2% volumetric.

DURABILITY

This is an extremely durable timber with estimates of service life when in contact with the ground being in excess of 25 years. It is highly resistant to attack from decay fungi and termites. Some species of *Tabebuia* in the Lapacho group, however, are not resistant to marine borers. The species *T. guayacan*, which is the principal Central American species, is reported to be resistant to these borers. The wood is extremely resistant to preservation treatments.

WORKABILITY

This hard heavy wood, often with wavy grain, is hard to work. It plays havoc with cutting blades on machine tools but not to a point that the wood should be abandoned. It turns and finishes well. Pre-boring before using nails or screws is advised.

USES

Uses include heavy construction, railroad crossties, turnery, industrial flooring and decorative veneer. It is currently finding ideal use as a substitute for the bland impregnated southern yellow pine for outside decks, walkways and railing in up-scale applications.

SUPPLIES

This timber is in the inventory of many wholesale lumber distributors at a medium high price. *Tabebuia rufescens* is marketed in the United Kingdom as pui.

Tamarindus indica

tamarind

by Jim Flynn

SCIENTIFIC NAME

Tamarindus indica. Derivation: The genus name *tamarind* derives from the Arabic *tamar-u'l-Hind*, which means "date of India". The specific epithet means the plant is "of India".

FAMILY

Fabaceae or Leguminosae, the legume family; (*Caesalpiniaceae*) the cassia group.

OTHER NAMES

tamarindo (Spanish), tamarin. The common name tamarind, also spelled tamerind, is applied to many other genera including *Albizia, Cojoba, Leucaena, Lysiloma, Niopa, Popanax*. Be careful when identifying the wood solely by its common name.

DISTRIBUTION

Tamarind is native to the dry savannahs of tropical Africa. Arab traders introduced it to Asia. It is believed that tamarind reached the New World with the first shipment of slaves. It is common throughout the tropics and well founded in the Caribbean, south Florida and Hawaii. Marco Polo recorded encountering it in 1298.

THE TREE

Tamarind has a short trunk and only grows to 40 feet or more in height. Its branches spread out to form a great dome-shaped crown. The pinnate leaves are 2 to 4.5 inches long with 10 to 18 pairs of oblong bluish-green leaflets. The leaves fold at night and in overcast weather. The long, flat, rust-colored pods contain a pulp consisting of 30% to 40% sugar and are a rich source of vitamins. The pods can be eaten fresh as well as used for seasoning in chutney, curries, etc. A well-liked carbonated drink is also made from the pods and extensively consumed throughout Latin America.

THE TIMBER

Tamarind is a grand wood that at one time was sold in North America as Madeira mahogany because of its mahogany-like appearance. The heartwood is dark purplish-brown and the well defined sapwood is dull yellow. This hard heavy wood has an average specific gravity of 0.90 (ovendry weight/green volume), equivalent to an air-dried weight of 73 pcf.

SEASONING

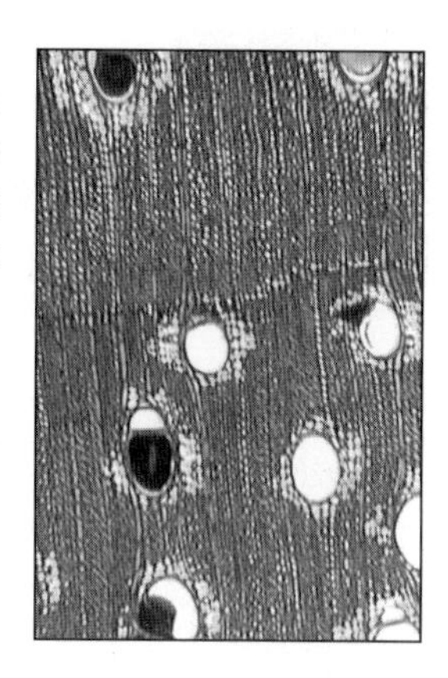

There does not appear to be any data on this characteristic and shrinkage values are not available. Because the tree has a short trunk with many branches, long clear planks are not generally obtainable. Wood slabbed in flitches would be apt to season erratically unless slowly and carefully dried.

DURABILITY

There is some conflicting data on whether termites like this wood. The U.S. National Academy of Sciences publication, *Tropical Legumes*, states that tamarind is "strong and termite proof". The USDA Agricultural Handbook, *Trees of Puerto Rico and the Virgin Islands*, states that "It is strong and durable, although very susceptible to attack by dry-wood termites".

WORKABILITY

The wood is hard, tough and difficult to work. Cabinet scrapers are very effective in smoothing flat surfaces. This wood is worth working as it takes a beautiful finish.

USES

Because the timber can be found worldwide, there are many applications where it finds use. In the Sudan, for example, it is used for general carpentry and boat building. Elsewhere it is used for construction timber, furniture, tool handles and turnery. It is a great fuel. In India, during WWII, it was used for gasogen units, which powered many cars and trucks.

SUPPLIES

Tamarind does not appear to be marketed commercially in the United States. Because the timber is so pervasive in the tropics, it is probably finding its way into bazaars and markets in other parts of the world. It is certainly not a rare species, but well worth finding and working.

Taxodium distichum

bald-cypress

by Jim Flynn

SCIENTIFIC NAME

Taxodium distichum. Derivation: The genus name is derived from the Latin *taxus* meaning yew and a suffix meaning like referring to the yew-like leaves. The specific epithet is Latin meaning two-ranked referring to the leaves being in two rows.

FAMILY

Taxodiaceae, the redwood family or the baldcypress family.

OTHER NAMES

cypress, southern-cypress, swamp-cypress, red-cypress, yellow-cypress, white-cypress, Tidewater red-cypress, gulf-cypress. The common name pond-cypress usually applies to the variety *nutans*.

DISTRIBUTION

The range extends from southern Delaware to southern Florida and along the lower Gulf Coast plain to southeastern Texas. Inland it grows along the many streams of the middle and upper coastal plains and northward through the Mississippi Valley to southeastern Oklahoma, southeastern Missouri, southern Illinois and southwestern Indiana.

THE TREE

Bald-cypress is a large deciduous conifer 100 to 120 feet in height with a diameter of 3 to 5 feet. It grows in swampy areas, along stream banks and rivers and can be found in pure stands. Once it was common to find trees 500 to 600 years old. Many trees are broadly buttressed, accompanied by root "knees" protruding from the swampy waters. It is conical-shaped when young, but the top flattens somewhat with age. The moderately thin bark is light reddish-brown with shallow furrows dividing the broad flat ridges that often separate into thin fibrous scales. The light yellowish-green deciduous leaves, two-ranked and spreading, have the appearance of feathers. The fruit, often growing in pairs, is a cone up to 1 inch in diameter and turns woody-brown when mature. Bald-cypress is the state tree of Louisiana.

THE TIMBER

The wide (1 inch or more) sapwood is pale yellowish-white merging into the heartwood which varies in color from yellowish to light or dark brown, reddish-brown or almost black. The color is darker in trees grown along the Gulf Coast and south Atlantic regions than in the north. The wood has a greasy feel with a rancid odor but does not impart taste, odor or color to food products. The grain varies from straight to even, and

the wood is coarse textured. Average reported specific gravity is 0.42 (ovendry weight/green volume), equivalent to an air-dried weight of 31 pcf. In terms of color, weight and durability, the wood is quite variable and is marketed under names such as red-cypress, tidewater red-cypress and white-cypress.

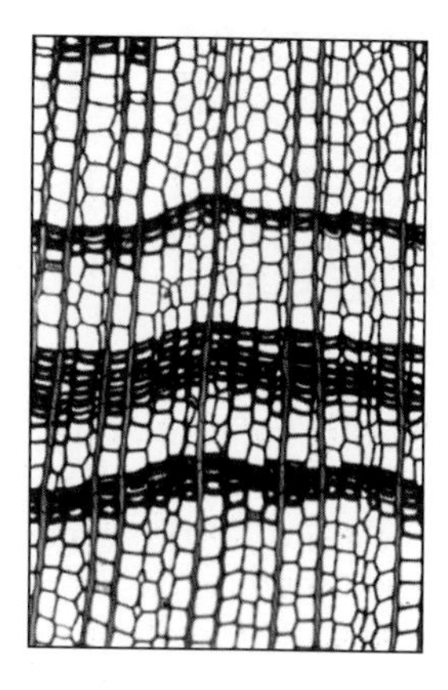

SEASONING

Freshly cut it contains a considerable amount of water, requiring more care in kiln drying than most other conifers. Under certain conditions, raised grain can be a problem but not discoloration by blue stain fungi. Average reported shrinkage values (green to ovendry) are 3.8% radial, 6.2% tangential and 10.5% volumetric.

DURABILITY

Bald-cypress has outstanding durability. Since virgin forest wood has a higher resistance to decay, second-growth timber should be treated with preservatives if used in situations that test durability. Pecky cypress contains pockets of wood attacked by fungus but decay is arrested when the tree is cut and poses no further problem.

WORKABILITY

It is an ideal wood to machine and shape and has very few adverse attributes. It glues, sands, nails and especially holds paint well.

USES

It is ideal for beams, posts, docks and bridges. It also is used for siding, sash, doors, paneling, trim and general millwork. Tanks, vats, greenhouse framing and others products subjected to moisture are ideal uses, as well as railroad crossties and shingles.

SUPPLIES

The peak cutting period was the early 1900s. Currently, adequate supplies of the wood are on the market at reasonable prices.

Taxus brevifolia

Pacific yew

by Max Kline

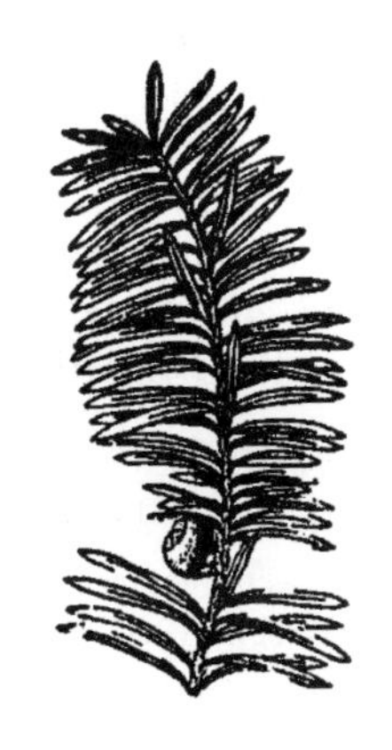

SCIENTIFIC NAME

Taxus brevifolia. Derivation: The genus name is the classical Latin name (Greek, *taxos*). The specific epithet means short-leaf which is in contrast to English yew, *T. baccata*.

FAMILY

Taxaceae, the yew family.

OTHER NAMES

western yew, yew.

DISTRIBUTION

Pacific yew is found from southern Alaska and British Columbia south to the coast ranges and Sierra Nevadas of central California to the western slopes of the Rocky Mountains in Montana and Idaho.

THE TREE

Pacific yew is found in small groups or singularly, scattered among other western conifers. The usual height is from 20 to 40 feet with a diameter of 12 to 15 inches. It can, however, reach heights of 60 to 75 feet and diameters of 18 to 30 inches. The bark is very thin, comparatively smooth and scaly.

THE TIMBER

The wood of Pacific yew is very hard and strong. It is dense and heavy with a very fine, close grain and fine texture. Average reported specific gravity is 0.60 (ovendry weight/green volume), equivalent to an air-dried weight of 46 pcf. The heartwood is bright orange to brown demarcated from the narrow white sapwood. It is odorless and tasteless and has a rather high luster. The grain is straight to variable.

SEASONING

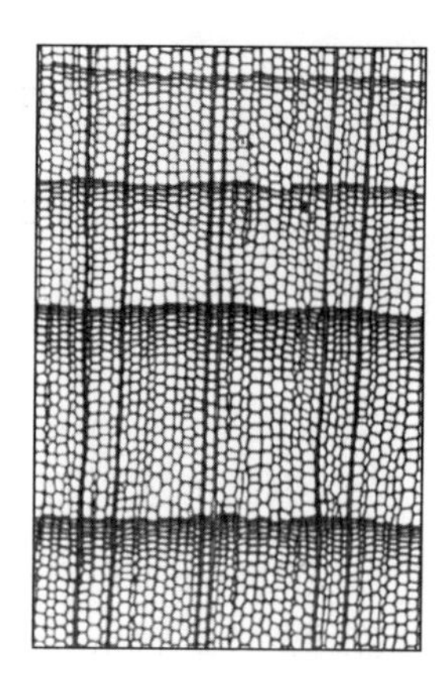

Pacific yew dries slowly and may develop severe shake. For this reason, it is desirable to slow the rate of drying through the use of end-grain coatings. Warping is not usually a problem. Kiln drying is very satisfactory provided a suitable drying schedule is used. Average reported shrinkage values (green to ovendry) are 4.0% radial, 5.4% tangential and 9.7% volumetric.

DURABILITY

The durability of Pacific yew is very high. It is resistant to all forms of insect and fungal attack and may be used for external purposes without a preservative treatment.

WORKABILITY

This wood may be worked satisfactorily with tools. It splits easily when nailing but holds screws well. It bends extremely well and is excellent for turnery. Some woodworkers find the dust toxic; a puffiness in the hands may develop and sneezing fits can occur while sawing Pacific yew. These effects are neither serious nor long lasting, but annoying to certain people. Non-oil finishes are preferred, as oil will turn the heartwood to a chocolate-tan color when applied. It finishes smoothly.

USES

The best known use of Pacific yew is for archery bows and other bent work. It is also a fine wood for turnery products and is occasionally used for canoe paddles and small cabinetwork. It is not of great commercial importance because of its scarcity. [Ed. note: This wood is often used for stringed musical instruments, especially the bodies of lutes. Further, the bark and leaves are a source of taxus, which has found use in the medical community as an agent for treating cancer.]

SUPPLIES

Since the tree is slow growing and not abundant, Pacific yew wood is considered to be rare and costly in either lumber or veneer form.

261 *Tectona grandis*

teak

by Max Kline

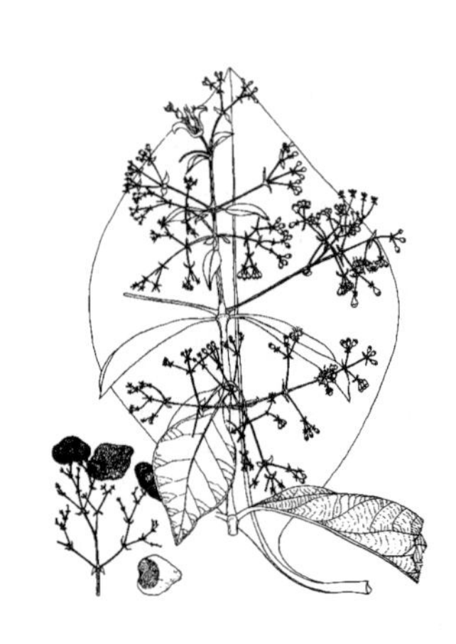

SCIENTIFIC NAME

Tectona grandis. Derivation: The genus name is from the Greek *tekton* meaning carpenter, referring to the use of the timber. The specific epithet is Latin meaning grand or large.

FAMILY

Lamiaceae, the mint family.

OTHER NAMES

Burma teak, Rangoon teak, Moulmein teak, India teak.

DISTRIBUTION

Myanmar (Burma), India, the Malaysian Peninsula, Thailand.

THE TREE

Teak trees are very variable in size and grow according to locality, soil and climate conditions. Commonly, teak grows 130 to 150 feet in height with a diameter of 4 to 5 feet. In Myanmar, there are reports of trees with diameters of 6 to 8 feet. The length of the clean bole can vary from 20 feet to as much as 80 to 90 feet. Teak is a hardwood and can be found in almost pure stands. Although it grows in a wide variety of soils, good drainage is essential. Teak is at home on the lower slopes and is rarely found at elevations higher than 3,000 feet. The leaves are 10 to 12 inches long and 7 to 14 inches wide.

THE TIMBER

True teak is dark golden yellow turning dark brown or almost black upon prolonged exposure to air. An irregular figuring of darker streaks or marks is quite common. The narrow sapwood is grayish or white. The wood has a distinctive leathery smell but no appreciable taste. The luster is dull. The timber has an oily surface when first cut that gives a sticky feel to the hand. Average reported specific gravity is 0.55 (ovendry weight/green volume), equivalent to an air-dried weight of 42 pcf. The grain can be a straight stripe to an occasional mottle. The wood has a coarse texture. Teak has excellent strength properties, which make the timber suitable for a wide range of structural uses.

SEASONING

Teak seasons well but rather slowly with very little shrinkage. Average reported shrinkage values (green to ovendry) are 2.5% radial, 5.8% tangential and 7.0% volumetric. Air or kiln drying can be done satisfactorily, and once dried the timber remains remarkably stable.

DURABILITY

Teak has outstanding durability in regards to alternate wet and dry conditions. This quality has long been appreciated in the ship building industry.

WORKABILITY

In general, teak works very well although it is rather hard on tools. Thin cutting edges are essential in all tool processes. Good quality steel tools perform best if the spindle speeds are somewhat reduced. Worked edges of the wood remain sharp. Mortising and drilling operations can be done quickly and cleanly. Teak can be varnished and polished effectively and takes nails and screws fairly well. Glue processes are best done on freshly machined or sawed surfaces because the oily nature of aged surfaces sometimes presents problems in bonding.

USES

Teak is one of the outstanding timbers of the world due to its valuable properties of durability, strength, weight, workability and good appearance. It is most useful in ship building for decking surfaces, while the veneer and plywood are used for paneling in fine homes and offices. Teak is in demand as a furniture wood. It is also used in products exposed to acids because of its chemical resistance. Carvings and turnings frequently are made from teak.

SUPPLIES

Teak is available in veneer or lumber forms but is higher in price than most imported timbers. [Ed. note: Because of the commercial value of this wood, there are many teak plantations operating in warmer climates and producing large quantities of teak. Reports indicate that the wood from the original growth areas is still preferred by discriminating users.]

Terminalia elliptica

Indian laurel

by Max Kline

SCIENTIFIC NAME

Terminalia elliptica. Derivation: The genus name refers to the clustered terminal leaves on the ends of branches. The specific epithet refers to the elliptical (twice the length of the width) shape of the leaves. A synonym is *T. tomentosa*.

FAMILY

Combretaceae, the combretum family.

OTHER NAMES

saj, sain, sein (Native Indian name), taukkyan (Myanmar name), East Indian walnut.

DISTRIBUTION

Indian laurel is widely distributed and common throughout India and Myanmar (Burma).

THE TREE

Indian laurel adapts itself to a wide variety of conditions, but the best tree development is reached on clay and moist soils where it may reach a height of 80 to 100 feet with a diameter at breast high of 25 to 40 inches.

THE TIMBER

Indian laurel is light brown with fine darker streaks or dark brown or brownish-black with bands of a darker color. The grain is often interlocked and has an attractive figure, especially on quartersawn surfaces. Average reported specific gravity is 0.73 (ovendry weight/green volume), equivalent to an air-dried weight of 58 pcf. The texture is medium, moderately open and uneven. The luster is dull to medium. Odor and taste are not distinct. Indian laurel is superior to teak (*Tectona grandis* WDS 261) in most strength properties. It is not recommended for steam bending treatments.

SEASONING

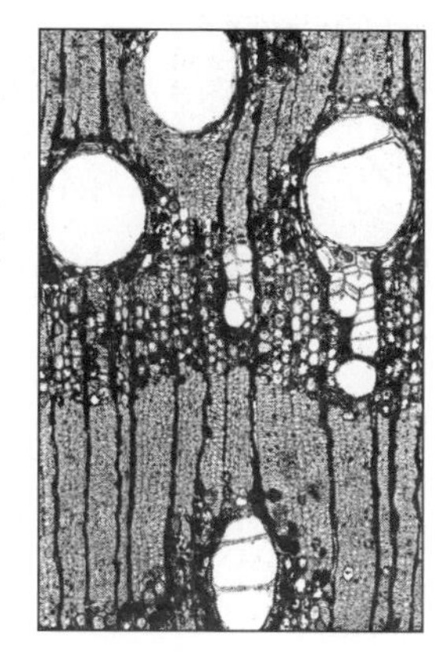

Indian laurel is reported to be a difficult timber to season, and degrade during drying may be excessive. If kept in the round, the wood is apt to split. Drying needs to be done slowly, and the most satisfactory results are obtained by kiln drying after a very short period of air drying or directly from the green condition. Average reported shrinkage values (green to ovendry) are 4.8% radial, 7.4% tangential and 13.2% volumetric.

DURABILITY

Indian laurel is moderately durable, but for sustained exposure to possible insect or fungal attack, a preservative treatment is advisable.

WORKABILITY

Indian laurel cannot be classified as easy to work as it is somewhat hard on tools. It works to a fibrous finish. No problems should be expected when mortising or drilling. Nail and screw holding properties are good, but pre-boring is suggested. It responds well to the normal finishing agents but a fair amount of filling is necessary prior to polishing.

USES

This timber has been used as a substitute for black walnut (*Juglans nigra* WDS 150) and is sometimes used for furniture, cabinets and gunstocks. It is more commonly used in the form of decorative veneer. Selected logs yield highly figured veneer of unusual attractiveness. In India, it finds uses for building and construction, harbor work, boat building and railway sleepers.

SUPPLIES

There is an abundant supply of Indian laurel in India, but due to the difficulty of seasoning, only specially selected material is exported which causes it to be scarce and expensive in the U.S. and European markets.

Terminalia ivorensis

idigbo

by Max Kline

SCIENTIFIC NAME

Terminalia ivorensis. Derivation: The genus name refers to the clustered terminal leaves on the ends of branches. The specific epithet means the plant is "of the Ivory Coast".

FAMILY

Combretaceae, the combretum family.

OTHER NAMES

emeri, framire, black afara, African teak, treme.

DISTRIBUTION

West tropical Africa including the Ivory Coast, Gold Coast and Nigeria.

THE TREE

Idigbo is a large tree that reaches 100 feet or more in height with diameters of 3 to 5 feet. The bole is straight and clear to 70 feet. It is frequently fluted and brittle heart is common.

THE TIMBER

The pale yellowish-brown sapwood and heartwood are not clearly defined. Usually the grain is straight, but occasionally may be interlocked which gives an attractive ribbon figure on quartersawn surfaces. The texture is moderately open, but uniform, and the wood has no distinctive smell. The luster is high. Average reported specific gravity is 0.43 (ovendry weight/green volume), equivalent to an air-dried weight of 32 pcf. The wood will corrode iron fastenings that are driven into it, and it contains a coloring agent that may, under damp conditions, leach out of the wood and stain anything with which it comes in contact. The dust may irritate the skin or respiratory tract. This timber is not rated as being a strong timber and is unsuitable for steam bending. If brittle heart is present, even lower strength properties are to be expected.

SEASONING

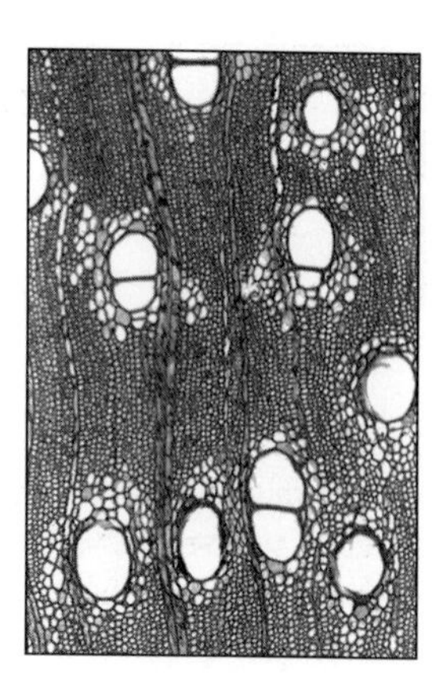

Idigbo seasons rapidly with little or no checking or distortion. It seasons equally well by kiln or air drying methods. Average reported shrinkage values (green to ovendry) are 3.5% radial, 5.2% tangential and 9.0% volumetric.

DURABILITY

The heartwood is rated as durable and moderately resistant to termite attack. The sapwood is attacked by powder post beetles. Idigbo is highly resistant to preservative treatment.

WORKABILITY

Idigbo works fairly easily with all hand and machine tools. Drilling or mortising should be done at relatively slow speeds to prevent a tendency to char at higher speeds. Good results are obtained with stains and polishes, but a fair amount of filling is necessary before a satisfactory surface is achieved. It has fairly good nail and screw holding properties and is said to glue without difficulty.

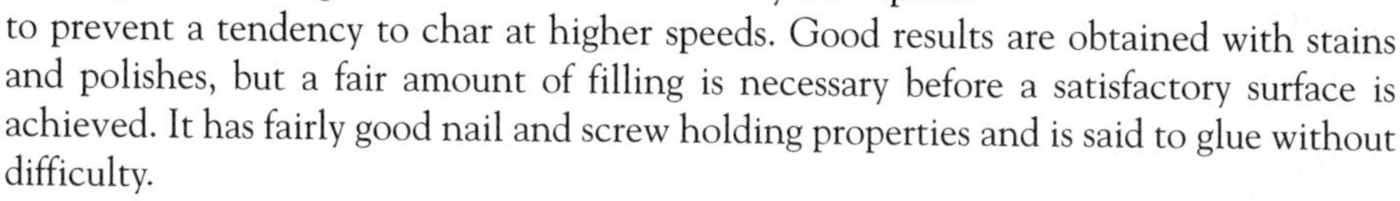

USES

Idigbo is a most useful lightweight, hard wood for up-scale joinery, such as window frames, doors and casings. Timbers that are free of knots are available in long lengths. It is also used for flooring, furniture components, veneer and light construction. Uses where there is contact with iron or with damp fabrics should be avoided to prevent corrosion and staining.

SUPPLIES

Considerable quantities are used in Europe for joinery, but a relatively small amount is imported into the United States, as other native species serve the same purposes. The cost is moderate.

Terminalia superba

limba

by Max Kline

SCIENTIFIC NAME

Terminalia superba. Derivation: The genus name refers to the clustered terminal leaves on the end of the branches. The specific epithet is Latin for superb or proud, which refers to the stature of the tree.

FAMILY

Combretaceae, the combretum family.

OTHER NAMES

korina, afara, ofram, akom, frake.

DISTRIBUTION

Occurring in rain and savanna forests, limba is widely distributed in west Africa from Sierra Leone to Angola and the Congo (formerly Zaire).

THE TREE

Limba is a tall tree with a straight bole that reaches 150 feet in height with a diameter of 3 to 5 feet above the large, thin 45° buttresses. The gray bark has long vertical fissures and patches from which flakes have fallen. The branching is whorled and confined to the upper parts of the tree. The leaves are crowded toward the end of the branchlets. They are arranged alternately and are simple measuring 4 inches long and 2 inches wide.

THE TIMBER

The color of limba is either uniformly creamy pale yellow or grayish-brown with irregular streaks of dark brown or black. The sapwood is not distinct from the heartwood. The texture is moderately coarse. The odor is mild. The luster is high and satiny, and the grain is straight to irregular or interlocked. Splinters can cause skin inflammation. Average reported specific gravity is 0.45 (ovendry weight/green volume), equivalent to an air-dried weight of 34 pcf. Brittle heart is sometimes present, rendering the wood appreciably weaker than normal wood and unsuitable for any use where strength is important.

SEASONING

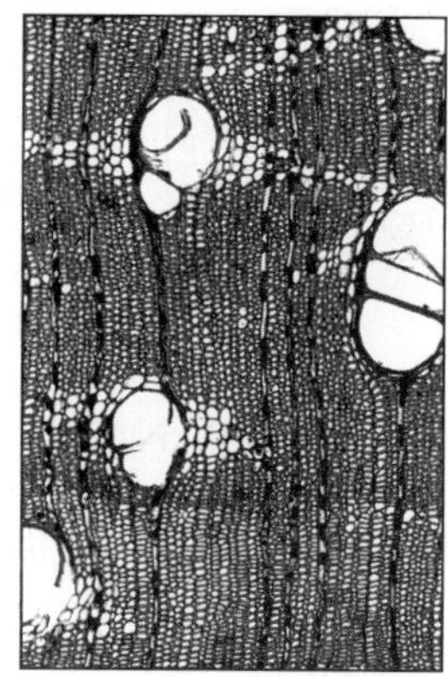

Limba seasons rapidly with little or no checking or warping. Average reported shrinkage values (green to ovendry) are 4.5% radial, 6.2% tangential and 10.8% volumetric. Once dried, it is rated as having only small movement in service.

DURABILITY

The timber is highly susceptible to damage by pinhole borer beetles, powder post beetles and subterranean termites. Special precautions should be taken immediately after the tree is felled to prevent deterioration.

WORKABILITY

Limba is easily worked with hand and machine tools. It has a tendency to split in nailing or screwing operations. It takes stain readily and polishes well after filling. Its gluing properties are satisfactory. The veneering characteristics are very good.

USES

Limba is used for light construction work, interior cabinetry, simulated traditional furniture, plywood and decorative moldings. The figured dark-colored heartwood is quite attractive for decorative plywood paneling.

SUPPLIES

Limba veneer is plentiful and lumber is available at a moderate price.

Testulea gabonensis

izombe

by Max Kline

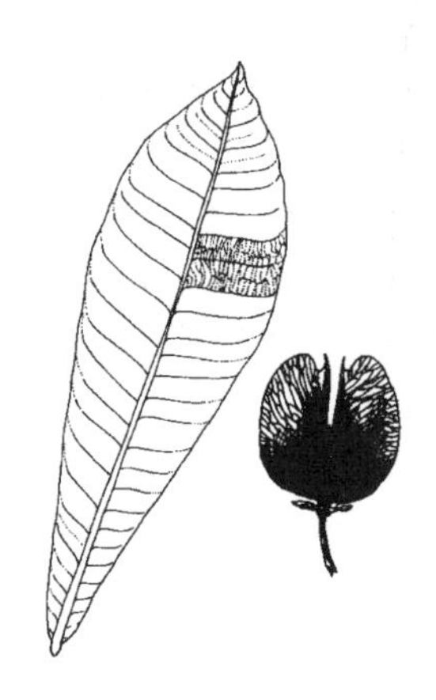

SCIENTIFIC NAME

Testulea gabonensis. Derivation: The genus name is Latin referring to the tortoise plant. The specific epithet means the plant is "of Gabon".

FAMILY

Ochnaceae, the ochna family.

OTHER NAMES

rone, ake, akewe.

DISTRIBUTION

Gabon and Cameroon.

THE TREE

Testulea is one of the 35 genera in the Ochnaceae family growing in subtropical to tropical climates. They are typified by evergreen leaves that are alternate and leathery in texture. The tree reaches 120 feet in height and has a straight bole, which is cylindrical and clear to 30 to 60 feet. Trunk diameters are 3 to 4 feet.

THE TIMBER

The color of izombe wood is light orange or orange-brown, usually with darker streaks. The sapwood is well demarcated from the heartwood. The odor is offensive when freshly cut but disappears after drying. The taste is not distinct. The texture is very fine and even. The grain is interlocked and produces a wavy or ribbon figure. The luster is shiny on smoothed surfaces. Average reported specific gravity is 0.60 (ovendry weight/green volume), equivalent to an air-dried weight of 46 pcf.

SEASONING

Izombe timber dries easily with little or no degrade. Average reported shrinkage values (green to ovendry) are 3.4% radial, 6.0% tangential and 10.4% volumetric.

DURABILITY

The heartwood has high durability and is resistant to termites. Weathering properties are satisfactory. Izombe is resistant to preservative impregnation.

WORKABILITY

Izombe saws well and works easily with both hand and power tools. It glues and nails well and takes a good finish.

USES

Izombe timber is used for door and window framing, furniture, flooring, turnery, carving, tool handles, sporting and athletic equipment, veneer and plywood.

SUPPLIES

Some izombe is imported into the United States by the large importers. It is available in lumber form at a moderately high price.

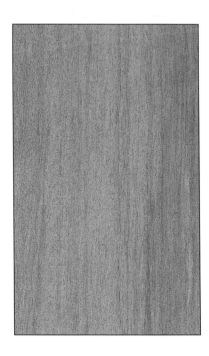

Tetraclinis articulata

thuya

by Jim Flynn

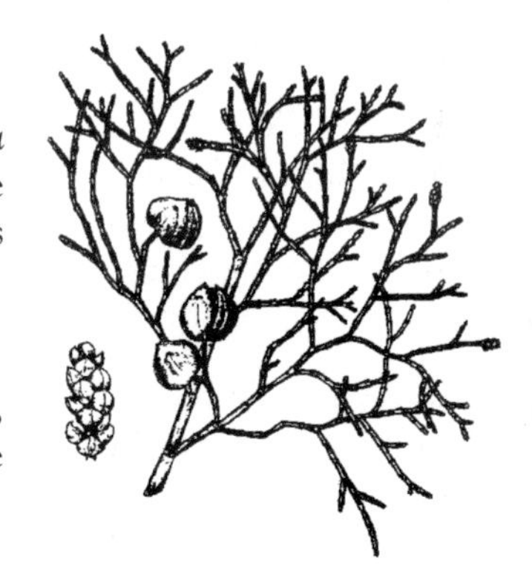

SCIENTIFIC NAME

Tetraclinis articulata. Derivation: The genus name *tetra* is Greek meaning four and *clinis* meaning angles. The specific epithet is Latin meaning articulated joints (branchlets). A synonym is *Callitris quadrivalvis*.

FAMILY

Cupressaceae, the cypress family. Within this family, the genus *Tetraclinis* is monotypic (consisting of one species).

OTHER NAMES

A wood that has been known to civilization as long as this probably has hundreds of common names. Ones that remain in the records are aratree, citrus wood (not related to the citrus species in the Rutaceae family), thyine wood, sandarac tree, thujawood, and the popular name thuya burl.

DISTRIBUTION

Thuya is native to northern Africa, primarily in the mid-elevations of the Atlas Mountains and in southern Spain.

THE TREE

This rare species of evergreen grows in dry areas and can withstand considerable periods of drought. It may reach 50 feet in height and grows along with Aleppo pine (*Pinus halepensis*) and the almost depleted Atlas cedar (*Cedrus atlantica*). The dense branches are ascending and terminate in flat, jointed, spray-like branchlets that are covered with scale-like leaves and are decurrent at the base and arranged in fours. The cones are solitary and rounded, about 0.375 inch in diameter and consist of four thick, woody and glaucous scales.

THE TIMBER

There is little reference to just thuya as it is always thuya burl. In the past, thuya was used for bridges, house construction and other utilitarian applications. Today, commercial interests are focused on thuya burl. These burls are formed on tree roots and require extensive digging to extract them. For centuries these trees have been subjected to fires kindled to open up small plots for agriculture. Some believe this process has encouraged the development of burls. The wood varies considerably but is reddish-tan darkening to a black hue. It is highly figured and single pieces can be found exhibiting grains patterns such as tiger (long and rippling), panther (spiral), peacock (eyes like peacock's tail) and parsley (a ruffled figure). While thuya burl may look somewhat like redwood burl (*Sequoia sempervirens* WDS 247), it is much denser with an average specific gravity of

0.72 (ovendry weight/green volume), equivalent to an air-dried weight of 57 pcf. It has a pleasant, sweet odor when worked.

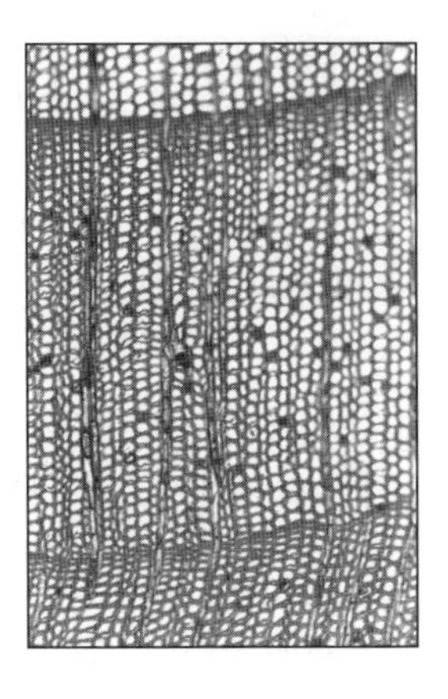

SEASONING

Shrinkage values are not available. The wood probably seasons well because of its high oil content which precludes rapid moisture loss.

DURABILITY

The high cost dictates that the wood be carefully used in applications not subjected to severe environmental conditions. The wood was used in temples and mosques in Spain and ancient Palestine.

WORKABILITY

Working with the grain of burls always requires special care and sharp tools. Thuya burl can be cut, milled and turned satisfactorily but planing and surfacing require particular skill. The most effective way of removing heavy sanding and saw marks is with cabinet scrapers. Spots where the nap of the grain abruptly changes can be encountered. Sanding with progressively finer grit is necessary and one should end up using 1200-grit paper. A fine polish will result with hardly the need for a finish. The wood will glue well if the joints are first wiped with a solvent, such as lighter fluid, to remove surface oil.

USES

Thuya burl was used in the Roman Empire for tabletops mounted on ivory legs and inlaid with gold. Artisans formed a special guild of wood joiners and ivory carvers, *citarii et eborarii*. The resin, sandarac, was used in making varnish and as a protective coating on paintings. Today, thuya burl is popular for veneered furniture. It is also used for small boxes, trinkets and bowls.

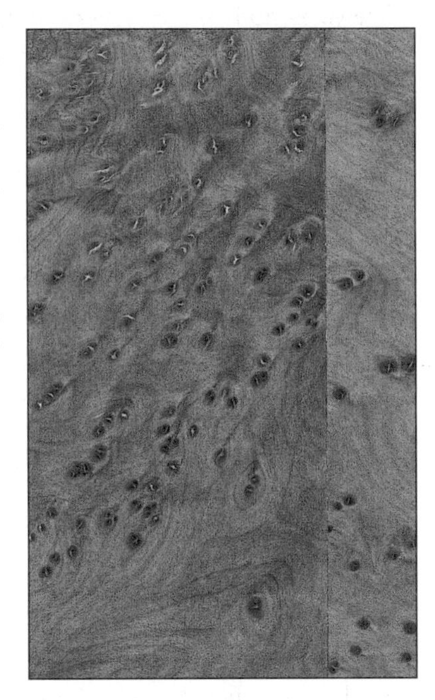

SUPPLIES

The predominant supplies are veneer but solid pieces are on the market. Expect to pay a premium price for them.

267

Thespesia populnea

portiatree

by Jim Flynn

SCIENTIFIC NAME

Thespesia populnea. Derivation: The genus name is from the Greek meaning divine, apparently because the species was considered a sacred tree in the islands of the South Pacific. The specific epithet refers to the poplar shape of the leaf.

FAMILY

Malvaceae, the mallow family.

OTHER NAMES

Milo is a popular common name in wide use. Other names include seaside mahoe, corktree, Spanish cork, emajaguilla, otaheita, rosewood of Seychelles, umbrella tree, false rosewood and cremon. There are many other names in the vernacular where the species is found.

DISTRIBUTION

Native to Asia, tropical Africa, Pacific islands (including Hawaii) and naturalized in other parts of the world including Florida, the Florida Keys and in the Caribbean region. It is widely distributed on the littoral of tropical landmasses due to germination of floating fruits and seeds.

THE TREE

Portiatree is an evergreen growing to 50 feet or more with a trunk diameter up to 12 inches. It has a dense crown and long lower branches that often spread and interlock with those of other trees to form a very dense thicket. The leaves are heart-shaped and pointed, 4 to 8 inches long and 2.5 to 5 inches wide. They are dark green and have long petioles. The flowers are large and bell-shaped and display some of the characteristics of the Hibiscus family of which they are a part. The tree often is planted for shade and as an ornamental. The bark contains a fiber often used for making crude ropes. Because of its small size, portiatree is not a great timber producer.

THE TIMBER

At first glance, the wood appears to be a rosewood (*Dalbergia* spp., specifically *D. retusa* WDS 098), but on closer examination, it lacks the shine of rosewood. The sapwood is light brown, clearly separated from the darkish heartwood that turns deep chocolate-brown near the center. It is attractively streaked with red and various shades of brown, often getting close to being black. The wood has a moderately fine texture and a specific gravity in the range of 0.61 (ovendry weight/green volume), equivalent to an air-dried weight of 47 pcf.

SEASONING

The wood seasons well and is not prone to warping or checking. Shrinkage values are not available. Portiatree can tolerate harsh drying treatments.

DURABILITY

The timber is classified as resistant to decay and dry-wood termites. Because portiatree is not harvested as large timbers, this characteristic is relatively unimportant.

WORKABILITY

This species is easy to saw and machine by both hand and machine tools. It is a choice wood for turnery and carving and can be worked in either the green or dry condition. It contains an oil that may impede the drying of varnish, nevertheless, it takes a high polish. Gluing may be troublesome unless precautions are taken to remove the surface oil from the joints to be glued.

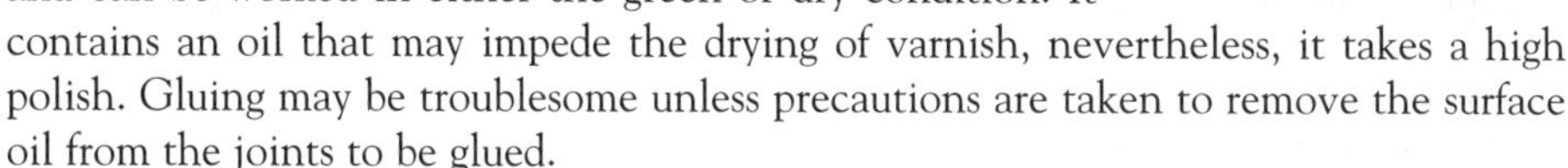

USES

Portiatree is an excellent hardwood for uses that do not require long lengths. Among the more important uses are flooring, furniture, turnery, carving and veneer. In Hawaii, the wood is considered the most easily worked of the species used for craft items offered in the tourist trade.

SUPPLIES

The large timber importers do not stock portiatree because of the limited availability. Local sawmills in the growth areas and specialty wood dealers are the best sources of supply. The wood often shows up at IWCS auctions, especially during the Southeast Regional Winter Meeting in Florida.

Thuja plicata

western redcedar

by Jon Arno

SCIENTIFIC NAME

Thuja plicata. Derivation: The genus name *thuja* is Greek for an aromatic wood, probably a juniper, that was highly prized in ancient times for choice durable furniture. The specific epithet means folded into plates, which refers to the flattened twigs with regularly arranged scale-like leaves.

FAMILY

Cupressaceae, the cypress family.

OTHER NAMES

giant arborvitae, canoe cedar, shinglewood, thuja geant.

DISTRIBUTION

Seldom forming pure stands, western redcedar is found in mixed forests with Sitka spruce (*Picea sitchensis* WDS 206), Douglas-fir (*Pseudotsuga menziesii* WDS 224) and other western species. Preferring moist soil, its range extends from southern Alaska to northern California. Absent in the arid central valleys, it reappears in the northern Rockies from Idaho and Montana into Canada where the mountains capture moisture from the prevailing winds.

THE TREE

Although sometimes dwarfed growing alongside redwood (*Sequoia sempervirens* WDS 247) and Douglas-fir, western redcedar would stand out as a giant with most other flora. On ideal sites, it can reach a height of 200 feet with a diameter of up to 10 feet. Several specimens in Washington, which are over 1,000 years old, have attained diameters in excess of 20 feet. The bark is stringy or fibrous and the fragrant, scale-like, evergreen leaves are similar to those of the common arborvitae cultivars. Western redcedar is the official tree of British Columbia, Canada.

THE TIMBER

The heartwood is reddish-brown, sometimes with a pinkish tinge and clearly demarcated from the narrow, almost white sapwood that seldom exceeds 1 inch in width on mature trees. The wood has a much coarser texture than redwood and has a distinct odor similar to aromatic cedar, but not quite so overpowering. It is weak, brittle and very easily dented. With an average reported specific gravity of 0.31 (ovendry weight/green volume), equivalent to an air-dried weight of 23 pcf, it is one of the least dense western softwoods and is substantially lighter than any of the domestic pines.

SEASONING

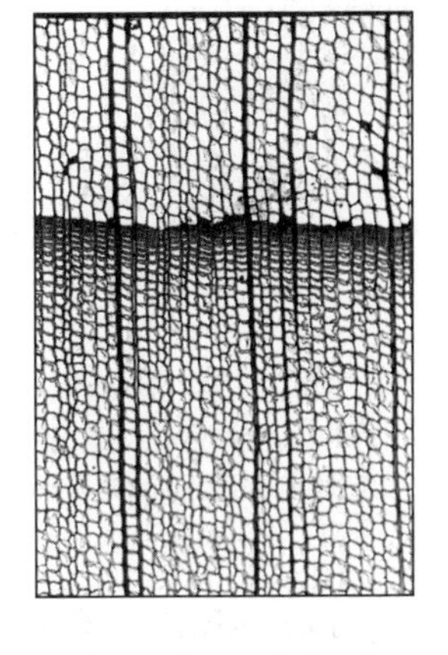

It dries easily with little degrade. Average reported shrinkage values (green to ovendry) are 2.4% radial, 5.0% tangential and 6.8% volumetric.

DURABILITY

It weathers extremely well and is one of the best North American softwoods for outdoor applications.

WORKABILITY

As one of the major lumber species in the United States and Canada, most woodworkers or homeowners have worked with or come in contact with western redcedar. Although it has a tendency to fray when crosscut, it provides little resistance to cutting tools. It glues and finishes well and takes nails without difficulty. Being extremely straight grained, it splits easily and predictably, making it one of the best species for shingles, shakes and shims. Not only is it durable when exposed to the elements, it develops a beautiful silver-gray patina that makes exterior maintenance unnecessary. This species is cited by several texts as being toxic to some individuals and excessive exposure to the dust can cause rashes and respiratory problems.

USES

Modern uses include siding, landscape decking, sashes, doors, boat building, fencing, shingles, shakes, shims, sheathing, telephone poles and veneer for exterior plywood panels. The tree is also used as a decorative planting. Historically, it played an important role in the economies of the Native American tribes of the Pacific Northwest where its fibrous bark was used for everything from rope to diapers. As its common name canoe cedar suggests, logs were used for dugout canoes.

SUPPLIES

Supplies are becoming less available. The species does not replenish itself quickly, and demand for it seems almost insatiable. Western redcedar is still readily available at moderate to low prices, but at the present rate of harvest clear grades may be virtually out of the market sometime in the first half of the 21st century.

Tieghemella heckelii

makore

by Max Kline

SCIENTIFIC NAME

Tieghemella heckelii. Derivation: The genus name is in honor of Philippe von Tieghem (1839 to 1914), the French naturalist. The specific epithet is in honor of Edouard Marie Heckel (1843 to 1916), the founder of the Botanical Garden of Marseille and author of several works on African flora. Synonyms are *Dumoria heckelii* and *Mimusops heckelii*.

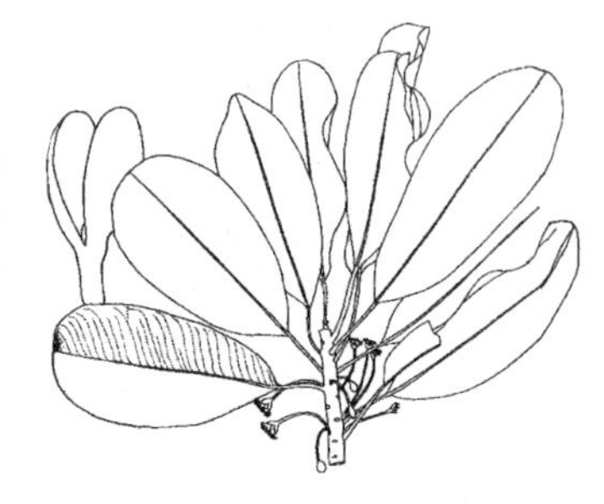

FAMILY

Sapotaceae, the sapodilla family.

OTHER NAMES

cherry mahogany, African cherry, baku, babu.

DISTRIBUTION

Widely distributed in west Africa from Sierra Leone to Gabon including the Ivory Coast and Ghana.

THE TREE

This typically tropical evergreen tree is one of about 30 similar species. The leaves are thick, shining, simple, entire and alternate. The small white flowers are borne in axillary fascicles. The tree reaches a height of 180 to 200 feet with straight cylindrical boles clear to 100 feet and diameters up to 9 feet. It is free from buttresses. In areas not subject to freezing, makore makes a very good ornamental.

THE TIMBER

Makore has a resemblance to close-grained mahogany or plainsawn sapele (*Entandrophyragma cylindricum* WDS 112). The heartwood varies from pinkish- or purplish-brown to dark blood red. The clearly demarcated sapwood is whitish or light pink and 2 to 3 inches wide. The grain is generally straight. The texture is fine to medium, and the wood is lustrous. Odor and taste are not distinct. It will stain when in contact with iron under damp conditions. The timber contains some silica. Strength properties resemble those of the denser grades of American mahogany. Straight-grained timber is classed as moderately good for steam bending. Average reported specific gravity is 0.55 (ovendry weight/green volume), equivalent to an air-dried weight of 42 pcf.

SEASONING

The timber seasons at a moderate rate with little degrade, and it is dimensionally stable in use. Average reported shrinkage values (green to ovendry) are 4.7% to 6.2% radial, 6.8% to 8.0% tangential and 10.6% to 11.0% volumetric.

DURABILITY

The heartwood is highly durable and usually resistant to all forms of fungus rot attack and insects including termites. No preservative treatments are used or required.

WORKABILITY

Makore causes rapid blunting of ordinary steel cutters and saws because of the silica present in the wood. Carbide-tipped saws are recommended for cutting. Pre-boring is advised when using nails or screws, as there is a tendency for splitting in these operations. Stains and polishes give excellent results and glued joints hold well. The fine dust produced in some finishing procedures may irritate the nose and throat and cause dermatitis.

USES

The wood compares favorably with mahogany for furniture and up-scale decorative work, including interior fittings and superior joinery. Makore can be used in the solid state or as veneer. Other common uses include flooring, instruments, tool handles, turnery, carving, ship building, textile industry rollers, laboratory benches and doors.

SUPPLIES

Makore is plentiful as veneer at a moderate price. Lumber can be obtained from importers.

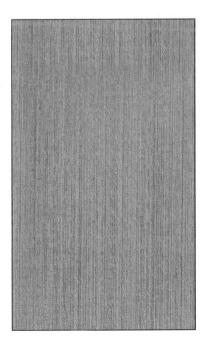

Tilia americana

American basswood

by Max Kline

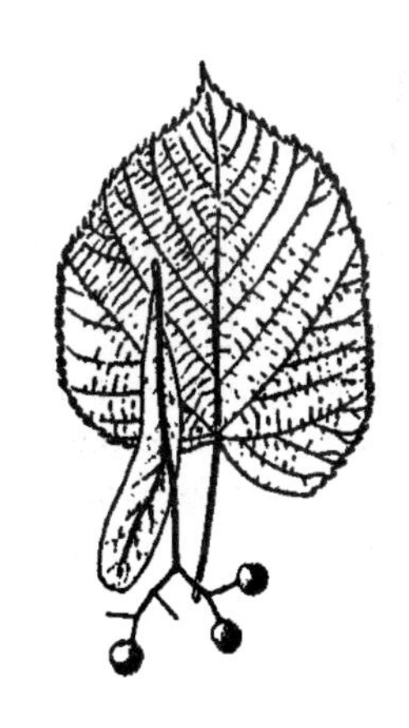

SCIENTIFIC NAME

Tilia americana. Derivation: The genus name is the classical Latin name, probably from the Greek *ptilon* meaning wing, referring to the wing-like bract of flower clusters. The specific epithet means the plant is "of America". A synonym is *T. glabra*.

FAMILY

Tiliaceae, the linden family.

OTHER NAMES

American linden, basswood, lime, linden, lime tree, American white wood.

DISTRIBUTION

The range extends from New Brunswick and southern Quebec to southern Manitoba and North Dakota, south to South Carolina, Tennessee, Missouri and Kansas.

THE TREE

American basswood is a beautiful tree reaching heights of 130 feet with diameters of 4 feet. Because of its dense foliage, hardiness and fast growth, it is widely used as a shade tree for street beautification. In the spring, the white to cream-colored flowers are profuse, developing into clusters of fruit. The tree thrives best in deep, rich, sandy soil.

THE TIMBER

The heartwood is creamy white to brownish and is not always easily distinguished from the wide, nearly white sapwood. The luster is medium. It is odorless and virtually tasteless. Average reported specific gravity is 0.32 (ovendry weight/green volume), equivalent to an air-dried weight of 24 pcf. The texture is rather fine and uniform. It is usually straight grained and does not split easily. American basswood is a soft wood with correspondingly low strength properties.

SEASONING

American basswood is easy to season and very dimensionally stable once dried. There is very little checking or twisting during seasoning. Average reported shrinkage values (green to ovendry) are 6.6% radial, 9.3% tangential and 15.8% volumetric.

DURABILITY

This timber has poor resistance to decay. It can be easily treated with preservatives, but for the usual purposes to which it is used, treatment is not normally done.

WORKABILITY

American basswood is an easy to work species, although sharp tools are recommended to achieve the best results. Because of its softness, nail holding properties are not the best. It is easily glued and cuts well with carving tools. Due to its soft texture, it does not stain satisfactorily but takes enamel easily.

USES

American basswood timber is preferred for products in the natural condition, especially when a clean, attractive appearance, light weight, and freedom from odor are essential such as in food containers including boxes, tubs, pails and baskets. It is favored for apiary supplies (bee hives and honey sections), slack cooperage, venetian blinds, toys and novelties. A considerable amount is manufactured into veneer and as core material for panels. Other uses are for paper pulp, excelsior, and ventriloquist dummy heads. Hobbyists find it to be one of the best carving woods for ship models, airplanes and wood sculpturing. [Ed. note: It is the choice wood for painting iconographic images in many of the old European churches.]

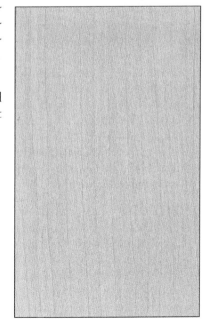

SUPPLIES

American basswood is listed in many wood catalogues and is plentiful in the United States at a relatively low price.

Toona ciliata

Australian red cedar

by Jim Flynn

SCIENTIFIC NAME

Toona ciliata. Derivation: The genus name is derived from the Hindi name *tune,* in turn derived from *tunna,* the Sanskrit name for the plant. The specific epithet refers to the *cilia* (small hairs) on the border of the flowers. A synonym is *T. australis*.

FAMILY

Meliaceae, the mahogany family.

OTHER NAMES

toon, Australian toon, thitkado, youhom, soeren, epi, kapere, tun.

DISTRIBUTION

This species is native to India, southeastern Asia and Queensland and New South Wales, Australia. Because it is a fast-growing tree, it is cultivated in many other places.

THE TREE

Because the species grows under differing climatic conditions, there is some variance in growth patterns. Native trees in Australia grow 120 feet in height with clear boles up to 80 feet and diameters as large as 5 feet. Because of extensive cuttings, trees of this size are now rare. It is a deciduous or semi-evergreen tree. The leaves are alternate, divided feather-fashion, about 12 inches long with nearly opposite, ovate, pointed leaflets obliquely cut at the base. Large clusters of small, fragrant green, white, yellow or reddish flowers grow on the branch tips. The smooth bark is light gray and finely fissured. The inner bark has a brownish outer layer, and the inner bark is streaked with pink and white and has a bitter taste. There is a tendency to confuse the *Toona* with *Cedrela odorata* (WDS 070). While both are in the same family, the seeds of *Toona* differ from *Cedrela* in that they are winged at the upper end or both ends while those of *Cedrela* are only winged at the lower end.

THE TIMBER

This lightweight wood has pale brown sapwood and reddish-brown, often described as brick red, strongly scented heartwood. The grain is straight and rarely interlocked. The texture is medium to coarse. The distinct growth rings show clearly on quartersawn surfaces often with a fiddleback or ribbon figure. Average specific gravity is 0.35 (ovendry weight/green volume), equivalent to an air-dried weight of 26 pcf. Tests by the U.S. Forest Products Laboratory reveal that the wood is similar to red alder (*Alnus rubra* WDS 018) in most properties.

SEASONING

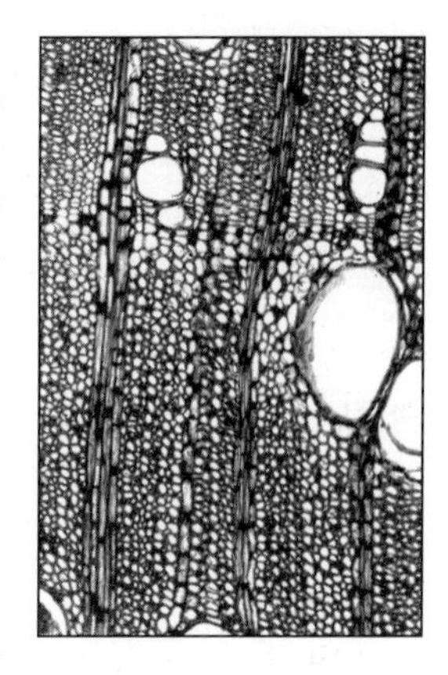

The timber will season quickly and well but is prone to cupping and warping. Proper spacing of stickers and weighting of stacks is important. Brittle heart and tension wood are occasionally encountered and require special attention when seasoning. Average reported shrinkage values (green to ovendry) are 3.2% radial and 5.5% tangential.

DURABILITY

Australian red cedar from native growth areas is much more durable than wood from plantation-grown trees. It is resistant to impregnation by preservation chemicals and is not stable in places where there is great variation in humidity.

WORKABILITY

This is an unquestionably easy wood to work with all kinds of tools. The density is sufficiently low to permit easy fabrication by machines and is easily steam bent. It glues well, can be stained effectively and is a good wood for painting. While it is easy to nail, it does not have a good holding properties. Sawdust may cause dermatitis.

USES

Australian red cedar is an extremely versatile timber. Uses include structural timbers, ship building, cabinetry, musical instruments, carvings, craft items, cigar boxes, fuel and turnery.

SUPPLIES

While supplies do not appear to be plentiful in the Western Hemisphere because of competing varieties of similar woods, there are extensive plantations in other parts of the world as well as a diminishing supply in areas where the species is native.

Triplochiton scleroxylon

obeche

by Max Kline

SCIENTIFIC NAME

Triplochiton scleroxylon. Derivation: The genus name *triplous* is Greek meaning triple and *chiton* meaning shirt which refers to the three covering (shirts) in the flower. The specific epithet *scleros* is Greek for cruel and *xylon* for wood.

FAMILY

Sterculiaceae, the stericulia or cacao family.

OTHER NAMES

ayous, African whitewood, soft satinwood, wawa, bush maple, African maple, African primavera, samba.

DISTRIBUTION

Widely distributed in West Africa. One of the most common trees of Nigeria, Ghana and the Ivory Coast.

THE TREE

Obeche is found in abundance and often in dense timber stands. The trees reach 150 feet in height with a diameter of 6 feet. The leaves resemble maple leaves. The bole is free from branches up to 80 feet.

THE TIMBER

Obeche is nearly white to a pale straw color with no clear distinction between the sapwood and the heartwood. It is the lightest utility hardwood in general use. Average reported specific gravity is 0.32 (ovendry weight/green volume), equivalent to an air-dried weight of 24 pcf. It has a moderately coarse but even texture. The grain is interlocked and when quartersawn presents a characteristic striped appearance. Odor and taste are not distinct. The luster is high and satiny. Obeche is fairly elastic and resilient but should not be used for purposes where strength is critical. The center of the log tends to be brittle. It is moderately good for steam bending.

SEASONING

The timber can and should be dried fairly rapidly to reduce the risk of fungal attack. It dries with little degrade. There is practically no tendency to split, but slight distortion may occur. Average reported shrinkage values (green to ovendry) are 3.0% radial, 5.4% tangential and 9.2% volumetric.

DURABILITY

The timber has a low standard of durability and is not resistant to decay or staining fungi. The heartwood resists preservative treatment.

WORKABILITY

Obeche works very easily with hand and machine tools and does not blunt the cutting edges very quickly. It is rather soft for turnery. Gluing is preferable to using nails or screws for jointed work. It stains and polishes well but first needs to be filled. The wood should be primed before being painted.

USES

Since this timber can be obtained in large sizes at a reasonable price, it finds applications for mass-produced cabinetry and kitchen furniture. It also is widely used for boxes and packing cases. Large logs are used for making commercial plywood. It is used as a substitute for American basswood (*Tilia americana* WDS 270), yellow-poplar (*Liriodendron tulipifera* WDS 162) and eastern white pine (*Pinus strobus* WDS 212).

SUPPLIES

Both obeche veneer and lumber are available and are inexpensive.

Turreanthus africanus

avodire

by Max Kline

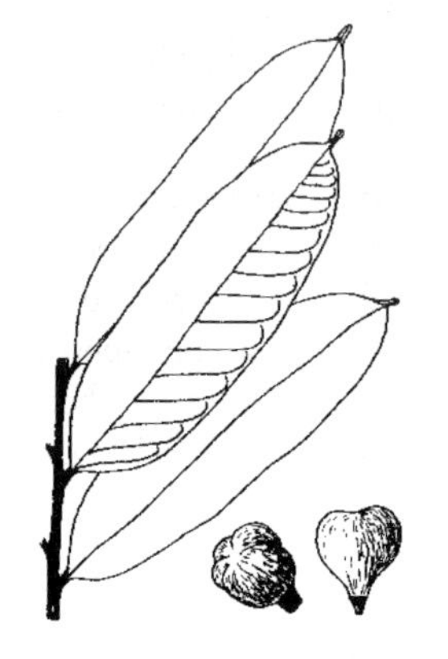

SCIENTIFIC NAME

Turreanthus africanus. Derivation: The genus name is in honor of Turra (1607 to 1688), a botanist in Padua, Italy and the Greek word *arithos* meaning flower. The specific epithet means the plant is "of Africa".

FAMILY

Meliaceae, the mahogany family.

OTHER NAMES

apapaye, wansenwa, apaya, engan, esu, songo, African satinwood.

DISTRIBUTION

Ranges from Sierra Leone to Angola with the Ivory Coast being the main source of timber for export.

THE TREE

Avodire is medium-sized reaching 100 feet in height with a diameter of 2 to 5 feet. The tree trunks are irregular or crooked and the low branching growth characteristics permit exportable logs from only about 25 to 50 feet of the trunk.

THE TIMBER

When freshly cut, avodire is pale creamy white but darkens to a golden yellow shade. The heartwood and sapwood cannot be distinguished. It has no odor or taste. It is capable of producing a satiny luster and a highly polished surface. Average reported specific gravity is 0.48 (ovendry weight/green volume), equivalent to an air-dried weight of 36 pcf. The grain is often wavy or interlocked with highly attractive figures including striped, mottled and curled. The texture is uniform. Avodire is a strong, tough and elastic timber in proportion to its weight, even approaching some properties of oak, a much heavier species.

SEASONING

Avodire timber can be seasoned rapidly, but with some tendency to cup or twist. Some splitting around knots may occur. Average reported shrinkage values (green to ovendry) are 4.6% radial, 6.7% tangential and 12.0% volumetric.

DURABILITY

Avodire is classified as a nondurable wood. The heartwood is extremely resistant to treatment with preservatives.

WORKABILITY

This timber may be quite easily worked with hand and machine tools with little dulling effect on the cutting edges. Avodire may be glued satisfactorily but tends to split when nailed therefore pre-boring is advisable. It will usually finish without difficulty although staining is somewhat uneven.

USES

Avodire is known as a decorative veneer. The wood is used for interiors, cabinets and furniture and is a superior joinery species.

SUPPLIES

Avodire is not available in quantity, but logs are exported from the Ivory Coast to world markets. Its price range is considered moderate.

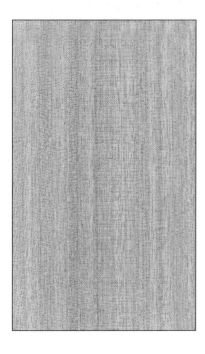

Ulmus americana

American elm

by Max Kline

SCIENTIFIC NAME

Ulmus americana. Derivation: The genus name is the classical Latin name. The specific epithet means the plant is "of America".

FAMILY

Ulmaceae, the elm family.

OTHER NAMES

white elm, water elm, soft elm, gray elm.

DISTRIBUTION

The growth range extends from southern Newfoundland across Canada to the Rocky Mountains, southward throughout the entire eastern United States to northern Florida and Texas.

THE TREE

American elm trees 2 to 4 feet in diameter and 80 to 100 feet in height are common, but diameters of 8 to 11 feet have been recorded. In the forest, trunk lengths of 30 to 60 feet are attained before the broad crown is formed. American elm was planted extensively in cities as shade trees. Dutch elm disease has wrought havoc as dead or dying trees can be seen almost everywhere in its growth area. American elm is the state tree of Massachusetts and North Dakota.

THE TIMBER

The heartwood is light grayish-brown, usually with a reddish tinge. The wide sapwood is lighter in color. The wood is hard, heavy, tough, difficult to split and usually odorless and tasteless. Average reported specific gravity is 0.46 (ovendry weight/green volume), equivalent to an air-dried weight of 35 pcf. The luster is low to medium. The texture is coarse, and the grain is straight to irregular. Growth rings are very conspicuous, and quartersawn surfaces show fine uniform rings. American elm is a strong timber and has excellent bending qualities. It also ranks high in toughness, elasticity, and wear resistance.

SEASONING

Care must be exercised in seasoning to prevent twisting and warping, especially when air drying. Average reported shrinkage values (green to ovendry) are 4.2% radial, 9.5% tangential and 14.6% volumetric.

DURABILITY

Resistance to fungal attack is not high, and the timber may suffer some attack from wood borers. Dutch elm disease is feared since it kills the trees and there is no known cure.

WORKABILITY

American elm works fairly well, but does not polish easily. Sawn surfaces can be woolly, but can be finished smoothly. Sharp tools are needed but they dull faster than is the case with many other wood species. American elm does not split well because cross-grain is frequently present. It takes nails and screws well and can be finished with the usual treatments.

USES

American elm has a large variety of uses including cooperage stays, hoops, baskets, ship building, boxes and crates, toys, woodenware, furniture, flooring, sporting goods and veneer for plywood.

SUPPLIES

The best stands of American elm are in the Great Lake states. It is generally available throughout the entire eastern United States from commercial dealers at a moderate price. Timber suppliers generally do not distinguish different species of elm in their inventories and will market a variety of species as elm.

Ulmus crassifolia

cedar elm

by Jim Flynn

SCIENTIFIC NAME

Ulmus crassifolia. Derivation: The genus name is the classical Latin name. The specific epithet means rough thick leaf.

FAMILY

Ulmaceae, the elm family.

OTHER NAMES

Texas elm, red elm, southern red elm, olmo (Spanish), scrub elm, lime elm.

DISTRIBUTION

The largest growth in its native range is on the Gulf coastal plain in eastern Texas, running in a wide swath from the Mexican border to the northern state line and beyond. It extends into southern Oklahoma, Louisiana and Arkansas and touches the southwestern corner of Tennessee and northwestern Mississippi. It appears in small outcroppings in northern Florida.

THE TREE

Cedar elm is the most common native elm tree in Texas. It can be found growing near streams in rich soil and on dry limestone hills. Occasionally, it will grow to 75 feet or more in height and develop a trunk about 24 inches in diameter with unusual specimens up to 36 inches. Cedar elm is a fast grower but is not known for having a long life. The trunk is straight, and the branches spread into a round-topped crown. The light brown bark is fairly thick (0.5 to 1 inch) and deeply furrowed. The leaves, with double-rowed teeth, are simple, alternate, pointed to slightly round at the tip, and broad and tapered unevenly at the base. The leaves are the smallest among the native elms. Another feature of cedar elm leaves is they can show the severity of air pollution by analysis of their sulfur content. The tree often is planted as an ornamental in southern climates.

THE TIMBER

Cedar elm wood is reddish-brown with wide, colorful yellow sapwood. It is hard, heavy, close grained, very strong and has exceptional shock resistance. It is denser and heavier than American elm (*Ulmus americana* WDS 274) or slippery elm (*U. rubra* WDS 276). Average reported specific gravity is 0.59 (ovendry weight/green volume), equivalent to an air-dried weight of 46 pcf. Cedar elm has an unusual cross-section close to the center that may be triangular, almost square or deeply and irregularly scalloped. The growth rings are indistinct.

SEASONING

There are no unusual seasoning problems recorded for cedar elm, though precautions should be taken to prevent warping when drying. Average reported shrinkage values (green to ovendry) are 4.7% radial, 10.2% tangential and 15.4% volumetric.

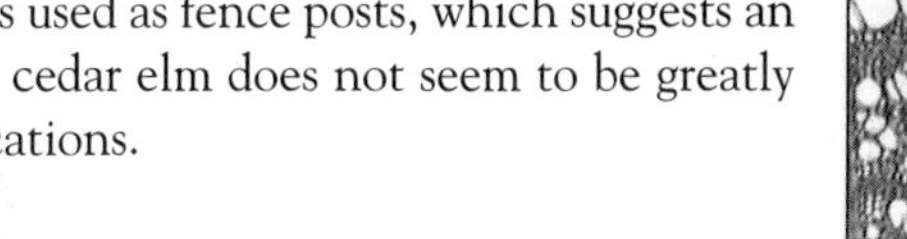

DURABILITY

Although the timber is used as fence posts, which suggests an attribute of durability, cedar elm does not seem to be greatly used in outdoor applications.

WORKABILITY

Unlike the less dense species of *Ulmus*, cedar elm is much more difficult to work. It requires more patience when working with hand tools and constant vigilance as to grain flow when working with machine tools, especially when planing to prevent tearout. It glues well, sands and finishes to a high luster and retains crisp edges. Cedar elm is a good wood for turnery.

USES

Cedar elm is an outstanding wood for steam bending and is used in the manufacture of furniture with bent components. Other uses include barrels, baskets, boxes and crates and caskets. It appears to have much potential for craft items and turnery.

SUPPLIES

It is problematic as to whether this wood can be obtained commercially outside its natural range. Although cedar elm has escaped the worst of the Dutch elm disease epidemic, the disease has taken its toll on many native elms and there is no great abundance of this timber. In addition, there are two classes of elm on the market, rock elm and soft elm. The rock elm group consists of rock elm (*Ulmus thomasii*), winged elm (*U. alata*), September elm (*U. serotina*) and cedar elm. The soft elm group consists of American elm and slippery elm. The only way to ensure you are getting the species you want is to do your own field collecting or check the anatomy of the wood.

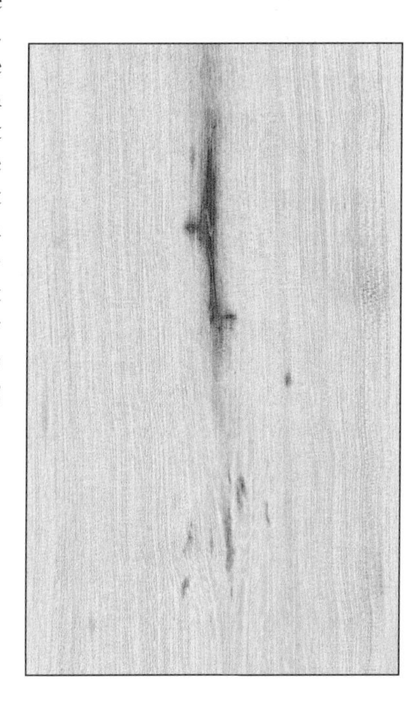

Ulmus rubra

slippery elm

by Max Kline

SCIENTIFIC NAME

Ulmus rubra. Derivation: The genus name is the classical Latin name. The specific epithet is Latin for red, which refers to the rusty or reddish-brown buds. A synonym is *U. fulva*.

FAMILY

Ulmaceae, the elm family.

OTHER NAMES

red elm, gray elm, soft elm, moose elm.

DISTRIBUTION

Slippery elm grows from Maine and the lower St. Lawrence Valley in Canada, westward across southern Ontario to eastern South Dakota and southward to western Florida and eastern Texas.

THE TREE

Slippery elm is a tree of medium size averaging from 40 to 60 feet in height with a comparatively short trunk 1 to 2 feet in diameter. The large spreading limbs branch haphazardly from the trunk and form an open flat-topped crown, irregular in outline. Frequently 1 inch thick, the bark is deeply furrowed. The inner bark is mucilaginous, which gives the tree its name.

THE TIMBER

The narrow sapwood is grayish-white to light brown with a faint characteristic odor. The heartwood is brown to dark brown frequently showing shades of red. Average reported specific gravity is 0.48 (ovendry weight/green volume), equivalent to an air-dried weight of 36 pcf. The wood is hard, strong, compact and durable. The luster is medium to low, and the texture is coarse. The grain is straight to irregular. Slippery elm is noted for resistance to wear and shock. The strength properties are good, but not exceptional. The timber has a reputation as an excellent species for steam bending.

SEASONING

During seasoning, slippery elm may warp or twist rather severely, but it does not split excessively. Slippery elm dries more quickly than most woods. When kiln drying, temperatures must be kept low. Average reported shrinkage values (green to ovendry) are 4.9% radial, 8.9% tangential and 13.8% volumetric.

DURABILITY

The durability of slippery elm is classified as fair to poor. Like other elms, this species is subject to Dutch elm disease.

WORKABILITY

Timber with wild grain will need some care in surfacing, but can be sanded well. Pre-boring before using nails is advised. The wood is rather difficult to split, usually saws "woolly" and is easy to bend. It responds well to the usual finishing treatments.

USES

Although not an important timber tree, slippery elm is used for furniture, wheel hubs, agricultural implements, railroad ties, ship building, fence posts and sills.

SUPPLIES

Slippery elm is often sold along with American elm (*Ulmus americana* WDS 274) and not as a distinct species. It is readily available at a moderate price.

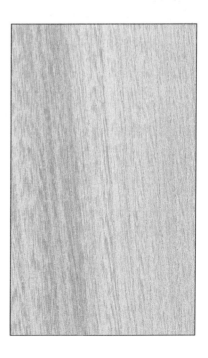

Umbellularia californica

California laurel

by Max Kline

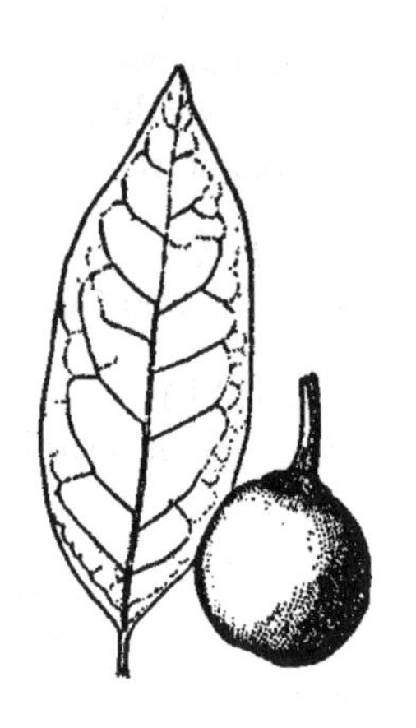

SCIENTIFIC NAME

Umbellularia californica. Derivation: The genus name *umbellula* is from the Latin meaning a small umbrella or small umbel, describing the inflorescence. The specific epithet means the plant is "of California" in recognition of where it was discovered.

FAMILY

Lauraceae, the laurel family.

OTHER NAMES

Oregon-myrtle, Pacific-myrtle, California-laurel, spice-tree, pepperwood, bay-laurel.

DISTRIBUTION

From the southwestern section of Oregon, southward along the coast to the southern border of California. Also in a narrow belt in central California for two-thirds of the state's length.

THE TREE

California laurel is a broadleaf evergreen easily distinguished by the strong, aromatic, pungent odor of its crushed leaves and its thick, dark reddish-brown, scaly bark. It may reach a height of 60 to 80 feet with a diameter or 2 to 3 feet and a straight clean trunk that is as much as 30 feet or more to the first branch. More commonly, however, the trunk is short and often divided near the ground.

THE TIMBER

The often variegated heartwood is yellowish-brown or olive and is not sharply defined from the wide, pale brown sapwood. When freshly cut, California laurel has a mild scent and taste. The luster is medium. Average reported specific gravity is 0.51 (ovendry weight/green volume), equivalent to an air-dried weight of 39 pcf. The texture is medium with straight to wavy grain. In Oregon, the wood is noted for its strikingly mottled pigment figures, varying from fine, delicate black lines to heavy splotches, occasionally streaked with gold and silver which are difficult to adequately describe.

SEASONING

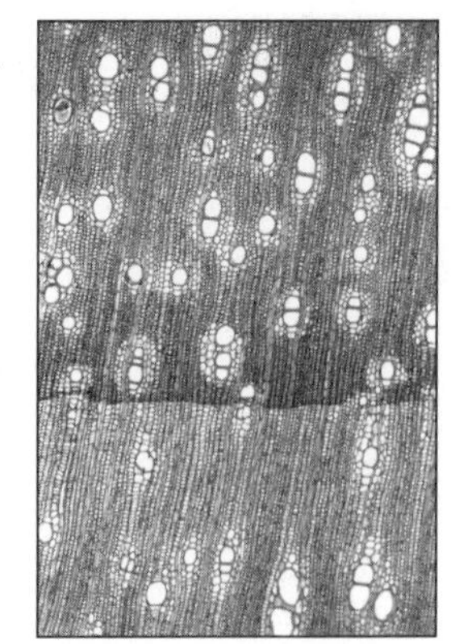

Generally the seasoning of California laurel takes years of painstaking care to bring out the finest color effects. When green, it is often submerged in water in log form so it develops the darker colors so highly esteemed by woodworkers. Great care must be taken when seasoning to prevent checking and warping. Average reported shrinkage values (green to ovendry) are 2.8% radial, 8.1% tangential and 11.9% volumetric.

DURABILITY

The heartwood of California laurel is quite durable.

WORKABILITY

California laurel is easily worked with tools and takes an exquisitely high polish.

USES

This wood is almost entirely consigned to making wood novelties of all descriptions and sizes. The burls are sliced into veneer for cabinetwork. California laurel is claimed by many to be the finest wood obtainable for novelty articles and is shipped to many parts of the world for this purpose.

SUPPLIES

Small pieces of California laurel can be purchased from dealers at high prices. Lumber is very scarce. Burls are one of the rarest figures and supplies are quite limited. When available, the burl wood sells for the highest price of all American woods. A favorite place to find this wood is along Oregon beaches tangled with a variety of other driftwood.

Zanthoxylum flavum

West Indian satinwood

by Jon Arno

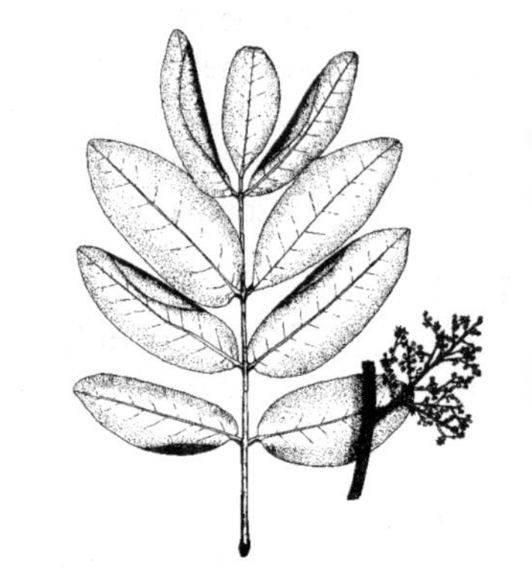

SCIENTIFIC NAME

Zanthoxylum flavum. Derivation: The genus name is Greek meaning yellow wood. The specific epithet also implies yellow. A synonym is *Fagara flava*.

FAMILY

Rutaceae, the rue or citrus family.

OTHER NAMES

yellow sanders, yellow wood, aceitillo, ayua, espinillo, noyer, yellowheart.

DISTRIBUTION

Native to the islands of the Caribbean from Cuba to St. Lucia including the Bahamas, it is found sparingly in the lower Florida Keys and was once abundant in Bermuda.

THE TREE

Like its citrus cousins, West Indian satinwood tends to develop a broad, low-branched crown when grown in the open. It seldom grows more than 40 feet in height with diameters ranging from 15 to 20 inches. The compound leaves are somewhat ash-like in appearance, and the small, egg-shaped fruit form in clusters, each containing a single shiny black seed.

THE TIMBER

The heartwood is creamy yellow, darkening on exposure to light orange or golden tan. The sapwood, almost white near the bark, becomes progressively darker until it blends into the heartwood. The diffuse-porous wood is fine textured and has high surface luster. The grain is often interlocked and wavy, producing very interesting figures. When freshly cut, this wood has a pleasant, coconut-like sweet scent. The wood is very strong but seldom used in applications where strength is seriously challenged. Average reported specific gravity is 0.73 (ovendry weight/green volume), equivalent to an air-dried weight of 58 pcf.

SEASONING

Shrinkage values are not available. Given its fine texture and relatively high density, it undoubtedly dries slowly.

DURABILITY

West Indian satinwood is rated as nondurable when exposed to the elements.

WORKABILITY

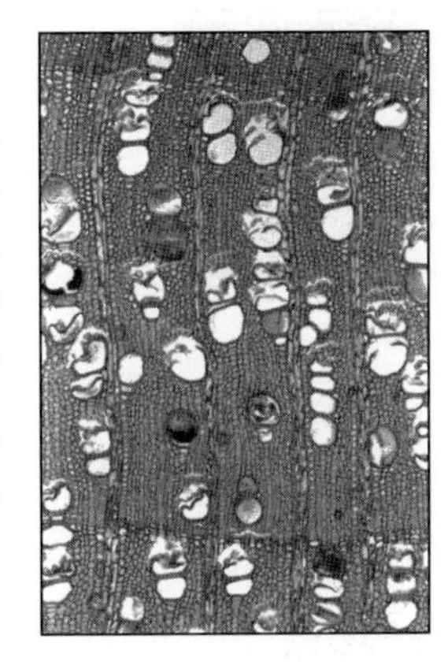

It is an excellent choice for turning, inlay work and small decorative items. When highly figured, it is a very showy wood and its yellowish-orange color compliments other darker woods. Its fine texture and high natural luster make it an easy wood to polish. Quickly appreciated when brought to Europe from the New World colonies, this wood has been a part of our cabinetmaking traditions for more than 300 years. Although its pleasant scent makes it enjoyable to work, over exposure to the dust can cause skin rashes and respiratory problems for some people.

USES

Uses include turnery, inlays, premium veneer, high-quality furniture and small novelty items and carvings.

SUPPLIES

Supplies are very scarce and expensive. Having been heavily exploited for so long, it is difficult to find mature specimens. From a commercial perspective, the wood is virtually out of distribution.

SPECIAL NOTE

The trade name satinwood can lead to confusion. The American species described is most frequently found in antique furniture of American or English origin. There are some acceptable substitutes. The two most often encountered in today's market are East Indian or Ceylon satinwood (*Chloroxylon swietenia* WDS 080) and East African satinwood (*Fagara macrophylla*), which belong to the Rutaceae family and are anatomically very similar to the American species. The East African variety can be quite variable in terms of weight and is often as much as 20% softer than the American species, while the East Indian species is normally about 10% heavier. These woods are similar in color, but the African species is usually slightly coarser in texture. Ayan (*Distemonanthus benthamianus* WDS 104) is sometimes called Nigerian satinwood but is a member of the legume family (Leguminosae). It is a very nice, golden yellow, fine textured, relatively stable cabinetwood with a low specific gravity of 0.58. Unfortunately, its rich yellow pigment is water-soluble, and it can contain very high levels (up to 1.3%) of silica which quickly dull blades. Finally, pau amarello (*Euxylophora paraensis*), a member of the Rutaceae family, is sometimes labeled satinwood because of its bright yellow color and soft luster.

Ziziphus thyrsiflora

ebano

by Jim Flynn

SCIENTIFIC NAME

Ziziphus thyrsiflora. Derivation: The genus name is derived from Zizouf, the Arabian name for this plant. The specific epithet is Greek meaning a plant having flowers in a thyrse (a dense cluster around the flower's axis [like a lilac]). The genus name was formerly spelled *Zizyphus*. The prestigious National Agrarian University of La Molina, Peru, where the species is native, continues to prefer the older spelling.

FAMILY

Rhamnaceae, the buckthorn family.

OTHER NAMES

Ebano is a Spanish word for ebony and is used in Peru as a common name because of the dark ebony-like color of the heartwood. Ebano is not a member of the true ebony family Ebenaceae. In Mexico some species of trees in the genus *Caesalpina* are also called ebano but are not related to the species described in this Wood Data Sheet. No other common names describing this species could be found. Again, this illustrates the futility of attempting to provide an accurate identification of wood solely based upon the common name.

DISTRIBUTION

Northern Peru and southwestern Ecuador.

THE TREE

This species of *Ziziphus* was one of the many plants collected in the Guayaquil area of western Ecuador by Richard Brinsley Hinds, the surgeon on H.M.S. Sulphur, during the ship's worldwide exploratory mission in 1836 to 1842. Ebano is not a large tree reaching an average height of approximately 40 feet. The leaves are 2 to 3 inches long and 2 to 2.5 inches wide with the prominent three-vein pattern of most trees in this genus.

THE TIMBER

When air dried, the sapwood is yellowish-cream. The heartwood varies from a deep brown to a blackish color, giving rise to the name ebony. The contour of the sapwood with the heartwood is irregular. The wood is straight grained with a fine texture and a high luster. Average specific gravity is 0.78 (ovendry weight/green volume), equivalent to an air-dried weight of 62 pcf. It is diffuse porous, and the mostly solitary and oval-shaped pores are barely visible to the naked eye. The growth rings are clearly shown in the cross-sectional axis thus highlighting dark streaks in the radial plane. There is an abundance of rhomboid-shaped crystals present. Neither peculiar taste nor odor can be determined.

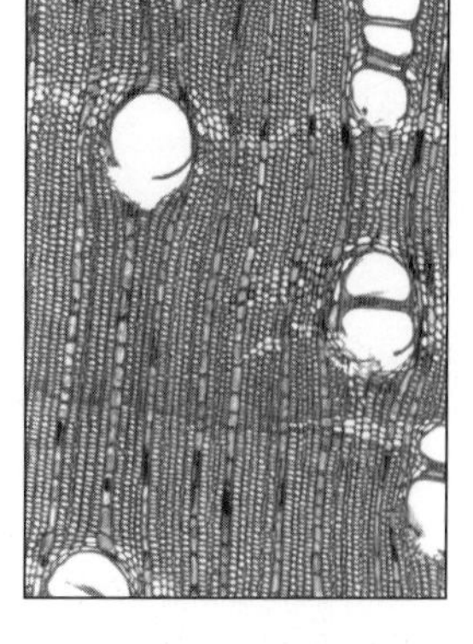

SEASONING

Shrinkage values are not available, but the appearance and workability of the wood seems to indicate that it cures well.

DURABILITY

Ebano is susceptible to fungi and insect attacks.

WORKABILITY

If you have ever worked with black walnut (*Juglans nigra* WDS 150) then you are close to the comparable workability of this wood. Moreover, if you have worked black walnut to a great degree you will have experienced the variable density in this species. Ebano is like the "heavy" walnut. Aside from the reported crystals present in the wood which may cause tool dulling, there are no unusual problems machining it. It squares off very nicely holding a sharp edge. It sands and finishes well and no problems are encountered when gluing.

USES

The attractive color, grain and crispness of the wood suggests that it is a very good candidate for cabinetry, furniture, craft items and turnery. This timber was one of a group of Peruvian species which was selected for special study and reporting by the National Agrarian University in Peru and the Nagoya University in Japan. The main purpose of the study was to acquaint timber users with the utility and diversity of Peruvian forest products.

SUPPLIES

Unfortunately, there does not seem to be much of this wood available in world commerce. Tourists visiting some of the historic places in Peru might obtain some timber from local lumber dealers. I had the good fortune of having a piece delivered to me directly from Peru. The wood was rather old and had many nail holes and scars and was obviously salvaged from a piece of furniture. Still, it made several beautiful samples.

APPENDICES

APPENDIX A

Biology and Taxonomy for Woodworkers

by Chuck Holder

This article will discuss woody plants in the context of their overall classification by science among all living things on Earth and within the Plant Kingdom. It will also provide a brief overview of taxonomy, a branch of science that deals with finding, describing, defining, naming and classifying living things. For the purpose of this book, however, the focus will largely be restricted to plant taxonomy, a part of systematic botany. Woodworkers and others interested in wood often ask, "How many different kinds of wood or species of trees or other woody plants are there?" It's a simple question with no known answer, but one that is discussed in this article. In addition, scientific and popular nomenclature for trees and wood is described along with examples of each for several kinds of trees.

TAXONOMY, THE SCIENCE OF CLASSIFICATION

Science defines a taxon (plural, taxa) as any related group of living things. It may be a whole kingdom or any lesser ranked grouping, such as an order, a family, a genus, a species or an even lower rank such as a sub-species or a variety. These groupings or taxa are the attempts of science to systematically describe, define, name and classify life on Earth, and taxonomy is the branch of science that is concerned with these efforts. Plant taxonomy attempts to define classifications that best reflect what is known about plants. As science advances, more techniques become available to study plants yielding more information that can be considered in making classifications. Over the centuries there has been a steady stream of more refined systems of classification. Initially, classification was based on growth form but now includes reproductive mechanisms, anatomy and chemical characteristics. Moreover, advances in the field of DNA and genetics could bring about wholesale reclassification of taxa in the future.

FIVE KINGDOMS

To approach an answer to the simple question "how many species of trees are there", consider the big picture of life on Earth. At first, only two kinds of life were recognized: plants and animals. Now, scientists have described five great kingdoms of life forms on Earth (Ref. 9). These additional three kingdoms are neither plants nor animals and comprise by far the most numerous, prolific and hardy life forms on the planet. The names of two, Kingdom Bacteria and Kingdom Fungi, are familiar to everyone, but the third, Kingdom Protoctista, is not so familiar. This kingdom is made up of the algae (including seaweeds and kelp), slimes and protozoa of the world. These kingdoms of living organisms can be excluded in our search for woody plants.

If we knew how many species of plants there were and if we knew how many of these plants had woody stems, we would have a rough approximation of how many kinds of wood there are. But most species of plants are not "woody". Plant forms are described as herbs, vines or lianas, shrubs and trees. All of these but herbs contain some material that most woodworkers would recognize and accept as wood. Even some herbs, such as bamboo, contain some hard material that one could easily accept and use as wood.

THE PLANT KINGDOM

To understand the Plant Kingdom as we know it today, it is useful to briefly trace the steps of plant evolution as one might encounter them on a stroll through Evolution House at Kew Gardens. Plant life began in the sea, and since water was a good supporter of plant structures, tree forms were unnecessary. At some point, much like their animal counterparts, green plants made a move onto land in tropical areas. Later, trees evolved as the competition for light and food rewarded those plants that reached up over their lesser rivals. Soon, every new type of plant that evolved on land tried to become a tree. The Carboniferous period saw club mosses that formed forests of very large, tree-like plants. Ferns that developed vascular systems (conduits) for moving water to their various parts superseded these. Later came the Cycads or early Gymnosperms, which introduced seeds to the competitive arsenal by giving their young embryos some built-in foodstuff for a head-start in life. The development of the Gymnosperms reached its peak with the Conifers, a group that is now widely used in commercial forestry and for ornamental purposes. Finally, the Angiosperms, or enclosed seeded flowering plants, evolved forging alliances with all kinds of animals for pollination and seed dispersal. Among the ranks of today's flowering plants, there are some herbs that are extremely competitive such as certain of the grasses that form bamboo "forests". Evolution has seen a succession of ever more efficient plants that have pushed their predecessors into extinction or into small survivor pockets. Many of those early "trees" have left a considerable legacy by way of the fossil fuels that now energize the modern world.

According to Margulis and Schwartz (Ref. 9), the Plant Kingdom is subdivided into twelve Divisions. The rank of Division is also called Phylum (plural, Phyla) in branches of biology other than botany. Of the twelve divisions, the first seven, noted below, contain no woody plants. The remaining five divisions, forming the ranks of the Gymnosperms and the Angiosperms, contain the woody plants we seek:

First 7 Divisions: Mosses, Liverworts, Hornworts, Club Mosses, Whisk-ferns, Horsetails and Ferns.

Next 4: Gymnosperms (naked seeded plants): the Cycads, Ginkgo, the Gnetophytes and the Conifers.

Last 1: Angiosperms (enclosed seeded or flowering plants): composed of monocotyledons and dicotyledons – the "monocots" and "dicots". Most, if not all, woody plants (trees, shrubs and woody vines or lianas) in the angiosperms are dicots although the monocots include some tree-like plants (i.e., among the palms and bamboos).

Since taxonomy is a living, changing, evolving science, new plants are being discovered continuously and species and higher taxa are being reclassified continuously, any attempt to show the number of taxa in the Plant Kingdom will be a "snapshot" at best.

The major taxa in the Plant Kingdom and estimated numbers are shown in **Table 1**.

Table 1.—Approximate number of taxa in the Plant Kingdom.

12 Divisions		Class	Order	Family	Genus	Species
Mosses to ferns: (1 to 7)		9	30	116	1,100	30,000
Gymnosperms: (8 to 11)		4	5	17	86	841
– Cycads, Gingko, Gnetophytes		3	3	8	16	241
– Conifers		1	2	9	70	600
Angiosperms: (12)		2	83	578	15,000	277,000
– Dicotyledons		1	64	469	11,700	212,000
– Monocotyledons		1	19	109	3,200	65,000
Total Taxa	**12 Divisions**	**15 Classes**	**118 Orders**	**711 Families**	**16,200± Genera**	**~310,000 Species**

It is interesting to note that the greatest commercial wood producers are the conifers or softwoods, which number approximately 600 species in only nine plant families. Today, the Plant Kingdom is composed of approximately 300,000 living species of plants that are known and described by science. These are grouped into approximately 16,000 genera that are in turn grouped into over 700 families based on various familial characteristics. To illustrate the range of uncertainty and state of flux in the field of taxonomy, different authoritative sources indicate that the number of plant species identified and known to science range from about 250,000 to 500,000 species and that another 500,000 species remain to be discovered.

So, how many different kinds of wood or species of woody plants are there? A definitive answer to this simple question is unknown. If one accepts the term wood in its broadest sense to mean "all plants with woody stems", that is one question that will result in one answer. If one limits the question to how many kinds of woody plants have arboreal form (i.e., are of tree shape and size) the answer would likely be much smaller. The question may further be limited to cover only those species of trees which are harvested and made available to wood users and consumers through commercial markets. However one measures it, the numbers are soft, non-definitive "guestimates" at best. It seems safe to say that no one has ever compiled "the list" of either all tree species or all species of woody plants, and therefore that number is unknown with any degree of certainty, just as the total number of plant species on Earth is unknown with certainty.

To try to summarize it in the universal context of the Plant Kingdom, some 300,000 species of now living plants are known to and identified by science. Of these, approximately 70,000 to 125,000 are thought to be woody species (i.e., species producing rigid xylem or wood, namely trees, shrubs and woody vines or lianas). Of these it is estimated that 20,000 to 40,000 species are arboreal (i.e., trees or tree-like plants depending on the definition of tree) and of the woody species, probably no more than 5,000 to 10,000 species are harvested for the wood used by local crafters *and* in commerce around the world. Of this group, it is thought that between 1,000 and 4,000 different species may provide wood that enters commercial markets throughout the world; and, of this number, probably no more than 300 different "kinds of wood" are available in any one national market at a given time. A "kind of wood" may include more than one species or genus marketed as a single "kind" (e.g., the SPF or spruce/pine/fir group, the *Shorea* spp./lauans groups, etc.) or conversely a single species sold as several "kinds of wood" (e.g., various figures such as bird's-eye or curly, various cuts such as flat or quartersawn and various grades).

Figure 1 represents several concepts of the "how many" question from "how many species are there" to "how many 'kinds of wood' are available commercially". The question must be clearly defined in terms of which concept it is addressing. Obviously, the answer, although not known precisely, should address the question in the same context. This book describes the "tree" and its wood characteristics for a sampling of 279 woody plants, both commercial and non-commercial species. Many of the most beautiful and interesting woods are available only locally, to those who harvest them on a small scale, and may not be available for sale in general commerce. The International Wood Collectors Society (IWCS) is an organization that makes such woods known and, through networking by its members, facilitates the trading of samples worldwide.

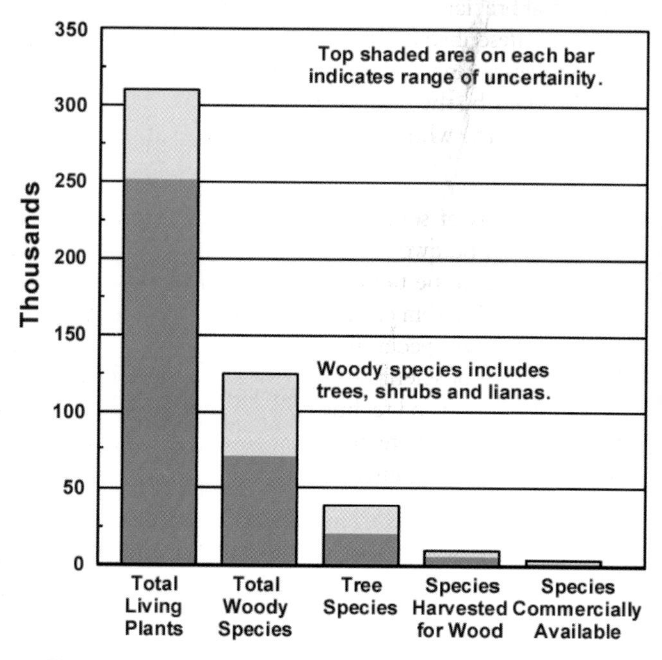

Figure 1.—Concepts for the number of woody plants.

NOMENCLATURE

Most trees and their woods usually have a number of common or vernacular names. This is because many species grow in different countries and are often known locally by a name related to their form, use or other characteristics. In addition, the same common name is often given to two or more possibly completely unrelated species, ironwood and redwood for example. Add to this the trade names often used by the lumber industry and the catchy marketing names designed to lure the buyer, and you have a recipe for confusion.

Science has adopted a system for naming living organisms that is accepted and recognized throughout the world. It assigns to every known species a binary or two-part name. There are strict rules, called the International Code of Botanical Nomenclature, governing plant names, which are administered and governed during meetings of the International Botanical Congress. The binary or two-part name for a species is sometimes inaccurately called a binomial. It consists of a genus name and a specific epithet and is, by convention, normally italicized in written form, with the genus or first part of the name capitalized. The specific epithet, or second part of the name, is commonly referred to as the species name. More correctly however, the species name is the full binary name consisting of both parts. In scientific or botanical literature, the names-initials-abbreviation that follows the species name refers to the botanist who first documentzed, described and published the plant and has the right to name the plant. Thus, eastern white pine is written as *Pinus strobus* L. indicating that this name was conferred on the plant by the great Swedish botanist Carl von Linné, better known as Linnaeus, (1707 to 1778) who is generally credited with instituting a consistent binary nomenclature.

Now let's look at some nomenclature and taxonomic classifications of a well-known softwood and hardwood and a not so familiar woody monocot. **Table 2** provides a step by step guide to the taxon classifications and the common and scientific names for an example species from each of the three main tree types. The Wood Data Sheet (WDS) for each of these species lists a number of other common names. The advantage of the binary system of scientific names is that it is accepted by dendrologists, botanists and other scientists, wood technologists and other serious students and users of wood around the world. While there are often numerous common names for the same tree, there is only one scientific name.

Table 2.—Nomenclature and taxonomy for three kinds of trees.

Kingdom	Plantae—The Plant Kingdom		
Division	Pinophyta (Gymnosperms)	Magnoliophyta (Angiosperms)	
Class	Pinopsida (conifers)	Magnoliopsida (dicots)	Liliopsida (monocots)
Order	Pinales	Proteales	Arecales
Family	Pinaceae (pine)	Proteaceae (protea)	Palmae (palm) also Arecaceae
Genus	*Pinus*	*Cardwellia*	*Roystonea*
Specific epithet	*strobus*	*sublimis*	*regia*
Species name	*Pinus strobus*	*Cardwellia sublimis*	*Roystonea regia*
Scientific name	*Pinus strobus* L.	*Cardwellia sublimis* F.Muell.	*Roystonea regia* (Kunth) O.F.Cook
Common name	eastern white pine	lacewood, silky oak	Cuban royal palm
Type of "tree"	softwood	hardwood	woody monocot
Reference	WDS 212	WDS 60	WDS 238

For example, in Australia there are several different species of silky oak (e.g., WDS 60 and WDS 131). The hardwood, northern silky oak, is shown in **Table 2** and is native to northern Queensland, Australia (WDS 60). In North America, the wood of this species is marketed and commonly known as lacewood. In Europe, however, lacewood refers to the quartersawn lumber of any number of species with a pronounced ray fleck. These were originally *Platanus* species (European plane tree, etc.) but later included any of the Australian silky oaks. Also in Europe, the wood of northern silky oak when not quartersawn is called silky oak. In spite of this interrelated web of common names, this tree and its wood is *Cardwellia sublimis* throughout the world, universally, uniquely and without confusion. The Wood Data Sheets in this book also provide interesting and historically significant descriptive notes on the derivation and history of the scientific names. These are often Latinized versions of the discoverer's name, the name of a deserving botanist, a Latin or

Greek description of part of the plant, or the location or use of the plant. It is entertaining and educational to make the effort to understand scientific names and their origins especially for those species with which one works and with which one wants to be familiar. One of the main purposes of IWCS is to encourage the proper identification and naming of wood. It is hoped that this book will contribute to that goal.

SOME CONCEPTS AND DEFINITIONS

Angiosperm: a seed-bearing plant with ovules/seeds that develop within an enclosed ovary and are born in a fruit.

Cotyledon: a leaf of the embryo of a seed that may evolve to become the first leaf of a seedling.

Dendrology: the study of trees. Its meaning has been expanded to cover those disciplines of botany and forestry that deal with taxonomy, nomenclature, morphology, phenology, ecology and the geographic range of trees.

Dicotyledons: one of two groups of Angiosperms. A plant usually has two cotyledons but may have zero to four. Observable characteristics are leaves with pinnate or palmate venation and floral parts that usually occur in multiples of four or five.

Family: a taxonomic group of related genera (singular, genus) that may contain only one genus but usually contains more than one.

Genus: (plural, genera) a taxonomic group of related species that may contain one species but usually contains more than one.

Gymnosperm: a seed bearing plant whose seeds are exposed (naked) and normally born on a seedcone.

Monocotyledons: one of two groups of Angiosperms. A plant that usually has one cotyledon, sometimes zero or two. Observable characteristics are leaves with parallel venations and floral parts that usually occur in multiples of three.

Morphology: the science of form and structure; that branch of biology that deals with the physical form, changing forms and comparative anatomy of living things.

Order: a taxonomic group of related families that may contain only one family but usually contains more.

Phenology: the study of the times of recurring natural phenomena such as flowering or fruiting.

Species: (plural, species) the basic taxonomic unit in the classification of living things; the basic evolutionary unit. A group of organisms that, left to their own devices, can or will breed to produce the next generation of organisms of the group.

Taxon: (plural, taxa) any group of related living things such as a species, a genus, a family or a kingdom.

Taxonomy: the science of finding, describing, defining, naming and classifying living things.

SELECTED REFERENCES

1. Bootle, Keith R. 1983. *Wood in Australia*. McGraw-Hill. Sydney, Australia.
2. Brummitt, R.K. 1992. *Vascular Plant Families and Genera*. Royal Botanic Gardens, Kew. Also access: www.kew.org.uk/data/vascplnt.html.
3. Cronquist, Arthur. 1968. *The Evolution and Classification of Flowering Plants*. Houghton Mifflin.
4. Flynn, Jr., J.H. 1997. "Primer on Proper Nomenclature of Botanical Species." *World of Wood*. January.
5. Harlow, W.M., E.S. Harrar and F.M. White. 1995. *Textbook of Dendrology*. 8th ed. McGraw Hill.
6. Hoadley, R. Bruce. 1990. *Identifying Wood: Accurate Results with Simple Tools*. The Taunton Press, Newtown, CT.
7. Hoadley, R. Bruce. 2000. *Understanding Wood: A Craftsman's Guide to Wood Technology*. 2nd ed. The Tauton Press, Newtown, CT.
8. Mabberley, D.J. 1997. *The Plant Book: A Portable Dictionary of the Vascular Plants*. 2nd ed. Cambridge University Press, Cambridge, U.K.
9. Margulis, L. and K. Schwartz. *Five Kingdoms: An Illustrated Guide to the Phyla of Life on Earth*. 3rd ed. W.H. Freeman & Co.
10. Stearn, William T. 1995. *Botanical Latin: History, Grammar, Syntax, Terminology and Vocabulary*. 4th ed. Timber Press.
11. Wiersema, John H. and Blanca León. 1999. *World Economic Plants: A Standard Reference*. CRC Press. Also access: GRIN Taxonomy: www.ars-grin.gov.

APPENDIX B

Insights on Wood Toxicity

by Jon Arno, Eugene Dimitriadis, Jim Flynn and Roy Tandy

The fact that some woods can be toxic is an absolute certainty. Pliny the Elder (23 to 79 A.D.), a Roman scholar and author of *Historia Naturalis*, described a case in which four soldiers died from drinking wine that was stored in flasks made of yew wood.

An examination of just how dangerous it might be to *work* with wood quickly raises many questions. The fact is our current state-of-the-art in medical science leaves us with far more questions than answers. Fear of the unknown is often worse than the underlying reality.

But caution is indeed warranted when working new woods. At the very least, the topic of wood toxicity can be dissected into broad terms, and it is helpful to do so. Based on their possible danger or potency, the hazardous substances found in or associated with cutting or handling wood can be divided into three main categories:

Irritants:
Substances that generally irritate exposed areas of the body of most individuals at some level of exposure. Their effect is usually proportional to exposure time and concentration.

Sensitizers:
Substances that can after repeated contact initiate a growing and potentially serious allergic reaction in susceptible individuals.

Poisons:
Substances that are universally lethal at some relatively low and often predictable dosage.

These categories are not mutually exclusive. Small doses of a poison may simply stimulate an allergic reaction in some individuals or be an irritant to others. On the other hand, a severe allergic reaction or long term or cumulative irritation in some instances may prove fatal.

POISONS

The very dangerous, and sometimes lethal, substances found in some woods can be further subdivided into two categories:

- *Natural compounds* which are produced by the tree itself or
- *Introduced substances* such as those naturally occurring from fungal spores, bacteria and their by-products, or those substances that are added by humans for a specific purpose.

Some woods which are relatively benign in terms of their natural chemistry become dangerous after they have been infected by fungi or bacteria, which in turn can produce toxins. In other cases, the fungi and bacteria themselves are pathogenic (dangerous if they invade our bodies). For example, wood from maple and birch, which are so chemically friendly their sap can be used to make sugar syrups, often appear on lists of toxic woods. Their inclusion is valid because there is clinical evidence supporting a potential problem, but it is very probable that the offending toxins are the result of microbial infestation and are not related to the wood's natural chemistry.

Also, wood is sometimes chemically treated to improve its decay resistance or to prevent attack by marine borers, termites or other wood-destroying agents. These intentionally introduced chemicals may be formulated with toxic metal salts (e.g., chromium copper arsenate) or organic materials such as tar extracts (creosote) or pentachlorophenols. In manufactured products, such as plywood and fiberboards, bonding agents containing formaldehyde can lead to long-term health and other effects. The boards can emit this substance over time by diffusion or by the heat produced from sawing.

Wood that contains natural compounds so potent as to be a potentially lethal poison at dosage levels the average individual might innocently ingest while working with them are very few. Also, these lethal compounds, which are often alkaloids but can be masked cyanides (e.g., cyanogenic glycosides, fluoroacetates and other naturally produced toxins), are usually present at higher levels only in the bark, sap, fruit and/or foliage. The levels can vary markedly depending on the growing season.

It has been suggested that these compounds originated as part of the tree's natural defense against browsing animals and insects. While these compounds can be very dangerous, the hazard to woodworkers is generally limited to instances when the manner in which the wood is used exposes the individual to an accumulation of the toxin. For example, there have been incidents where the use of oleander (*Nerium oleander*) branches as meat skewers and stirring sticks in food preparation have caused fatalities. The lesson is that heat, moisture and time stimulate the release of the toxins found in some wood and the woodworker should be very careful in selecting a species for making kitchen utensils, food storage containers or items such as toys on which an infant might chew.

Also, woodworkers should consider personal health circumstances that might make a particular wood dangerous. For example, yew (*Taxus* spp.) contains a cardiovascular depressant that is so powerful its use as a heart medicine has been explored. While the fruit and foliage of yew contains this compound in potentially lethal quantities, the wood has been used for centuries in cabinetry and archery bows. The compound taxol found in Pacific yew (*Taxus brevifolia*) is currently being used to treat cancer patients. Woodworkers with cardiovascular problems should be alert to the risk and may want to either avoid the wood altogether or at the very least use it with extreme caution and certainly not use the wood for items that come into contact with food. A similar caution can be made about Australian conkerberry and other *Carissa* spp. which have similar cardiac compounds and toxins. Those with an aspirin allergy also need to take care and should avoid using the wood from willows and birches (*Salix* spp. and *Betula* spp.). Species of these genera can contain appreciable concentrations of salicylic acid, the essential ingredient in aspirin.

There is a host of examples we might cite to illustrate the point that the trees we look to as the source of some of the finest cabinetry woods in the world have a darker side when it comes to potential toxicity. The latex of sandbox tree (*Hura crepitans*) contains the poisonous alkaloid huratoxin. Other species contain well known toxic alkaloids like strychnine, atropine, berberine, cocaine, etc. But not all potentially poisonous species are tropical exotics. Many members of the rose family (Rosaceae) yield beautiful cabinetry woods, while their seeds and foliage can be very dangerous. For example, the foliage of black cherry (*Prunus serotina*) contains a precursor of cyanide. In fact, the wilted leaves of black cherry have been known to poison livestock, yet the wood is used without incident by thousands of woodworkers.

This is not to say that natural toxins are found exclusively in the foliage and never in the wood. Many species protect their woody tissue from decay by producing antiseptic compounds. One example is Tasmanian Huon pine (*Largarostrobos franklinii*), a yellow wood rich in phenolic materials that are similar to clove oil. Another example is camphor-tree (*Cinnamomum camphora*) which has high concentrations of terpenes. Like camphor, it is pleasant in low quantities but in high concentrations on a hot day in a confined place can lead to health problems. In addition, the heat from mechanical actions (cutting or planing) or friction (sanding) can cause the emission of the volatile substances contained in the wood. If these substances are inhaled for extended periods, they may cause health problems.

These compounds, referred to as wood extractives, are typically stored in the interior cells of the wood. Since extractives often oxidize and polymerize into colorful pigments in the heartwood, there seems to be some moderate correlation between vividly colored woods and their potential toxicity. But it is irrational to conclude that dark wood equates to deadly and pale wood is safe. These extractives are often sensitizers and/or irritants depending upon the susceptibility of the individual, but the dosage required to make them lethal poisons is not likely obtainable by occasional, low-level exposure to sawdust.

Interestingly, the common name for the tree can give a clue to its potential toxicity. Australian Cooktown ironwood or red ironwood (*Erythrophleum chlorostachys*) is also called camel poison tree and could well be called toxic as all parts of the plant, especially the leaves, twigs and underbark, contain toxic alkaloids. In one recorded case, hungry cattle held in a truck under a red ironwood tree died after digesting just a handful of leaves. While there is no evidence that the wood of this species is toxic, like all durable woods, it should be handled and worked with caution. Two species of *Excoecaria* are called blind-your-eye or river poison tree by Aborigines as the sap can cause temporary blindness if it gets into the eyes (when chopping for example), as well as producing blisters on the skin and throat and causing a headache. The imported species from Central America called poisonwood or Honduras walnut (*Metopium* spp.) derives its name from the irritating and toxic sap even though the wood contains little or no toxicity.

In addition, some spores from moldy wood can be toxic or pathogenic. Care should be exercised when cutting spalted (mold infected) and rotting wood, even if it is for fire-wood. For example, *Cryptostroma corticale* is frequently found growing in the bark/sapwood interface of maples and birches. The widespread river red gum (*Eucalyptus camaldulensis*) from Australia has recently been found to host a *Cryptococcus* species of fungus in its rotting heartwood. The spores of this organism enter the body by inhalation and can cause serious symptoms of meningitis leading to a few deaths every year.

SENSITIZERS

Perhaps the least understood aspect of wood toxicity relates to the sensitizing properties of some wood extractives. Our research in this field progresses daily, but it is a tremendous challenge, not only from the perspective of the wood chemistry involved, but also because of the subtleties of the human organism. The extractives involved are compounds of diverse and sometimes complex molecular structure including phenols, polyphenols, terpenes, glycosides, quinones, alkaloids, flavonoids, resins, saponins and many others. Not all individuals react to these classes of compounds in the same way. Like a bee sting to most of us, the first low-level exposure may produce little if any reaction, while for other individuals this exposure activates a cumulative allergic response. The first-time reaction can be delayed and mild. Thereafter, the individual increasingly becomes sensitized to the same chemical or class of chemical so that subsequent exposures produce a greatly increased immune response. This can lead to fatal results if avoidance measures are not practiced. More insidiously, woodworkers may stop using a wood that might be a problem (e.g., ipe which contains desoxylapechol) and turn to teak or white peroba only to be faced with the same sensitizer in these woods and thus a recurring health problem.

Allergic reactions are not common, but they affect enough individuals to make them a significant potential problem and one of which all woodworkers should be wary. Statistically, about 2% to 5% of woodworkers develop an allergic response to compounds in wood. If many species are handled, as is the case with wood collectors, the chances for allergic reaction may increase. Unfortunately, some woods will bother most everyone eventually; other woods will rarely cause problems to anyone.

The variables behind this response are many:

- The wood (species, wet or dry, with bark or without, etc.)
- Nature of particles (fine, coarse, wet or dry, vapor emitting)
- Nature of contact (size of contact area, dust, vapor, where and how)
- Environment (dust extraction, temperature, air movement)
- Person (immunity, sensitivity, degree of care exercised, contact frequency).

That an organic substance like wood should be counted among the potential allergens or even be considered a danger in our environment is certainly not surprising. Many common foods, fruits, dairy products and seafoods have a similar capacity to initiate allergies. It is probably correct to assume that individuals with a history of immune system problems or known allergic sensitivities to pollen or other common allergens should adopt a less adventurous attitude when it comes to experimenting with strange or new species. Nonetheless, an allergic response to an unfamiliar wood can happen to anyone at any time. This is because many of the compounds found in wood are quite unique. In other words, they are not always chemically similar enough to the many common organic substances we encounter every day so as to make them unthreatening to our immune systems.

When adding a new wood to your repertoire, it is always wise to carefully limit your exposure to the dust for the first several times you work the wood. Allergic reactions are seldom immediate in that the body often tolerates the first exposure without producing symptoms. Symptoms may take days, weeks or even months to develop. Basically, what happens is the immune system uses the first contact as a stimulus to gear up for fighting the allergen. More unpleasant and serious misery comes only with subsequent exposure.

IRRITANTS

Wood can be both a physical and a chemical irritant. Physically, the fine dust of any substance, no matter how inert it might be, can act as an irritant to the skin and especially the mucous membranes of the respiratory tract. This is because fine dust particles tend to desiccate any surface they come in contact with by absorbing moisture and oils. The sawdust of many woods further complicates the problem of physical irritation in that they also contain chemicals that are irritating or caustic to sensitive human tissue, e.g., mucous membranes. Even some of our most popular and otherwise least threatening cabinetry woods present this problem. For example, many woodworkers find that the oaks are especially irritating to work, because they contain high levels of tannic acid. The volatile terpenes in numerous softwoods also prove troublesome. To some, the ebony chin rest of a violin or even wood pencils can produce allergies.

Just how irritating a wood may be is often a function of the intensity of exposure. At very low levels the fine dust of many woods actually produces a very pleasant scent. Rosewoods (of various genera) around the world often have a rose-like scent and the sandalwoods are famous for their perfumed wood. In fact, many woodworkers consider these pleasant aromas to be an important, intangible yet pleasant additional benefit of working with wood. Some of the most pleasantly scented species, however, are potentially toxic, walnut (*Juglans* spp.) and rosewood (*Dalbergia* spp.) to name just two. Walnut contains juglone, which in adequate doses is both a laxative and an anaesthetic, while a substance called dalbergione in Central and South American rosewood is a rather potent irritant and potential sensitizer.

Other interesting effects have emerged more recently. Certain woods, like the myriad of chemicals in our everyday lives, have been found to even have hormone-like (phytoandrogenic) effects on us. The health effects have yet to be confirmed.

One might conclude from this that the substances in wood which constitute simple irritants are not all that dangerous. This is probably true at low levels of exposure, but there is considerable evidence suggesting that continuous, long-term irritation is potentially life-threatening. The chance of developing nasal cancer is known to be 5 to 40 times greater if you work with wood. Similarly, the relationship between cigarette smoking and both emphysema and lung cancer, pipe smoking and lip cancer, heavy consumption of alcohol and both gastrointestinal cancer and liver disease all confirm that too much of anything that irritates human tissue anywhere in the body carries with it an increased health risk.

It would appear that wood dust is no exception. There is clinical evidence suggesting a probable correlation between prolonged and heavy exposure to wood dust and the occurrence of nasal cancer. Even though the incidence of this cancer is rare at 7 in 10,000, it is much higher in certain high-risk areas.

But be forewarned, 80% of all nasal cancers are associated with woodworkers. It is, however, a question of degree: degree of exposure to certain woods and the degree of response by each individual. In addition, there is reason for some concern that the cumulative effects of irritation caused by multiple irritants may compound the risk. In other words, the destructive pleasures many of us engage in, such as smoking and drinking, may tend to combine with long-term exposure to wood dust, solvent fumes and other pollutants in our modern world to 'stack the deck' against longevity. We do not yet have certain confirmation of these relationships, but it is probable that they exist.

PRECAUTIONS WHEN HANDLING WOOD

In many respects the problem of wood toxicity for the woodworker is one that requires individual management. The magnitude of the risk we each face is not just a function of the wood species we elect to use and how we use them, but also our individual sensitivities, our metabolism and our physical condition. There is no one correct approach for all woodworkers, but there are some sensible guidelines that are universally appropriate:

1. Take particular care working with new or unknown species, especially when making food or toys or items subject to extended contact (e.g., handles) until you have researched the species (see listed woods). Read the literature or talk to suppliers or other woodworkers who have experience with the wood in question. Avoid putting wood in your mouth to test its taste even if described as pleasant or unusual.

2. Always "introduce" yourself to a new species slowly and cautiously. Limit your exposure to the dust for at least the first several times you work with it until you are reasonably sure you do not have an allergic sensitivity to the wood's chemistry.

3. If you experience an unusual reaction of any kind (usually in the form of skin rash, watery eyes, respiratory difficulty, headache, dizziness or nausea) stop working with the wood and seek qualified medical advice as soon as possible.

4. As a general practice, keep the shop clean, well ventilated and as dust free as possible. Wearing a respirator or dust mask and/or installing a dust collection system (locating the dust extractor outside if possible) makes good sense, especially if woodworking is your vocation or an avocation to which you dedicate a significant amount of your spare time. The U.S. threshold limit for oak and beech wood dust is a mere 1 mg/cu meter and for softwoods it is 5 mg/cu meter. Even if you feel the dust is not particularly objectionable, long-term exposure to it presents unnecessary long-term health risks. Furthermore, remember that wood dust in a confined space is a potential explosion hazard as well as fire hazard. Vacuum rather than sweep.

5. Don't overlook personal hygiene. When working with potentially toxic woods don't take your dusty work clothes to the bedroom and transfer the dust there

as well. Change work clothes frequently. Strip off work clothes, take them directly to the washing machine and immediately shower.

6. Wear sensible clothes with long sleeves, covering exposed skin. Take care with long hair, rings and loose clothing when working with machinery so they do not get caught and drag you to them.

REFERENCES

References used for the tabulated information are:

Albert Forest Products. *Toxicity of Wood.* www.city-net.com/albertfp/toxic.htm.

Anderson, E. 1993. *Plants of Central Queensland.* DPI. Queensland, Australia.

Australian Forest Industries Journal. 1984. March.

Bolza, E. 1978. *Timber and Health.* CSIRO Report. Reprinted by CSIRO Publications. Melbourne, Australia.

Cartwright, L. 1988. *Plants Used in Herbal Remedies.* Australian Journal of Pharmacy. Vol. 69.

Collins, D.J., C.J. Culvenor, et al. 1990. *Plants for Medicines.* CSIRO Publications. Melbourne, Australia.

Concannon, J. 1989. *Herbal Hazards.* Australian Journal of Pharmacy. Vol. 70.

Everist, S.L. 1986. *Use of Fodder Trees and Shrubs.* DPI. Queensland, Australia.

Guidelines on Occupational Cancer. 1988. Health and Safety Bulletin. Vol. 56. Oct. Australian Council of Trade Unions. Victoria, Australia.

Keating, W.G. and E. Bolza. 1982. *Characteristics, Properties and Uses of Timbers, SE Asia, Northern Australia and the Pacific.* Vol 1. Inkarat Press. Melbourne, Australia.

Lassak E.V. and T. McCarthy. 1986. *Australian Medicinal Plants.* Mandarin, Australia.

McBarron, A.J. 1976. *Medical and Veterinary Aspects of Plant Poisons in NSW.*

McCann M. and A. Babin. *Woodworking Hazards.*

Personal communications with researchers, woodworkers and personal experience.

Woods, B. and C.D. Calnan. 1976. *Toxic Woods.* British Journal of Dermatology, 94, suppl. 13.

ABOUT THE AUTHORS

Jon Arno is a well known writer and consultant specializing in wood technology, the properties of commercial timbers and their use. He has written numerous articles for U.S. wood publications and is the author of *The Woodworker's Visual Handbook.* Jon graduated from the University of Michigan and resides in Michigan, USA.

Eugene Dimitriadis is an associate editor and writer for *World of Wood* and other wood publications. He graduated from Adelaide University with a PhD in Organic Chemistry (plant extracts). He has researched medicinal chemistry, toxic plants and wood extractives at universities and institutions in Canada and Australia. He has published widely and remains active in these fields of interest.

Jim Flynn is an associate editor of *World of Wood* and frequently researches and writes technical descriptions in the Wood of the Month column. He specializes in tree/wood taxonomy in which capacity he assisted with species names. He is the author of *A Guide to Useful Woods of the World* and was a stringed musical instrument maker who has worked with a great many woods.

Roy Tandy retired as a Captain after 32 years of service in the Navy as a Medical Service Corps Officer. He was Chief of Staff to the Surgeon General and Director of Internal Review of the Medical Department. Roy, an avid wood turner, occasionally encountered health problems with some woods. Following an extensive literary search, he published a paper on toxic woods in *World of Wood* in 1995 reporting a distinct lack of published material on the subject.

Australian Timbers Known to Produce Some Health Problems

The wood species in this list were compiled to highlight the diversity of "toxic timbers" endemic to Australia. This list has been kept separate from the "Worldwide Timbers" list to focus on this point. Since some of these woods are exported from Australia, the list has value to wood handlers worldwide.

Common Name	Scientific Name	Some reported health problems
alpine ash	*Eucalyptus delegatensis*	Irritating to mucous membranes, eyes, nose and throat; dermatitis
blackbean	*Castanospermum australe*	Irritating to mucous membranes, nose, mouth, throat, genitals, armpits
blackwood	*Acacia melanoxylon*	Dermatitis and skin reactions; asthma
blue gum (Tasmania)	*Eucalyptus globulus*	Dermatitis
bottlebrush	*Callistemon* spp.	Dermatitis and skin eruptions
brigalow	*Acacia harpophylla*	Dermatitis
coolabah/coolibah	*Eucalyptus coolibah, E. microtheca*	Skin irritation from bark and dust; other coolabah species produce similar effects
coolamon tree, shitwood	*Gyrocarpus americanus*	Can cause blindness
Crows ash	*Flindersia australis*	Dermatitis
cypress pine, white	*Callitris glaucophylla*	Dermatitis; swollen eyelids; asthma; nasal cancer; irritation of mucous membranes; furanculosis
dead finish	*Acacia tetragonophylla*	Wood splinters and thorns (phyllodes?) cause skin irritation; wood dust causes dermatitis
grass tree, black boy	*Xanthorrhoea* spp.	Dermatitis; potentially carcinogenic
grey box	*Eucalyptus microcarpa*	Eczema; irritating to mucous membranes
grey myrtle	*Diospyros* spp.	Causes skin eruptions (especially splinters)
gutta percha	*Excoecaria parvifolia*	Temporary blindness; has a milky irritating sap
jarrah	*Eucalyptus marginata*	Irritating to nose, throat, eyes
lemon scented gum	*Corymbia citriodora (Euc. citriodora)*	Dermatitis
messmate	*Eucalyptus obliqua*	Dermatitis; asthma; sneezing
milky mangrove	*Excoecaria agallocha, E. dallachyana*	Sap irritating and can cause temporary blindness, headaches, burning throat, skin blistering

miva mahogany	*Dysoxylum mollisimum* Congestion of lungs, eyes and irritation of mucous membranes; headaches; nosebleeds; appetite loss
mountain ash	*Eucalyptus regnans* Irritating to nose, throat and eyes; dermatitis
mulga	*Acacia aneura* Headaches; vomiting; irritation; may contain a poison
myrtle beech	*Nothofagus cunninghamii* Irritates mucous membranes
poison walnut/laurel	*Cryptocarya pleurosperma* Bark very irritating, causes breathing problems, vomiting, giddiness, dermatitis
Queensland maple	*Flindersia brayleana* Dermatitis
red cedar	*Toona ciliata* (syn. *Toona australis*) Violent headaches and earaches; giddiness; stomach cramps; asthma; dermatitis
red siris/Mackay cedar	*Albizia toona* Irritating to eyes, nose and throat; sneezing; conjunctivitis; nosebleeds; dermatitis
rose butternut	*Blepharocarya involucrigera* Dermatitis; conjunctivitis
saffron heart	*Halfordia scleroxyla* Splinters cause infections; dust causes dermatitis and lung congestion
silky oak, northern	*Cardwellia sublimis* Skin irritation and eruptions; worse when wood is green
silky oak, southern	*Grevillea robusta* Skin eruptions and blistering; irritating to eyelids and mucous membranes; dermatitis; worse when wood is green
spotted gum	*Corymbia maculata (Euc. maculata)* Dermatitis
sugar gum	*Eucalyptus cladocalyx* Irritating to eyes (especially sap); saw chips are sharp and irritating
tea tree	*Leptospermum* spp. Dermatitis; skin eruptions
thorny yellowwood	*Zanthoxylum brachycanthum*, Z. spp. Slow-healing lesions from thorns and splinters; dermatitis; cramps; irritating to eyes and throat; visual disturbances
yellow gum	*Eucalyptus leucoxylon* Irritating to nose and throat

WARNING

This list is not exhaustive and does not necessarily represent all of the woods discussed in this book. Care should be taken when working with all unfamiliar timbers, especially when wet or if you have respiratory or known allergic tendencies. If you experience health problems, record the timbers you have worked with in past few days and see your doctor promptly!

Worldwide Timbers Known to Produce Some Health Problems

Common Name	Scientific Name Some reported health problems
abura	*Hallea ledermannii, H. stipulosa* Nausea; eye irritation; giddiness
African blackwood	*Dalbergia melanoxylon* Acute dermatitis; sneezing; conjunctivitis
African boxwood	*Gonioma kamassi* Irritates nose and throat; headaches; fainting; shortness of breath; asthma; erythema
afrormosia	*Pericopsis elata* Dermatitis; rhinitis; asthma
agba	*Gossweilerodendron balsamiferum* Dermatitis
ailanthus	*Ailanthus altissima* Dermatitis
albizia	*Falcataria moluccana* Irritates eyes, nose, alimentary tract; nausea; dermatitis
alder	*Alnus* spp. Dermatitis
angelique	*Dicorynia guianensis* Headaches; respiratory problems; contains a poisonous alkaloid
ash, white	*Fraxinus americana, F. excelsior* Occupational asthma/rhinitis
avodire	*Turraeanthus africanus* Sawdust is a skin and respiratory irritant; cause unknown
ayan	*Distemonanthus benthamianus* Dermatitis
bald-cypress	*Taxodium distichum* Respiratory irritant and sensitizer
balsam fir	*Abies* spp. Eye and skin irritant (especially bark)
baltic, white	*Picea abies* Asthma; skin irritation
beech, American	*Fagus grandifolia* Eye, skin and respiratory sensitizer and suspected carcinogen
beech (European)	*Fagus sylvatica* Nasal and sinus cancers; dermatitis
bibu	*Holigarna arnottiana* Sap causes skin blistering, eye irritation
birch	*Betula* spp. Sensitizer, those allergic to aspirin need to beware
boxwood (Europe, Asia minor)	*Buxus sempervirens* Dermatitis; irritates nose, eyes and throat
boxwood, Knysna	*Gonioma kamassi* Irritates nose and throat; headaches; fainting; shortness of breath; asthma; erythema
bubinga	*Guibourtia* spp. Dermatitis; probably caused by sensitizing quinones
cabreuva	*Myrocarpus fastigiatus* May cause asthma and systemic toxic reactions

calophyllum	*Calophyllum* spp. Dermatitis; kidney damage; irritates nose and throat
camphor-tree	*Cinnamomum camphora* Can cause dermatitis and shortness of breath especially when wood is green or hot and in hot weather
caroba	*Jacaranda caroba* Dermatitis
cashew nut	*Anacardium occidentale* Sap causes blisters; dermatitis
cedar, Atlas	*Cedrus atlantica* Irritates mucous membranes
cedar, deodar	*Cedrus deodara* Irritates mucous membranes
cedar of Lebanon	*Cedrus libani* Respiratory irritant; probably terpenes
cedar, red (Aust., SE Asia, PNG, India)	*Toona ciliata, T. calantas* Violent earaches; giddiness; stomach cramps; asthma; dermatitis; irritation of mucous membranes
cedar (S. America)	*Cedrela* spp. Dermatitis; asthma; nasal cancer; irritates nose and throat
cedar, white	*Thuja occidentalis* Dermatitis
chechem (black poison wood)	*Metopium brownei* Respiratory eye and skin irritation; sap very irritating
cherry, black	*Prunus serotina* Wheezing; giddiness
cocobolo	*Dalbergia granadillo, D. retusa* Acute dermatitis; sneezing; conjunctivitis; asthma; nausea
cocuswood	*Byra ebenus* Dermatitis; contact sensitization
cordia	*Cordia milleni* Dermatitis; contains a sensitizing quinone
courbaril	*Hymenaea courbaril* Skin irritant; probably a quinone
cypress, Monterey	*Cupressus macrocarpa* Volatile oils can cause headaches; wet wood can irritate skin in hot weather
dahoma	*Piptadeniastrum africanum* Dermatitis; coughing; sneezing; nose bleeds
difundu	*Vitex congolensis* Dermatitis
djohar	*Senna siamea, Senna* spp. Skin discoloration; eye irritation; keratitis
Douglas-fir (Oregon)	*Pseudotsuga menziesii* Dermatitis; nasal cancer; irritates eyes and throat; leaves and bark are prone to irritate eyes and skin
ebony, various	*Diospyros crassiflora, D.* spp. Acute dermatitis; sneezing; conjunctivitis; skin inflammation
ekki	*Lophira alata* Itching; dermatitis; cause not reported
elm (Eurpoean)	*Ulmus* spp. Dermatitis; irritates nose and throat; nasal cancer
eyum	*Dialium dinklagei* Irritates nose and throat; dermatitis
gaboon	*Aucoumea klaineana* Skin itching; irritates nose and eyes; bronchial asthma
goncalo alves	*Astronium fraxinifolium* Dermatitis

greenheart (S. America) *Chlorocardium rodiei*
Headaches; wheezing; shortness of breath; visual disturbances; diarrhea; erythema; splinters

guarea *Guarea* spp.
Dermatitis; asthma; nausea; viusal disturbances; headaches

hemlock, western *Tsuga heterophylla*
Dermatitis; eczema

idigbo *Terminalia ivorensis*
Skin and respiratory irritant; contains triterpenes

imbuya *Ocotea porosa*
May cause occupational asthma

incense-cedar *Calocedrus decurrens*
Dermatitis; eczema; contains thymoquinone

ipe (lapacho) *Tabebuia* spp.
Dermatitis; irritating to mucous membranes; systemic symptoms; shortness of breath; headaches

iroko *Milicia excelsa*
Dermatitis; possible irritation of mucous membranes; oedema of eyelids; respiratory difficulties; giddiness

jelutong (SE Asia) *Dyera costulata*
May cause contact allergy

juniper, Phoenician *Juniperus phoenicea, J. sabina*
Nausea; headaches; skin irritation

katon *Wallichia disticha*
Respiratory irritant

kingwood *Dalbergia cearensis*
Eye and skin irritant

kirundu *Antiaris toxicaria*
Headaches; cardiac effects; probably a glycoside

lancewood, red (S. America) *Manilkara huberi*
Dermatitis

larch *Larix decidua*
Dermatitis; respiratory irritant; probably terpenes

lauan *Shorea* spp.
Dermatitis; irritation of eyes, nose, throat

laurel, California *Umbellularia californica*
Wood and dust common respiratory sensitizers

lignumvitae *Guaiacum* spp.
Dermatitis

limba *Terminalia superba*
Splinters can cause festering, skin irritation and inflammation

lingue *Afzelia africana*
Irritates eyes, nose and throat

locust, black *Robinia pseudoacacia*
Wood dust, the bark especially can be a skin and eye irritant and can cause nausea

logwood *Haematoxylon campechianum*
Dermatitis; cause not reported

Macassar ebony *Diospyros celebica*
Dermatitis

mahogany, African *Khaya* spp.
Dermatitis; rhinitis; nasal cancer

mahogany, American *Swietenia* spp.
Dermatitis; giddiness; vomiting; "mahogany cough"; furunculosis

makore *Tieghemella heckelii*
Dermatitis; nose bleeds; asthma; sneezing; nausea; headaches; giddiness

mansonia	*Mansonia altissima* Dermatitis; nausea; vomiting; nose bleeds; sneezing; giddiness; asthma; mucosal irritation, cardiac effects
manzanillo, Cuba	*Hippomane mancinella* Resin is very caustic causing skin eruptions and vomiting
maple (spalted)	*Acer* spp. Skin irritation and pneumonitis likely
Maria	*Calophyllum brasiliense* Dermatitis; appetite loss; kidney damage; fainting; insomnia
marking nut	*Semecarpus australiensis, S. anacardium* Sap and wood dust can cause skin lesions, eye irritation, nose bleeds
massaranduba	*Pouteria procera* Dermatitis
merbau	*Instia bijuga, I. palembanica* Dermatitis; rhinitis
mesquite	*Prosopis juliflora* Respiratory irritant; probably an alkaloid; wood smoke can cause respiratory problems
moabi	*Baillonella toxisperma* Respiratory irritant; cause not reported
molompangady	*Breonia* spp. Dermatitis; sores
muhuhu	*Brachylaena huillensis* Dermatitis
muninga	*Pterocarpus angolensis* Bronchitis; asthma; dermatitis
narra (PNG)	*Pterocarpus indicus* Dermatitis; asthma
needlewood (India and SE Asia)	*Schima wallichii* Bark causes skin irritation
niangon	*Heritiera utilis* Skin irritant; cause not reported
nkassa	*Erythrophleum le-testui* Headaches; giddiness; vomiting; bowel and stomach disorders; nose and throat irritation
oak, English (Europe and Japan)	*Quercus robur, Q.* spp. Nasal cancer; dermatitis; sneezing
oak (U.S.)	*Quercus* spp. Nasal cancer; dermatitis; sneezing; occupational asthma
obeche	*Triplochiton scleroxylon* Asthma; sneezing; congestion of lungs; irritation of mucous membranes; possibly dermatitis
odoum	*Milicia regia* Dermatitis; nose and throat irritation
okume	*Aucoumea klaineana* Skin itch; irritates nose and eyes; bronchial asthma
oleander	*Nerium* spp., *Thevetia* spp. Sap and other plant parts should all be treated as very poisonous
olivewood	*Olea* spp. Eye, skin and respiratory irritant and sensitizer
orangewood	*Citrus* spp. Respiratory irritant
opepe	*Nauclea trillesii* Dermatitis; irritates mucous membranes
padauk	*Pterocarpus* spp. Asthma; dermatitis

palisander	*Jacaranda brasiliana*	
		Dermatitis
parinari	*Parinari* spp.	
		Dermatitis
pau armarello	*Euxylophora paraensis*	
		Dermatitis
pau marfim	*Balfourodendron riedelianum*	
		Dermatitis; occupational rhinitis; asthma
pau santo	*Zollernia paraensis*	
		Dermatitis; cause not reported
peppercorn	*Schinus molle*	
		Sap and wet wood irritates eyes and mucous membranes
Pernambuco	*Caesalpinia echinata*	
		Headaches; nausea; painful swelling of arms; visual disturbances
peroba, pink	*Aspidosperma polyneuron*	
		Dermatitis; irritates mucous membranes; nausea; sweating; asthma; cramps; fainting; drowsiness
peroba, white	*Paratecoma peroba*	
		Dermatitis; irritates mucous membranes
pine, eastern white	*Pinus strobus*	
		May be cytotoxic and genotoxic
pine, western white	*Pinus monticola*	
		Volatile extractives may be cytotoxic and genotoxic
pine, white (New Zealand)	*Dacrycarpus dacrydioides*	
		Dermatitis; irritates nose and throat
poplar	*Populus* spp.	
		Asthma; dermatitis; bronchitis
Port-Orford-cedar (U.S.)	*Chamaecyparis lawsoniana*	
		Dermatitis; can irritate eyes, skin and respiratory system (monoterpenes?)
purpleheart	*Peltogyne* spp.	
		Can cause nausea
quebracho	*Schinopsis quebracho-colorado*	
		Respiratory and nasal irritant; potential carcinogenic
ramin	*Gonystylus bancanus*	
		Dermatitis
redwood (U.S.)	*Sequoia sempervirens*	
		Dermatitis; asthma; wood dust respiratory irritant and carcinogen
redcedar, western (U.S.)	*Thuja plicata*	
		Mucosal irritation; dermatitis; asthma; nasal cancer; nausea; nose bleeds; giddiness; stomach pain
rengas (SE Asia)	*Gluta* spp. (syn. *Melanorrhoea* spp.)	
		Occupational allergic contact dermatitis; blistering; chronic ulcers
rimu (New Zealand)	*Dacrydium cupressinum*	
		Irritating to nose and eyes
rosewood (Indian)	*Dalbergia latifolia, D. sissoo, D. oliveri*	
		Dermatitis
rosewood (S. America)	*Dalbergia nigra, D. stevensonii*	
		Dermatitis
sandalwood (Asia, India)	*Santalum album*	
		Respiratory irritant; cause not reported
sandbox tree	*Hura crepitans*	
		Latex contains a strong irritant, huratoxin, potentially poisonous
Santos rosewood (pao ferro)	*Machaerium scleroxylon*	
		Can cause contact dermatitis
sapele	*Entandophragma cylindricum*	
		Skin irritation; sneezing

sassafras (U.S.) — *Sassafras albidum*
Irritates skin and respiratory passages; suspected carcinogen

satine (bloodwood) — *Brosimum* spp.
Salivation; thirst; nausea and respiratory tract irritation

satinwood, Ceylon — *Chloroxylon swietenia*
Dermatitis; headaches; swollen scrotum; irritates mucous membranes

satinwood, West Indian — *Zanthoxylum flavum*
Dermatitis

snakewood — *Brosimum guianense*
Salivation; thirst; nausea; respiratory tract irritation

spindle tree — *Euonymus europaeus*
Occupational sensitivity to wood dust

spruce — *Picea* spp.
Occasionally respiratory irritant

sucupira — *Bowdichia nitida*
Can cause allergic contact dermatitis

sumac — *Rhus* spp.
Bark causes blisters, dermatitis

sweetgum — *Liquidambar styraciflua*
Dermatitis

talh — *Acacia seyal*
Dermatitis; coughing; irritates mucous membranes

taun — *Pometia pinnata*
Dermatitis; rhinitis

teak — *Tectona grandis*
Dermatitis; conjunctivitis; contact urticaria; over sensitivity to light; swollen scrotum; nausea

teak, Rhodesian — *Baikiaea plurijuga*
Respiratory irritant; cause not reported

tigerwood — *Lovoa trichilioides*
Irritates mucous membranes and alimentary tract; nasal cancer

tonka — *Dipteryx odorata*
Respiratory irritant; "toxic effects" reported

walnut, black (U.S.) — *Juglans nigra*
Irritates eyes, skin

walnut (European) — *Juglans regia*
Dermatitis; irritates nose and throat; nasal cancer

walnuts — *Juglans* spp.
Nasal cancer; dermatitis; irritation of mucous membranes

wenge — *Millettia laurentii*
Irritates eyes, skin and respiratory system

willow — *Salix* spp.
Sensitizer, those allergic to aspirin need to beware

yang — *Dipterocarpus alatus*
Dermatitis; inflames mucous membranes; furunculosis; splinters cause festering

yew — *Taxus baccata*, *Taxus* spp.
Headaches; lung congestion; nausea; fainting; visual disturbances; irritation of alimentary canal

zebrawood — *Microberlinia brazzavillensis*
Irritates eyes and skin

WARNING

This list is not exhaustive and does not necessarily represent all of the woods in this book Care should be taken when working with all unfamiliar timbers, especially when wet or i you have respiratory or known allergic tendencies. If you experience health problems record the timbers you have worked with in past few days and see your doctor promptly

APPENDIX C

Selected References

There are many resources available, printed and electronic, which cover the numerous aspects of wood reported in this book including taxonomy and the proper naming of plants and wood, dendrology or the study of trees and wood technology, the characteristics and appropriate uses of various kinds of wood. This list is not exhaustive but provides the reader with additional paths to follow. Additional references can be found in Appendix A and Appendix B.

TAXONOMY AND NOMENCLATURE

Brummitt, R.K. 1992. *Vascular Plant Families and Genera.* www.rbgkew.org.uk/data/genfile.html. Royal Botanic Gardens, Kew.

Little, Jr., E.L. 1979. *Checklist of United States Trees (Native and Naturalized).* Agric. Handbook 541. Washington DC: U.S. Department of Agriculture, Forest Service.

Reveal, James L. *Vascular Plant Family Nomenclature: Names in Current Use.* www.inform.umd.edu/PBIO/fam/ncu.html. Norton-Brown Herbarium, University of Maryland.

U.S. Department of Agriculture. GRIN Taxonomy. www.ars-grin.gov/npgs/tax/index.html.

U.S. Department of Agriculture. www2.fpl.fs.fed.us/CommNames2000.html (a searchable common name database).

DENDROLOGY AND TREE IDENTIFICATION

Boland, J.H. et al. 1992. *Forest Trees of Australia.* CSIRO Publications, Australia.

Discovery Channel. 2000. *Trees.* Discovery Books. New York, NY (an international sampling of significant trees).

Johnson, Hugh. 1993. *The International Book of Trees.* Mitchell Beazley Publishers.

Turner, Jr., R.J. and Ernie Wasson. 1999. *Botanicia: The Illustrated A-Z of Over 10,000 Garden Plants* (includes CD ROM). Welcome Rain.

WOOD, WOOD IDENTIFICATION AND WOOD TECHNOLOGY

Chudnoff, Martin. 1984. *Tropical Timbers of the World.* Agric. Handbook 607. Washington DC: U.S. Department of Agriculture, Forest Service.

Forest Products Laboratory. 1999. *Wood Handbook: Wood as an Engineering Material.* General Technical Report FPL-GTR-113 U.S. Department of Agriculture, Forest Service. Reprinted with permission by the Forest Products Society, Madison, WI.

Ives, Ernie. 2001. A *Guide to Wood Microtomy, Making quality microslides of wood sections*. 63 Church Lane, Sproughton, Ipswich, Suffolk, IP8 3AY, England.

Lincoln, William A. 1986. *World Woods in Color*. Linden Publishing Co., Fresno, California.

Panshin, A.J. and Carl de Zeeuw. 1980. *Textbook of Wood Technology: Structure, Identification, Properties and Uses of the Commercial Woods of the United States and Canada*. McGraw-Hill.

Record, Samuel J. and Robert W. Hess. 1943. *Timbers of the New World*. Reprinted 1986. Yale University Press.

WEBSITES

FPL, Forest Products Laboratory, USDA: www.fpl.fs.fed.us/about.htm.

FPS, Forest Products Society: www.forestprod.org.

IWCS, International Wood Collectors Society: www.woodcollectors.org.

NeHoSoC, Dutch Wood Collectors Society: www.nehosoc.nl.

INDICES

COMMON NAME INDEX

If a common name is known for a species, look it up in this alphabetic list for a reference to the Wood Data Sheet (WDS) number for the species. The WDS are published in numerical order in the book.

A

A Tree Grows in Brooklyn, 15
aba, 164
aborzok, 171
aboudikro, 112
Abyssinian boxwood, 50
acacia, 237
acacia, catclaw, 4
acacia, devil's-claw, 4
acacia, false, 237
acacia, Gregg, 4
acacia, thorny, 126
acacia, three-thorned, 126
acaja, 251
acajou, 155, 254
aceite Maria, 55
aceitillo, 278
acero, 157
acetillo, 136
achi, 130
acuapa, 144
aderno, 29
adza, 32
afara, 264
afara, black, 263
African black walnut, 175
African blackwood, 96
African cedar, 133
African cedar, pink, 133
African cherry, 269
African coralwood, 228
African (East) yellowwood, 12
African mahogany, 155
African maple, 272
African oak, 164, 181
African padauk, 228
African pearwood, 32
African primavera, 272
African rosewood, 135
African sandalwood, 118
African satinwood, 200, 273
African teak, 181, 263
African-teak, 200
African walnut, 166
African whitewood, 272
afrormosia, 200
afzelia, 13
agba, 130
agoho, 67
ahuejote, 114
ailanthus, 15
ake, 265
akewe, 265
akom, 264
akondoc, 185
akume, 135
alageta, 201
alamo blanco, 219
Alaska-cedar, 78
Alaska cypress, 78
Alaska yellow-cedar, 78
alder, Oregon, 18
alder, red, 18
alder, western, 18
alderleaf cercocarpus, 76
algarobo, 221
algarrobo, 146, 241
alligator-pear, 201
alligator tree, 161
alligator wood, 161
almacigo, 49
almon, 249
alona wood, 166
alstonia, Palimira, 19
amansa-mujer, 220
amapa, 86, 257
amaranth, 199
amarello, 28, 215
amargosa, 28
amasisa, 114
amazakone, 135
ambahu, 69
amboyana, 226
American ash, 121
American basswood, 270
American beech, 119
American black walnut, 150
American chestnut, 65
American coffeebean, 137
American elm, 274
American holly 147
American hornbeam, 61
American laurel, 153
American linden, 270
American mahogany, 137, 254
American mountain-ash, 250
American plane tree, 214
American redbud, 75
American rowantree, 250
American (South) locust, 146
American smoketree, 89
American sumac, 236
American (tropical) ebony, 45
American walnut, 150
American white wood, 270
American yellowwood, 83
amole de bolita, 243
anacaguita, 252
ancient pine, 207
Andaman padauk, 225
Andaman redwood, 225
andira, 21
andiroba, 59
angelim, 21
angelin, 21
angico, 197
angsana, 226
anis, 187

anume, 179
apapaye, 273
apaya, 273
apes-earring, 108
apple, 170
apple, common, 170
apple, custard, 27
apple, hedge, 167
apricot, 201
aprono, 175
aquacate, 201
aragan, 256
arariba, 73
aratree, 266
araucaria, 24
arborvitae, giant, 268
Arbuckle white cedar, 151
arbuti tree, 25
argento, 116
Arizona-ironwood, 193
Arkansas oak, 231
Arkansas water oak, 231
aroeira, 29
arrow-wood, 88
arrowwood, 196
aru, 67
asamela, 200
ash, 122
ash, American, 121
ash, American mountain-, 250
ash, basket, 123
ash, Biltmore, 121
ash, black, 123
ash, brown, 123
ash, canary, 116
ash, cane, 121
ash, common, 122
ash, European, 122
ash, false, 48
ash, French, 122
ash, hoop, 72, 123
ash, Hungarian, 122
ash, Italian olive, 122
ash, Japanese, 124, 154
ash leafed maple, 7
ash, maple-, 7
ash, mountain, 116
ash, mountain-, 250
ash, smallseed white, 121
ash, swamp, 123
ash, Turkey, 122
ash, white, 121
ash, yellow, 83
Ashe juniper, 151
aspen, golden, 219
aspen, mountain, 219
aspen, quaking, 219
aspen, quiverleaf, 219
aspen, trembling, 219
assacu, 144
aukkyu, 40
Australian blackwood, 6
Australian laurel, 110
Australian maple, 120
Australian oak, 116
Australian pine, 67
Australian red cedar, 271
Australian silky oak, 60
Australian tea tree, 160
Australian toon, 271
Australian walnut, 110
avocado, 201
avodire, 273
axemaster, 157
ayan, 104
ayanran, 104
ayous, 272
ayua, 278
azobe, 164

babu, 269
badi, 185
Bahia wood, 51
baku, 269
bakundu, 164
balata, 173
bald-cypress, 259
balm, 217
balm-of-Gilead, 217
balong ayam, 141
balsa, 189
balsam, 1
balsam, Canada, 1
balsam fir, 1
balsam of Peru, 184
balsam poplar, 217
balsamaria, 55
balsamo, 184
bamboo, 203
bamboo, moso, 203
bamboo, tortoise-shell, 203
banana, 27
banana, false, 27
barauna, 246
bari, 55
baria, 86
bariaco, 157
barre, 104
barwood, 228
basket ash, 123
basswood, 270
basswood, American, 270
bastard elm, 72
bat tree, 169
bata, 201
bateo, 59
bay, 129
bay, bull, 169
bay, holly-, 129
bay-laurel, 277
bay, loblolly-, 129
bay, tan, 129
beaded locust, 253
beadtree, 177
bean, mahogany, 13
bean tree, 66
bean, walnut, 110
bear-berry, 253
bear, Indian, 68
bebeere, 79
beech, American, 119
beech, blue, 61
beech, Chilean, 187
beech, red, 119
beech, stone, 119
beech, water, 61, 214
beech, white, 119
beech, winter, 119
beefwood, 67, 173
bellota, 252
belltree, 139
benge, 134
Benin mahogany, 133
Benin satinwood, 200
Benin walnut, 166
Benin wood, 155
bera cuchivaro, 47
bete, 175

bethabara, 257
bibira, 79
big laurel, 169
big-leaved ivy, 153
bigleaf mahogany, 254
bigtree, 248
bilinga, 185
billetwood, Lagos, 102
bilsted, 161
Biltmore ash, 121
birch, black, 38
birch, canoe, 39
birch, cherry, 38
birch, curly, 37
birch, grey, 37
birch, hard, 37
birch, mahogany, 38
birch, paper, 39
birch, red, 38
birch, silver, 37, 39
birch, swamp, 37
birch, sweet, 38
birch, white, 39
birch, yellow, 37
birchleaf cercocarpus, 76
bird cherry, 222
bird cherry, European, 222
bishopwood, 40
black afara, 263
black ash, 123
black birch, 38
black cherry, 223
black ebony, 102
black ironwood, 157
black laurel, 129
black locust, 237
black maple, 10
black mulberry, 183
black-olive, 46
black pine, 209
black poison wood, 178
black poplar, 217
black spruce, 204
black titi, 90
black tupelo, 188
black walnut, 150
black walnut, African, 175
black walnut, eastern, 150
black wattle, Darwin, 3
black willow, 240
black willow, western, 240
blackbead, ebony, 108
blackbean, 66
blackgum, 188
blackwood, 6
blackwood, African, 96
blackwood, Australian, 6
blackwood, Bombay, 95
blackwood, Tasmanian, 6
bloodwood, 43
blue beech, 61
blue catalpa, 198
blue mahoe, 143
bodark, 167
bodock, 167
boessi papaja, 69
bog oak, 232
bog spruce, 204
bois cochon, 91
bois-d'arc, 167
bois de fer, 157
bois de lance, 195
bois mulatre, 178
bois rouge, 228
boise de rose, 94
bokoli, 171
Bombay blackwood, 95
bompegya, 171
bonga, 71
bongossi, 164
bonsamdua, 104
Borneo camphorwood, 106
boro, 114
bosse, 133
bottlebrush, 176
bouleau a papier, 39
bouleau juane, 37
bowwood, 167
box, Circassian, 50
box, Persian, 50
box-elder, 7
boxelder, 7
boxwood, 50, 88
boxwood, Abyssinian, 50
boxwood, false, 88
boxwood, Turkey, 50
branquilho, 136
brauna, 246
Brazil redwood, 43
Braziletto, 51
Brazilian louro, 22
Brazilian mahogany, 215
Brazilian pinkwood, 94
Brazilian rosewood, 97
Brazilian walnut, 191
Brazilnut, 36
Brazilwood, 51
bristlecone pine, 207
bristlecone pine, Colorado, 207
bristlecone pine, Great Basin, 207
bristlecone pine, intermountain, 207
bristlecone pine, Rocky Mountain, 207
bristlecone pine, western, 207
broad-leaved paperbark, 176
broom tea tree, 160
broussonet, 44
brown ash, 123
brown ebony, 45, 52
brushbox, 165
buaja, 128
bubinga, 135
bucida, 46
bucida, oxhorn, 46
buckeye, 11
buckeye, fetid, 11
buckeye, Ohio, 11
buckeye, stinking, 11
buckwheat-tree, 90
bull-bay, 169
bulletwood, 173
bully tree, 173
burkea, 48
Burma padauk, 227
Burma rosewood, 226
Burma teak, 261
buruta, 80
bush, calico, 153
bush maple, 272
butternut, 149
buttonball, 214
buttonwood, 214

C

cabbage bark, red, 21
cabbage-palm, 239
cabbage palmetto, 239

cabbage, swamp, 239
cabinet cherry, 223
cadmia, 57
cajeput-tree, 176
caju assu, 20
calaba tree, 55
calabar, 102
calambuca, 55
calico bush, 153
calicowood, 139
California chestnut oak, 163
California incense-cedar, 54
California laurel, 277
California-laurel, 277
California redwood, 247
California sugar pine, 210
calumban, 17
cam-lai, 93
camagon, 101
camajura, 252
camaruca, 252
camibar, 220
camirio, 17
campeche, 138
camphor-tree, 82
camphorwood, Borneo, 106
Canada balsam, 1
canalete, 86
canaletta, 86
canang odorant, 57
cananga, 57
canary ash, 116
canary wood, 162
canarywood, 73
candela, 215
candelon, 235
candle tree, 91
candleberry, 17
candlenut, 17
candlenut-tree, 17
cane ash, 121
canella imbuia, 191
canime, 220
canoe birch, 39
canoe cedar, 268
canoe wood, 162
caoba, 254
Cape Damson ebony, 96
Cape laurel, 190
carabali, 241
caracoli, 20
caragana, 58
caragana, common, 58
carate, 49
cardinalwood, 43
Caribbean pine, 208
carocaro, 113
Carolina hickory, 64
Carolina palmetto, 239
Carolina poplar, 218
Carolina silverbell, 139
castana del maranon, 36
castanheiro, 36
castano, 252
castor aralia, 154
casuarina, 67
catahua, 144
catalpa, blue, 198
catapla, northern, 68
catclaw, 4
catclaw acacia, 4
catclaw, Gregg, 4
cativo, 220
cautivio, 220
caviuna, 97
cecropia, 69
cedar, African, 133
cedar, Alaska-, 78
cedar, Alaska yellow-, 78
cedar, Arbuckle white, 151
cedar, Australian red, 271
cedar brake, 151
cedar, California incense-, 54
cedar, canoe, 268
cedar, cigar-box, 70
cedar elm, 275
cedar, incense- 54
cedar, Java, 40
cedar, mountain, 151
cedar, Oregon-, 77
cedar, Ozark white, 151
cedar, pencil, 152
cedar, pink African, 133
cedar, Port-Orford-, 77
cedar, Port-Orford white-, 77
cedar, post-, 151
cedar, rock-, 151
cedar, Spanish-, 70
cedar, Texas, 151
cedar, white, 177
cedar, yellow, 151
cedar, yellow-, 78
cedro, 70, 151
cedro macho, 59
cedro prieto, 178
cedron, 133
ceiba, 71
cenicero, 241
cercocarpus, alderleaf, 76
cercocarpus, birchleaf, 76
cercocarpus, curlleaf, 76
cercocarpus, hairy, 76
cetico, 69
Ceylon satinwood, 80
chaca, 49
chachin, 178
chaguaramo, 238
chanfuta, 13
chanfuti, 13
chechem, 178
chechen, 178
cheesewood, white, 19
chene rouge, 233
cherioni, 243
cherry, African, 269
cherry birch, 38
cherry, bird, 222
cherry, black, 223
cherry, cabinet, 223
cherry, European bird, 222
cherry mahogany, 269
cherry, rum, 223
cherry, wild, 223
cherry, wild black, 223
chestnut, 65
chestnut, American, 65
chestnut, Moreton Bay, 66
chestnut oak, 163
chestnut oak, California, 163
chestnut, sweet, 65
chica, 252
chicle, 174
chicot, 137
Chilean beech, 187
chimtoc, 157
chinaberry, 177
chinatree, 177
chinatree, wild, 243
Chinese sumac, 15
Chinese umbrella tree, 177

chinkapin, 81
chinkapin, evergreen-, 81
chinkapin, giant, 81
chinkapin, golden, 81
chinkapin, goldenleafed, 81
chinkapin, western, 81
chittamwood, 89
Christmas holly, 147
Christmas-tree, 24
chumprak, 141
chun, stinking, 15
cigar-box cedar, 70
cigar tree, 68
cinnamon-wood, 244
Circassian box, 50
citrus wood, 266
coachwood, 74
coast madrone, 25
coast redwood, 247
coast spruce, 206
cockspur coral bean, 114
cockspur coralbean, 114
coco, 100
cocobolo, 98
coconut, 84
cocus, 45
cocuswood, 45
coffee bean, American, 137
coffeenut, 137
coffeetree, Kentucky, 137
coffeewood, 52
coigue, 187
coihue, 187
Colorado bristlecone pine, 207
colorado, mangle, 235
common apple, 170
common ash, 122
common caragana, 58
common coral tree, 114
common mesquite, 221
common palmetto, 239
common pear, 229
common persimmon, 103
comwood, 228
Confederate pintree, 126
Congo wood, 166
Congowood, 96
copachu, 220
copal, 256
copal tree, 15
coracy, 199
corail, 228
coral sumac, 178
coral tree, common, 114
coralbean, 253
coralbean, cockspur, 114
coralwood, African, 228
corcho, 189
cordia wood, 87
cordovan, 245
cork pine, 212
cork, Spanish, 267
corktree, 267
cornel, 88
cornel, white, 88
coronel, 157
Cortez, 256, 257
cottontree, silk-, 71
cottontree, silk-, white, 71
cottonwood, eastern, 218
cottonwood, southern, 218
courbaril, 146
coyan, 187
cra huang, 93
crabwood, 59, 136
cremon, 267
cresta de gallo, 114
cry baby tree, 114
cuangare, 100
Cuban oysterwood, 136
Cuban royal palm, 238
cuchi, 29
cucumber magnolia, 168
cucumbertree, 168
curacai, 220
curlleaf cercocarpus, 76
curly birch, 37
curtidor, 145
custard apple, 27
cut leafed maple, 7
cypress, 53, 259
cypress, Alaska, 78
cypress, bald-, 259
cypress, gulf-, 259
cypress, Lawson, 77
cypress, Murray River, 53
cypress, Nootka, 78
cypress-pine, 53
cypress pine, white, 53
cypress, pond-, 259
cypress, red-, 259
cypress, Sitka, 78
cypress, southern-, 259
cypress, swamp-, 259
cypress, Tidewater red-, 259
cypress, yellow, 78
cypress, yellow-, 259
cypress, western, 53
cypress, white-, 259
cyrilla, swamp, 90

D

dalchini, 82
damo, 124
damoni, 105
danta, 186
dante, 186
dao, 105
Darwin black wattle, 3
date plum, 103
dead-tree, 137
degame, 56
degema, 155
demerara, 79
desert-ironwood, 193
devil-tree, 19
devil's-claw acacia, 4
devil's tree, 200
dibetou, 166
dikela, 182
dilly, 174
dilly, wild, 173
dimpampi, 32
dogberry, 250
dogwood, Florida, 88
dogwood, flowering, 88
dogwood, Jamaica, 213
dorea, 105
dormilon, 241
Douglas-fir, 224
Douglas-fir, yellow, 224
Douglas spruce, 224
Douglas yew, 224
dragon's tooth, 114
Dudley willow, 240
durango, 256
durmast oak, 232

E

ear-pod wattle, 3
eartree, 113
East African yellowwood, 12
East Indian rosewood, 95
East Indian sandalwood, 242
East Indian satinwood, 80
East Indian walnut, 16, 262
eastern black walnut, 150
eastern cottonwood, 218
eastern fir, 1
eastern hophornbeam, 194
eastern juniper, 152
eastern red oak, 233
eastern redbud, 75
eastern redcedar, 152
eastern spruce, 204, 205
eastern white oak, 230
eastern white pine, 212
eban, 135
ebano, 52, 279
ebony, black, 102
ebony blackbead, 108
ebony, brown, 45, 52
ebony, Cape Damson, 96
ebony ebano, 108
ebano, ebony, 108
ebony, Gabon, 102
ebony, golden, 101
ebony, green, 45
ebony, Indian, 101
ebony, Jamaican, 45
ebony, Macassar, 101
ebony, Maracaibo, 52
ebony, Mozambique, 96
ebony, Nigerian, 102
ebony, Senegal, 96
ebony, Texas, 108
ebony, tropical American, 45
edabalii, 55
edoucie, 133
ejen, 104
ekki, 164
ekor belangkas, 111
el siebo, 114
elder, box-, 7
elm, American, 274
elm, bastard, 72
elm, cedar, 275
elm, gray, 274, 276
elm, lime, 275
elm, moose, 276
elm, red, 275, 276
elm, scrub, 275
elm, slippery, 276
elm, soft, 274, 276
elm, southern red, 275
elm, Texas, 275
elm, water, 274
elm, white, 274
emajaguilla, 267
embuia, 191
emeri, 263
empress tree, 198
encino, 234
endivi, 164
enea, 189
enebro, 151
engan, 273
English oak, 232
epi, 271
epinette noire, 204
epinette rouge, 205
eprou, 186
escore, 164
espave, 20
espavel, 20
espinillo, 278
estoraque, 184
esu, 273
European ash, 122
European bird cherry, 222
European oak, 232
Eve's necklace, 253
evergreen-chinkapin, 81
evergreen holly, 147

F

false acacia, 237
false ash, 48
false banana, 27
false boxwood, 88
false-lignumvitae, 136
false-mamey, 55
false rosewood, 267
Fernambuco, 51
fetid buckeye, 11
fetid shrub, 27
fir, balsam, 1
fir, Douglas-, 224
fir, eastern, 1
fir, red, 224
fir, yellow Douglas-, 224
fireman's cap, 114
fish-poison tree, 213
fishfuddletree, 213
fishpoison-tree, Florida, 213
flamewood, 93
flor amarillo, 257
floresa, 220
Florida dogwood, 88
Florida fishpoison-tree, 213
Florida royal palm, 238
flowered satinwood, 80
flowering dogwood, 88
forked-leaf white oak, 230
frake, 264
framire, 263
freijo, 87
French ash, 122
French oak, 232
fromagier, 71

G

Gabon ebony, 102
gargoran, 245
garu buaja, 128
gateado, 30
gelam, 176
genuine mahogany, 254
ghost tree, 214
giant arborvitae, 268
giant chinkapin, 81
giant gum, 116
giant sequoia, 248
ginkgo, 125
gintungan, 40
gmelina, 127
gnulgu, 117
Gold Coast mahogany, 155
golden aspen, 219
golden chinkapin, 81
golden ebony, 101
golden walnut, Nigerian, 166
goldenleafed chinkapin, 81
gommier, 91
gommier blanc, 91

gommier montagne, 91
goncalo alves, 30
Gooding willow, 240
gopherwood, 83
gordonia, 129
granadillo, 45, 52, 98
grand bassam, 155
grandillo, 136
gray elm, 274, 276
gray oak, 233
Great Basin bristlecone pine, 207
great laurel magnolia, 169
great maple, 8
green ebony, 45
greenheart, 79
greenheart, Surinam, 257
Gregg acacia, 4
Gregg catclaw, 4
grenadillo, 96
grevillea, 131
grey birch, 37
gri-gri, 46
grignon, 46
guaiacum wood, 132
guanacaste, 113
guanandi, 55
guao, 178
guapinol, 146
guarea, 133
guarea, scented, 133
guarumo, 69
guatambu, 33
guayacan, 47, 52, 132, 257
guayacan plovillo, 257
gubas, 111
gulf-cypress, 259
gum, 161
gum, giant, 116
gum, sour, 188, 196
gum, star-leafed, 161
gum, stringy, 116
gum, swamp, 116
gum tree, yellow, 188
gum, tupelo, 188
gumbo-limbo, 49
gumhar, 127
gunwood, 150
gusi, 31

H

Haag tree, 222
habillo, 144
habin, 213
hackberry, 72
hackmatack, 159
hacktree, 72
hairy cercocarpus, 76
hairy sumac, 236
hard birch, 37
hard maple, 10
hardtack, 76
harewood, 8
hari-gara, 154
Hawaiian mahogany, 5
haya prieta, 195
hazel pine, 161
he-balsam, 205
he-huckleberry, 90
heaven, tree-of-, 15
heavenwood, 15
hedge apple, 167
hembra, 69
hickory, Carolina, 64
hickory, nutmeg, 63
hickory pine, 207, 211
hickory poplar, 162
hickory, red heart, 64
hickory, shagbark, 64
hickory, swamp, 63
hickory, upland, 64
hickory, white, 64
hobo, 251
hog plum, 251
holly, American, 147
holly-bay, 129
holly, Christmas, 147
holly, evergreen, 147
holly, prickly, 147
holly, white, 147
Honduras mahogany, 254
Honduras rosewood, 99
honey locust, 126
honey mesquite, 221
honey mesquite, western, 221
honey pod, 221
honeyshucks, 126
hoop ash, 72, 123
hophornbeam, 194
hophornbeam, eastern, 194
hornbeam, 194
hornbeam, American, 61
horseflesh, 173
horsetail tree, 67
huacamayo-chico, 241
hualo, 187
huayra caspi, 252
Humboldt redwood, 247
Hungarian ash, 122
Huon pine, 158
hura wood, 144
huynh, 141

I

ico, 61
idigbo, 263
ilang-ilang, 57
imbauba, 69
imbuia, 191
imbuia, canella, 191
imbuya, 191
immortelle, 114
impas, 156
incense-cedar, 54
incense-cedar, California, 54
incienso, 184
India teak, 261
Indian bear, 68
Indian (East) rosewood, 95
Indian (East) sandalwood, 242
Indian (East) satinwood, 80
Indian (East) walnut, 16, 262
Indian ebony, 101
Indian laurel, 262
Indian soap-plant, 243
Indian-walnut, 17
Indian (West) satinwood, 278
intermountain bristlecone pine, 207
ipe, 257
iphane, 140
ipil, 148
iroko, 181
ironwood, 61, 132, 193, 194
ironwood, Arizona-, 193
ironwood, black, 157
ironwood, desert-, 193
ironwood, red, 164
ironwood, Sonora, 193

ironwood, South-Sea, 67
ironwood, swamp, 90
Italian olive ash, 122
Ivory Coast mahogany, 155
ivory, pink, 35
ivory, red, 35
ivorywood, red, 35
ivy, 153
ivy, big-leaved, 153
ivy leaf laurel, 153
ivywood, 153
izombe, 265

J

jaboncillo, 243
jacaranda, 97
jacareuba, 55
Jamaica dogwood, 213
Jamaica plum, 251
Jamaican ebony, 45
Japanese ash, 124, 154
jarina, 113
jarrah, 115
jasmine, 27
jasminier, 27
jatoba, 146
Java cedar, 40
javillo, 144
jelutong, 107
jelutong bukit, 107
jelutong paya, 107
jenisero, 113
Jenny wood, 87
jobo, 251
jucaro, 46
Judas-tree, 75
June bud, 75
juniper, 152
juniper, Ashe, 151
juniper, eastern, 152
juniper, Mexican, 151
juniper, Virginia, 152
jutahy, 146
juvia, 36

K

kaede, 8
kaikumba, 171
kalingag, 82
kalmia, 153
kambala, 181
kanze, 141
kanzo, 141
kapa, 44
kapere, 271
kapoktree, 71
kapur, 106
kauri, 14
kauri-pine, 14
kauri, New Zealand, 14
kaurikopal, 14
kauvula, 111
kayu, 82
keladan, 106
kelobra, 113
kembang, 141
kempas, 156
Kentucky coffeetree, 137
keruntum, 165
kevazingo, 135
khaya, 155
kiboto, 182
kikko-chiku, 203
kilingi, 185
kingwood, 30, 92
koa, 5
koa-ka, 5
koka, 40
kokko, 16
kokrodua, 200
kolobra, 113
konigspalme, 238
korina, 264
kotibe, 186
koul, 175
krapa, 59
kssingang, 135
kukui, 17
kurahara, 55
kusia, 185
kusonoki, 82
kwabohoro, 133
kwila, 148

L

lacewood, 60, 131
lagarto caspi, 55
Lagos billetwood, 102
Lagos wood, 155
lamio, 105
lana, 189
lancewood, 56, 195
lancewood, red, 173
lanutan-bagio, 128
lapacho, 257
lapacho negro, 257
lara, 241
larch, 159
larch, Montana, 159
larch, mountain, 159
larch, western, 159
large flowered evergreen magnolia, 169
lauan, 249
laurel, American, 153
laurel, Australian, 110
laurel, bay-, 277
laurel, big, 169
laurel, black, 129
laurel, California, 277
laurel, California-, 277
laurel, Cape, 190
laurel, Indian, 262
laurel, ivy leaf, 153
laurel magnolia, great, 169
laurel, mountain-, 153
laurel, poison, 153
laurel, sheep, 153
laurel, swamp, 129
lauro negro, 86
lauro pardo, 86
lauro rosa, 22
Lawson cypress, 77
leadwood, 157
leatherwood, 90
leatherwood, southern, 90
leatherwood, swamp, 90
lebbek, 16
leche de Maria, 55
lechillo, 61
lemonwood, 56
lengue, 187
lentisco, 76
leopardwood, 42
letterwood, 42
lettok, 19
libuyu, 112
Lichtnussbaum, 17
life of man, 250
lignumvitae, 132

lignumvitae, false-, 136
lignumvitae, Marcaibo, 47
lilac, 255
lilac, Persian, 177
lilly-of-the-valley-tree, 196
limba, 264
lime, 270
lime elm, 275
lime tree, 270
linden, 270
linden, American, 270
liquidambar, 161
liso, pau, 33
live oak, 163, 234
live oak, sand, 234
live oak, Texas, 234
live oak, Virginia, 234
loblolly-bay, 129
locust, beaded, 253
locust, black, 237
locust, honey, 126
locust, pink, 253
locust, South American, 146
locust, sweet, 126
locust, thorny, 126
locust, white, 237
locust, yellow, 83, 237
lodgepole pine, 209
lodgepole pine, Rocky Mountain, 209
logwood, 138
loup, 105
louro, Brazialian, 22
Lovoa wood, 166
lumban, 17
lumbangtree, 17
lumbayan, 141
lumpa, 148
lumpho, 148
lusia, 185

M

Macassar ebony, 101
macaya, 21
madera negra, 257
madrona, 25
madrone, 25
madrone, coast, 25
madrone, Pacific, 25
madrono, 25
magnolia, cucumber, 168
magnolia, great laurel, 169
magnolia, large flowered evergreen, 169
magnolia, mountain, 168
magnolia, southern, 169
mahoe, 143
mahoe, blue, 143
mahoe, mountain, 143
mahoe, seaside, 143, 267
mahogany, African, 155
mahogany, American, 137, 254
mahogany bean, 13
mahogany, Benin, 133
mahogany, bigleaf, 254
mahogany birch, 38
mahogany, Brazilian, 215
mahogany, cherry, 269
mahogany, genuine, 254
mahogany, Gold Coast, 155
mahogany, Hawaiian, 5
mahogany, Honduras, 254
mahogany, Ivory Coast, 155
mahogany, mountain-, 76
mahogany, Nigerian, 155
mahogany, Philippine, 249
mahogany, pink, 133
mahogany, pod-, 13
mahogany, rose, 74
mahogany, Santos, 184
mahogany, sapele, 112
mahogany, scented, 112
mahogany, true, 254
mahogany, Western Australian, 115
mahogany, white, 256
mai pradoo, 227
maidenhair tree, 125
maidon, 225
makarati, 48
makore, 269
malabar, 95
malaleuca, 176
manako, 172
manga, 172
mangabe, 245
mangle colorado, 235
mango, 172
mangot, 172
mangrove, 235
mangrove, red, 235
mangue, 172
Manitoba maple, 7
mansonia, 175
manuka, 160
manzanita, 26
manzanita, whiteleaf, 26
mao zhu, 203
maple, African, 272
maple-ash, 7
maple, ash leafed, 7
maple, Australian, 120
maple, black, 10
maple, bush, 272
maple, cut leafed, 7
maple, great, 8
maple, hard, 10
maple, Manitoba, 7
maple, negundo, 7
maple, Queensland, 120
maple, red, 9
maple, Red River, 7
maple, rock, 10
maple, scarlet, 9
maple silkwood, 120
maple, soft, 9
maple, sugar, 10
maple, swamp, 9
maple, sweet, 10
maple, sycamore, 8
maple, three leafed, 7
maple, water, 9
Maracaibo ebony, 52
Maracaibo lignumvitae, 47
maranon, 20
Maria, 55
masarakana, 118
Mayday tree, 222
mazabalo, 59
mbenge, 134
melabau, 165
melawis, 128
melima, 141
mengkulang, 141
mengris, 156
Menzies spruce, 206
meranti, 249
merbau, 148
mesquite, 221

mesquite, common, 221
mesquite, honey, 221
mesquite, velvet, 221
mesquite, western honey, 221
Mexican juniper, 151
mgando, 48
miguelario, 100
milky pine, 19
milkwood, 19
milo, 267
mimosa, 241
mimosa, Texas, 4
mist tree, 89
mkarati, 48
mkehli, 13
moabi, 32
moca, 21
mockorange, 167
mombin, 251
monkey-pod, 241
Montana larch, 159
moose elm, 276
mopane, 140
mopanie, 140
Moreton Bay chestnut, 66
morototon, 245
moso bamboo, 203
Moulmein teak, 261
mountain ash, 116
mountain-ash, 250
mountain-ash, American, 250
mountain aspen, 219
mountain cedar, 151
mountain larch, 159
mountain-laurel, 153
mountain magnolia, 168
mountain mahoe, 143
mountain-mahogany, 76
mountain pine, 211
mountain red oak, 233
mountain silverbell, 139
mountain sumac, 250
mousou-chiku, 203
movingui, 104
Mozambique ebony, 96
mpingo, 96
msangala, 48
mse, 12
mshenzi, 41
mtolo, 118
mubuubu, 41
muenge, 228
muermo, 117
muhugu, 41
muhuhu, 41
muirapiranga, 43
mukarati, 48
mukshi, 31
mukusi, 31
mulberry, black, 183
mulberry, paper-, 44
mulberry, red, 183
mulberry, silkworm, 183
mulga, 2
munhiti, 118
mura, 30
muramo, 220
Murry pine, 53
Murry River cypress, 53
Murry River pine, 53
musaru, 140
muscarala, 48
musclewood, 61
musenene, 12
mushunga, 12
mutenye, 134
mutivoti, 118
mutomboti, 118
muyovu, 112
mvule, 181
mvumvo, 41
mwani, 140
myrtle, Oregon-, 277
myrtle, Pacific-, 277

N

nakora, 154
nambar, 98
nancito, 145
naranjo chino, 167
narra, 226
narra, red, 225
narra, yellow, 225
negundo maple, 7
nettle tree, 72
New Guinea rosewood, 226
New Guinea walnut, 105
New Guineawood, 105
New Zealand kauri, 14
New Zealand tea tree, 160
nhoi, 40
Nicaragua rosewood, 98
Nigerian ebony, 102
Nigerian golden walnut, 166
Nigerian mahogany, 155
Nigerian satinwood, 104
Nigerian teak, 181
nimwood, 177
nire, 187
nispero, 174
njabi, 32
Nookta cypress, 78
Norfolk-Island-pine, 24
northern catalpa, 68
northern red oak, 233
northern silky oak, 60
northern white pine, 212
noyer, 278
noyer de bancoul, 17
noyer des Moloques, 17
noz da India, 17
nutmeg hickory, 63

O

o-heh-yah-tah, 65
oak, African, 164, 181
oak, Arkansas, 231
oak, Arkansas water, 231
oak, Australian, 116
oak, Australian silky, 60
oak, bog, 232
oak, California chestnut, 163
oak, chestnut, 163
oak, durmast, 232
oak, eastern red, 233
oak, eastern white, 230
oak, English, 232
oak, European, 232
oak, forked-leaf white, 230
oak, French, 232
oak, gray, 233
oak, live, 163, 234
oak, mountain red, 233
oak, northern red, 233
oak, northern silky, 60
oak, peach, 163
oak, peduncalate, 232
oak, pollard, 232
oak, ridge white, 230

oak, sand live, 234
oak, sessile, 232
oak, she-, 67
oak, silky, 60, 131
oak, silky, northern, 60
oak, silky, southern, 131
oak, southern silky, 131
oak, stave, 230
oak, tan, 163
oak, tanbark-, 163
oak, Tasmanian, 116
oak, Texas live, 234
oak, Virginia live, 234
oak, water, 231
oak, white, 230
obeche, 272
obobo, 133
oboto, 171
obuina, 97
ocote, 208
odum, 181
ofram, 264
ofun, 175
ohez, 82
ohia, 179
Ohio buckeye, 11
oilnut, 149
okwen, 180
olbaum, 192
oleaster, 109
oleo vermelho, 184
olive, 192
olive ash, Italian, 122
olive, black-, 46
olive, Russian-, 109
olive, wild, 139
olivier, 192
olivo, 192
olmo, 275
ologbomodu, 171
opepe, 185
opossum-wood, 139
Oregon alder, 18
Oregon-cedar, 77
Oregon-myrtle, 277
Oregon pine, 224
orejero, 113
oriental wood, 110
oro, 186
oroko, 181
Osage-orange, 167
osan, 181
oscuro, 55
otaheita, 267
otoba, 100
otutu, 186
ovoue, 186
oxhorn bucida, 46
oysterwood, 136
oysterwood, Cuban, 136
Ozark white cedar, 151

P

Pacific madrone, 25
Pacific-myrtle, 277
Pacific yew, 260
padauk, 225
padauk, African, 228
padauk, Andaman, 225
padauk, Burma, 227
paldao, 105
Palimira alstonia, 19
palisandro, 98
palisandro de Honduras, 99
pallissandre, 182
palm, cabbage-, 239
palm, Florida royal, 238
palm, royal, 238
palm, royal Cuban, 238
palma real de Cuba, 238
palmeir real, 238
palmetto, 239
palmetto, cabbage, 239
palmetto, Carolina, 239
palmetto, common, 239
palmier royal, 238
palo barranco, 61
palo blanco, 243, 256
palo colorado, 90
palo de hierro, 193
palo de sangre, 43
palo de tinta, 138
palo fierro, 108
palo rosa, 28
palo santo, 132
palo silo, 61
palta, 201
pana cimarrona, 245
Panama, 252
paniala, 40
paper-bark, 176
paper birch, 39
paper-mulberry, 44
paperbark, broad-leaved, 176
Para rubbertree, 142
paradise flower, 4
paradise-tree, 15
paraiso, 177
Parana-pine, 23
partridge wood, 21, 52
pau Brasil, 51
pau de balsa, 189
pau ferro, 199
pau liso, 33
pau marfim, 33
pau preto, 97
pau rainha, 43
pau rosa, 22, 94
pau roxo, 199
paulownia, royal, 198
pava, 245
pavo, 245
pawpaw, 27
payoong, 93
payung, 93
pea-shrub, 58
pea-shrub, Siberian, 58
pea-tree, 58
pea-tree, Siberian, 58
peach oak, 163
pear, alligator-, 201
pear, common, 229
pear tree, 229
pear tree, wild, 188
pearwood, 229
pearwood, African, 32
pecan, 62, 63
pecan nut, 62
pecanier, 62
pedunculate oak, 232
pelawan, 165
pencil cedar, 152
penda, 256
penkwa, 112
pepperidge, 188
pepperwood, 277
perfume tree, 57
Pernambuco, 51
peroba, pink, 28
peroba, red, 28

peroba rosa, 28
perota, 113
persea, 201
Persian box, 50
Persian lilac, 177
persimmon, 103
persimmon, common, 103
peulmahonia, 13
peuplier faux-tremble, 219
pheasantwood, 21
Philippine mahogany, 249
pianowood, 97
pilon, 145
pin blanc, 212
pin tordu, 209
pine ancient, 207
pine, Australian, 67
pine, bristlecone, 207
pine, black, 209
pine, California sugar, 210
pine, Caribbean, 208
pine, Colorado bristlecone, 207
pine, cork, 212
pine, cypress-, 53
pine, eastern white, 212
pine, Great Basin bristlecone, 207
pine, hazel, 161
pine, hickory, 207, 211
pine, Huon, 158
pine, intermountain bristlecone, 207
pine, kauri-, 14
pine, lodgepole, 209
pine, milky, 19
pine, mountain, 211
pine, Murray, 53
pine, Murray River, 53
pine, Norfolk-Island-, 24
pine, northern white, 212
pine, Oregon, 224
pine, Parana, 23
pine, prickly, 211
pine, Puget Sound, 224
pine, pumpkin, 212
pine, Rocky Mountain bristlecone, 207
pine, Rocky Mountain lodgepole, 209
pine, sugar, 210
pine, Table Mountain, 211
pine, western bristlecone, 207
pine, Weymouth, 212
pine, white cypress, 53
pink African cedar, 133
pink ivory, 35
pink locust, 253
pink mahogany, 133
pink peroba, 28
pink sophora, 253
pink tree, 176
pinkwood, Brazilian, 94
pino, 208
pino blanco, 23
pino parana, 23
pintree, Confederate, 126
piqua, 133
plane, sycamore, 8
plane tree, 8
plane tree, American, 214
plaqueminier, 103
plum, date, 103
plum, hog, 251
plum, Jamaica, 251
plum, Spanish, 251
plum, Yucatan, 251
po'a'aha, 44
pochota, 71
pod-mahogany, 13
podo, 12
poison laurel, 153
poison wood, 178
poison wood, black, 178
poisonwood, 136
polladro, 232
pollard oak, 232
pond-cypress, 259
poplar, balsam, 217
poplar, black, 217
poplar, Carolina, 218
poplar, hickory, 162
poplar, trembling, 219
poplar, tulip, 162
poplar, yellow-, 162
popple, 219
porcupinewood, 73
Port-Orford-cedar, 77
Port-Orford white-cedar, 77
portiatree, 267
possumwood, 103, 144
post-cedar, 151
poui, 257
pradoo, 227
prickly holly, 147
prickly pine, 211
pride-of-India, 177
primavera, 256
primavera, African, 272
princesstree, 198
Puget Sound pine, 224
pui, 257
pulai, 19
pumpkin pine, 212
punktree, 176
purga, 235
purpleheart, 199
putumuju, 73

Q

quaking aspen, 219
quebracho, 246
quebracho macho, 246
quebracho moro, 246
Queensland maple, 120
Queensland walnut, 34, 110
quiverleaf aspen, 219

R

raintree, 241
rakuda, 144
ramin, 128
ramon, 76
Rangoon teak, 261
rattlebox, 139
rauli, 187
red alder, 18
red beech, 119
red birch, 38
red cabbage bark, 21
red cedar, Australian, 271
red-cypress, 259
red-cypress, Tidewater, 259
red elm, 275, 276
red fir, 224
red heart hickory, 64
red ironwood, 164
red ivory, 35
red ivorywood, 35

red lancewood, 173
red mangrove, 235
red maple, 9
red mulberry, 183
red narra, 225
red oak, eastern, 233
red oak, mountain, 233
red oak, northern, 233
red peroba, 28
Red River maple, 7
red sassafras, 244
red spruce, 205, 224
red syringa, 48
red tea tree, 160
red titi, 90
redbark, 200
redbay, 202
redbay, smooth, 202
redbay, swamp, 202
redbud, American, 75
redbud, eastern, 75
redcedar, eastern, 152
redcedar, western, 268
redgum, 161
redwood, 247
redwood, Andaman, 225
redwood, Brazil, 43
redwood, California, 247
redwood, coast, 247
redwood, Humboldt, 247
redwood, Sierra, 248
redwood, Zambesi, 31
Rhodesian teak, 31
Rhodesiese kiaat, 31
ridge white oak, 230
roble, 187, 256
roble de Chile, 117
roble ruili, 187
rock-cedar, 151
rock maple, 10
Rocky Mountain bristlecone pine, 207
Rocky Mountain lodgepole pine, 209
rode kabbes, 21
rone, 265
rooisering, 48
rose mahogany, 74
rosewood, African, 135
rosewood, Brazilian, 97
rosewood, Burma, 226
rosewood, East Indian, 95
rosewood, false, 267
rosewood, Honduras, 99
rosewood, New Guinea, 226
rosewood, Nicaragua, 98
rosewood of Seychelles, 267
rosewood of Siam, 93
rosewood, Thailand, 93
roundwood, 250
rowan berry, 250
rowantree, American, 250
royal palm, 238
royal palm, Cuban, 238
royal palm, Florida, 238
royal paulownia, 198
ru, 67
rubbertree, 142
rubbertree, Para, 142
rubberwood, 142
Ruk-attana, 19
rum cherry, 223
Russian-olive, 109

S

saba, 100
sabino, 151
sain, 262
saj, 262
samaguare, 241
saman, 241
samba, 272
sand live oak, 234
sandalwood, 242
sandalwood, African, 118
sandalwood, East Indian, 242
sandalwood, yellow, 242
sandarac tree, 266
sandbox tree, 144
Santa-Maria, 55
Santos mahogany, 184
sapta, 12
sapele, 112
sapele mahogany, 112
sapgum, 161
sapin baumier, 1
sapodilla, 173, 174
sapote, 174
sassafras, 244
sassafras, red, 244
satin walnut, 161
satine, 43
satine rubane, 43
satinwood, African, 200, 273
satinwood, Benin, 200
satinwood, Ceylon, 80
satinwood, East Indian, 80
satinwood, flowered, 80
satinwood, Nigerian, 104
satinwood, scented, 74
satinwood, soft, 272
satinwood, West Indian, 278
satinwood, yellow, 200
saunders, white, 242
saunders, yellow, 242
sauz, 240
saxifrax-tree, 244
scarlet maple, 9
scented guarea, 133
scented mahogany, 112
scented satinwood, 74
scholar-tree, 19
scrub elm, 275
seaside mahoe, 143, 267
sebo, 100
sein, 262
selano, 60
selunsur, 165
sempervirens, 247
sen, 154
sendok sendok, 111
Senegal ebony, 96
sepira, 79
sequoia, giant, 248
seraya, 249
sericote, 85
service tree, 250
sesendok, 111
sessile oak, 232
shagbark hickory, 64
shantsi, 140
she-oak, 67
sheep laurel 153
shell bark, 64
shinglewood, 268
shioji, 124
shisham, 95
shorebay, 202
Siberian pea-shrub, 58
Siberian pea-tree, 58

Sierra redwood, 248
siete pisos, 24
silk-cottontree, 71
silk-cottonwood, white, 71
silkbay, 202
silkwood, 120
silkwood, maple, 120
silkworm mulberry, 183
silky oak, 60, 131
silky oak, Australian, 60
silky oak, northern, 60
silky oak, southern, 131
silver birch, 37, 39
silverbell, Carolina, 139
silverbell, mountain, 139
simmon, 103
sipiroe, 79
siris, 16
Sitka cypress, 78
Sitka spruce, 206
sitsal, 95
slippery elm, 276
smallseed white ash, 121
smoketree, 89
smoketree, American, 89
smooth redbay, 202
snakewood, 42
snowdrop tree, 139
soap-plant, Indian, 243
soapberry, western, 243
soeren, 271
soft elm, 274, 276
soft maple, 9
soft satinwood, 272
songo, 273
Sonora ironwood, 193
sophora, Texas, 253
sophore, pink, 253
sorrel-tree, 196
sour gum, 188, 196
sourwood, 196
South American locust, 146
South American walnut, 87
South-Sea ironwood, 67
southern cottonwood, 218
southern-cypress, 259
southern leatherwood, 90
southern magnolia, 169
southern red elm, 275
southern silky oak, 131
Spanish-cedar, 70
Spanish cork, 267
Spanish plum, 251
Spanish walnut, 220
spearwood, 2
specklewood, 42
spice-tree, 277
spoonwood, 153
spruce, black, 204
spruce, bog, 204
spruce, coast, 206
spruce, Douglas, 224
spruce, eastern, 204, 205
spruce, Menzies, 206
spruce, red, 205, 224
spruce, Sitka, 206
spruce, swamp, 204
spruce, tideland, 206
spruce, yellow, 205, 206
staghorn sumac, 236
star-leafed gum, 161
stave oak, 230
stinkhout, 190
stinking buckeye, 11
stinking chun, 15
stinkwood, 188, 190
stone beech, 119
stringy gum, 116
stump tree, 137
sugar maple, 10
sugar pine, 210
sugar pine, California, 210
sugarberry tree, 72
sumac, 236
sumac, American, 236
sumac, Chinese-, 15
sumac, coral, 178
sumac, hairy, 236
sumac, mountain, 250
sumac, staghorn, 236
sumac, velvet, 236
sumauma, 71
sunsun, 252
suradan, 145
Surinam greenheart, 257
surra, 67
swamp ash, 123
swamp birch, 37
swamp cabbage, 239
swamp-cypress, 259
swamp cyrilla, 90
swamp gum, 116
swamp hickory, 63
swamp ironwood, 90
swamp laurel, 129
swamp leatherwood, 90
swamp maple, 9
swamp redbay, 202
swamp spruce, 204
swamp willow, 240
sweet birch, 38
sweet chestnut, 65
sweet locust, 126
sweet maple, 10
sweetbay, 202
sweetbrush, 76
sweetgum, 161
sycamore, 8, 214
sycamore maple, 8
sycamore plane, 8
syringa, red 48

T

tabanuco, 91
tabasara, 220
Table Mountain pine, 211
tabonuco, 91
tacamahac, 217
tachuari, 257
tacuna, 69
taito, 220
tamarack, 159
tamarack, western, 159
tamarin, 258
tamarind, 258
tamarindo, 258
tambooti, 118
tambotie, 118
tambouti, 118
tamerind, 258
tami, 189
tamo, 124
tan bay, 129
tan-oak, 163
tanbark-oak, 163
tanoak, 163
tapa, 44
tapana, 145
tascate, 151
Tasmanian blackwood, 6

Tasmanian oak, 116
tat-talun, 148
taukkyan, 262
tawa, 34
tayokthe, 40
tea tree, 160
tea tree, Australian, 160
tea tree, broom, 160
tea tree, New Zealand, 160
tea tree, red, 160
teak, 261
teak, African, 181, 263
teak, African-, 200
teak, Burma, 261
teak, India, 261
teak, Moulmein, 261
teak, Nigerian, 181
teak, Rangoon, 261
teak, Rhodesian, 31
teraling, 141
terbulan, 111
term, 40
tesota, 193
tetoka, 36
Texas cedar, 151
Texas ebony, 108
Texas elm, 275
Texas live oak, 234
Texas mimosa, 4
Texas sophora, 253
Thailand rosewood, 93
thitkado, 271
thorny acacia, 126
thorny locust, 126
three leafed maple, 7
three-thorned acacia, 126
thuja geant, 268
thujawood, 266
thuya, 266
thuya burl, 266
thyine wood, 266
tideland spruce, 206
Tidewater red-cypress, 259
tigerwood, 30, 166
tinto, 138
tiss-wood, 139
titi, 90
titi, black, 90
titi, red, 90
titi, white, 90
tocary, 36
tokiwakaede, 8
tola, 130
tola branco, 130
toog, 40
toon, 271
toon, Australian, 271
torchwood, 45
torito, 145
tortoise-shell bamboo, 203
tortoiseshell wood, 42
totara, 216
trac, 93
tracwood, 93
tree-of-heaven, 15
trembling aspen, 219
trembling poplar, 219
treme, 263
trementino, 220
trompillo, 145
trompy, 69
tropical American ebony, 45
true mahogany, 254
trumpet-tree, 69
trumpet-wood, 69
tsanya, 186
tsomvori, 118
tuai, 40
tualang, 156
tule moreira, 181
tulip poplar, 162
tuliptree, 162
tulipwood, 94
tun, 271
tung, 17
tupelo, 188
tupelo, black, 188
tupelo gum, 188
Turkey ash, 122
Turkey boxwood, 150
turpentine tree, 49
tutury, 36

U

ubande, 118
ubatan, 29
ucar, 46
ulmo, 117
umbrella tree, 267
umbrella tree, Chinese, 177
umgoloti, 35
umgusi, 31
umkusu, 31
umnini, 35
umnondo, 48
umnukane, 190
umthombothi, 118
una-de-gato, 4
upland hickory, 64
urunday, 29
urundel, 29

V

varilla, 55
varnishtree, 17
velvet mesquite, 221
velvet sumac, 236
verawood, 47
vermilion wood, 225
vermillion, 228
vesi, 148
vinegar tree, 236
vinhatico, 215
violete, 92
violetwood, 92, 199
virgilia, 83
Virginia juniper, 152
Virginia live oak, 234
virola, 100
vuga, 179

W

wait-a-bit, 4
walnut, African, 166
walnut, African black, 150
walnut, American, 150
walnut, Australian, 110
walnut bean, 110
walnut, Benin, 166
walnut, black, 150
walnut, Brazilian, 191
walnut, East Indian, 16, 262
walnut, eastern black, 150
walnut, Indian-, 17
walnut, New Guinea, 105
walnut, Nigerian golden, 166
walnut, Queensland, 34, 110
walnut, satin 161
walnut, South American, 87
walnut, Spanish, 220

walnut, white, 149
wansenwa, 273
water beech, 61, 214
water elm, 274
water maple, 9
water oak, 231
water oak, Arkansas, 231
watho, 41
wattle, Darwin black, 3
wattle, ear-pod, 3
wauke, 44
wawa, 272
Wellingtonia, 248
wenge, 182
West Indian satinwood, 278
western alder, 18
Western Australian mahogany, 115
western black willow, 240
western bristlecone pine, 207
western chinkapin, 81
western cypress, 53
western honey mesquite, 221
western larch, 159
western redcedar, 268
western soapberry, 243
western tamarack, 159
western yew, 260
Weymouth pine, 212
white ash, 121
white ash, small seed, 121
white beech, 119
white birch, 39
white cedar, 177
white cheesewood, 19
white cornel, 88
white-cypress, 259
white cypress pine, 53
white elm, 274
white hickory, 64
white holly, 147
white locust, 237
white mahogany, 256
white oak, 230
white oak, eastern, 230
white oak, forked-leaf, 230
white oak, ridge, 230
white pine, eastern, 212
white pine, northern, 212
white-poplar, 162
white saunders, 242
white silk-cottontree, 71
white titi, 90
white walnut, 149
white wood, American, 270
whiteleaf manzanita, 26
whitewood, 162
whitewood, African, 272
wild black cherry, 223
wild cherry, 223
wild chinatree, 243
wild dilly, 173
wild olive, 139
wild pear tree, 188
willow, black, 240
willow, Dudley, 240
willow, Gooding, 240
willow, swamp, 240
willow, western black, 240
wine tree, 250
winter beech, 119
wiri, 12
woman's tongue, 16

Y

yachidamo, 124
yagrumbo, 69
yagrume, 245
yagrumo, 69
yagrumo macho, 245
yarumo, 245
yaya, 195
ye-padauk, 40
yellow ash, 83
yellow birch, 37
yellow cedar, 151
yellow-cedar, 78
yellow-cedar, Alaska, 78
yellow cypress, 78
yellow-cypress, 259
yellow Douglas-fir, 224
yellow gum tree, 188
yellow locust, 83, 237
yellow narra, 225
yellow-poplar, 162
yellow sandalwood, 242
yellow sanders, 278
yellow satinwood, 200
yellow saunders, 242
yellow spruce, 205, 206
yellow wood, 278
yellowheart, 278
yellowwood, 12, 83, 89, 215
yellowwood, American, 83
yellowwood, East African, 12
yemane, 127
yew, 260
yew, Douglas, 224
yew, Pacific, 260
yew, western, 260
yin-hsing, 125
ylang-ylang, 57
ylang-ylang-tree, 57
ylang-ylangbaum, 57
yomo, 228
youhom, 271
Yucatan plum, 251

Z

Zambesi redwood, 31
zapotillo, 174
zebrano, 180
zebrawood, 180
zingana, 180
ziricote, 85, 86
zorrowood, 30

FAMILY NAME INDEX

This index contains the names of all plant families represented by species in this book. Each plant family is listed alphabetically by its scientific name and a common family name or names related to it. Then, for each family, the species members in this book are shown by preferred common name, Wood Data Sheet number and alphabetically by scientific name.

Family Scientific and Common Name

Common Name	WDS No.	Scientific Name
Abietaceae, see Pinaceae, the pine family		
Asteraceae, see Compositae, the composite or daisy family		
Aceraceae, the maple family		
boxelder	7	*Acer negundo*
sycamore maple	8	*Acer pseudoplatanus*
red maple	9	*Acer rubrum*
sugar maple	10	*Acer saccharum*
Anacardiaceae, the cashew family		
espave	20	*Anacardium excelsum*
urunday	29	*Astronium balansae*
goncalo alves	30	*Astronium fraxinifolium*
American smoketree	89	*Cotinus obovatus*
paldao	105	*Dracontomelon dao*
mango	172	*Mangifera indica*
chechem	178	*Metopium brownei*
staghorn sumac	236	*Rhus typhina*
quebracho	246	*Schinopsis quebracho-colorado*
jobo	251	*Spondias mombin*
Annonaceae, the annonia or custard-apple family		
pawpaw	27	*Asimina triloba*
ylang-ylang	57	*Cananga odorata*
lancewood	195	*Oxandra lanceolata*
Apocynaceae, the dogbane family		
white cheesewood	19	*Alstonia scholaris*
pink peroba	28	*Aspidosperma polyneuron*
jelutong	107	*Dyera costulata*
Aquifoliaceae, the holly family		
American holly	147	*Ilex opaca*
Araliaceae, the ginseng family		
sen	154	*Kalopanax septemlobus*
yagrumo macho	245	*Schefflera morototoni*
Araucariaceae, the araucaria family		
kauri	14	*Agathis australis*
Parana-pine	23	*Araucaria angustifolia*
Norfolk-Island-pine	24	*Araucaria heterophylla*

Arecaceae, see Palmae, the palm family

Asteraceae, see Compositae, the composite family

Betulaceae, the birch family, includes Carpinaceae and Corylaceae

red alder 18 *Alnus rubra*
yellow birch 37 *Betula alleghaniensis*
sweet birch 38 *Betula lenta*
white birch 39 *Betula papyrifera*
American hornbeam 61 *Carpinus caroliniana*
eastern hophornbeam 194 *Ostrya virginiana*

Bignoniaceae, the bignonia or trumpet creeper family

northern catalpa 68 *Catalpa speciosa*
primavera 256 *Tabebuia donnell-smithii*
ipe 257 *Tabebuia* spp.

Bombacaceae, the bombax family

ceiba 71 *Ceiba pentandra*
balsa 189 *Ochroma pyramidale*

Boraginaceae, the borage family

sericote 85 *Cordia dodecandra*
canalete 86 *Cordia gerascanthus*
freijo 87 *Cordia goeldiana*

Burseraceae, the bursera or torchwood family

gumbo-limbo 49 *Bursera simaruba*
gommier 91 *Dacryodes excelsa*

Buxaceae, the box family

boxwood 50 *Buxus sempervirens*

Caesalpiniaceae, see Fabaceae or Leguminosae, the legume family

Carpinaceae, see Betulaceae, the birch family

Casuarinaceae, the casuarina or beefwood family

casuarina 67 *Casuarina* spp.

Cecropiaceae, the cecropia family

trumpet-tree 69 *Cecropia peltata*

Clusiaceae, see Guttiferae, the mangosteen family

Combretaceae, the combretum family

jucaro 46 *Bucidia buceras*
Indian laurel 262 *Terminalia elliptica*
idigbo 263 *Terminalia ivorensis*
limba 264 *Terminalia superba*

Compositae, the composite or daisy family; also Asteraceae

muhuhu 41 *Brachylaena huillensis*

Cornaceae, the dogwood family

flowering dogwood 88 *Cornus florida*
blackgum 188 *Nyssa sylvatica*

Corylaceae, see Betulaceae, the birch family

Cunionaceae, the cunonia family

coachwood 74 *Ceratopetalum apetalum*

Cupressaceae, the cypress family

white cypress pine 53 *Callitris glaucophylla*
incense-cedar 54 *Calocedrus decurrens*
Port-Orford-cedar 77 *Chamaecyparis lawsoniana*
Alaska-cedar 78 *Chamaecyparis nootkatensis*
Ashe juniper 151 *Juniperus ashei*
eastern redcedar 152 *Juniperus virginiana*

thuya	266	*Tetraclinis articulata*
western redcedar	268	*Thuja plicata*
Cyrillaceae, the cryilla family		
swamp cyrilla	90	*Cyrilla racemiflora*
Dipterocarpaceae, the dipterocarp family		
kapur	106	*Dryobalanops aromatica*
lauan	249	*Shorea* spp.
Ebenaceae, the ebony family		
Macassar ebony	101	*Diospyros celebica*
Gabon ebony	102	*Diospyros dendo*
common persimmon	103	*Diospyros virginiana*
Elaeaganaceae, the elaeaganus or oleaster family		
Russian-olive	109	*Elaeagnus angustifolia*
Ericaceae, the heath family		
Pacific madrone	25	*Arbutus menziesii*
whiteleaf manzanita	26	*Arctostaphylos viscida*
mountain-laurel	153	*Kalmia latifolia*
sourwood	196	*Oxydendrum arboreum*
Eucryphiaceae, the eucryphia family		
ulmo	117	*Eucryphia cordifolia*
Euphorbiaceae, the spurge family		
kukui	17	*Aleurites moluccana*
bishopwood	40	*Bischofia javanica*
kauvula	111	*Endospermum macrophyllum*
tamboti	118	*Excoecaria africana*
oysterwood	136	*Gymnanthes lucida*
rubbertree	142	*Hevea brasiliensis*
sandbox tree	144	*Hura crepitans*
pilon	145	*Hyeronima alchorneoides*
Fabaceae or Leguminosae, the legume family; (*Caesalpiniaceae*) the cassia group		
chanfuta	13	*Afzelia quanzensis*
Rhodesian teak	31	*Baikiaea plurijuga*
makarati	48	*Burkea africana*
Pernambuco	51	*Caesalpinia echinata*
partridge wood	52	*Caesalpinia granadillo*
eastern redbud	75	*Cercis canadensis*
mopane	140	*Colophospermum mopane*
ayan	104	*Distemonanthus benthamianus*
honey locust	126	*Gleditsia triacanthos*
agba	130	*Gossweilerodendron balsamiferum*
benge	134	*Guibourtia arnoldiana*
bubinga	135	*Guibourtia tessmannii*
Kentucky coffeetree	137	*Gymnocladus dioicus*
logwood	138	*Haematoxylum campechianum*
mopane	140	*Hardwickia mopane*
courbaril	146	*Hymenaea courbaril*
merbau	148	*Intsia* spp.
kempas	156	*Koompassia malaccensis*
zebrawood	180	*Microberlinia brazzavillensis*
purpleheart	199	*Peltogyne paniculata*
cativo	220	*Prioria copaifera*
tamerind	258	*Tamarindus indica*

Fabaceae or Leguminosae, the legume family; (*Mimosaceae*) the mimosa group

mulga	2	*Acacia aneura*
ear-pod wattle	3	*Acacia auriculiformis*
Gregg acacia	4	*Acacia greggii*
koa	5	*Acacia koa*
Australian blackwood	6	*Acacia melanoxylon*
lebbek	16	*Albizia lebbeck*
ebony blackbead	108	*Ebenopsis ebano*
guanacaste	113	*Enterolobium cyclocarpum*
angico	197	*Parapiptadenia rigida*
vinhatico	215	*Plathymenia reticulata*
honey mesquite	221	*Prosopis glandulosa*
monkey-pod	241	*Samanea saman*

Fabaceae or Leguminosae, the legume family; (*Papilionaceae*) the pea or pulse group

partridge wood	21	*Andira inermis*
cocuswood	45	*Brya ebenus*
caragana	58	*Caragana arborescens*
blackbean	66	*Castanospermum australe*
canarywood	73	*Centrolobium ochroxylon*
canarywood	73	*Centrolobium orinocense*
canarywood	73	*Centrolobium paraense*
canarywood	73	*Centrolobium robustum*
canarywood	73	*Centrolobium tomentosum*
yellowwood	83	*Cladrastis lutea*
kingwood	92	*Dalbergia cearensis*
flamewood	93	*Dalbergia cochinchinensis*
tulipwood	94	*Dalbergia decipularis*
East Indian rosewood	95	*Dalbergia latifolia*
African blackwood	96	*Dalbergia melanoxylon*
Brazilian rosewood	97	*Dalbergia nigra*
cocobolo	98	*Dalbergia retusa*
Honduras rosewood	99	*Dalbergia stevensonii*
cockspur coralbean	114	*Erythrina crista-galli*
wenge	182	*Millettia laurentii*
balsamo	184	*Myroxylon balsamum*
desert-ironwood	193	*Olneya tesota*
afrormosia	200	*Pericopsis elata*
Florida fishpoison-tree	213	*Piscidia piscipula*
Andaman padauk	225	*Pterocarpus dalbergoides*
narra	226	*Pterocarpus indicus*
Burma padauk	227	*Pterocarpus macrocarpus*
African padauk	228	*Pterocarpus soyauxii*
black locust	237	*Robinia pseudoacacia*
Texas sophora	253	*Styphnolobium affine*

Fagaceae, the beech family

American chestnut	65	*Castanea dentata*
giant chinkapin	81	*Chrysolepis chrysophylla*
American beech	119	*Fagus grandifolia*
tanoak	163	*Lithocarpus densiflorus*
white oak	230	*Quercus alba*
Arkansas oak	231	*Quercus arkansana*
English oak	232	*Quercus robur*

northern red oak	233	*Quercus rubra*
live oak	234	*Quercus virginiana*
Ginkgoaceae, the ginkgo family		
ginkgo	125	*Ginkgo biloba*
Gramineae, the grass family; also Poaceae		
moso bamboo	203	*Phyllostachys edulis*
Guttiferae, the mangosteen family; also Clusiaceae		
Maria	55	*Calophyllum brasiliense*
oboto	171	*Mammea africana*
Hamamelidaceae, the witch-hazel family		
sweetgum	161	*Liquidambar styraciflua*
Hippocastanaceae, the horsechestnut family		
Ohio buckeye	11	*Aesculus glabra*
Juglandaceae, the walnut family		
pecan	62	*Carya illinoinensis*
nutmeg hickory	63	*Carya myristiciformis*
shagbark hickory	64	*Carya ovata*
butternut	149	*Juglans cinerea*
black walnut	150	*Juglans nigra*
Lamiaceae, the mint family		
gmelina	127	*Gmelina arborea*
teak	261	*Tectona grandis*
Lauraceae, the laurel family		
pau rosa	22	*Aniba rosaeodora*
tawa	34	*Beilschmiedia tawa*
greenheart	79	*Chlorocardium rodiei*
camphor-tree	82	*Cinnamomum camphora*
Queensland walnut	110	*Endiandra palmerstonii*
stinkwood	190	*Ocotea bullata*
imbuya	191	*Ocotea porosa*
avocado	201	*Persea americana*
redbay	202	*Persea borbonia*
sassafras	244	*Sassafras albidum*
California laurel	277	*Umbellularia californica*
Lecythidaceae, the Brazilnut family		
Brazilnut	36	*Bertholletia excelsa*
Leguminosae, the legume family, see Fabaceae/Leguminosae		
Magnoliaceae, the magnolia family		
yellow-poplar	162	*Liriodendron tulipifera*
cucumbertree	168	*Magnolia acuminata*
southern magnolia	169	*Magnolia grandiflora*
Malvaceae, the mallow family		
blue mahoe	143	*Hibiscus elatus*
portiatree	267	*Thespesia populnea*
Meliaceae, the mahogany family		
andiroba	59	*Carapa guianensis*
Spanish-cedar	70	*Cedrela odorata*
sapele	112	*Entandrophragma cylindricum*
guarea	133	*Guarea cedrata* and *G. thompsonii*
African mahogany	155	*Khaya* spp.
tigerwood	166	*Lovoa trichilioides*
chinaberry	177	*Melia azedarach*

bigleaf mahogany	253	*Swietenia macrophylla*
Australian red cedar	271	*Toona ciliata*
avodire	273	*Turraeanthus africanus*
Mimosaceae, see Fabaceae or Leguminosae, the legume family		
Moraceae, the mulberry family		
snakewood	42	*Brosimum guianense*
satine	43	*Brosimum rubescens*
paper-mulberry	44	*Broussonetia papyrifera*
Osage-orange	167	*Maclura pomifera*
iroko	181	*Milicia excelsa*
red mulberry	183	*Morus rubra*
Myristicaceae, the nutmeg family		
virola	100	*Dialyanthera* spp.
Myrtaceae, the myrtle family		
jarrah	115	*Eucalyptus marginata*
mountain ash	116	*Eucalyptus regnans*
tea tree	160	*Leptospermum scoparium*
brushbox	165	*Lophostemon confertus*
cajeput-tree	176	*Melaleuca quinquenervia*
ohia	179	*Metrosideros collina*
Nothofagaceae, the southern beech family		
coigue	187	*Nothofagus dombeyi*
Ochnaceae, the ochna family		
ekki	164	*Lophira alata*
izombe	265	*Testulea gabonensis*
Oleaceae, the olive family		
white ash	121	*Fraxinus americana*
European ash	122	*Fraxinus excelsior*
black ash	123	*Fraxinus nigra*
tamo	124	*Fraxinus sieboldiana*
olive	192	*Olea europaea*
lilac	255	*Syringa vulgaris*
Palmae, the palm family; also Arecaceae		
coconut	84	*Cocos nucifera*
Cuban royal palm	238	*Roystonea regia*
cabbage palmetto	239	*Sabal palmetto*
Papilionaceae, see Fabaceae or Leguminosae, the legume family		
Pinaceae, the pine family; also Abietaceae		
balsam fir	1	*Abies balsamea*
western larch	159	*Larix occidentalis*
black spruce	204	*Picea mariana*
red spruce	205	*Picea rubens*
Sitka spruce	206	*Picea sitchensis*
bristlecone pine	207	*Pinus aristata*
Caribbean pine	208	*Pinus caribaea*
lodgepole pine	209	*Pinus contorta*
sugar pine	210	*Pinus lambertiana*
intermountain bristlecone pine	207	*Pinus longaeva*
Table Mountain pine	211	*Pinus pungens*
eastern white pine	212	*Pinus strobus*
Douglas-fir	224	*Pseudotsuga menziesii*

Platanaceae, the sycamore family
sycamore 214 *Platanus occidentalis*

Poaceae, see Gramineae, the grass family

Podocarpaceae, the podocarp family
podo 12 *Afrocarpus falcata*
Huon pine 158 *Lagarostrobus franklinii*
totara 216 *Podocarpus totara*

Proteaceae, the protea family
lacewood 60 *Cardwellia sublimis*
silky oak 131 *Grevillea robusta*

Rhamnaceae, the buckthorn family
pink ivory 35 *Berchemia zeyheri*
leadwood 157 *Krugiodendron ferreum*
ebano 279 *Ziziphus thyrsiflora*

Rhizophoraceae, the mangrove family
mangrove 235 *Rhizophora mangle*

Rosaceae, the rose family
mountain mahoganies 76 *Cercocarpus* spp.
birchleaf cercocarpus 76 *Cercocarpus betuloides*
hairy cercocarpus 76 *Cercocarpus breviflorus*
curlleaf cercocarpus 76 *Cercocarpus ledifolius*
alderleaf cercocarpus 76 *Cercocarpus montanus*
apple 170 *Malus sylvestris*
European bird cherry 222 *Prunus padus*
black cherry 223 *Prunus serotina*
pearwood 229 *Pyrus communis*
American mountain-ash 250 *Sorbus americana*

Rubiaceae, the madder family
degame 56 *Calycophyllum candidissimum*
opepe 185 *Nauclea diderrichii*

Rutaceae, the rue or citrus family
pau marfim 33 *Balfourodendron riedelianum*
Ceylon satinwood 80 *Chloroxylon swietenia*
Queensland maple 120 *Flindersia brayleana*
West Indian satinwood 278 *Zanthoxylum flavum*

Salicaceae, the willow family
balsam poplar 217 *Populus balsamifera*
eastern cottonwood 218 *Populus deltoides*
trembling aspen 219 *Populus tremuloides*
black willow 240 *Salix nigra*

Santalaceae, the sandalwood family
sandalwood 242 *Santalum album*

Sapindaceae, the soapberry family
western soapberry 243 *Sapindus saponaria*

Sapotaceae, the sapodilla family
moabi 32 *Baillonella toxisperma*
bulletwood 173 *Manilkara bidentata*
sapodilla 174 *Manilkara zapota*
makore 269 *Tieghemella heckelii*

Scrophulariaceae, the figwort family
royal paulownia 198 *Paulownia tomentosa*

Simaroubaceae, the ailanthus or quassia family

ailanthus	15	*Ailanthus altissima*

Sterculiaceae, the stericulia or cacao family

mengkulang	141	*Heritiera simplicifolia*
mansonia	175	*Mansonia altissima*
danta	186	*Nesogordonia papaverifera*
chica	252	*Sterculia apetala*
obeche	272	*Triplochiton scleroxylon*

Styracaceae, the storax family

Carolina silverbell	139	*Halesia carolina*

Taxaceae, the yew family

Pacific yew	260	*Taxus brevifolia*

Taxodiaceae, the redwood or baldcypress family

redwood	247	*Sequoia sempervirens*
giant sequoia	248	*Sequoiadendron giganteum*
baldcypress	259	*Taxodium distichum*

Theaceae, the tea family

loblolly-bay	129	*Gordonia lasianthus*

Thymelaeaceae, the mezereon family

ramin	128	*Gonystylus bancanus*

Tiliaceae, the linden family

American basswood	270	*Tilia americana*

Ulmaceae, the elm family

hackberry	72	*Celtis occidentalis*
American elm	274	*Ulmus americana*
cedar elm	275	*Ulmus crassifolia*
slippery elm	276	*Ulmus rubra*

Zygophyllaceae, the caltrop family

verawood	47	*Bulnesia arborea*
lignumvitae	132	*Guaiacum officinale*

SCIENTIFIC NAME INDEX

In addition to providing a link to the Wood Data Sheet number, this index will identify synonyms for scientific names. If a scientific name is in current use, it will appear in straight type and will be associated with the proper Wood Data Sheet. Scientific names appearing in *italics* are synonyms. They will be linked to the Wood Data Sheet bearing the current name.

Scientific Name	WDS No.
A	
Abies balsamea	1
Acacia aneura	2
Acacia auriculiformis	3
Acacia greggi	*4*
Acacia greggii	4
Acacia koa	5
Acacia melanoxylon	6
Acer negundo	7
Acer pseudoplatanus	8
Acer rubrum	9
Acer saccharum	10
Achras zapota	*174*
Aesculus glabra	11
Afrocarpus falcata	12
Afrormosia elata	*200*
Afzelia quanzensis	13
Agathis australis	14
Ailanthus altissima	15
Albizia lebbeck	16
Albizia saman	*241*
Aleurites moluccana	17
Alnus rubra	18
Alstonia scholaris	19
Anacardium excelsum	20
Andira inermis	21
Andromeda arborea	*196*
Aniba duckei	*22*
Aniba rosaeodora	22
Araucaria angustifolia	23
Araucaria heterophylla	24
Arbutus menziesii	25
Arctostaphylos viscida	26
Asimina triloba	27
Aspidosperma peroba	*28*
Aspidosperma polyneuron	28
Astronium balansae	29
Astronium fraxinifolium	30
B	
Baikiaea plurijuga	31
Baillonella toxisperma	32
Balfourodendron riedelianum	33
Beilschmiedia tawa	34
Berchemia zeyheri	35
Bertholletia excelsa	36
Betula alleghaniensis	37
Betula lenta	38
Betula lutea	*37*
Betula papyrifera	39
Bischofia javanica	40
Bombax pentandrum	*71*
Brachylaena huillensis	41
Brachylaena hutchinsii	*41*
Brosimum aubletii	*42*
Brosimum guianense	42
Brosimum paraense	*43*
Brosimum rubescens	43
Broussonetia papyrifera	44
Brya ebenus	45
Bucida buceras	46
Bulnesia arborea	47
Burkea africana	48
Bursera simaruba	49
Buxus sempervirens	50
C	
Caesalpinia echinata	51
Caesalpinia granadillo	52
Callitris glaucophylla	53
Callitris quadrivalis	*266*
Calocedrus decurrens	54
Calophyllum brasiliense	55
Calophyllum lucidum	*55*
Calycophyllum candidissimum	56
Cananga odorata	57
Canangium odoratum	*57*
Caragana arborescens	58
Carapa guianensis	59

Cardwellia sublimis 60
Carpinus caroliniana 61
Carya illinoensis *62*
Carya illinoinensis 62
Carya myristiciformis 63
Carya ovata 64
Castanea americana *65*
Castanea dentata 65
Castanopsis chrysophylla *81*
Castanospermum australe 66
Casuarina spp. 67
Catalpa speciosa 68
Cecropia peltata 69
Cedrela mexicana *70*
Cedrela odorata 70
Ceiba pentandra 71
Celtis occidentalis 72
Centrolobium ochroxylon 73
Centrolobium orinocense 73
Centrolobium paraense 73
Centrolobium robustum 73
Centrolobium tomentosum 73
Ceratopetalum apetalum 74
Cercis canadensis 75
Cercocarpus betuloides 76
Cercocarpus breviflorus 76
Cercocarpus ledifolius 76
Cercocarpus montanus 76
Chamaecyparis lawsoniana 77
Chamaecyparis nootkatensis 78
Chlorocardium rodiei 79
Chlorophora excelsa *181*
Chloroxylon swietenia 80
Chrysolepis chrysophylla 81
Cinnamomum camphora 82
Cladrastis kentukea *83*
Cladrastis lutea 83
Cocos nucifera 84
Colophospermum mopane 140
Cordia dodecandra 85
Cordia gerascanthus 86
Cordia goeldiana 87
Cornus florida 88
Cotinus obovatus 89
Cybistax donnell-smithii *256*
Cyrilla racemiflora 90

D

Dacrydium franklinii *158*
Dacryodes excelsa 91
Dalbergia cearensis 92
Dalbergia cochinchinensis 93
Dalbergia decipularis 94
Dalbergia latifolia 95
Dalbergia melanoxylon 96
Dalbergia nigra 97
Dalbergia retusa 98
Dalbergia stevensonii 99
Dialyanthera spp. 100
Didymopanax morototoni *245*
Diospyros celebica 101
Diospyros dendo 102
Diospyros virginiana 103
Distemonanthus benthamianus 104
Dracontomelon cumingianum *105*
Dracontomelon dao 105
Dracontomelon edule *105*
Dryobalanops aromatica 106
Dryobalanops sumatrensis *106*
Dumoria heckelii *269*
Dyera costulata 107

E

Ebenopsis ebano 108
Elaeagnus angustifolia 109
Endiandra palmerstonii 110
Endospermum macrophyllum 111
Entandrophragma cylindricum 112
Enterolobium cyclocarpum 113
Erythrina crista-galli 114
Eucalyptus marginata 115
Eucalyptus regnans 116
Eucryphia cordifolia 117
Excoecaria africana 118

F

Fagara flava *278*
Fagus americana *119*
Fagus grandifolia 119
Flindersia brayleana 120
Fraxinus americana 121
Fraxinus excelsior 122
Fraxinus mariesii *124*
Fraxinus nigra 123
Fraxinus sieboldiana 124

G

Ginkgo biloba 125
Gleditsia triacanthos 126
Gmelina arborea 127
Gonystylus bancanus 128
Gordonia lasianthus 129
Gossweilerodendron balsamiferum 130
Grevillea robusta 131
Guaiacum officinale 132
Guarea cedrata 133
Guibourtia arnoldiana 134
Guibourtia tessmannii 135
Guilandina echinata *51*

Gymnanthes lucida 136
Gymnocladus dioicus 137

H

Haematoxylum campechianum 138
Halesia carolina 139
Hardwickia mopane 140
Heritiera simplicifolia 141
Hevea brasiliensis 142
Hibiscus elatus 143
Hura crepitans 144
Hyeronima alchorneoides 145
Hymenaea courbaril 146

I

Ilex opaca 147
Intsia spp. 148

J

Juglans cinerea 149
Juglans nigra 150
Juniperus ashei 151
Juniperus mexicana *151*
Juniperus virginiana 152

K

Kalmia latifolia 153
Kalopanax pictus *154*
Kalopanax septemlobus 154
Khaya spp. 155
Koompassia malaccensis 156
Krugiodendron ferreum 157

L

Lagarostrobos franklinii 158
Larix occidentalis 159
Laurus camphora *82*
Leptospermum scoparium 160
Libocedrus decurrens *54*
Liquidambar macrophylla *161*
Liquidambar styraciflua 161
Liriodendron tulipifera 162
Lithocarpus densiflorus 163
Lophira alata 164
Lophira procera *164*
Lophostemon confertus 165
Lovoa trichilioides 166
Lyonia arborea *196*

M

Maclura pomifera 167
Magnolia acuminata 168
Magnolia grandiflora 169
Malus sylvestris 170
Mammea africana 171
Mangifera indica 172
Manilkara bidentata 173
Manilkara zapota 174
Manilkara zapotilla *174*
Mansonia altissima 175
Melaleuca quinquenervia 176
Melia azedarach 177
Melia toosendan *177*
Metopium brownei 178
Metrosideros collina 179
Microberlinia brazzavillensis 180
Milicia excelsa 181
Millettia laurentii 182
Mimusops heckelii *269*
Morus rubra 183
Myroxylon balsamum 184

N

Nauclea diderrichii 185
Nectandra rodiei *79*
Nesogordonia papaverifera 186
Nothofagus dombeyi 187
Nyssa sylvatica 188

O

Ochroma lagopus *189*
Ochroma pyramidale 189
Ocotea bullata 190
Ocotea porosa 191
Ocotea rodiei *79*
Olea europaea 192
Olneya tesota 193
Ostrya virginiana 194
Oxandra lanceolata 195
Oxydendrum arboreum 196

P

Parapiptadenia rigida 197
Paulownia tomentosa 198
Peltogyne paniculata 199
Pericopsis elata 200
Persea americana 201
Persea borbonia 202
Phoebe porosa *191*
Phyllostachys edulis 203
Picea mariana 204
Picea rubens 205
Picea sitchensis 206
Pinus aristata 207
Pinus caribaea 208
Pinus contorta 209
Pinus lambertiana 210

Pinus longaeva 207
Pinus pungens 211
Pinus strobus 212
Piptadenia rigida 197
Piratinera guianensis 42
Piscidia erythrina 213
Piscidia piscipula 213
Pithecellobium flexicaule 108
Pithecellobium saman 241
Platanus occidentalis 214
Plathymenia reticulata 215
Podocarpus falcatus 12
Podocarpus totara 216
Populus balsamifera 217
Populus deltoides 218
Populus tremuloides 219
Populus trichocarpa 217
Prioria balsamifera 130
Prioria copaifera 220
Prosopis glandulosa 221
Prunus padus 222
Prunus serotina 223
Pseudotsuga menziesii 224
Pterocarpus dalbergioides 225
Pterocarpus indicus 226
Pterocarpus macrocarpus 227
Pterocarpus soyauxii 228
Pyrus communis 229

Q

Quercus alba 230
Quercus arkansana 231
Quercus robur 232
Quercus rubra 233
Quercus virginiana 234

R

Rhamnus zeyheri 35
Rhizophora mangle 235
Rhus typhina 236
Robinia pseudoacacia 237
Roystonea elata 238
Roystonea floridana 238
Roystonea regia 238

S

Sabal palmetto 239
Salix nigra 240
Samanea saman 241
Santalum album 242
Sapindus drummondii 243
Sapindus indicus 243
Sapindus saponaria 243
Sapota achras 174
Sapota zapotilla 174
Sassafras albidum 244
Schefflera morototoni 245
Schinopsis lorentzii 246
Schinopsis quebracho-colorado 246
Sequoia sempervirens 247
Sequoiadendron giganteum 248
Shorea spp. 249
Siphonia brasiliensis 142
Sophora affinis 253
Sorbus americana 250
Spirostachys africana 118
Spondias lutea 251
Spondias mombin 251
Sterculia apetala 252
Sterculia carthagensis 252
Styphnolobium affine 253
Swietenia macrophylla 254
Syringa vulgaris 255

T

Tabebuia donnell-smithii 256
Tabebuia spp. 257
Tamarindus indica 258
Taxodium distichum 259
Taxus brevifolia 260
Tectona grandis 261
Terminalia elliptica 262
Terminalia ivorensis 263
Terminalia superba 264
Terminalia tomentosa 262
Testulea gabonensis 265
Tetraclinis articulata 266
Thespesia populnea 267
Thuja plicata 268
Tieghemella heckelii 269
Tilia americana 270
Tilia glabra 270
Toona australis 271
Toona ciliata 271
Triplochiton scleroxylon 272
Tristania conferta 165
Turraeanthus africanus 273

U

Ulmus americana 274
Ulmus crassifolia 275
Ulmus fulva 276
Ulmus rubra 276
Umbellularia californica 277

Z

Zanthoxylum flavum 278
Ziziphus thyrsiflora 279